The Role of Protein and Amino Acids in Sustaining and Enhancing Performance

Committee on Military Nutrition Research

Committee on Body Composition, Nutrition and Health

Food and Nutrition Board

Institute of Medicine

NATIONAL ACADEMY PRESS
Washington, D.C. 1999

NATIONAL ACADEMY PRESS • 2101 Constitution Avenue, N.W. • Washington, DC 20418

NOTICE: The project that is the subject of this report was approved by the Governing Board of the National Research Council, whose members are drawn from the councils of the National Academy of Sciences, the National Academy of Engineering, and the Institute of Medicine. The members of the committee responsible for the report were chosen for their special competences and with regard for appropriate balance.

The Institute of Medicine was established in 1970 by the National Academy of Sciences to enlist distinguished members of the appropriate professions in the examination of policy matters pertaining to the health of the public. In this, the Institute acts under both the Academy's 1863 congressional charter responsibility to be an adviser to the federal government and its own initiative in identifying issues of medical care, research, and education. Dr. Kenneth I. Shine is president of the Institute of Medicine.

This report was produced under grant DAMD17-94-J-4046 between the National Academy of Sciences and the U.S. Army Medical Research and Materiel Command. The views, opinions, and/or findings contained in chapters in Parts II through VIII that are authored by U.S. Army personnel are those of the authors and should not be construed as official Department of the Army positions, policies, or decisions, unless so designated by other official documentation. Human subjects who participated in studies described in these chapters gave their free and informed voluntary consent. Investigators adhered to U.S. Army regulation 70-25 and United States Army Medical Research and Development Command regulation 70-25 on the use of volunteers in research. Citations of commercial organizations and trade names in this report do not constitute an official Department of the Army endorsement or approval of the products or services of these organizations. The chapters are approved for public release; distribution is unlimited.

Library of Congress Catalog Card No. 98-89750

International Standard Book Number 0-309-06346-9

Additional copies of this report are available from:

National Academy Press
2101 Constitution Avenue, N.W.
Lock Box 285
Washington, D.C. 20055

Call (800) 624-6242 or (202) 334-3313 (in the Washington metropolitan area), or visit the NAP's on-line bookstore at *http://www.nap.edu*.

Copyright 1999 by the National Academy of Sciences. All rights reserved.

Printed in the United States of America.

The serpent has been a symbol of long life, healing, and knowledge among almost all cultures and religions since the beginning of recorded history. The image adopted as a logotype by the Institute of Medicine is based on a relief carving from ancient Greece, now held by the Staatliche Museen in Berlin.

COMMITTEE ON MILITARY NUTRITION RESEARCH

ROBERT O. NESHEIM (*Chair, through June 30, 1998*), Salinas, California
JOHN E. VANDERVEEN (*Chair*), Rockville, Maryland
LAWRENCE E. ARMSTRONG, Departments of Physiology and Neurobiology, and Exercise Science, University of Connecticut, Storrs
WILLIAM R. BEISEL, Department of Molecular Microbiology and Immunology, The Johns Hopkins University School of Hygiene and Public Health, Baltimore, Maryland
GAIL E. BUTTERFIELD, Nutrition Studies, Palo Alto Veterans Affairs Health Care System and Program in Human Biology, Stanford University, Palo Alto, California
WANDA L. CHENOWETH, Department of Food Science and Human Nutrition, Michigan State University, East Lansing
JOHN D. FERNSTROM, Department of Psychiatry, Pharmacology, and Neuroscience, University of Pittsburgh School of Medicine, Pennsylvania
G. RICHARD JANSEN, (*through August 31, 1997*), Department of Food Science and Human Nutrition, Colorado State University, Fort Collins
ROBIN B. KANAREK, Department of Psychology, Tufts University, Boston, Massachusetts
ORVILLE A. LEVANDER, Nutrient Requirements and Functions Laboratory, U.S. Department of Agriculture Beltsville Human Nutrition Research Center, Beltsville, Maryland
ESTHER M. STERNBERG, Neuroendocrine Immunology and Behavior Section, National Institute of Mental Health, Bethesda, Maryland
DOUGLAS W. WILMORE, Department of Surgery, Brigham and Women's Hospital, Boston, Massachusetts

Food and Nutrition Board Liaison
JOHANNA T. DWYER, Frances Stern Nutrition Center, New England Medical Center Hospital and Departments of Medicine and Community Health, Tufts Medical School and School of Nutrition Science and Policy, Boston, Massachusetts

U.S. Army Grant Officer Representatives
LTC KARL E. FRIEDL, Military Operational Medicine Program, U.S. Army Medical Research and Materiel Command, Fort Detrick, Frederick, Maryland

Staff
MARY I. POOS (*from May 23, 1998*), Study Director
SYDNE J. CARLSON-NEWBERRY, Program Officer
MARIZA SILVA (*from August 31, 1998*), Project Assistant

SUBCOMMITTEE ON BODY COMPOSITION, NUTRITION, AND HEALTH OF MILITARY WOMEN

BARBARA O. SCHNEEMAN (*Chair*), College of Agricultural and Environmental Sciences, University of California, Davis
ROBERT O. NESHEIM (*Vice Chair*), Salinas, California
JOHN P. BILEZIKIAN, Department of Medicine, College of Physicians and Surgeons, Columbia University, New York, New York
NANCY F. BUTTE, Children's Nutrition Research Center, Baylor College of Medicine, Houston, Texas
STEVEN B. HEYMSFIELD, Human Body Composition Laboratory, Weight Control Unit, and Obesity Research Center, St. Luke's-Roosevelt Hospital Center, New York, New York
ANNE LOOKER, Division of Health Examination Statistics, National Center for Health Statistics, Hyattsville, Maryland
GORDON O. MATHESON, Division of Sports Medicine, Department of Functional Restoration, Stanford University School of Medicine, Stanford, California
BONNY L. SPECKER, The Martin Program in Human Nutrition, South Dakota State University, Brookings

Committee on Military Nutrition Research Liaison
GAIL E. BUTTERFIELD, Nutrition Studies, Palo Alto Veterans Affairs Health Care System and Program in Human Biology, Stanford University, Palo Alto, California

Food and Nutrition Board Liaison
JANET C. KING, U.S. Department of Agriculture Western Human Nutrition Research Center, San Francisco and University of California, Berkeley

U.S. Army Grant Representative
LTC KARL E. FRIEDL, USA, Army Operational Medicine Research Program, U.S. Army Medical Research and Materiel Command, Fort Detrick, Frederick, Maryland

Staff
REBECCA B. COSTELLO (*through May 22, 1998*), Project Director
MARY I. POOS (*from May 23, 1998*), Project Director
SYDNE J. CARLSON-NEWBERRY, Program Officer
SUSAN M. KNASIAK-RALEY (*through April 3, 1998*), Research Assistant
MELISSA L. VAN DOREN (*through August, 1998*), Project Assistant

FOOD AND NUTRITION BOARD

CUTBERTO GARZA (*Chair*), Division of Nutrition, Cornell University, Ithaca, New York
JOHN W. ERDMAN, JR. (*Vice Chair*), Division of Nutritional Sciences, College of Agriculture, University of Illinois at Urbana-Champaign
LARRY R. BEUCHAT, Center for Food Safety and Quality Enhancement, University of Georgia, Griffin
BENJAMIN CABALLERO, Center for Human Nutrition, The Johns Hopkins School of Hygiene and Public Health, Baltimore, Maryland
FERGUS M. CLYDESDALE, Department of Food Science, University of Massachusetts, Amherst
ROBERT J. COUSINS, Center for Nutritional Sciences, University of Florida, Gainesville
JOHANNA T. DWYER, Frances Stern Nutrition Center, New England Medical Center Hospital and Departments of Medicine and Community Health, Tufts Medical School and School of Nutrition Science and Policy, Boston, Massachusetts
SCOTT M. GRUNDY, Center for Human Nutrition, University of Texas Southwestern Medical Center at Dallas
CHARLES H. HENNEKENS, Boca Raton, Florida
ALFRED H. MERRILL, JR., Department of Biochemistry, Emory Center for Nutrition and Health Sciences, Emory University, Atlanta, Georgia
LYNN PARKER, Child Nutrition Programs and Nutrition Policy, Food Research and Action Center, Washington, D.C.
ROSS L. PRENTICE, Division of Public Health Sciences, Fred Hutchinson Cancer Research Center, Seattle, Washington
A. CATHERINE ROSS, Department of Veterinary Science, The Pennsylvania State University, University Park
ROBERT M. RUSSELL, USDA Human Nutrition Research Center on Aging, Tufts University, Boston, Massachusetts
ROBERT E. SMITH, R. E. Smith Consulting, Inc., Newport, Vermont
VIRGINIA A. STALLINGS, Division of Gastroenterology and Nutrition, The Children's Hospital of Philadelphia, Pennsylvania
STEVE L. TAYLOR, Department of Food Science and Technology and Food Processing Center, University of Nebraska, Lincoln

Staff
ALLISON A. YATES, Director
GAIL SPEARS, Administrative Assistant
GARY WALKER, Financial Associate

Preface

This publication, *The Role of Protein and Amino Acids in Sustaining and Enhancing Performance*, is the latest in a series of reports based on workshops sponsored by the Committee on Military Nutrition Research (CMNR) of the Food and Nutrition Board (FNB), Institute of Medicine, National Academy of Sciences. Other workshops or symposia have included such topics as nutrition and immune function; emerging technologies in nutrition research; food components to enhance performance; nutritional needs in hot, cold, and high-altitude environments; body composition and physical performance; and fluid replacement and heat stress. These workshops form part of the response that the CMNR provides to the Commander, U.S. Army Medical Research and Materiel Command, regarding issues brought to the committee through the Military Nutrition and Biochemical Division of the U.S. Army Research Institute of Environmental Medicine (USARIEM) at Natick, Massachusetts and the Military Operational Medicine Program at Fort Detrick, Maryland.

HISTORY OF THE COMMITTEE

The CMNR was established in October 1982 following a request by the Assistant Surgeon General of the Army that the Board on Military Supplies of the National Academy of Sciences set up a committee to advise the U.S. Department of Defense on the need for and conduct of nutrition research and related issues. The committee was transferred to the FNB in 1983. The committee's current tasks are

- to identify nutritional factors that may critically influence the physical and mental performance of military personnel under all environmental extremes;
- to identify deficiencies in the existing database;
- to recommend research that would remedy these deficiencies, as well as approaches for studying the relationship of diet to physical and mental performance; and
- to review and advise on standards for military feeding systems.

Within this context, the CMNR was asked to focus on nutrient requirements for performance during operational missions rather than on requirements for military personnel in garrison (the latter were judged not to differ significantly from those of the civilian population).

Although the membership of the committee has changed periodically, the disciplines represented consistently have included human nutrition, nutritional biochemistry, performance physiology, food science, and psychology. For issues that require broader expertise than exists within the committee, the CMNR has convened workshops or utilized consultants. The workshops provide additional state-of-the-art scientific information and informed opinion for the consideration of the committee.

FOCUS OF THE REPORT

The request for this review originated with scientists at USARIEM who were concerned about the unique nutritional demands placed on soldiers during combat. They were particularly concerned about the role that dietary protein may play in controlling muscle mass and strength, response to injury and infection, and cognitive performance.

Past reports of the CMNR usually have focused on issues of current concern to the military. Traditional methods of research, data gathering, and analysis have provided the factual base for study of a problem and recommended solutions. Several previous CMNR reports have focused on issues of protein nutriture and performance. In 1992, the CMNR noted in an evaluation of Army Ranger training that trainees experienced significant loss of muscle mass during periods of intense physical exertion (IOM, 1992). A follow-up report

(IOM, 1993b) found that increases in energy intake only partially prevented such losses. The report *Food Components to Enhance Performance* (IOM, 1994) briefly considered the influences of protein and amino acids on physical and cognitive performance and response to stress. The most recent CMNR report, *Military Strategies for Sustainment of Nutrition and Immune Function in the Field* (IOM, 1999), considered the effects of diet, including protein and individual amino acids such as glutamine, on immune response. This report looks further into the many questions regarding the optimal level of protein intake in a high-stress field environment. How to measure protein balance and estimate protein requirements accurately; how these requirements are affected by physical activity, gender, hormonal factors, and stress; and whether muscle function and cognition are influenced by protein intake and by individual amino acids are all active areas of research.

The CMNR decided that the best way to review the state of knowledge in this area was through a workshop. The purpose of this workshop was to bring together leading scientists in the field of protein metabolism to seek their assessment of the current state of knowledge and to determine, based on these assessments, on a careful reading of the literature, and on the expertise of the committee members themselves, whether the recommended intakes of protein or individual amino acids for soldiers should be modified.

In May 1996, CMNR and USARIEM personnel met to frame a series of questions, outline the workshop, and identify qualified speakers. A follow-up planning meeting was held in January 1997 and included several members of the Subcommittee on Body Composition, Nutrition, and Health of Military Women. Invited workshop speakers were asked to prepare a paper for presentation and publication that described the key issues of protein metabolism. USARIEM scientists also participated in the workshop, which resulted in a well-rounded group. At the one-day workshop, held in Washington, D.C. in March 1997, each speaker gave a formal presentation, which was followed by questions and a brief discussion period. The proceedings were tape recorded and professionally transcribed. At the end of each group of presentations, a general discussion of the overall topic was held. Immediately after the workshop, the CMNR met in executive session to review the issues, to draft summaries of the presentations, and to provide responses to the sponsor's task questions. Committee members subsequently met with staff in June 1997 and worked separately and together using the authored papers, additional reference materials provided by the staff through limited literature searches, and personal expertise and experience to draft the overview, summary, conclusions, and recommendations.

ORGANIZATION OF THE REPORT

A project overview and a summary of the CMNR's responses to the task questions, conclusions, and recommendations, constitute Part I of this volume. Part II includes papers contributed by speakers at the workshop. Part I has been reviewed by an outside group with expertise in the topic areas and experience in military issues. For the most part, the authored papers in Part II appear in the order in which they were presented at the workshop (see workshop agenda in Appendix A). These chapters have undergone limited editorial changes, have not been reviewed by the outside group, and represent the views of the individual authors. Selected questions directed toward the speakers and the speakers' responses are included when they provide a flavor of the workshop discussion. The invited speakers also were requested to submit a brief list of selected background papers prior to the workshop. These recommended readings, relevant citations collected by CMNR staff prior to the workshop, and selected citations from each chapter are included in the selected bibliography (see Appendix D).

ACKNOWLEDGMENTS

It is my pleasure as chairman of the CMNR to acknowledge the help of Institute of Medicine President Kenneth I. Shine; Food and Nutrition Board Director Allison Yates and former Acting Director Carol Suitor; Study Director Mary Poos; and former FNB staff—Rebecca Costello, study director, and Sydne J. Newberry, staff officer. Also, I wish to acknowledge the excellent editorial efforts and able assistance of Mariza Silva, CMNR project assistant; Susan M. Knasiak-Raley, former research assistant to the CMNR; and Melissa Van Doren, former CMNR project assistant, in word processing and preparing the camera-ready copy for this report. Finally, I wish to acknowledge the assistance of managing editor Michael A. Edington and Reports and Information Office associate Claudia M. Carl; National Academy of Sciences Librarian Susan Fourt and assistant Patricia Kaiser; and editors Judy Grumstrup-Scott and Florence Poillon.

I wish to acknowledge as well the excellent contributions by the speakers and their commitment to participating in the workshop and preparing papers on their assigned areas with relatively short notice.

Finally, I express my appreciation to the members of the CMNR who have participated in the proceedings of the workshop and the discussions and preparation of summaries and recommendations in this report. Their continued dedication to providing sound, timely recommendations on issues brought to our attention is commendable. I also wish to acknowledge the many years of fine work by Richard Jansen who rotated off the committee prior to the final preparation of this report, but nevertheless continued to provide his expertise. I especially want to acknowledge the dedicated leadership of the CMNR provided by Robert O. Nesheim who stepped down as chair in July, 1998. Thank you all for your commitment to the success of this program.

This report has been reviewed by individuals chosen for their diverse perspectives and technical expertise, in accordance with procedures approved by the Report Review Committee of the National Research Council. The purpose of this independent review is to provide candid and critical comments that will assist the authors and the Institute of Medicine in making the published report as sound as possible and to ensure that the report meets institutional standards for objectivity, evidence, and responsiveness to the study charge. The review comments and draft manuscript remain confidential to protect the integrity of the deliberative process. The CMNR wishes to thank the following individuals for their participation in the review of this report: Elsworth K. Buskirk, Pennsylvania State University; Gerald Combs, Sr. (retired); Melvin Grumbach, University of California, San Francisco; John M. Kinney, Rockefeller University; and T. Peter Stein, University of Medicine and Dentistry of New Jersey. While the individuals listed above have provided many constructive comments and suggestions, it must be emphasized that responsibility for the final content of this report rests entirely with the authoring committee and the IOM.

JOHN E. VANDERVEEN, *Chair*
Committee on Military Nutrition Research

Contents

PREFACE vii

EXECUTIVE SUMMARY 1

I   COMMITTEE SUMMARY AND RECOMMENDATIONS

     1   Committee Overview 19
     2   Responses to Questions, Conclusions, and Recommendations 77

II   AUTHORED PAPERS AND WORKSHOP DISCUSSIONS

     3   Protein and Amino Acids: Physiological Optimization for Current and Future Military Operational Scenarios 85
        LTC Karl E. Friedl

     4   Overview of Garrison, Field, and Supplemental Protein Intake by U.S. Military Personnel 93
        LTC (ret) Alana D. Cline and John P. Warber

5	The Energy Costs of Protein Metabolism: Lean and Mean on Uncle Sam's Team .... 109 *Dennis M. Bier*	
6	Regulation of Muscle Mass and Function: Effects of Aging and Hormones .... 121 *Niels Moller and K. Sreekumaran Nair*	
7	Effects of Protein Intake on Renal Function and on the Development of Renal Disease .... 137 *Mackenzie Walser*	
8	Infection and Injury: Effects on Whole Body Protein Metabolism	155
	Douglas W. Wilmore	
9	Inherent Difficulties in Defining Amino Acid Requirements .... 169 *D. Joe Millward*	
10	Amino Acid Flux and Requirements: Counterpoint Tentative Estimates Are Feasible and Necessary .... 217 *Vernon R. Young*	
11	Physical Exertion, Amino Acid and Protein Metabolism, and Protein Requirements .... 243 *Michael J. Rennie*	
12	Skeletal Muscle Markers .... 255 *Dympna Gallagher, Steven B. Heymsfield, and Zi-Mian Wang*	
13	Alterations in Protein Metabolism Due to the Stress of Injury and Infection .... 279 *Robert R. Wolfe*	

Discussion I 285

14	Amino Acid and Protein Requirements: Cognitive Performance, Stress, and Brain Function .... 289 *Harris R. Lieberman*
15	Supplementation with Branched-Chain Amino Acids, Glutamine, and Protein Hydrolysates: Rationale for Effects on Metabolism and Performance .... 309 *Anton J. M. Wagenmakers*

16 Dietary Supplements Aimed at Enhancing Performance: Efficacy and Safety Considerations 331
Timothy J. Maher

Discussion II 341

APPENDIXES

A Workshop Agenda 349
B Biographical Sketches 353
C Acronyms and Abbreviations 369
D Proteins and Amino Acids—A Selected Bibliography 373
E Protein and Energy Content of Selected Operational Rations 411

INDEX 413

The Role of Protein and Amino Acids in Sustaining and Enhancing Performance

Executive Summary

As the U.S. military faces the millennium and the changing nature of modern warfare, it must anticipate physical and mental challenges never encountered before. Longer periods of intense physical exertion and possible food deprivation; advanced weaponry requiring maximum attention, precision, and decision-making ability; and greater threats of infection, injury, and exposure to environmental stressors are quickly becoming the reality that soldiers face. Military scientists charged with maintaining and optimizing the health and performance of their personnel are looking to the role that nutrition may play in this process, and have expressed particular interest in the body of current research suggesting the importance of protein and amino acids.

Proteins catalyze virtually all chemical reactions in the body, regulate gene expression, comprise the major structural elements of all cells, regulate the immune system, and form the major constituents of muscle. Individual amino acids, the components of proteins, also serve as neurotransmitters, hormones, and modulators of various physiological processes. Every aspect of physiology involves proteins. The relationships between dietary protein and bodily protein metabolism are a major focus of research. Many questions remain regarding the validity of methods for assessment of protein balance; thus, the question of how best to assess dietary protein requirements remains unanswered. In addition, the influences of genetic factors, hormones, physical activity, injury and infectious processes, and environmental stresses on protein metabolism and protein requirements continue to be explored. Another major focus of research is the role of protein and amino acid balance in modulating physiological function and behavior, as measured for example by physical and mental performance. The possibility that protein or individual amino acids in quantities that exceed those required to maintain protein balance may have the potential to contribute to performance optimization is of great interest.

COMMITTEE'S TASK

In 1992, the Committee on Military Nutrition Research (CMNR) was asked by the Military Nutrition Division of the U.S. Army Research Institute of Environmental Medicine to conduct an investigation into the fate of a group of soldiers in special forces training. The committee found that these soldiers lost significant amounts of muscle mass during the training period and recommended changes in their intakes of energy and protein. In 1994, the CMNR was asked to conduct a workshop and prepare a report on the performance-enhancing potential of individual food components. The committee recommended further research on the ability of the amino acid tyrosine to enhance several measures of cognitive performance in stressful situations. The CMNR conducted a similar workshop in 1996 to examine the influence of dietary factors, including protein and amino acids, on immune function. Finally, in 1997, the CMNR was asked by the U.S. Army Medical Research and Materiel Command to update its earlier findings and to conduct a workshop on the potential role of protein and amino acids in maintaining and enhancing the physical and cognitive performance of soldiers. Specifically, the committee was asked to respond to three questions:

1. Do protein requirements increase with military operational stressors, including high workload with or without energy deficit? Are there gender differences in protein requirements in endurance exercise?
2. What is the optimal protein content (and protein–energy ratio) for standard operational rations, and specifically, is the Military Recommended Dietary Allowance (MRDA) for operational rations (100 g/d for men and 80 g/d for women) appropriate? Is the protein MRDA for women appropriate during pregnancy and lactation?
3. Is there evidence that supplementation with specific amino acids (AAs) or modification of dietary protein quality would optimize military performance, either cognitive or physical, during high workload, psycho-

logical stress, or energy deficit? What are the risks of amino acid supplements and high-protein diets?

METHODS

In considering the questions posed by the Army, the CMNR collaborated with the Subcommittee on Body Composition, Nutrition, and Health of Military Women. The two committees convened a workshop in March 1997, in Washington, D.C., to bring together experts in protein metabolism, exercise physiology, and cognitive psychology, as well as military nutritionists familiar with historical and recently completed nutritional assessment surveys. Each of the workshop speakers was asked to prepare a review paper. After the workshop, the committees summarized the presentations, and drawing upon their summaries of the workshop, the commissioned papers, background materials provided by the speakers, focused literature searches by the staff, and their own expertise in the field of protein metabolism, they prepared an overview of the pertinent issues, responses to the sponsor's questions, conclusions, and recommendations. These form chapters 1 and 2 of this report. The authored papers, which can be found in Chapters 3 through 16, were not subject to review.

FINDINGS

Effect of Stressors on Protein Requirements of Military Personnel

Baseline Protein and Amino Acid Requirements

Current estimates of protein and amino acid requirements are based on studies employing the technique of nitrogen balance assessment. A 1985 report of the Food and Agriculture Organization, in collaboration with the World Health Organization and the United Nations University (FAO/WHO/UNU, 1985) proposed a protein requirement of 0.625 g per kilogram of body weight per day (g/kg BW/d) for egg or beef protein, so-called high-quality proteins. The current recommended dietary allowance (RDA) for protein in the U.S. diet is 0.8 g/kg BW/d for adults (Table ES-1; NRC, 1989).

Consensus exists for the current adult protein requirement but not the adult requirement for total essential (indispensable) amino acids (IAAs). Based on nitrogen balance data, the recommendation for IAAs as a percentage of total protein intake is 43 percent for children and only 11 percent for adults (FAO/WHO/UNU, 1985). Since the 1985 report, Young and colleagues have presented data showing that the FAO/WHO/UNU pattern cannot maintain

TABLE ES-1 Recommended Dietary Allowances for Protein

Age (years) or Condition	Weight (kg)	RDA g/d	RDA g/kg BW/d
Males			
19–24	72	58	0.8
25–50.	79	63	0.8
51+	77	63	0.8
Females			
19–24	58	46	0.8
25–50	63	50	0.8
51+	65	50	0.8
Pregnant		60	
Lactating (first 6 months)		65	
Lactating (second 6 months)		62	

SOURCE: Adapted from NRC (1989).

amino acid homeostasis. From these data they conclude that the adult IAA requirement is 31 percent of the total protein requirement, approximately 3 times the FAO/WHO/UNU estimate (Marchini et al., 1993; see also Young, Chapter 10). The importance of this debate lies in the definition of dietary protein quality, which is an assessment of the efficiency of protein utilization, determined in part by how closely the IAA content of a given protein resembles an IAA scoring pattern for the age group in question. The scoring pattern is based on the estimated IAA requirements for that age group. If the 1985 FAO/WHO/UNU estimate of adult IAA requirements is correct, protein quality is no longer an issue for adults, either here in the United States or worldwide.

Young's hypothesis regarding adult IAA requirements is based on estimations of obligatory indispensable amino acid oxidation computed from obligatory nitrogen (N) losses and studies of amino acid oxidation using stable isotopically labeled amino acid infusion. This hypothesis has been challenged most persistently by Millward (1994; see also Chapter 9). Millward's assertion that Young's data are flawed is based on several arguments, the most serious of which are that the amount of tracer infused in Young's experiments was great enough to alter amino acid balance significantly and that the true precursor amino acid enrichment was not measured accurately. In addition, Millward argues that the real metabolic demand for IAA is based only on net protein synthesis, not turnover. There is also disagreement on the nutritional status of subjects in previous studies that assessed protein and AA requirements and the need to employ subjects who have adapted to the experimental protein intake.

The practical implications of the debate between Millward and Young center around the requirements for the IAA lysine. The lysine content of cereal grain proteins, which are considered low-quality proteins, is too low to support the growth of infants. However, if Millward is correct in his assessment of IAA requirements, all dietary proteins, whether animal or plant in origin, contain enough lysine to provide adequate amino acids to adults if these proteins are consumed in quantities that meet overall protein requirements. Both Millward and Young agree that further experiments are necessary to clarify adult IAA requirements.

Physical Activity

Despite a commonly held belief that athletes, particularly body builders, have greater requirements for dietary protein than sedentary individuals, the evidence in support of this contention is controversial. A major function of amino acids in muscle during exercise is to supply intermediates to the tricarboxylic acid (TCA) cycle so that the energy needs of the contractile apparatus of the muscle are met. The extent of this amino acid oxidation depends on exercise intensity, nutritional status (energy stores), training status, and gender. However, it is not clear that the magnitude of this amino acid oxidation is sufficient either to increase whole-body protein requirements when energy intake matches output or to warrant increasing the MRDA for protein.

Efforts to investigate the possibility that protein requirements are increased by physical activity have been complicated by several factors. First, increases in protein intake result in increases in the catabolic processes that oxidize amino acids. This adaptation has significant implications for individuals who habitually consume high-protein diets: these individuals may face the risk of significant loss of protein stores if circumstances such as field operations suddenly force them to curtail protein intake. Another factor confounding the effort to examine the influence of physical exertion on protein requirements is the finding that energy deficit results in an increased protein requirement. Finally, a number of studies have suggested that regular moderate physical activity may have a protective effect on protein retention. Some studies have suggested that the protein requirements of athletes may be as high as 2–2.5 g/kg BW/d. The most carefully controlled studies have shown that endurance exercise increases protein requirements in men. However, in male weight lifters, the maintenance of body mass requires no increase over the requirements of sedentary individuals. The small number of studies of physical activity and protein requirements that have included women show that the needs of women appear to be similar to those of men. Thus, the evidence to date suggests that sustained endurance exercise, such as that done by some military personnel, may increase protein requirements over the amounts required by sedentary individuals; however, typical intakes (1.5 g/kg BW/d) are usually well within the range of requirements as long as energy intake is adequate to meet energy expenditure.

Infection, Injury, and Illness

Systemic infections and serious injuries trigger alterations in body protein metabolism that lead to increases in protein requirements. The contractile proteins of skeletal muscles undergo rapid breakdown to supply amino acids for energy and for specific immune responses in the liver and immune tissues. Protein requirements are increased to approximately 1.5 g/kg BW/d in almost all trauma patients except burn patients, whose requirements are elevated to 2–2.5 g/kg BW/d (although treatment advances are slowly lowering these requirements). However, the provision of high levels of protein to severely burned patients fails to induce repletion of muscle protein stores. The administration of anabolic hormones, such as growth hormone, insulin-like growth factor, and testosterone, in conjunction with nutritional support is the subject of considerable research at the present time, and a number of positive studies have been reported indicating accelerated wound healing. However, two recent trials of recombinant human growth hormone administration to postsurgical and posttrauma patients in several intensive care units were discontinued due to higher mortality rates among the treated patients.

Other Stressors

The influences of temperature extremes and high altitude on protein requirements have been discussed in earlier CMNR reports. Although sweat losses of nitrogen can be considerable, the need for protein does not appear to increase in hot climates. Similarly, protein needs are not increased in cold temperatures. The need to increase fluid intake when consuming a high-protein diet has led to recommendations that excesses in dietary protein intake be avoided in environments where access to drinking water may be a problem. Decreases in lean body mass observed during acclimatization to high altitude appear to be due to the overall decrease in energy intake, rather than to an increased requirement for protein. The effects of combined stressors, such as intense physical activity superimposed on change in climate or altitude, on protein requirements have not been investigated systematically.

The MRDA for Protein

The current RDA for protein for men and women (the nonmilitary population) is 0.8 g/kg BW/d (see Table ES-1). In comparison, the MRDA for protein is 100 g/d for men and 80 g/d for women. These recommendations are for physically active individuals in temperate climates with energy intakes of 3600 kcal/d for men and 2000–2800 kcal/d for women. Thus, for active-duty men and women of current mean weight 78 kg and 63 kg, respectively, the MRDA is approximately 1.3 g/kg BW/d. The formulation of operational rations is based on the MRDAs.

Studies conducted in the 1940s and throughout the past decade have shown that military personnel maintain relatively high protein intakes both in the field and in garrison. Protein intake in the field averages 1.3 g/kg BW/d for men and 1.2 g/kg BW/d for women, exceeding the estimated protein requirement for sedentary individuals and the increased requirement observed by several investigators for athletes. Further, dietary surveys indicate that soldiers in the field preferentially eat the higher-protein components of rations, resulting in mean protein intakes of 86–132 g/d for men and 68–96 g/d for women.

Because protein metabolism is influenced by energy intake and expenditure, these factors must be considered in judging the adequacy of soldiers' protein intake. Studies conducted on Ranger trainees revealed that although these soldiers experienced an average daily energy deficit of 1200 kcal (over 8 weeks), protein intake averaged more than 100 g/d. Nevertheless, the loss of lean tissue mass was significant and was inversely proportional to initial body fat stores, suggesting that negative energy balance contributed to this loss.

Thus, in the absence of additional data on energy balance, the MRDA for protein and the protein intake of the average soldier appear to be more than adequate. Data are insufficient to establish an ideal protein–energy ratio.

Pregnancy and Lactation

Currently, there are no MRDAs for pregnancy or lactation. As shown in Table ES-1, the RDA for protein is increased by 10 g/d for pregnant women and 15 g/d for lactating women. The recommended protein intakes for women in the weight range, of 46–63 kg would be 44–57 g/d during pregnancy and 60–72 g/d during lactation. Some studies have suggested that the RDA for protein during lactation is inadequate and have shown that actual protein requirements may be as high as 1.5 g/kg BW/d (69–94 g/d) (Motil et al., 1996). The MRDA of 80 g/d would therefore be sufficient to meet the apparent protein requirements of pregnant or lactating women.

Benefits and Risks of Protein and Amino Acid Supplements and Alternative Sources of Dietary Protein

Evidence from recent Army surveys suggests that in addition to the high dietary protein intake of the average soldier, many soldiers use protein and amino acid supplements in the belief that these products will improve performance (Warber et al., 1996). In addition, evidence suggests that increasing numbers of soldiers, particularly women, are consuming diets largely or completely lacking in animal products, with the result that dietary protein is increasingly of plant origin. The potential health benefits and risks of protein and amino acid supplements, high-protein diets, and vegetarian diets were considered.

Protein and Amino Acid Supplements and Cognitive Performance

It has been known for some time that the synthesis in brain of several neurotransmitters (serotonin and the catecholamines, dopamine and norepinephrine) is influenced by the levels of their precursor amino acids (tryptophan and tyrosine, respectively) in the diet. In laboratory animals, the rates of synthesis of these transmitters respond to changes in tryptophan and tyrosine concentrations that occur following single meals, as well as to chronic changes in dietary protein content. Of greatest potential relevance to the military, brain tyrosine and tryptophan concentrations in animals can decrease when protein intake falls below the required level, leading to reductions in synthesis of serotonin and the catecholamines. Under field conditions, which are typically associated with increased physical activity and stress (conditions that stimulate the turnover of serotonin and the catecholamines), such changes may be relevant. Reduced food intake could impede serotonin and catecholamine synthesis at a time when increased amounts of these neurotransmitters may be required to support important cognitive functions.

However, relatively little is known about how reduced serotonin synthesis might impede brain function. Serotonin neurons are most active when animals are awake and physically active and are important in channeling sensory information to the brain. Thus, it is possible that some aspects of sensorimotor function might be diminished under conditions of extended energy and protein deprivation. No studies have been conducted to evaluate the possible benefit of protein or tryptophan supplementation under such conditions. Catecholamine neurons increase their activity in response to a variety of stressful conditions and appear to regulate behavioral parameters such as attention and state of arousal. A 1994 CMNR report presented preliminary evidence that tyrosine supplements might improve some types of cognitive performance under stressful conditions and recommended further research to explore the potential value of tyrosine supplements for enhancing soldier performance (IOM, 1994). Only one subsequent study has shown that tyrosine administration appears to restore marksmanship performance degraded by fatigue and cold, but the effects of tyrosine are even less evident under normal circumstances. Thus, although earlier studies appeared to show promise, there have been no more recent studies to reevaluate the earlier findings.

There is little evidence at present, therefore, to justify the administration of either tryptophan or tyrosine for the purpose of modifying behavior. Further research is necessary on the behavioral effects of such supplements as well as the effects of meal-related changes in brain serotonin and catecholamines.

Protein, Amino Acids, Muscle Mass, and Physical Performance

Current methods for routine assessment of muscle mass remain inadequate, particularly when applied to the measurement of changes in muscle mass. In addition, the relationship between muscle mass and measures of function such as strength and endurance remains unclear. Strength appears to be related to the synthesis of both myosin heavy chain and the mitochondrial enzyme cytochrome C oxidase. Apparent gender differences in endurance disappear with appropriate adjustment for differences in muscle mass.

The impact of dietary protein intake and physical activity on muscle protein synthesis is modulated by the hormones insulin, growth hormone, insulin-like growth factor I (IGF-I), thyroid hormone, cortisol, and particularly testosterone. Testosterone, growth hormone, and IGF-I all stimulate muscle protein synthesis, whereas insulin appears to inhibit protein breakdown. Cortisol, which is secreted in response to stress, stimulates the breakdown of muscle protein, although its mechanisms of action are not precisely known.

Although there is considerable research interest in the potential for supplemental protein or amino acids to increase muscle protein synthesis or physical performance, little evidence supports such a role in the absence of anabolic steroids. The "central fatigue" theory, which hypothesized that the administration of branched-chain amino acids would decrease or delay the central nervous system and muscle fatigue that resulted from increased brain tryptophan, is not supported by evidence. In contrast, a postexercise feeding regimen combining carbohydrate and protein appears to increase the endurance of individuals performing strenuous exercise over several days.

Protein, Amino Acids, and Immune Function

In addition to the increased requirement for protein in individuals suffering systemic infection or major injury, growing evidence appears to suggest a potentially beneficial role for the amino acid glutamine in modulating immune function. Studies of hospitalized, critically ill patients show that glutamine supplementation of total parenteral nutrition solutions results in increased survival and decreased length of hospital stay (Wilmore et al., 1997a). However, no effect of glutamine supplementation on the parameters of immune function was observed in Special Forces trainees, whose increased rate of infection and depressed immune function may be linked to the high level of physical exertion during training (Shippee et al., 1995). In addition, the influence of glutamine supplementation on infection rate in athletes (the so-called overtraining effects) has not been studied extensively.

Issues of Protein Quality and Timing of Consumption

Considerable debate currently surrounds the requirement of adults for high-quality protein. Consumption of plant proteins such as soy protein or a combination of cereal and legume proteins in place of all or part of the animal protein in the diet, which results in a diet lower in total fat, saturated fat, and cholesterol and higher in soluble fiber and other potentially beneficial substances, has been shown to lower serum cholesterol and triglyceride levels. Among the plant proteins, soy protein has a balance of amino acids that most closely resembles that of animal proteins, and substitution of this protein in rations would not markedly decrease their nutritional value. However, the taste might be affected; thus, product development and testing are required to ensure the acceptance of soy products in rations. As mentioned above, research suggests that protein and carbohydrate meals consumed shortly after exercise may improve glycogen storage and maintenance of lean mass, resulting in increased endurance on subsequent days.

Dietary Protein and Renal Function

Chronic high levels of dietary protein have been associated with decline in renal function in aging individuals; however the evidence is not sufficient to conclude that dietary protein itself is the causative factor. In contrast, high-protein diets can stimulate nephrolithiasis (kidney stone formation) in otherwise healthy individuals. Because a restriction of fluid intake can increase the likelihood of stone formation, deployed military personnel are at especially high risk if water supplies are inadequate.

Dietary Protein and Calcium Status

The potential calciuretic effect of high-protein diets has been recognized for some time in men and women. For each 50 g increment in dietary protein, an estimated 60 mg additional urinary calcium is lost. This loss may be modulated by other dietary factors such as phosphorus. At present, there is limited evidence to suggest that high dietary protein intake may be a risk factor for osteoporosis. Calcium balance appears to remain close to equilibrium with protein intakes up to 74 g/d and calcium intakes of 500–1400 mg/d. In addition, there is evidence to suggest that regular weight-bearing physical activity contributes to bone strength. Thus, protein intake would not appear to be a major risk factor for calcium loss in the average military woman.

Amino Acid Toxicity

The risk of deleterious interactions between pharmacological doses of single amino acid supplements and over-the-counter or prescription medications may be considerable. In addition, supplement purity is a major consideration, as demonstrated by the harm that resulted from widespread use of tryptophan supplements that were later believed to have contained a low-level contaminant. Lack of safety data on the consumption of high levels of individual amino acids by normal, healthy individuals suggests that their use be limited until further research is performed.

Protein and Amino Acid Supplements and Pregnancy

Evidence from a major nutritional supplementation study suggests that the use of high-protein supplements (to achieve a protein intake that represents 34 percent of the total daily energy intake, in contrast to the more usual 11–14 percent) may be harmful to fetal development. Studies in laboratory animals suggest that supplementation of the diet with single amino acids also may be dangerous to the developing fetus.

CONCLUSIONS AND RECOMMENDATIONS

Effects of Stressors on Protein Requirements

At the present time, considerable controversy exists regarding the validity of estimations of protein and amino acid requirements. In addition, the evidence is insufficient to conclude that high levels of physical activity increase protein requirements for individuals whose energy intake matches their output. Continuous excessive intake of protein may cause increased protein catabolism, resulting in greater risk of negative protein balance when protein intake is reduced.

As emphasized in earlier IOM reports (IOM, 1992, 1995), the importance of adequate energy intake (sufficient to match output and to avoid weight loss) and protein intake should be emphasized to soldiers as the primary means of maintaining lean tissue mass. Research is needed to resolve the controversy regarding the adult requirement for indispensable amino acids and to quantitate more precisely the effect of energy deficit on protein and indispensable amino acid requirements.

Systemic infections and serious injuries clearly increase protein requirements. However data suggest that in patients recovering from burns, the elevation of dietary protein intake alone does not permit the recovery of muscle mass to begin immediately, because of the development of the acute phase response, which is accompanied by changes in hormonal status. Research on the effects of treatment with anabolic hormones is ongoing.

Military researchers and physicians should pay careful attention to civilian research on the effects of treatment with anabolic hormones on recovery from burns and other injuries. Where appropriate, military-specific models should be developed.

Military Recommended Dietary Allowances for Protein

Without more data on the functional implications of varying protein intakes, it is not possible to define with accuracy the optimal protein content of standard operational rations. However, based on currently available data, the use of the MRDA for operational rations is appropriate and provides a generous level of protein intake. The MRDA covers the protein requirements of pregnant and lactating women.

Current MRDAs for protein should be maintained. Provided that energy intake is adequate, no increase in MRDAs is necessary for pregnant or lactating women.

Benefits and Risks of Supplemental Protein, Amino Acids, and Alternative Sources of Dietary Protein

Research fails to support the use of protein supplements to facilitate muscle building under conditions of adequate energy and protein intake.

Given adequate nutritional intake, soldiers should not use protein supplements for muscle building. Military researchers and physicians should pay careful attention to civilian research on the use of anabolic hormones to increase muscle or lean tissue mass.

At the present time, considerable debate surrounds the adult requirement for indispensable amino acids and thus high-quality proteins.

Protein supplied in operational rations should be of high quality and digestibility. Energy intakes should be adequate, and sources

of energy should be consumed within 2 h of an intense bout of endurance exercise, to replace depleted muscle glycogen.

There is a lack of safety data on the consumption of high levels of individual amino acids by normal, healthy individuals. Furthermore, research supporting the use of tyrosine supplements to enhance cognitive performance under field conditions is inconclusive.

Single amino acid supplements should not be used to modify cognitive performance, due to potential toxicity and insufficient evidence of efficacy.

Supplemental glutamine and arginine have yet to show conclusively beneficial effects on immune function. The MRDA, if consumed, provides adequate protein and energy to sustain immune function under normal field conditions.

The military should test the ability of supplemental glutamine and arginine to enhance the immune response and decrease rates of infection under field conditions and in seriously injured hospitalized patients.

Current intakes of protein among military populations are high and show no apparent harmful effects, provided fluid intake is adequate. There is no evidence of increased health risks from a high intake of dietary protein; however, an amino acid imbalance may be created with use of single amino acid or protein supplements. Although no data are available from groups similar in age and fitness characteristics to military personnel, a review of the information available shows that high protein intake is not associated with direct effects on renal dysfunction. However, high-protein diets may indirectly stimulate renal stone formation and result in an increased renal workload, due to the need to concentrate urine. High protein intake has been shown to increase urinary calcium loss, but there is no definitive evidence that the level of protein intake observed in Army women in field conditions represents a risk factor for osteoporosis.

Given the high protein content of operational rations, adequate fluid intake should be emphasized, as recommended by the "Fluid Doctrine" (IOM, 1994).

BOX ES-1 Major Recommendations

- As recommended in earlier IOM reports (IOM, 1992, 1995), the importance of adequate nutrient intake (with sufficient energy to match output and to avoid weight loss) should be emphasized to soldiers as the primary means of maintaining lean tissue mass.
- Given adequate nutritional intake, soldiers should not use protein supplements for muscle building.
- Single amino acid supplements should not be used to modify cognitive performance, due to potential toxicity and insufficient evidence of efficacy.
- Current MRDAs for protein should be maintained. Provided that energy intake is adequate, no increase in MRDAs is necessary for pregnant or lactating women.
- Protein supplied in operational rations should be of high quality and digestibility.
- Energy intakes should be adequate, and source of energy should be consumed within 2 h of an intense bout of endurance exercise, to replace depleted muscle glycogen.
- Military researchers and physicians should pay careful attention to civilian research on the effects of treatment with anabolic hormones on recovery from burns and other injuries. Where appropriate, military-specific models should be developed.
- The military should test the ability of supplemental glutamine and arginine to enhance the immune response and decrease rates of infection under field conditions and in seriously injured hospitalized patients.
- Given the high protein content of operational rations, adequate fluid intake should be emphasized, as recommended by the Fluid Doctrine (IOM, 1994).

REFERENCES

FAO/WHO/UNU (Food and Agriculture Organization of the United Nations/World Health Organization/United Nations University). 1985. Energy and protein requirements. Report of a joint expert consultation. World Health Organization Technical Report Series No. 724. Geneva: World Health Organization.
IOM (Institute of Medicine). 1992. A Nutritional Assessment of U.S. Army Ranger Training Class 11/91. March 23. Washington, D.C.: National Academy Press.
IOM. 1994. Food Components to Enhance Performance, An Evaluation of Potential Peformance-Enhancing Food Components for Operational Rations, B.M. Marriott, ed. Washington, D.C.: National Academy Press.
IOM. 1995. Not Eating Enough, Overcoming Underconsumption of Military Operational Rations, B.M. Marriott, ed. Washington, D.C.: National Academy Press.
Marchini, J.S., J. Cortiella, T. Hiramatsu, T.E. Chapman, and V.R. Young. 1993. Requirements for indispensable amino acids in adult humans: Longer term amino acid kinetic study with support for the adequacy of the Massachusetts Institute of Technology amino acid requirement pattern. Am. J. Clin. Nutr. 58:670–683.
Millward, D.J. 1994. Can we define indispensable amino acid requirements and assess protein quality in adults? J. Nutr. 124:1509S–1516S.
Motil, K.J., T.A. Davis, C.M. Montandon, W.W. Wong, and P.D. Klein. 1996. Whole-body protein turnover in the fed state is reduced in response to dietary protein restriction in lactating women. Am. J. Clin. Nutr. 64:32–39.
NRC (National Research Council). 1989. Recommended Dietary Allowances, 10th ed. Washington, D.C.: National Academy Press.
Shippee, R.L., S. Wood, P. Anderson, T.R. Kramer, M. Neita, and K. Wolcott. 1995. Effects of glutamine supplementation on immunological responses of soldiers during the Special Forces Assessment and Selection Course [abstract]. FASEB J. 9:731.
Warber, J.P., F.M. Kramer, S.M. McGraw, L.L. Lesher, W. Johnson, and A.D. Cline. 1996. The Army Food and Nutrition Survey, 1995–97. Technical Report. Natick, Mass.: U.S. Army Research Institute of Environmental Medicine.
Wilmore, D.W. 1997a. Glutamine saves lives! What does it mean? Nutrition 13(4):375–376.

I

Committee Summary and Recommendations

Part I of this report provides the Committee on Military Nutrition Research's (CMNR) overview and summary of key issues in protein metabolism, its response to specific questions posed by the Army, and the committee's conclusions and recommendations. The CMNR was requested by the U.S. Army Medical Research and Materiel Command and the Military Nutrition and Biochemistry Division of the U.S. Army Research Institute of Environmental Medicine to review the state of knowledge on protein requirements and determine if the Military Recommended Dietary Allowance (MRDA) need to be revised.

In Chapter 1, the committee presents an overview of the project using relevant background materials and the proceedings of the workshop held in March 1997 to provide a summary of key issues in determining protein requirements, various militarily relevant stressors that may influence protein requirements, and the benefits and risks of supplemental protein or individual amino acids.

The committee's response to the three questions posed by the Army, listed below, and its conclusions and recommendations are presented in Chapter 2.

1. Do protein requirements increase with military operational stressors, including high workload with or without energy deficit? Are there gender differences in protein requirements in endurance exercise?

2. What is the optimal protein content (and protein–energy ratio) for standard operational rations, and specifically, is the protein MRDA for operational rations (100 g/d for men and 80 g/d for women) appropriate? Is the protein MRDA for women appropriate during pregnancy and lactation?

3. Is there evidence that supplementation with specific amino acids (AAs) or modification of dietary protein quality would optimize military performance, either cognitive or physical, during high workload, psychological stress, or energy deficit? What are the risks of amino acid supplements and high-protein diets?

1

Committee Overview

INTRODUCTION

Proteins form the major constituents of muscle, catalyze virtually all chemical reactions in the body, regulate gene expression, and comprise the major structural elements of all cells. Individual amino acids, the components of proteins, also serve as neurotransmitters, hormones, and modulators of various physiological processes. Every aspect of physiology involves proteins. According to Bier (see Chapter 5), credit for the name "protein" is given to the Dutch chemist Gerardus Johannes Mulder, who wrote an article in French that was published in a Dutch journal on July 30, 1838. In this article, he asserted that this material was the essential general principle of all animal body constituents and defined it by the Greek word *proteus* (which he translated to the Latin, *primarius*, meaning primary). Mulder appears to have taken this word directly from a letter sent to him by the Swedish chemist Jacques Bursailleus on July 10, 1838, in which the name *protein* had been suggested. Aside from the amazing fact of a Dutch chemist borrowing a Latin word from a Swedish chemist, which he defined in Greek in an article written in French for a Dutch journal, the entire sequence of events appears to have occurred in a period of 20 days, demonstrating the efficiency of both mail service and scientific publication in those days.

The relationship between dietary protein and bodily protein metabolism is a major focus of research today. Many questions remain regarding the validity of methods for assessing protein balance; thus, the question of how best to assess dietary protein requirements remains unanswered. In addition, the influence of genetics, hormones, physical activity, infectious processes, and environmental stresses on protein metabolism and protein requirements continues to be explored. Another major focus of research that is of great interest is the role of dietary protein and amino acids in modulating physiological function, as measured for example by physical and mental performance. The possibility that protein or individual amino acids in quantities that exceed those required to maintain protein balance may have the potential to contribute to performance optimization is of great interest.

THE ARMY'S INTEREST IN DIETARY PROTEIN AND PROTEIN BALANCE

Because of the unique demands placed on soldiers in combat, the military is particularly concerned about the role that dietary protein may play in controlling muscle mass and strength; response to injury, infection, and environmental stress; and cognitive performance. As described in Chapter 3 by Karl Friedl, the longer, more isolated deployments and maneuvers that are becoming more commonplace may limit access to rations. The nutritional studies of Ranger trainees conducted by the U.S. Army Research Institute of Environmental Medicine (USARIEM) (IOM, 1992, 1993b) identified losses of up to 30 percent in lean body mass (including organs, such as the liver, plasma, and proteins) after 3 weeks of limited food intake and high energy expenditure. Although increased energy intake offset these losses somewhat, they were still significant, suggesting the need for additional energy, protein, or both. In these studies, the observed decrease in lean body mass was accompanied by changes in serum levels of several hormones including testosterone, insulin-like growth factor I (IGF-I), and triiodothyronine (T_3) (Friedl, 1997; Nindl et al., 1997), the significance of which is unclear. Because the administration of these hormones is known to stimulate protein synthesis under some conditions, the Army maintains considerable interest in exploring their potential both to ameliorate the losses in lean body mass sustained by troops under conditions of extreme negative energy balance and to stimulate an increase in muscle mass and physical performance. In contrast to the limited intakes of protein and energy measured in the Ranger studies, a more recent study, in which soldiers subsisted on the field ration known as Meals, Ready to Eat (MREs) for 30 days, showed average energy intakes of 2500 kcal/d and protein intakes of 103 g/d (Thomas et al., 1995), raising questions about the optimum protein-to-energy ratio for performance and health.

As the typical battlefield scenario becomes more automated, soldiers must attend to increasing numbers of signals in the face of increasing amounts and sources of noise, with increasingly dangerous consequences for failure. Thus, the possibility that cognitive performance may depend on diet and that performance optimization may be achievable through dietary modifications such as amino acid supplements is of considerable interest to the military. A 1994 report by the Committee on Military Nutrition Research (CMNR), entitled *Food Components to Enhance Performance*, briefly considered the influence of protein and amino acids (and all other dietary components) on physical and cognitive performance and response to stress (IOM, 1994). Data were presented on the effect of protein-to-carbohydrate ratio on mental alertness, the effect of physical activity on protein requirements, and the influence of branched-chain amino acids, tyrosine, and tryptophan in pharmacological amounts on cognitive function. The report concluded that the potential ability of tyrosine supplements to sustain alertness and cognitive performance in the face of environmental stress merited further investigation.

Finally, the risk of injury and infection faced by soldiers in the field is extremely high. At the same time, conditions of sleep and food deprivation, environmental extremes, and heightened emotional stress all exert a negative impact on the immune system. The CMNR report *Military Strategies for Sustainment of Nutrition and Immune Function in the Field* (IOM, 1999), considered the effects of diet, including protein and individual amino acids such as glutamine, on immune response and concluded that although the role of energy intake in immune function is probably more significant than that of protein, individual amino acids such as glutamine and arginine appear to play crucial roles in modulating immune function. The effects of these amino acids are considered in greater detail in this report.

ESTIMATION OF PROTEIN REQUIREMENTS

Current estimates of protein requirements for mature humans and the methods used to assess these requirements are being scrutinized by the research community and are a source of considerable disagreement.

Protein Metabolism

The requirement for protein arises from growth, from the need to replace obligatory losses, and from the need to respond to environmental stimuli. The

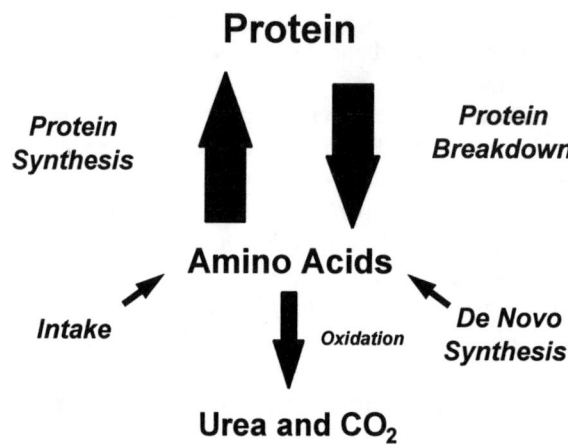

FIGURE 1-1 Pathways of protein turnover. SOURCE: Young and Marchini, 1990.

breakdown products of protein—amino acids—enter the free amino acid pool of the body (distributed among body fluids and tissues) from four sources: (1) dietary proteins; (2) so-called dispensable (nonessential) amino acids, which can be synthesized in the body; (3) breakdown products of body protein (particularly skeletal muscle, the largest tissue in the body and the site of protein storage); and (4) the products of recycling by intestinal microbes (Figure 1-1). In mature humans, a homeostatic mechanism maintains the balance between tissue protein synthesis and breakdown by drawing on the free amino acid pool.

Methods for Assessment of Protein Requirements

Because the majority of nitrogen in the body is associated with protein and amino acids, nitrogen has been used as a marker for assessing whole-body and tissue protein flux and status. The traditional method for assessing whole-body protein metabolism is nitrogen balance, where nitrogen (N) intake and output (in feces, sweat, and urine, as well as other miscellaneous sources) are measured and the difference $[N_{bal} = (N_{in} - N_{out})]$ is expressed in grams of nitrogen per day (g N/d). Total body protein loss or retention is then calculated using the conversion factor of 6.25 g N/g protein (Munro and Crim, 1994).

A state of positive nitrogen balance exists when the total nitrogen output is less than the total nitrogen ingested. Positive nitrogen balance requires adequate protein and energy intake plus a stimulus for synthesis. A state of positive nitrogen balance (an anabolic state) exists for the synthesis of new tissues during the growth observed in childhood, adolescence, and pregnancy. When dietary protein or energy intake is inadequate or an individual experiences an acute-phase response, nitrogen excretion exceeds nitrogen intake and a state of negative nitrogen balance exists (net protein catabolism). When dietary protein is adequate or more than adequate and energy intake matches energy output, a state of nitrogen equilibrium exists ($N_{in} = N_{out}$ for any intake above the required level). A state of nitrogen equilibrium is required to maintain total body protein mass. Altering protein or energy intake or physical activity may alter nitrogen balance.

Despite the usefulness of nitrogen balance assessment in estimating the adequacy of protein intake, there are significant limitations to its use, including overestimation of nitrogen intake and incomplete collections of urine, feces, or sweat, which result in an underestimation of nitrogen output. The net outcome is an overestimate of nitrogen retention and of the body's ability to adapt to inadequate protein intakes; these overestimates limit the ability of the nitrogen balance technique to assess nitrogen requirement. The limitations of nitrogen balance assessment are discussed further below and by Millward and Young in Chapters 9 and 10, respectively.

In the mid-1940s, stable isotopes of hydrogen (^2H) and nitrogen (^{15}N), were made available for use in biomedical research. However, the mass spectrometry technology that would use these isotopes for rapid analysis of biological specimens was not widely available until the late 1970s. With the improvement of this technology and the widespread availability of stable isotope-labeled metabolites, amino acid kinetic studies have come to augment nitrogen balance in examining the effects of dietary protein, energy, and physical activity on overall protein metabolism.

Amino acids labeled with stable isotopes of hydrogen (^2H), nitrogen (^{15}N), and carbon (^{13}C) have been administered orally and intravenously. With the use of the primed continuous infusion technique, amino acid turnover can be studied in subjects of all ages under many physiological conditions (Munro and Crim, 1994). The calculation of amino acid flux (Q) is based on the following assumptions: (1) the body's free amino acid pool is a homogeneous mixture that can be sampled from the plasma pool; (2) the only sources of the target amino acids entering the body pool are dietary protein (I) and intracellular protein breakdown (B); and (3) amino acid removal from this pool occurs by irreversible oxidation (E) or synthesis into protein (Z). In reality, a large quantity of recycled amino acids is derived daily from the breakdown of body proteins. In addition to the sizable turnover of blood cells, mucosal cells of gastric and intestinal villi are continuously moved toward villus tips where they slough off and undergo digestion; released free amino acids are then reabsorbed into the plasma pool (Munro and Crim, 1994). Thus, the equation $Q = B + I = E + Z$ describes the steady-state relationship in which the total entry of amino acids into the free amino acid pool ($B + I$) is equal to the total exit of amino acids from the free amino acid pool ($E + Z$). Rates of protein synthesis and protein breakdown can be calculated from this equation (Picou and Taylor-

Roberts, 1969). At isotopic steady state, total amino acid turnover (Q) is measured, and the rate of protein breakdown can be calculated knowing the rate of amino acid intake. Likewise, the rate of protein synthesis can be calculated when the rate of amino acid disappearance is known. If a ^{13}C-labeled amino acid is used, oxidation can be measured from $^{13}CO_2$ excretion rates.

Because of its unique role as an amino acid that is oxidized in skeletal muscle and not converted to a tricarboxylic acid (TCA) cycle intermediate, leucine (in the ^{13}C form) has been the amino acid of choice for many amino acid kinetic studies. However, due to its unique metabolism, it may not be representative of the entire pool of amino acids. Glycine labeled with ^{15}N has also been used extensively to study protein synthesis and breakdown because it has the advantage of ubiquitous utilization.

FAO/WHO/UNU Requirements and RDAs: Current Estimates of Average Protein Intake

Estimations of protein and amino acid requirements are currently based on nitrogen balance studies. The 1985 report of the Food and Agriculture Organization (FAO), World Health Organization (WHO), and United Nations University (UNU) proposed a protein requirement of 0.625 g per kilogram of body weight per day (g/kg BW/d) for egg or beef protein and a "safe" level of 0.75 g/kg BW/d for mixed protein if the protein is as digestible as egg or beef (FAO/WHO/UNU, 1985). The current recommended dietary allowance (RDA) for protein in the U.S. diet (which is derived by adding two standard deviations to the estimated requirement) is 0.8 g/kg BW/d for adult men and women (Table 1-1) (NRC, 1989).

Also based on nitrogen balance data, the recommendation for total essential or indispensable amino acids (IAAs) as a percentage of protein intake is 43 percent for infants and 11 percent for adults (FAO/WHO/UNU, 1985). Essential (indispensable) and nonessential (dispensable) amino acids are traditionally distinguished on a nutritional basis because essential amino acids cannot be synthesized by the body and must be part of the diet to permit growth or to maintain nitrogen balance, whereas nonessential amino acids can be synthesized by the body. Metabolically, however, the distinctions are less clear because a number of essential amino acids can be formed by transamination (at least in laboratory animals). By this criterion, only the amino acids lysine and threonine appear not to be synthesized by transamination and are therefore indispensable (as discussed further below, the concentrations of these two amino acids in cereal proteins are so low as to limit their ability to sustain growth). By this same argument, glutamic acid and serine are the only truly dispensable amino acids because they can be synthesized by reductive amination of ketoacids. A

TABLE 1-1 Recommended Dietary Allowances for Protein

Age (years) or Condition	Weight (kg)	RDA g/d	RDA g/kg BW/d
Males			
19–24	72	58	0.8
25–50	79	63	0.8
51+	77	63	0.8
Females			
19–24	58	46	0.8
25–50	63	50	0.8
51+	65	50	0.8
Pregnant		60	
Lactating (first 6 months)		65	
Lactating (second 6 months)		62	

SOURCE: Adapted from NRC (1989).

third class—the conditionally essential amino acids—is synthesized from other amino acids. However, this synthesis is confined to particular organs and may be limited by certain physiological factors such as age or disease state (Reeds and Beckett, 1996). As knowledge increases and techniques improve, the distinction between essential and nonessential amino acids becomes less clear. Adding to this lack of clarity are observations such as the one by Stucky and Harper (1962), who found that if rats were fed a diet adequate in nitrogen but lacking in nonessential amino acids, the growth rate of the animals was significantly decreased.

Importance of the Debate over Indispensable Amino Acid Requirements

Although consensus exists at present for the adult protein requirement this is not the case for the adult requirement of indispensable amino acids. Since the 1985 FAO/WHO/UNU report, Young and coworkers have presented data that contradict the findings of the report; based on these data, Young suggests that the adult requirement for total IAAs is 31 percent of the protein requirement, or about three times the FAO/WHO/UNU estimate (McLarney et al., 1996; Young, 1987, 1994; Young and El-Khoury, 1995a; Young and Marchini, 1990; Young et al., 1989; see also Chapter 10). This contention of the group at Massachusetts Institute of Technology (MIT) for higher indispensable amino acid needs has been countered by Millward and colleagues (Millward, 1994; Millward and Rivers, 1988, 1989; see also Chapter 9), who find significant methodological problems in the studies of Young and coworkers. This debate is important, because it influences whether or not protein quality is an issue to be considered in setting protein requirements. Protein quality, a measure of the efficiency with which dietary protein is utilized, can be assessed by comparison of the amino acid profile of a given protein to various amino acid scoring patterns such as those developed by the FAO/WHO for various age groups. If the requirement for IAAs is low (as proposed by FAO/WHO and Millward), the pattern is easily matched by most proteins, and protein quality ceases to be an issue in setting protein requirements for adults. However, if the FAO estimates are incorrect and indispensable amino acids are required in the higher amounts proposed by Young, individual protein sources may duplicate the scoring patterns poorly, and protein quality may then become a significant determinant of protein requirements.

Argument for Higher Indispensable Amino Acid Requirements

Young has based his argument for higher indispensable amino acid requirements on two related measures: the obligatory oxidative losses of these amino acids and the calculated obligatory losses based on daily nitrogen loss. In the latter calculation, Young assumes that the efficiency of dietary protein use is about 70 percent and that the lost protein has the composition of mixed body protein. Indispensable amino acid requirements calculated in these two ways (the MIT pattern) are approximately the same. In 1991, an expert panel of FAO/WHO also agreed that the IAA needs for adults are greater than those in the 1985 report and proposed that the amino acid pattern for preschool children (FAO/WHO, 1991), a pattern similar to the MIT pattern, be recommended for adults. Young argues that protein and indispensable amino acid intakes have to be high enough to provide sufficient flux for optimum "metabolic control." This concept proposes that a high flux rate of amino acids or other substrates provides a kinetic basis for a sensitive control mechanism to ensure adequate provision of metabolic intermediates. In the case of protein, these important intermediates would be amino acids such as glutamine, tyrosine, and tryptophan, which have important physiological roles to play independent of their incorporation into protein.

To prove their point, Young and colleagues carried out a long-term study to compare the effects of the FAO (FAO/WHO/UNU, 1985), MIT, and egg patterns of indispensable amino acids on amino acid balance in healthy young adults (Marchini et al., 1993). After a week on the egg pattern (high in IAAs), 20 young men were placed on diets resembling either the FAO, the MIT, or the egg pattern for three weeks. Based on a negative leucine balance while the subjects were on the FAO (compared with the MIT) pattern and changes in serum amino acid profiles, Marchini et al. (1993) concluded that the FAO pattern is not capable of maintaining amino acid homeostasis.

Since the 1991 FAO/WHO meeting, several groups have reevaluated the existing data and concluded that the original FAO recommendations were likely to be underestimates but stopped short of endorsing the MIT pattern (Fuller and Garlick, 1994; Waterlow, 1996). In 1994, an expert panel met to consider the issue. After the meeting, the panel recommended that the entire question of how amino acid requirements are determined be reexamined but that, in the interim, the MIT pattern be accepted (Clugston et al., 1996). However, as subsequently pointed out by Millward and Waterlow (1996), this recommendation was not the consensus of the attendees but was inserted during postmeeting editing.

Argument Against Higher Indispensable Amino Acid Requirements

Millward and colleagues have challenged Young's data point by point (Millward, 1994; Millward and Rivers, 1988, 1989; see also Chapter 9). They suggest first that Young's stable isotope amino acid oxidation data, derived from stable isotope-labeled amino acid infusion studies, are flawed for two reasons. First, the amount of tracer used in the infusion studies is itself high enough to influence the oxidation of the amino acid and thus the balance determined. Second, the enrichment of the amino acid precursors being oxidized is not accurately measured, a critical issue in the interpretation of stable isotope research.

Next, Millward argues that there is no valid basis for assuming that the obligatory amino acid losses (as calculated from obligatory nitrogen loss) resemble the pattern of body protein, because some of the amino acids released during normal turnover are known to be preferentially recycled (lysine and threonine). In addition, he believes that the metabolic demand for IAAs is determined not by the need for high flux rates, but by the obligatory losses and the relative ability of the body to adapt on a diurnal basis to varying levels of these amino acids in the diet (he notes that digestive enzymes secreted in response to a meal can, over the short run, assist in meeting the indispensable amino acid needs by breaking down themselves). Finally, Millward points out that in the longer-term study mentioned above (Marchini et al., 1993), nitrogen balance did not differ significantly between the MIT and the FAO patterns; this finding suggests that both patterns support overall body protein economy.

The Rebuttal

Young agrees with Millward that there are inherent difficulties in defining requirements for indispensable amino acids. The two most serious and difficult-to-resolve problems are (1) accounting for the mass of stable isotope infused, which is large enough to affect nitrogen balance, and (2) determining the true precursor enrichment rate of the amino acid being infused and under study. On the first point, the agreement between IAA requirements calculated from oxidation rates and from nitrogen balance leads Young to conclude that the mass of stable isotope infused does not "profoundly" affect the calculation of amino acid oxidation rates. He agrees, however, that this issue deserves more attention.

On the question of true precursor amino acid enrichments in the stable isotope experiments, Young points out that this is a problem primarily for lysine since measurements in experiments with branched-chain amino acids are made from keto acids derived intracellularly from the infused amino acid. Studies using L-[1-^{13}C]phenylalanine as an indicator amino acid for determining the lysine requirement have yielded a requirement of 40 mg/kg BW/d (Duncan et al.; 1996, Zello et al., 1993), an estimate close to Young's own tentative new requirement for lysine (50 mg/kg BW/d). In this technique, the indicator amino acid (labeled phenylalanine) is infused at graded levels of lysine intake, and the "breakpoint" in $^{13}CO_2$ excretion is measured, under the assumption that the uptake of phenylalanine into protein will be sharply decreased and its oxidation sharply increased at the point where lysine intake becomes inadequate.

Young's definition of the maintenance amino acid pattern for adults is generally similar to the amino acid pattern in body protein, except for lysine, threonine, and methionine, whose patterns were derived more from the results of his tracer studies. Young agrees that the body has significant ability to conserve lysine under conditions of inadequate intake. His calculations suggest that the lysine requirement is 30 percent lower than that found in mixed body protein, due to lysine conservation that results from diurnal cycling.

Resolution of the Debate

The practical implications of the debate between Young and Millward revolve primarily around lysine: the lysine content of cereal proteins is limiting for growth. If Millward is correct, then all dietary proteins, whether plant or animal, contain enough lysine and other amino acids to support adequate protein nutriture of adults if consumed in amounts that meet the protein requirement (although some military personnel in the 18–22-year age group are still growing, a factor that might influence the requirement for some amino acids). Millward has shown that wheat protein, a protein that is particularly low in lysine, is well utilized in adults in the postprandial period, even when net protein synthesis occurs. He suggests that the low level of lysine in this protein is supplemented by the tissue free amino acid pools. However, older data from Longenecker (Longenecker, 1961, 1963; Longenecker and Hause, 1959, 1961) show that the ingestion of wheat protein by dogs or humans may result in decreased plasma lysine levels accompanied by increased levels of other indispensable amino acids. Such data support the contention that a postprandial breakdown of body protein may supply the indispensable amino acids necessary for synthesis. However, under such circumstances, other IAAs may be used less efficiently for protein synthesis when lysine is limiting in the protein consumed, which supports Young's belief that the indispensable amino acid requirement is higher than currently recommended. Thus, the controversy over requirements for IAAs is still unresolved.

The implications of this debate for the current state of knowledge of protein and amino acid requirements for the military depend in part on the current intake of dietary protein and amino acids by military personnel and in part on other factors influencing protein requirements in these individuals, as discussed below.

STRESSORS THAT INFLUENCE PROTEIN REQUIREMENTS

As discussed by Friedl in Chapter 3, the stressors encountered most frequently by military personnel are high levels of physical activity with or without energy restriction; illness, injury, and infection; and environmental extremes. Although each of these stressors may somehow influence protein metabolism and protein requirements directly, they also produce changes in hormonal status that can influence protein metabolism as well. The impact of each of these factors on protein metabolism and requirements has been the subject of intense investigation in the civilian research community. A brief summary of relevant findings is presented here.

Physical Activity and Energy Restriction

The question of whether individuals who routinely engage in intensely physical occupational or athletic activities have increased requirements for dietary protein appears to have arisen from the observations that during exercise, muscle protein is utilized for fuel and that exercise can lead to an increase in muscle mass. However, whether protein requirements are in fact increased by physical activity is unclear and a subject of intense controversy. In Chapter 11, Rennie reviews the role of protein and its breakdown products, amino acids, in exercising muscle and discusses changes in protein metabolism induced by energy deficit.

Exercise and Amino Acid Catabolism

A major function of amino acid breakdown in muscle during periods of exercise is to supply tricarboxylic acid intermediates (anaplerosis) so that the oxidation of acetyl coenzyme A (CoA) can proceed at rates appropriate to the energy needs of the contractile apparatus. The exercise-induced increase in muscle alanine production may be a marker for this process. Specifically, glutamate can react with pyruvate, via the action of alanine-aminotransferase, to produce alanine and α-ketoglutarate. The latter then feeds into the TCA cycle, and the former provides a mechanism for shuttling nonacidic gluconeogenic precursors to the liver. In addition, valine and isoleucine can be deaminated and enter the TCA cycle via succinyl CoA. Leucine, in contrast, may deplete TCA intermediates (cataplerosis) by promoting the transamination of α-ketoglutarate to glutamate. Rennie proposes that a major function of circulating glutamine, an amino acid that decreases in the circulation under circumstances of severe stress or trauma, may be to augment glutamate stores. Glutamine crosses the cell membrane easily and can be converted by the enzyme glutaminase to glutamate or by transaminase to α-ketoglutaramide, which can then be converted to α-ketoglutarate and ammonia. According to Rennie, these proposed mechanisms await validation, and the magnitude of the conversions has not been shown to limit energy availability. In addition, a threshold glutamate concentration for TCA cycle intermediate generation has not been established.

A different but possibly related question is why obese individuals placed on starvation diets experience potentially fatal muscle wasting despite adequate stores of energy. Owen and coworkers (1998) recently conducted a series of experiments to address this question and to test the hypothesis that in spite of adequate energy stores, muscle must be broken down to supply TCA cycle intermediates, so that energy can be used, and to supply precursors for hepatic gluconeogenesis. The administration of phenylacetate, which binds to plasma glutamine resulting in its excretion as phenylacetylglutamine, to obese subjects for the last 3 days of a 3-week fast resulted in no change in urinary creatinine, urea, uric acid, ammonium, or ketone body excretion and no decrease in hepatic or renal gluconeogenesis. These observations are consistent with a continuous demand for amino acid oxidation, presumably to supply TCA cycle intermediates and gluconeogenic precursors.

The extent of any increase in amino acid oxidation during exercise depends on several variables. These include exercise intensity, availability of energy from glucose and fat, nutritional status, training status (trained athletes experience less amino acid oxidation in muscles), and gender (Millward et al., 1994). In Chapter 11, Rennie concludes that although the oxidation of amino acids in muscle supplies energy, the magnitude of this oxidation does not appear to be sufficient to allow protein to be considered a major metabolic fuel during periods of exercise. Although it is theoretically possible that repeated bouts of intense physical activity in the face of limited nutrient intake might deplete glutamate and glutamine to an extent that would affect the TCA cycle, this possibility has not been tested.

Contractile Activity and Muscle Protein Turnover

Muscle protein synthesis appears to decrease during exercise and to rebound after exercise; thus, any net change in nitrogen balance or muscle protein is observed only over a period of several days. Muscle protein breakdown during normal exercise is limited to soluble or membrane proteins degraded via lysosomal proteases. The degradation of myofibrillar protein appears to increase only with eccentric exercise (exercise such as walking downhill, in which muscle is forced to contract as it is stretched) (Fielding et al., 1991); the consequences of this damage to muscle tissue are discussed later. Little evidence exists to suggest that exercise leads to significant accumulation of muscle protein in the absence of steroid-induced growth (Forbes, 1985, cited in Millward et al., 1994).

According to Rennie (1996), any negative balance of protein synthesis and breakdown created during exercise is rapidly reversed by protein intake. Attempts to investigate the possibility that protein requirements are increased by physical activity are confounded by the observation, discussed by Millward (Millward et al., 1994; see also Chapter 9), that increases in protein intake result in increases in the catabolic processes leading to the oxidation of amino acids. A potentially significant implication of this adaptation is that individuals who habitually consume high-protein diets may face the risk of significant losses of protein stores if suddenly forced to curtail protein intake.

Energy Balance and Protein Requirements

The direct influence of energy intake on nitrogen balance has been recognized for many years (Cuthbertson and Munro, 1937). In an attempt to quantify this effect, Calloway (1975) fed groups of men diets in which first the levels of protein were varied while energy was held constant and then the levels of protein were held to the level nearest individual need (determined by N balance) while energy was varied. When 85 percent of maintenance energy was provided, nitrogen balance fell to –0.61 g/d. In contrast, when 115 percent of maintenance energy was provided, nitrogen balance rose to 0.59 g/d, with the greater increase occurring when energy intake rose from 85 to 100 percent of maintenance. Marginal protein intake appeared to have less effect than marginal energy intake. In a similar set of experiments, Kishi and coworkers (1978) fed diets of increasing energy content to groups of men for 3-week periods and found that at energy intakes of 40, 45, 48.2, and 57 kcal/kg, the estimated protein requirements to maintain nitrogen balance were 0.78, 0.56, 0.51, and 0.42 g/kg BW per day, respectively. This effect of energy intake on apparent maintenance protein requirements is observed only when protein intake is not limiting. Moreover, although it is clear that experimentally determined nitrogen balance is the result of both protein and energy intake, energy intake has not been considered in determining the protein requirements of various groups. If the results of Kishi and coworkers were applied to a 70-kg man, the protein requirement would be 54.6 g/d at an energy intake of 2800 kcal/d (the equivalent of two MREs). It is not known whether the relationship between energy intake and protein intake can be extrapolated to lower energy intakes (such as those of Ranger trainees, who in past studies have been observed to consume approximately 1500–1700 kcal/d, including approximately 50 g protein). Furthermore, because the studies of Kishi and coworkers and Calloway relied on the technique of nitrogen balance, more reliable and precise results would be expected if the studies were repeated using more up-to-date techniques. Finally, it is possible that the estimation of resting energy expenditure and energy requirements could be improved by the use of newer techniques such as magnetic resonance imaging (MRI) derived organ-tissue mass measurement (Gallagher et al., 1998).

The mechanism by which energy intake exerts its effect on protein balance is not completely understood. In the resting state, carbohydrate intake inhibits protein catabolism. Two possible mechanisms have been proposed. One possibility is that the carbohydrate-stimulated increase in insulin secretion increases synthesis and decreases breakdown of protein. An alternative possibility is that the provision of carbohydrate simply decreases liver gluconeogenesis, so that amino acids are no longer drawn away from muscle. The provision of carbohydrates prior to and during exercise inhibits protein breakdown and oxidation of amino acids. Although it is known that carbohydrate and fatty acids are the primary fuels during exercise, the effect of triglycerides and medium-chain fatty acids on protein turnover is less well understood.

The issue of whether or not physical activity increases protein requirements is complicated. Results of studies depend in part on the energy intake of the participants, the kind and intensity of the exercise studied, and the length of time given for adaptation to varying protein intakes (Butterfield, 1987). It has long been held that individuals experiencing an energy deficit—whether it is the result of a decrease in energy intake, an increase in energy expenditure, or both—exhibit an increase in protein breakdown and protein requirements (Calloway and Spector, 1954; Calloway, 1975). However, the effect of chronic increases in energy expenditure and negative energy balance on muscle mass of exercising individuals is unclear. Several groups of investigators have observed that athletes undergoing strength and endurance training require greater protein intake to maintain nitrogen balance than do sedentary individuals (Lemon et al., 1992; Meredith et al., 1989; Phillips et al., 1993; Tarnopolsky et al., 1988, 1990a). However, the results of these studies have been questioned, in part because no differences were observed in performance or lean tissue mass between athletes consuming high-protein diets and those on lower-protein diets. In addition, many of these studies were too short to allow for adaptations to changes in protein intake (about 10 days) or exercise regimen (about 2 weeks) (Gontzea et al., 1975), which makes their interpretation difficult.

The possibility that moderate (approximately 50–60 percent of maximum capacity) exercise may have a protective effect on muscle protein in the negative energy balance situation has been suggested by a small number of studies (Carraro et al., 1990; Stein et al., 1989; Stroud et al., 1996; Todd et al., 1984); however insufficient research has been conducted to confirm this observation. The classic study of college athletes and soldiers by Chittenden (1907) showed that a gradual 50 percent reduction in protein intake (to 0.75 g/kg BW/d) over 5 months did not decrease and in fact increased strength while decreasing fatigue. A study by Butterfield and Calloway (1984) also showed that moderate physical activity optimized protein utilization by previously untrained individuals who consumed lower-than-average amounts of protein. Buskirk (1996) has suggested that although controversy remains with regard to protein requirements and optimal protein intakes for athletes, a protein intake of 1.5 g/kg BW/d (the usual protein intake for most Americans) should be sufficient to preserve muscle mass and nitrogen balance if energy intake is sufficient.

Although the protein requirements of women have been studied inadequately, the small amount of available data suggests that the needs of nonpregnant women may be similar to those of men. Kurzer and Calloway (1986) showed variations in urea excretion in sedentary women, corresponding to the phases of the menstrual cycle, but also showed that nitrogen balance could be maintained on a protein intake approximating 0.8 g/kg BW/d. As mentioned above, protein requirements are related to energy intake. Because energy intake in active women is often lower than predicted from activity pattern (Mulligan and Butterfield, 1990), these individuals may be in negative energy balance and, therefore, would have increased requirements for protein. The reported tendency of military women to restrict energy intake intentionally because of concerns about weight and body mass (IOM, 1998) may further interfere with their ability to meet protein requirements during field operations.

In summary, the evidence to date suggests that sustained physical activity does not result in increased protein requirements for appropriately trained (and possibly even sedentary) individuals who are in energy balance. Data on physically active women are insufficient to determine whether their protein requirements are higher than those of sedentary women.

Infection, Injury, and Illness

Systemic infections and severe injuries trigger complex but predictable alterations in body protein metabolism that lead to increases in protein requirements. These effects, which are a component of acute-phase reactions now known to be initiated by the proinflammatory cytokines, include a rapid catabolic destruction of skeletal muscle proteins and, simultaneously, an equally rapid synthesis of many other body proteins.

These shifts in protein metabolism appear to be purposeful. The responses to infection and injury provide the body with newly formed proteins and the immune system elements needed to mount a successful and effective defense, but at considerable nutritional costs.

In the response of protein metabolism to injury and infection, the contractile proteins of skeletal muscle serve as an amino acid "bank." These muscle proteins undergo rapid catabolic breakdown, and the free amino acids liberated are released into plasma and used in visceral organs for a number of purposes. Some amino acids (especially the branched-chain group) are metabolized in situ within the muscle cell to provide energy, and their nitrogen components are immediately reutilized to create new amino acids (for example, glutamine and alanine) that are released into the circulation. Both glutamine and alanine are metabolized in visceral tissues, as sources of the additional energy needed to sustain the body-wide hypermetabolic state characteristic of acute-phase reactions.

Under the influence of proinflammatory cytokines, the liver takes up large quantities of plasma free amino acids. Hepatic uptake of amino acids is so large that their plasma concentrations may decline despite the massive quantities of amino acids being made available by muscle protein catabolism (Beisel, 1992; Kinney and Elwyn, 1995; Wilmore, 1991).

In addition to their metabolic degradation as sources of fuel, the resulting amino acids are used by the liver to manufacture numerous enzymes, metallothioneins, lipoproteins, and the large array of acute-phase reactant plasma proteins (α_1-antitrypsin, α_1-acid glycoprotein, haptoglobin, C-reactive protein, fibrinogen, the third component of complement, ceruloplasmin, amyloid, and orosomucoid). Protein synthesis is also enhanced in lymphocytes and immunologic tissues (Beisel, 1992).

In the liver, excess phenylalanine is converted to tyrosine, and excess tryptophan is metabolized via the kynurenine pathway, resulting in the creation of diazo reactants that are excreted in the urine. The hepatic enzymes needed to metabolize excess phenylalanine and tryptophan are induced rapidly as part of the body's cytokine-induced acute-phase response (Beisel, 1992). Nitrogen components of other degraded amino acids are converted into urea and excreted in the urine (Beisel, 1992; Wilmore, 1991). Losses of body nitrogen are thus markedly increased with infection or trauma.

Changes in dietary protein and energy requirements to sustain muscle protein synthesis during recovery from infections and severe injury are of considerable importance to the military situation (Wolfe, Chapter 13). A full discussion of the impact of illness and injury on protein and energy requirements is beyond the scope of this report; the reader is referred to a comprehensive review such as that by Souba and Wilmore (1994). In Chapter 8, Wilmore notes that protein requirements are approximately 1.5 g/kg BW/d in almost all patients recovering from serious injury or systemic infection, with the exception of burn patients, whose requirements are elevated to 2–2.5 g/kg BW/d. In addition, the postinjury energy requirements of burn victims are among the highest known (Sakurai et al., 1995). However, the evolution of care for the injured, including those with thermal injuries, has greatly decreased the duration of hypermetabolic states and attenuated the peak response (D.W. Wilmore, Los Angeles, personal communication). The provision of intravenous energy sources and large quantities of amino acids to severely burned patients is unable to induce effective production of muscle protein (Sakurai et al., 1995). As discussed later, hormonal therapies have been used in combination with nutritional support in an attempt to improve this situation. In badly burned children, growth hormone has demonstrated positive effects on muscle protein synthesis (Gore et al., 1991). Despite the insulin resistance that develops during infection and trauma, insulin (along with adequate glucose) markedly increases the synthesis of muscle protein (Sakurai et al., 1995); however, protein breakdown increases as well, which negates the effect. Wolfe (see Chapter 13) speculates that testosterone, which stimulates protein synthesis under normal circumstances and is known to be suppressed in male burn patients, might have a synergistic effect with insulin on muscle protein synthesis.

Other Stressors

There is no question that deployed soldiers in field operations experience the simultaneous effects of multiple stressors. In addition to an interest in the effects of physical activity, injury, and infection, the military maintains considerable interest in the effects of temperature extremes and high altitude on protein requirements. Therefore, several previous reports of the CMNR have reviewed this topic.

Heat

As reviewed by Buskirk (1993), Mitchell and Edman (1949, 1951) postulated that protein requirements would increase slightly in hot environments as the result of increased sweat losses of nitrogen or tissue catabolism secondary to hyperthermia. Studies by Consolazio and coworkers (Consolazio and Shapiro, 1964) found that the increased protein intake of soldiers in the heat appeared to be due to the increase in energy intake, rather than to some innate need for more protein in hot environments. Calloway and colleagues (1971) estimated dermal losses of protein due to exercise-induced sweating and found that such losses could add approximately 0.5 g/d to previously calculated protein losses in active individuals, but they proposed that sweat-induced losses would likely be less in sedentary individuals. Paul (1989) observed that because protein contributes to energy needs during prolonged exercise and because sweat losses of nitrogen increase during intense prolonged exercise in hot weather, protein requirements might be increased under these circumstances (the effects of acclimatization are not known). Nevertheless, Buskirk (1993) concluded that no evidence exists for an increase in the apparent protein requirement of soldiers (beyond the Military Recommended Dietary Allowance [MRDA] for protein) in hot environments. He further cautioned that an increase in dietary protein would raise fluid requirements (see discussion of protein and renal function later in this chapter) in individuals whose fluid requirements were already elevated by the effects of exercise and heat and whose intake might already be restricted by logistical factors.

Cold and High Altitude

In 1996, the CMNR reported on the effects of cold and high-altitude conditions on nutrient requirements (IOM, 1996). According to studies summarized by Jones and Lee (1996) and by LeBlanc (1996), there is no conclusive evidence that protein needs are increased by brief or prolonged exposure to cold temperatures. In a classic study cited by Rennie in Chapter 11, Stroud and coworkers (1996) showed that among two subjects who trekked across the Antarctic while in profound negative energy balance, whole-body protein turnover was slightly increased in one and slightly decreased in the other.

The MRDAs do not include a higher protein allowance for cold weather; however, they do include a higher energy allowance. Thus, the recommended contribution of protein to total energy intake in the cold (and the ratio of protein to total energy in the Ration, Cold Weather) is actually lower than that in moderate temperatures (and in the MRE). Because the metabolism of extra dietary protein results in increased urea excretion and, therefore, an increased fluid requirement, this lowered protein-to-energy ratio may be advantageous in cold environments where the availability of drinking water may be limited (Askew, 1989).

Butterfield (1996) and Hoyt and Honig (1996) reviewed the influence of high altitude on metabolism and on macronutrient requirements, and reported significant losses of lean tissue mass, body weight, and fat mass among most subjects. However, as demonstrated by Butterfield and coworkers (1992), if subjects consumed sufficient energy to maintain energy balance, weight losses were corrected and nitrogen balance was maintained. These findings suggest that the loss of lean tissue mass and the negative nitrogen balance often experienced at high altitude are due entirely to the negative energy balance caused by altitude-induced anorexia and increased metabolic rate, rather than to an increase in protein requirements. The CMNR concluded that there appeared to be no rationale for increasing the MRDA for protein for individuals working at high altitudes; however, the committee also suggested the need for further research on the effects of intense physical exertion on protein requirements at high altitudes.

Combined Stressors

It is generally recognized that at any given time, deployed soldiers and those in field training face a combination of stressors that may include energy imbalance (secondary to undereating, intense physical activity, or both), severe injury, systemic infection, climatic extremes, and changes in altitude. As described above, attempts to identify the effects of such stressors, alone or in combination, on protein requirements have thus far been inconclusive. Additional stressors, such as exposure to unidentified environmental contaminants and the emotional consequences of the battlefield and of separation from a familiar environment, have also been recognized, but far less is understood about their impact on physiology and nutritional status and how this might influence protein requirements (Friedl, 1997; IOM, 1995).

THE MRDA FOR PROTEIN

The protein (and energy) content of military operational rations was formulated during World War II on the basis of Food and Nutrition Board recommendations (NRC, 1941). Based on data for energy consumption and expenditure of soldiers during the war, the initial standard of 70 g protein per 3000 kcal for the 70 kg reference man was increased in 1947 to 100 g protein per day (based on 3600-kcal total energy intake) for physically active military men in temperate climates. The protein MRDA for women is 80 g/d based on a daily energy intake of 2000–2800 kcal (AR40-250, 1947). This amount of protein is equivalent to 11 percent of total recommended energy intake (results of recent national nutritional surveys, including the National Health and Nutrition Examination Survey (NHANES) III, have shown that protein intake in the U.S. population averages 14 to 16 percent of total food energy for both males and females).

Since the 1940s, the mean weight of male soldiers has increased from 68 to 78 kg; similarly, the mean weight of female soldiers has increased from 61 to 63 kg. Thus, the MRDA for protein expressed on a g/kg BW basis is 1.3 for both men and women. Based on observations suggesting that protein requirements may be increased for individuals engaged in specific types of exercise, the question arises whether recommendations established for protein intake by soldiers during World War II are still appropriate for military personnel today.

Protein Intake Studies of Military Personnel

Studies conducted in the 1940s and over the past decade have shown that military personnel maintain relatively high protein intakes during field operations as well as in garrison. Table 1-2 summarizes the average and range of intakes measured in studies reviewed by Cline and others (Baker-Fulco et al., 1995; King et al., 1993; see also Cline, Chapter 4). Estimated protein intakes in the field and in garrison for both men and women met or exceeded the MRDA of 100 g/day and 80 g/day, respectively. Relative to body weight, the average daily protein intake for men in the field was 1.3 g/kg BW and for women, 1.2 g/kg BW, exceeding the estimated protein requirement of 0.8 g/kg BW and even the increased requirement suggested by Tarnopolsky and coworkers (1990b). Both energy and protein intake appear to have decreased between the 1940s and more recent times, whereas average body weight has increased over this period. The significance of this apparent decrease is not known; the observation may be attributable to changes in the method of dietary intake assessment or may be related to an observation by Cline and Warber (see Chapter 4) that approximately one-fourth of soldiers report trying to lose weight during field training exercises. Nevertheless, protein consumption by soldiers is still high relative to the RDA.

Operational rations include the individual combat rations such as the Meal, Combat, Individual (MCI); the Meal, Ready-To-Eat; and other rations (including A, B, K, T, and unitized group ration [UGR]; see Appendix B) used to support operations in the field. The energy and protein contents of operational rations are set at a level of 3600 kcal and 100 g/d, respectively. Because of their high energy content, operational rations theoretically contain 2.8 g protein per 100 kcal. According to Cline and Warber (see Chapter 4), the actual energy content of operational rations varies from 2794 kcal in C rations to 4300 kcal in B rations. The protein content varies from 79 g in K rations to 142 g in the UGR. Therefore, the protein-to-energy ratio varies from 2.8 to 4.3 g protein per 100 kcal (11–17 percent of energy), depending on the type of operational ration.

In comparing macronutrient intake in the field and in garrison, it appears that protein and energy intakes in the field did not exceed and may have been lower than intakes in garrison, even though energy expenditure would be expected to be higher in the field than in garrison. In the small number of studies in which energy expenditure was estimated, soldiers in the field tended to be in negative energy balance (as evidenced by weight loss), although their protein intake met or exceeded the MRDA (none of these studies included women). Dietary surveys indicate that soldiers preferentially eat the high-protein entrees of the rations (Baker-Fulco, 1995). As a result, mean protein intakes of soldiers in the field as a percentage of total dietary energy are higher than suggested by the ration contents.

Pregnancy, Lactation, and the MRDA for Protein

At the present time, there are no MRDAs for pregnancy or lactation. As shown in Table 1-1, the RDA for protein is increased by 10 g/d for pregnant women and 15 g/d for lactating women (NRC, 1989). Thus, the recommended protein intakes for women in the weight range of 46–63 kg would be 44–57 g/d during pregnancy and 60–72 g/d during lactation. Some studies have suggested that the RDA for protein during lactation may be insufficient to meet the requirements of lactating women and have shown that actual protein requirements may be as high as 1.5 g/kg BW/d or 69–94 g/d (Motil et al., 1990, 1996). The MRDA of 80 g/d would therefore be sufficient to meet the apparent protein requirements of most pregnant or lactating women.

Summary

In summary, the protein intake of soldiers in both garrison and field situations appears adequate relative to the current MRDA. Because energy expenditures have not been measured for average soldiers in the field, it is not possible to determine their risk for negative energy balance, which could increase requirements for protein. MRDAs for women appear to be adequate to support pregnancy and lactation.

PERFORMANCE BENEFITS AND HEALTH RISKS OF SUPPLEMENTAL PROTEIN, AMINO ACIDS, AND PLANT PROTEINS

Supplement Use Among Army Personnel

The use of various protein and amino acid supplements by military personnel reflects the current public interest in strength training and body building. Such products are readily available at military commissaries, exchanges, and fitness centers, according to Cline and Warber (see Chapter 4). In a recent Army survey of protein supplement use, 36 percent of Army personnel less than 30 years old reported having used amino acids, and 30 percent had used protein powders. Supplement use was lower in persons more than 30 years of age: 26 percent reported having used amino acids, and 19 percent reported using protein powders. The highest use was reported among combat arms personnel (this category includes infantry, armor, field artillery, air defense, and Special Forces), followed by combat service support personnel (includes ordnance, quartermaster, and transportation), and combat support personnel (includes engineer, chemical, military intelligence, military police, signal, aviation, and civil affairs). Supplement use among men was appproximately double that among women, and individuals required to eat in military dining facilities reported a greater likelihood of supplement use than those who received food allowances to eat off-site. Reasons given for supplement use included the desire to increase energy, improve athletic performance, increase muscle mass, engage in strength training, and gain weight (Warber et al., 1996).

TABLE 1-2 Dietary Protein Intake of Military Personnel (g/d)

Reference	Location	Protein Intake (% of total energy)	N	Duration and Type
Kretsch et al., 1986	USMA	84 (13.7) 125 (13.4)	54 women 13 men	5 d training
Rose and Carlson, 1986	Ft. Sill	129	31 men	8 d field
USACDEC/USARIEM, 1986	Hawaii	67 (14.6) 98–113[a]	40 women 33–38 men	44 d field 16, 48 d field
Askew et al., 1987	Camp E Allen[b]	112	17 men	30 d training
Carlson et al., 1987	NCO Academy	124 (15.5)	43 men	8 d training
Szeto et al., 1987	Ft. Lewis, WA	75 (16.4) 103 (15.1)	12 women 31 men	6 d garrison
Szeto et al., 1988	Ft. Devens	114 (14.0)	54 men	7 d garrison
Szeto et al., 1989	Ft. Devens	126 (16.1)	51 men	8 d garrison
Rose RW et al., 1989	Ft. Jackson, SC (BCT)	96 (15.6) 125 (15.5)	40 women 41 men	7 d training
Rose MS et al., 1989	Ft. Hood	82 (14.0) 113[c]	N/A women N/A men	8 d field
Edwards et al., 1991	Bolivia	68 (9.0) 97[d] 100[d]	13 women 35 men 32 men	15 d field
Baker-Fulco et al., 1992	Marine OCS	160 (14.5)	12 men	5 d training
Klicka et al., 1993	USMA	79 (13.7) 130 (14.6)	86 women 11 men	5 d training
Baker Fulco et al., 1994	Marine OCS	169 (14.5)	16 men	5 d training
King et al., 1994	Ft. Jackson	82 (12.7)	49 women	7 d training

TABLE 1-2 Continued

Reference	Location	Protein Intake (% of total energy)	N	Duration and Type
Thomas et al., 1995	Ft. Chaffee[e]	125 (17.2) 103 (16.7)	32 men 34 men	30 d garrison and training
Cline et al., 1998	Ft. Sam Houston	75 (14.7)	50 women	56 d basic training
Baker-Fulco, 1997 personal communication	Camp Mackall	43.7 (9.8)[f] 59.4 (14.1)[g] 65.6 (10.3)[f] 83.5 (14.9)[g]	34 women 33 women 35 men 95 men	7 d training
Hirsch et al., in press	Camp Parks	103.5 (19.9)[h] 81.8 (14.8)[g] 57.1 (9.8)[i] 135.4 (20.1)[h] 103.7 (14.9)[g] 69.6 (10.3)[i]	19 women 18 women 13 women 13 men 16 men 11 men	4 d

NOTES: BCT, basic combat training; N/A, not available; NCO, noncommissioned officer training; OCS, officer candidate school; USMA, U.S. Military Academy, West Point.

a Wt. loss 1.3–2.1%.
b 2 A rations + 1 MRE or 3 MRE: wt loss 4.8%.
c Wt. loss 1.0%.
d Wt. loss 2.0%.
e Wt. loss 2.2%.
f Concept ration.
g MRE.
h High-protein MRE.
i High-carbohydrate MRE.

No data are available on the voluntary use of specific amino acids by military personnel or on the effects of protein supplement use on performance of military personnel. The effects of tyrosine and tryptophan supplementation on soldier performance have been examined by Lieberman, Askew, and coworkers and are reviewed in the next section.

Protein and Amino Acid Supplements and Cognitive Performance

Although it is well known that several amino acids are precursors to neurotransmitters or neurotransmitters themselves, the brain's precise need for these amino acids is not well known. The concentrations of these precursor amino acids in blood influence their availability to brain neurons and, as a result, the ability of neurons to synthesize neurotransmitter products (because the rate of production of some transmitters is directly influenced by local concentrations of their precursors). The accessibility of amino acids to the brain is controlled by specific transport carriers located at the blood–brain barrier, the physical site of which is the endothelial cell of the brain capillaries (Pardridge, 1977). These sites can have a major influence on whether or not the brain meets its amino acid needs. It is thus essential to understand how this transport carrier operates. In Chapter 14, Lieberman notes that these carriers are selective for groups of amino acids based on size and charge. Of particular relevance to neurotransmitter production in brain is the large neutral amino acid (LNAA) transporter, which is responsible for the uptake of tryptophan (the precursor of serotonin) and of tyrosine and phenylalanine (precursors of the catecholamines). Several other amino acids are also transported into brain by this carrier. The LNAA transporter has been found to be competitive at normal plasma levels of these amino acids; consequently, phenomena that change plasma levels of one or more LNAAs indirectly modify their uptake into the brain. According to Lieberman (see Chapter 14), when the availability of these transmitter precursors to the brain changes, brain functions are modified, including cognitive performance and affective state (mood). Since the availability of one or more of these amino acids to brain becomes compromised during periods of undernutrition or stress, he argues that the resulting reduction in transmitter production could diminish brain function. This issue is highly relevant to the performance of soldiers in the field, who often operate under conditions of food restriction and increased exposure to multiple stressors (Lieberman, 1994).

Tryptophan

The rate of serotonin synthesis and release by brain neurons is directly influenced by the tryptophan concentration in brain (Fernstrom, 1990; Sharp et al., 1992) and, thus, by the uptake of tryptophan from the circulation. The consumption of a protein-free meal increases the concentration of tryptophan in blood relative to that of the other LNAAs. The mechanism of action of this influence involves the stimulation of insulin release by dietary carbohydrate. Insulin stimulates the release of nonesterified fatty acid (NEFA) molecules from the serum carrier protein, albumin, and their subsequent uptake by adipocytes. The amino acid tryptophan, which has an affinity for albumin, binds in place of the NEFA. As a result, the total serum tryptophan concentration remains constant (unlike the concentrations of other amino aids, which decrease in response to insulin), and the bound tryptophan is able to enter the brain. In contrast, consumption of a high-protein meal results in a decrease in brain tryptophan and serotonin levels because the other LNAAs, which compete with tryptophan for binding to the LNAA transporter and thus for transport to the brain, are present in higher concentrations in dietary protein than is tryptophan (Fernstrom, 1990). As a result, serotonin synthesis is modified in several brain regions, including the hypothalamus and cerebral cortex. Thus, brain serotonin synthesis may be modulated by the protein-to-carbohydrate ratio of the overall diet or a recent meal.

To date, it is not known whether meal-related changes in serotonin production influence brain functions, but a number of studies have linked the changes in serotonin synthesis that follow the administration of tryptophan or its LNAA competitors to functional effects. In laboratory animals, serotonin neurons are most active when animals are awake and physically active (Jacobs and Fornal, 1993) and play an important role in channeling sensory information to the brain (Messing and Lytle, 1977; Walters et al., 1979).

Of particular relevance to the military, Lieberman notes, is the observation that tryptophan administration produces "mental fatigue" and has been used to promote sleep. Administration of tryptophan is also reported to reduce pain sensitivity (Lieberman et al., 1983; Seltzer et al., 1983). A decade ago, significant toxicity was attributed to tryptophan supplements, apparently due to a contaminant that survived the purification process for the amino acid (Hartmann and Greenwald, 1984; Lieberman et al., 1985).

Relatively little is known about how a reduction in serotonin synthesis might influence performance. The administration of an amino acid mixture that should reduce brain tryptophan levels and serotonin synthesis has been found to promote aggressiveness (Cleare and Bond, 1994) and depression (Delgado et al., 1990) in human subjects. These findings suggest that the changes in brain serotonin synthesis that accompany the ingestion of normal foods may produce similar, though less remarkable, effects on these behaviors (because meals cause smaller changes in serotonin than those produced by amino acid treatments). At present, no data exist that evaluate the magnitude of such effects.

Chronic, substantial reductions in protein intake can reduce brain tryptophan levels and serotonin production in laboratory animals (Fernstrom and Wurtman, 1971), with the ingestion of proteins naturally low in tryptophan having the most pronounced effects on serotonin (Fernstrom and Hirsch, 1975) and behavior (Fernstrom and Lytle, 1976). These data are insufficient to allow an estimate to be made of a brain tryptophan "requirement" in humans in relation to habitual dietary protein intake. Some human populations (for example, those that subsist on a corn-based diet) may experience chronic tryptophan deficiency (Wurtman and Fernstrom, 1979). However, although such groups might be expected to experience the behavioral effects predicted from animal studies (insomnia, increased pain sensitivity, depression), no data currently exist to support such a possibility. At least several weeks would be required for such a deficiency state to develop in humans. Nevertheless, the combat situation is typically associated with increased physical activity and stress, which stimulate serotonin turnover (Chaouloff, 1989; IOM, 1994). Thus, serotonin production may be impeded at a time when the requirement for this neurotransmitter is increased to support crucial brain functions. It is possible that aspects of sensory and motor function might become diminished under field conditions in which food intake (protein intake, in particular) is reduced for an extended period. Under such conditions, protein and/or tryptophan supplementation could have potential benefit. However, no research has been conducted to evaluate this possibility in animals or humans under the conditions of intended use.

Tyrosine

The only other amino acid studied in any detail in relation to its conversion to a brain neurotransmitter is tyrosine, another LNAA and a catecholamine precursor. Lieberman notes that when catecholamine neurons are actively firing in brain, the rate at which they synthesize catecholamines (primarily dopamine or norepinephrine) increases and becomes responsive to the local tyrosine concentration (Fernstrom, 1990; Lieberman, 1994; Wurtman et al., 1981).

Catecholamine neurons typically increase their activity under a variety of stressful conditions (Fernstrom, 1994; Stone, 1975), and appear to be important regulators of such behavioral parameters as attention, state of arousal, and mood (Lieberman and Shukitt-Hale, 1996). Stress-induced increases in catecholamine production can be magnified in animals by the administration of tyrosine, which may result in beneficial effects on spatial reference and working memory of laboratory animals under stressful conditions (Lehnert et al., 1984a; Shukitt-Hale et al., 1996; Shurtleff et al., 1993). Tyrosine administration also appears to improve memory and learning in humans under adverse environmental circumstances (Banderet and Lieberman, 1989; Deijen and Orlebeke, 1994). For example, Marine Corps sharpshooters who received tyrosine supplements experienced a partial reversal of the cold-induced deterioration in their marksmanship performance (Shurtleff et al., 1994). In another study, tyrosine administration was associated with an amelioration of the sleep deprivation-induced decline in psychomotor function and vigilance (Neri et al., 1995). However, such effects are less evident under normal circumstances, and no recent studies have updated or reevaluated these findings.

Finally, in studies of soldiers examined under field conditions and consuming military rations, deficits in energy intake as well as in mental performance often develop along with reductions in plasma levels of tyrosine and tryptophan. A study by Askew and coworkers (1987) showed that the development of behavioral changes under these circumstances correlated most closely with decreases in plasma tryptophan, suggesting that plasma tryptophan levels may be most closely predictive of the development of a state of tryptophan deficiency in brain and changes in cognitive function.

In summary, very little is known regarding the influence of stressful conditions on brain requirements for the amino acids that are neurotransmitter precursors, an issue of particular relevance to the military. In addition, data on tyrosine supplementation are insufficient to demonstrate conclusive effects on cognitive performance.

Protein, Amino Acids, Muscle Mass, and Physical Performance

Measurement of Muscle Mass

Although skeletal muscle (SM) contains more than half of the body's protein, methods that quantitate the amount and quality of skeletal muscle remain underinvestigated and inadequately validated. The methods, which range from atomic to whole-body levels of body composition, are reviewed by Heymsfield in Chapter 12.

Since methods for determination of human body composition are by necessity indirect, all methodologies are dependent on mathematical models. Heymsfield suggests that it is possible to subdivide these body composition methods into (1) descriptive methods and (2) model-based methods involving the type of mathematical models used in the quantitation of SM.

Descriptive Methods. Empirical methods include anthropometry, bioelectric impedance analysis (BIA), ultrasound, and urinary biochemical markers (3-methylhistidine, creatinine) and are based on limited theory. These methods rely on the use of linear regression models based on a criterion method for the quantitation of SM. Each descriptive method would measure a quantity related to skeletal muscle, and this measurable quantity could be used in a linear regression model to predict SM: $SM = a \times (\text{measurable quantity}) + b$. Based on a brief review of these descriptive methods, Heymsfield noted that anthropometry is perhaps the most limited in accurately predicting SM.

Segmental BIA can be used to quantitate specific skeletal muscle groups. In addition to the conventional placement of electrodes on the hands and feet, the electrodes can be placed leg to leg for an estimation of lean tissue in the legs.

Likewise, placement of electrodes in an arm-to-arm pattern (or some segment of the arm) results in an estimation of arm lean tissue. Further developments in BIA methodology may increase its usefulness for quantitating muscle mass (Baumgartner et al., 1989; Heymsfield et al., 1995; Nuñez et al., 1995, 1997; Wang et al., 1996). However, the use of BIA in this way will require determination of the actual composition of various body parts from which to interpret the BIA data.

Methods cited by Heymsfield based on the quantitation of urinary metabolites include the use of urinary creatinine and 3-methylhistidine. Measurement of 24-h urinary creatinine excretion has been used as a surrogate for measurement of muscle mass, based on a well-known association between muscle mass and creatinine metabolism. The large majority of creatinine is produced from creatine within skeletal muscle (and, thus, in proportion to muscle mass), which is followed by quantitative urinary excretion of the creatinine. Creatinine excretion in humans varies by gender, due to gender differences in muscle mass, and is affected by total dietary protein and the amount of muscle-containing foods (red meat) consumed. Thus, its use as a measure of muscle mass requires some dietary control.

A second urinary metabolite that has been used as a marker of muscle metabolism is the post-translationally modified amino acid 3-methylhistidine (3-MeHis), which is released as a breakdown product from myofibrillar muscle protein. While urinary 3-MeHis excretion is an indicator of muscle mass, 3-MeHis is also produced in the gut as a by-product of meat digestion and therefore must also be measured on a meat-free diet. High correlations were found when computerized tomography (CT) measurements of total body skeletal muscle were compared with urinary 3-methylhistidine excretion or muscle mass predicted from 3-methylhistidine (Lukaski et al., 1981; Wang et al., 1998).

The development of prediction equations for muscle mass would benefit from additional research into the use of urinary metabolites as biomarkers to predict muscle mass.

Model-Based Methods. Model-based methods, which include (1) the imaging techniques of computerized tomography and magnetic resonance imaging, (2) dual-energy x-ray absorptiometry, and (3) in vivo neutron activation and whole-body counting, are based on sound and well-developed theory. These methods rely on the use of a biological model based on in vivo quantitation of SM using phantoms or cadavers. Each model-based method would measure a quantity related to skeletal muscle and this measurable quantity could be used in a mathematical model to predict SM:

$$SM = a \times (measurable\ quantity).$$

The most significant advance in the field of estimating muscle mass has been in the use of imaging methods; CT was introduced in the late 1970s and MRI within the past 15 years. The advance is based in part on the ability to perform accurate calibrations of these methods using phantoms and human cadavers, resulting in accurate estimates of muscle areas (CT) or volumes (MRI). The coefficient of variation associated with muscle mass determination by CT or MRI is 2–3 percent when the instruments are calibrated by comparing the results obtained with excised cadaver tissue or filled balloons to the actual weights of these standards (Heymsfield et al., 1997).

CT methodology in vivo relies on measurement of the CT number. Each type of tissue has a particular CT number, with the number for adipose tissue being lower than that for muscle tissue, which in turn is lower than that for bone. By using the CT number and algorithms, it is possible to separate in each slice the pixels that belong to adipose tissue and bone, while the remaining pixels, which belong to muscle, are determined by difference.

The CT number is also directly related to the composition of muscle and adipose tissue and enables the researcher and clinician to distinguish between anatomic muscle and nonadipose muscle tissue. The nonadipose skeletal muscle would contain the actual myofibrillar protein, while the anatomic muscle contains both intramuscular adipose tissue and nonadipose tissue skeletal muscle. Thus, within an anatomic muscle, the actual mass of functional contractile protein can be distinguished from the noncontractile fat tissue.

The advantage of MRI over CT is the absence of radiation exposure. The disadvantage of MRI is the time involved in its use. Although it is currently possible to scan the entire body from head to toe over 40 to 50 slices, in 20 minutes, subsequent analysis is time-consuming, requiring at least one full day per person to analyze each slice using sophisticated software programs for adipose tissue and lean tissue.

With the advent of stronger magnets, it has become possible to use magnetic resonance spectroscopy for analysis of carbon, hydrogen (protons), phosphorus, and sodium in vivo; from these measurements, tissue compartments can be calculated. This type of spectroscopy not only measures the amount of tissue present but also monitors in real time in vivo metabolic processes such as glycogenolysis, lipogenesis, and water balance. This powerful tool, magnetic resonance spectroscopy, is a major resource for future research.

Dual-energy x-ray absorptiometry (DXA) is a method originally designed for the quantitation of bone and bone density and therefore has been most useful in studying diseases such as osteoporosis. More recently, algorithms have been developed to use in interpreting total body scans for other body components. The differential attenuation of the energies from the two x-rays by soft tissue and bone is used to quantitate the amount of these tissues. Further partitioning of the soft tissue into lean and fat is similarly accomplished. DXA software then allows the separation of the body into discrete regions. Research in

Heymsfield's laboratory has estimated that 75 percent of total body muscle mass is found in the arms and legs. Other regions of interest such as the lower leg also can be identified and isolated using the software. This approach has been validated by comparing total body muscle mass determined by CT to appendicular skeletal muscle estimated by DXA. The correlation coefficient between the two is 0.95, which indicates a strong association and the feasibility of using DXA for the quantitation of muscle mass in humans in vivo.

The primary limitations on the accuracy of using imaging techniques to estimate muscle mass are those imposed by the technology employed. As mentioned above, even the most accurate methods of CT and MRI have a minimal error of measurement of 2 to 3 percent. Thus, for estimating changes in muscle mass, the expected change would have to be greater than 6 percent to make the measurement feasible. Such a change in muscle mass could be seen only under drastic circumstances, such as a change in total body weight of more than 10 percent. Current research includes the further development of BIA and imaging techniques, improvement of the urinary creatinine method, and perhaps most importantly, developments in the measurement of dynamic in vivo changes in muscle metabolism, composition, and subsequent muscle function.

Control of Muscle Mass and Function

The function and regulation of skeletal muscle mass have been reviewed by Nair in Chapter 6. Muscle mass constitutes 40 to 45 percent of body weight and accounts for approximately 70 percent of body cell mass. Nair presents a model that links muscle mass and function (in the form of contractility) with metabolic processes. Skeletal muscle has important locomotive and metabolic functions. Many of the metabolic functions that occur in skeletal muscle depend on muscle mass as well as interrelated factors such as circulating hormone levels, training status, and age. Thus, indicators of muscle function such as strength and endurance are influenced by these factors as well.

Strength. Although muscle mass and strength are significantly correlated, muscle strength may be regulated independently from muscle mass. Strength is a functional indicator of overall muscle "quality." Sarcopenia syndrome, which refers to the age-related decline in muscle mass, is characterized by loss of strength, power, speed, and endurance, as well as by poor balance, resulting in an increased potential for bone injury due to falls. According to a model presented by Nair, adjustment for the age-related decline in muscle mass reveals an apparent functional impairment, which points to a disturbance in the quality of muscle with age and inactivity.

Endurance. Evaluation of endurance with measures such as maximal oxygen consumption reveals gender differences; however these differences essentially disappear after appropriate adjustment for differences in body composition. In contrast, the age-related decline in endurance persists even after such adjustments for body composition are made, supporting the observation of an apparent change in muscle quality with aging. Endurance is compromised by myocardial problems such as decreased cardiac output, reduction in maximal exercise heart rate and stroke volume, and increased peripheral resistance. Each of these factors influences the delivery of oxygen to working muscle.

Muscle Metabolism. According to Nair, muscle protein quality is maintained by a remodeling process involving the replacement of old and damaged protein with new protein. Both mitochondrial and sarcoplasmic proteins undergo this process, which declines with age. The decline in the fractional synthetic rate of myosin heavy chain (a sarcoplasmic protein) is correlated with a decline in leg curl strength. The decline in mitochondrial protein synthesis is associated with a decrease in the activity of the mitochondrial enzyme cytochrome C oxidase, which is involved in energy utilization and storage. This decline may explain both a decrease in the efficiency of substrate metabolism and an increase in the fatigability of aging muscle. These changes in mitochondrial and sarcoplasmic proteins taken together are associated with an age-related decline in endurance capacity. The functional significance of these age-related changes is unknown for younger muscle.

Hormonal Interactions

Anabolic Hormones. Muscle is metabolically sensitive to insulin action; insulin administration decreases protein breakdown but does not appear to increase muscle protein synthesis in humans except when preceded by a large infusion of amino acids. In insulin-dependent diabetes, muscle protein breakdown is increased, with little effect on muscle protein synthesis. As a result, insulin deprivation is associated with increased net muscle protein loss. Insulin deprivation also is associated with increased glucagon levels, which have been shown to increase oxidation of the essential amino acid leucine. The insulin resistance observed in severely injured or postoperative patients and those with systemic infections may contribute to the muscle catabolism observed in these individuals (Black et al., 1982). Administration of insulin is associated with an increase in muscle protein synthesis in catabolic patients given amino acid-containing formulas (Pearlstone et al., 1994; Sakurai et al., 1995).

Growth hormone (GH) and IGF-I also stimulate muscle protein synthesis over the short term. Growth Hormone replacement in GH-deficient children and adults results in increased lean body mass or muscle mass (Collipp et al., 1973; Cuneo et al., 1991) and improvement in muscle function (endurance and strength) (Jorgensen et al., 1989). GH supplementation of elderly subjects, who normally have low GH concentrations, also increases lean body mass and muscle mass (Schwartz, 1995). The effects of GH administration on skeletal muscle protein synthesis appear to depend on experimental factors such as study duration and route of administration. For example, local infusion of growth hormone resulted in an increase in forearm protein synthesis with no change in protein breakdown; these effects may be mediated by a stimulation of IGF-I synthesis (Fryburg et al., 1991). Systemic infusion of GH for 3 days resulted in increased synthesis and decreased breakdown of forearm protein (Fryburg and Barrett, 1993; Wolf et al., 1992). Butterfield and coworkers (1997) observed increases in skeletal muscle protein synthesis, nitrogen balance, whole-body protein synthesis, and net protein synthesis after 1 month of daily injections of GH given to postmenopausal women. However, systemic infusion of men with GH to achieve the same range of serum concentrations, for varying lengths of time, resulted in no change in protein synthesis or degradation in the vastus lateralis muscle of the leg, as measured by the methods of continuous infusion or balance (Copeland and Nair, 1994; Welle et al., 1996; Yarasheski et al., 1995).

Administration of IGF-I, a paracrine hormone that is believed to mediate the effects of GH, also has demonstrated contradictory effects on muscle protein synthesis. Systemic infusion of IGF-I produced no effect on whole-body protein synthesis or breakdown (Elahi et al., 1993), but local infusion increased forearm protein synthesis and, at higher doses, decreased breakdown (Fryburg 1994, 1996; Fryburg et al. 1995). Injection of IGF-I twice daily for 1 month in postmenopausal women resulted in increased nitrogen balance, whole-body protein synthesis, skeletal muscle protein synthesis, and net protein synthesis and in decreased protein breakdown (Butterfield et al., 1997). The question of whether IGF-I mediates all or some of the effects of GH on muscle protein synthesis is unresolved, in part because the effects of IGF-I are influenced not only by its binding to IGF-I receptors but also by the secretion of its binding proteins, the control of which is not fully understood. Because of the influence of IGF-binding proteins and the apparent localization of at least some IGF-I effects, determining an effective dose of IGF-I is difficult. In addition, measurement of plasma levels of IGF-I may indicate trends but has limited functional significance.

Efforts to determine the effects on skeletal muscle protein synthesis of exogenous GH or IGF-I combined with resistance training have shown no additional effects of GH beyond those of the training itself in young and elderly sedentary men and weight lifters (Yarasheski et al., 1995), although GH increased fat-free mass and whole-body protein synthesis in young men and fat-free mass in elderly men (suggesting the possibility that some body tissues may be more sensitive to the effects of GH than skeletal muscle) (Rooyackers and Nair, 1997).

Administration of GH and IGF-I to critically ill patients has demonstrated positive effects on nitrogen balance and preservation of lean body mass in some studies but not in others. In badly burned children, GH has demonstrated positive effects on muscle protein synthesis (Gore et al., 1991). IGF-I administered to HIV patients improves nitrogen balance and protein turnover transiently (Lieberman et al., 1994). In contrast, Wolf and coworkers (1992) found no effect of GH on skeletal muscle protein synthesis in catabolic cancer patients. Sandstrom and coworkers reported no effect of IGF-I administration on nitrogen balance or protein breakdown in postoperative patients receiving dextrose with no added amino acid source. Nevertheless, it is believed that the combination of GH and IGF-I may increase nitrogen balance and prevent skeletal muscle protein breakdown in critically ill patients if the GH can counteract the hypoglycemic effects of high doses of IGF-I and if amino acid concentrations are adequate (Rooyackers and Nair, 1997). Recently, however, two large clinical trials of recombinant human GH with intensive care patients (recovering from any of several types of surgery or trauma) were terminated prematurely due to an unexpected and as yet unexplained increase in the mortality rate among the GH-treated groups (B. Lippe, Los Angeles, personal communication, 1998). Thus, further study is needed before GH can be considered for the treatment of critically ill patients.

Testosterone is undoubtedly the anabolic hormone most closely associated with building muscle mass. Levels of free testosterone in the blood decline with age and are associated with the rate of synthesis of myosin heavy chain. Testosterone administration to hypogonadal men substantially increases muscle mass, muscle strength, and muscle protein synthesis. Administration of testosterone in supraphysiological doses has recently been shown to increase muscle mass and muscle strength (Bhasin et al., 1996); however the effects of moderate doses on athletes in training have been inconsistent. Nair (see Chapter 6) indicates that the mechanism by which testosterone increases muscle mass and muscle strength remains unknown. Wolfe (see Chapter 13) speculates that testosterone, which stimulates protein synthesis under normal circumstances and is known to be suppressed in male burn patients, might have a synergistic effect with insulin on muscle protein synthesis.

Catabolic Hormones

Catabolic hormones, such as the glucocorticoids and glucagon, increase muscle protein breakdown and net catabolism of amino acids. The catabolic effect of glucocorticoids on skeletal muscle is evident in individuals suffering from Cushing's disease, which is characterized by an excess production of corticosterone. Glucocorticoid-induced protein breakdown has been shown to be inhibited by GH alone and completely reversed by coadministration of GH and

IGF-I (Berneis et al., 1997). The catabolic effect of glucagon is observed primarily in patients with Type-I (insulin-dependent) diabetes.

Effects of Protein, Energy, and Amino Acid Supplementation on Physical Performance

Protein and Energy Intake in Long-Distance Cyclists

Recent metabolic studies of cyclists engaged in the Tour de France bicycle race may provide some insight into nutritional factors that can contribute to the ability of an individual to perform sustained strenuous endurance activity. Saris and coworkers (1997) studied four cyclists during the race, which involves covering a distance of approximately 2640 miles (4000 km) over a period of 22 days, with 1 day of rest. During this race, mean daily energy expenditure ranged from 12.9 to 32.7 MJ/d (3071–7786 kcal/d), with an overall mean of 25.4 (6048 kcal/d); the duration of work was as much as 8–9 h/d, and altitudes frequently reached more than 2 km; however, the cyclists lost no weight or body fat. Mean daily energy intake matched or slightly exceeded energy expenditure, with protein representing 15.4 percent, carbohydrates 60.6 percent, and fat 22.4 percent of energy consumed (a high proportion of the energy consumed was in the form of carbohydrate-rich drinks). Further analysis revealed that intake closely followed expenditure, so that energy balance was maintained on a day-to-day basis with a maximum lag of 2 days (glycogen repletion was found to be complete within 16 hours). Although the contribution of protein to total energy was normal, total protein intake was very high compared with the RDA. Resting total plasma amino acid concentrations did not change during the race, although a large number of amino acids decreased or increased over the course of the race. This finding suggests that the daily recovery periods were too short to restore amino acid balance, but it was not possible to interpret the significance of individual changes or to draw conclusions about the effect of strenuous exercise on amino acid metabolism. Nevertheless, the results of this study demonstrate that individuals can sustain high levels of endurance activity over an extended period with no loss of lean body mass when energy intake is sufficient to match output.

The Fatigue Theory

Branched-chain amino acid (BCAA) supplements could theoretically benefit physical performance in several ways. They could supply TCA cycle intermediates, decrease the use of other energy sources (sparing glycogen), inhibit muscle protein breakdown, or limit the transport of tryptophan into the brain. A major mechanism by which the BCAAs are hypothesized to affect performance positively is via the last mechanism, known as the "central fatigue theory" of Newsholme (see Wagenmakers, Chapter 14; see also Blomstrand et al., 1991). This hypothesis states that exercise fatigue is in part a result of the amount of serotonin produced in the brain and that it is the availability of the precursor tryptophan that regulates the synthesis of serotonin. As discussed above, tryptophan levels in the brain are dependent on the activity of the large neutral amino acid transporter (see Lieberman, Chapter 14), which responds to the relative concentration(s) of free tryptophan and the BCAAs. If the ratio of tryptophan to BCAAs increases, the transporter will carry more tryptophan and serotonin synthesis in the brain will similarly increase.

In the exercising individual, the free tryptophan concentration in the blood increases due to competition with free fatty acids (FFAs) for binding sites on circulating albumin. The FFAs liberated from fat stores by the hormonal milieu generated by exercise essentially displace tryptophan from the albumin molecule. At the same time, BCAA concentrations in the blood are decreasing due to their use as a metabolic fuel by muscle. Thus, the theory proposes that exercise increases free tryptophan availability to the transporter and thus the brain, and the increase in brain serotonin shortens the time to fatigue. According to this theory, manipulation of the tryptophan-to-BCAA ratio by increasing BCAA intake during exercise should theoretically decrease serotonin synthesis, lengthen the time to fatigue, and improve performance.

However, studies have shown that the administration of BCAA during strenuous exercise results in virtually no change in time to fatigue or improvement in performance (Blomstrand et al., 1991; Davis et al., 1992; Kreider et al., 1993), and according to Wagenmakers (see Chapter 15), BCAA administration may actually decrease performance in circumstances of intense, long-duration exercise where glycogen stores are depleted. Wagenmakers and Rennie describe the importance of TCA cycle intermediates in the provision of energy substrates to the mitochondria and point out that, in fact, high levels of circulating leucine may deplete α-ketoglutarate, one of these intermediates. When α-ketoglutarate is converted to glutamate by the enzyme branched-chain α-keto acid dehydrogenase, the leucine carbon "skeleton" is left to be burned for energy. This hypothesis is borne out by work in individuals with McArdle's disease, who cannot release glucose from glycogen and who respond to BCAA administration with a significant decrease in maximum oxygen consumption (Sahlin et al., 1995). Similarly, research in healthy individuals infused with α-keto acids of the BCAAs, who showed an improvement in performance that was theoretically due to augmentation of the TCA intermediate pool, actually experienced a depletion in the TCA cycle intermediate pool (Katz et al., 1986; Sahlin et al., 1990). Thus, Wagenmakers cautions against the use of BCAA supplements as ergogenic aids for military troops, suggesting that these supplements may deplete TCA cycle intermediates when exercise is intense and of long duration and could thus hasten the time to fatigue.

Based on the observation that protein and carbohydrate combined together in a postexercise feeding results in a high insulin response and high rates of glycogen synthesis (Zawadzki et al., 1992), Wagenmakers conducted work on the effect of glutamine supplementation on glycogen synthesis. To date he has found no positive effect of glutamine but has confirmed the synergistic effect of carbohydrate and protein, and concludes that implementation of a postexercise feeding regimen including these two nutrients may be useful for promotion of the continued endurance of troops performing strenuous exercise over several days.

Thus, there appears to be no evidence at the present time to suggest that amino acid supplementation would optimize physical performance in healthy individuals who consume the MRDA for protein. There may, however, be situations in which such supplementation would theoretically be beneficial. For example, with exercise during moderate protein restriction, exogenous glutamine may enhance skeletal muscle protein synthesis and aid general maintenance of acid–base homeostasis. However, it should be emphasized that no studies are available to confirm these hypotheses. What data are available confirm that when adequate energy and protein are consumed, supplements will not be beneficial.

Amino Acids and Immune Function

The role of protein and specific amino acids in immune function has been reviewed in detail in a recent CMNR report (IOM, 1999). It is fully evident that every aspect of immune function and all innate host defensive mechanisms are entirely dependent upon the body's ability to synthesize new proteins. It is also evident that immunologically related proteins comprise a vast array of unique, highly specific individual molecules, each with its own purpose and function. Such proteins include antibodies, cytokines, thymic hormones, acute-phase reactant proteins, metal-binding proteins, the numerous proteins involved in complement, kinin and coagulation systems, lipoproteins, cell surface protein receptors and other components of newly synthesized lymphocytes and phagocytic cells, as well as numerous enzymes and structural proteins distributed throughout the body.

The synthesis of all these immunologically important proteins depends upon a complete and balanced array of free amino acids in body cells, and a lack of adequate dietary protein, as seen worldwide in subjects with protein–energy malnutrition (PEM), is the most common cause of nutritionally acquired immune dysfunction syndromes (NAIDS) and cachexia-related deaths. Young children, aged individuals, and patients with severe debilitating illness or trauma are the most frequent victims of NAIDS. Although healthy, well-fed military personnel should not experience protein-related immune dysfunction, the stress of severe Ranger training, with its associated dietary deprivation and extreme loss of body weight and muscle mass (IOM, 1992), did produce early evidence of both PEM and NAIDS.

Although, as noted above, the structure and function of immunologically important proteins is dependent upon the balanced availability of amino acids, there is growing evidence that the supplemental administration of certain amino acids can produce immunological benefits. Glutamine is the best example of this, as noted in an earlier CMNR report (IOM, 1999). Plasma glutamine concentrations decline with strenuous exercise and in "overtrained" individuals (Rowbottom et al., 1995), who are at an increased risk for infection. Wagenmakers (see Chapter 15) discusses the role of the conditionally essential amino acid glutamine in immune function. This amino acid is made in large amounts (20–25 g/d) in muscle and liver; it provides fuel for cells lining the intestine (VanderHulst et al., 1993) and may be important for the maintenance of function in immune cells (mucosal effects of glutamine may play a role in reducing the incidence or severity of intestinal infection).

Under circumstances of extreme stress and trauma, as well as intense exercise, the circulating level of glutamine diminishes, reaching a minimum attained 2 hours after the cessation of an exercise bout. Under normal conditions, skeletal muscle produces more glutamine than any other amino acid. Preliminary studies by Castell and colleagues (1996) of infection rate in very active athletes suggest that glutamine supplementation may decrease the rate of infection (that is, "overtraining effects"). However, Wagenmakers points out design flaws in the experiment that make this conclusion suspect. Some evidence suggests that the proliferation of lymphocytes and other parameters of immune function are boosted by glutamine (Jacobi et al., 1997; O'Riordain et al., 1994; Ziegler et al., 1994). In studies of hospitalized, critically ill patients, glutamine supplementation of total parenteral solutions resulted in increased survival and decreased length of hospital stay (Griffiths et al., 1997; Wilmore, 1997a; Ziegler et al., 1992).

The amino acid arginine also may serve as an immunostimulant and as an immune system modulator (Barbul et al., 1990; Wilmore, 1991). These effects may not be as great as those seen with glutamine; however, arginine is the sole substrate for the nitric oxide synthase-catalyzed formation of nitric oxide (NO), with citrulline as the other product.

Growing evidence now indicates that NO is highly important as a microbicidal and tumoricidal molecule, with an effectiveness that may surpass that of better known-oxidative mechanisms. Nitric oxide appears to have many other actions throughout the body (Albino and Mateo, 1995), some of which support mechanisms of host defense. However, more evidence is needed to determine if arginine supplements might reduce the incidence of infectious illness in healthy subjects.

Issues of Protein Quality and Timing of Consumption

Diets consumed by people worldwide typically supply protein at 9 to 14 percent of energy. The greatest differences in the protein value of human diets are in protein quality, not quantity. Diets in the developing world are based largely on cereals, with small contributions from legumes and little if any animal protein; thus, these diets are typically of low quality. Protein quality is a measure of the efficiency with which dietary protein is converted to body protein. Proteins of higher quality are needed in lesser amounts, and proteins of lower quality in greater amounts to synthesize body proteins. Protein quality is assessed in relation to a reference amino acid pattern with adjustment made for digestibility. All cereals contain less of the amino acid lysine than suggested by the reference pattern limitations, and corn is also lower than the reference in the amino acid tryptophan. Legumes and animal proteins supply excess lysine (with respect to the reference pattern) but are limiting in methionine. Thus, cereals, on the one hand, and legumes, meat, or milk, on the other, are known as complementary proteins since their amino acid compositions "complement" each other when they are consumed together. Since the cyclical nature of protein metabolism makes it possible for an amino acid deficiency in a meal to be made up from body stores, it is not necessary that complementary proteins be consumed in the same meal, but consuming them during the same day is highly desirable (Young et al., 1989). Thus, timing the consumption of complementary proteins is not an issue for operational rations employed by the military as these rations are currently formulated.

Considerable debate surrounds the need for high-quality protein in adults. If the estimated indispensable amino acid requirements were as low as indicated by classic nitrogen balance experiments—that is, about 15 percent of the total protein requirement—even poor-quality cereal diets would meet this requirement. On the other hand, if the need for IAAs is as high as indicated by recent stable isotope experiments (that is, about 50 percent), then protein quality is as important a consideration in diets consumed by adults as it is in those of infants and preschool children.

In considering this issue with respect to operational rations, it is desirable that the protein supplied be of high quality. During recovery from infection or trauma, including but not limited to blood loss, lost body protein must be replaced and this protein requirement is high for indispensable amino acids, especially lysine. Moreover, higher-quality proteins are used with greater efficiency, resulting in the excretion of less urea and a decreased renal solute load. For military women who become pregnant or who are lactating, the proteins synthesized by the body (breast milk and the products of conception) are high-quality animal proteins that are high in IAAs. The 1991 FAO/WHO working group on protein and amino acid requirements suggested that protein quality be assessed using amino acid requirements appropriate for the preschool child as the scoring pattern (Clugston et al., 1996).

Potential Benefits of Plant and Legume Proteins

Plant proteins are generally associated with less total fat, less saturated fat, and more polyunsaturated fat compared to most animal proteins. In addition, plants do not contain cholesterol, whereas red meats, poultry, and some shellfish do. The consumption of plant proteins such as soy-derived protein or a combination of plant proteins in place of part or all of the animal protein in the diet would therefore decrease the level of total fat, saturated fatty acids, and cholesterol in the diet—changes that are known to lower serum cholesterol and saturated fat. In addition, some plant protein sources contain other substances not found to be associated with animal proteins, such as soluble fiber, which also decrease the levels of serum cholesterol and saturated fatty acids. Plant proteins, such as the cereal proteins, are low in one or more of the indispensable amino acids. If Young were correct in his assessment of the IAA needs of adults, substitution of these proteins for animal proteins in military rations would decrease their nutritional value. However, among the plant proteins, soy has a better balance of the essential amino acids necessary for maintenance of lean mass (thus, it is higher-quality protein). A meta-analysis of the effects of soy-based diets (Anderson et al., 1995) supports the beneficial influence of these diets on blood lipids, although these data require confirmation and the responsible factor(s) must be determined. Substitution of soy protein for animal proteins in whole or in part would not decrease the nutritional value of military rations. The use of soy has the potential to improve the long-term health benefits of the diet. However to achieve these benefits, it might be desirable for foods made with soy to have the organoleptic (sensory) qualities of traditional animal protein sources. In addition, although soy-based foods have found acceptance in civilian markets, testing must be accomplished to ensure acceptance of these foods in military rations.

Effects of Timing of Protein Intake

The effect of timing of protein intake on exercise and cognitive performance has not been investigated extensively. However, available data suggest that postexercise feeding may be beneficial to maintenance of glycogen and protein stores. Early work by Cuthbertson and Munro (1937) showed that feeding immediately after an exercise bout decreased nitrogen loss in the urine over the subsequent 24 h. These authors suggested that the provision of protein immediately after exercise conserved body protein as an energy source. More recent work by Zawadzki and coworkers (1992) and others has shown that feeding protein in conjunction with carbohydrate (CHO) immediately after exercise resulted in an increase in circulating insulin, in comparison to feeding CHO alone or feeding nothing. This increase in circulating insulin translates into an improvement in glycogen storage, a factor shown by some to be important in the ability to sustain long bouts of strenuous exercise. Tipton and Wolfe (1998) have shown an improvement in protein synthesis when amino acids are provided immediately after exercise, compared to no feeding. Borchers and Butterfield (1992) showed that feeding either CHO, protein and CHO, or protein alone in equicaloric amounts immediately after exercise resulted in a diminution of urinary urea in the 24 h after the exercise bout, compared with no meal at all (reflecting improved protein utilization), even when total energy intake over the day was constant.

Thus, postexercise feeding, especially of CHO and protein, may result in an improvement in glycogen storage and maintenance of lean mass, thereby potentially increasing the ability to continue strenuous activity on a subsequent day. It should be noted, however, that all these data were derived from men, and the results may be different from women. Tarnopolsky and coworkers (1995) have shown that active women respond differently to glycogen loading than do men.

Risks Associated with High-Protein Diets and Supplements

Data from national surveys on food consumption demonstrate that protein intake is highly variable among both male and female adults. Variation also is evident for the intake of some individuals on a day-to-day basis. Chronic high levels of protein intake may increase amino acid catabolism and foster the body's adaptation to higher protein intakes. As noted by Bier in Chapter 5, this adaptation may be detrimental in situations where stress increases need or intake is diminished. The individual who has adapted to an increased amino acid catabolism may show a greater deficit in protein balance when a more moderate or a low intake of protein occurs. Further, high-protein diets lead to the generation of excess nitrogenous end products, which in turn require a greater intake of fluids to permit their excretion. This extra fluid requirement may be an added stress in hot environments.

Protein and Renal Function

Evidence from several studies has suggested that chronic high protein intake may contribute to the deterioration of renal function that is observed with aging. However, few studies have included considerations of energy intake or physical activity, and even fewer studies have included women specifically.

Nevertheless, available data do not suggest that dietary protein per se is a causative factor in the age-related deterioration of renal function in humans. Tobin and Spector (1986) measured renal function on two occasions separated by 10 to 18 years in 198 healthy men who were participants in a longitudinal study on aging. No correlation could be found between the observed changes in renal function and the levels of protein intake. Moreover, no relationship was identified between the decline in creatinine clearance with age and the level of protein intake. Other studies in humans have led to similar conclusions (Kerr et al., 1982). Animal studies also fail to support the hypothesis that high-protein diets compromise renal function; chronic high-protein diets fed to rodents for 2 years had no effect on glomerular filtration rate (GFR) or renal pathology (Collins et al., 1990).

In Chapter 7, Walser reviews the well-known adverse effects of a high protein intake in patients with impaired renal function that had reached end stage (Walser, 1992) and the therapeutic value of reducing protein intake of such patients, but these concerns do not seem applicable to healthy young adults. Similarly, the potentially deleterious effects of high protein intakes in aged individuals (who are losing renal function in association with senescence) are not applicable to individuals of military age. In fact, as noted by Walser, the renal clearances of both inulin (a chemical used to measure clearance) and creatinine in healthy subjects were increased by higher-protein diets.

The one potential renal danger of high protein intakes in healthy individuals, as cited by Walser, is nephrolithiasis (Robertson et al., 1979b). Renal stone formation has a highly complex pathogenesis that involves the excretion of calcium, sodium, sulfate, oxalate, and purines, all of which are likely to be increased in subjects consuming a high-protein diet (Tschope and Ritz, 1985). A major consequence of high-protein diets is increased excretion of urea, which results from the amino groups of oxidized amino acids. A restricted intake of water may increase the work required of the kidney to excrete these by-products in a concentrated urine, resulting in a compromise to the kidney and predisposition to nephrolithiasis. Adverse effects on renal function were not mentioned among the potential dangers cited by Maher (FASEB/LSRO, 1992; see also Chapter 16) from high intakes of individual amino acids, however he cited the need for additional studies.

Thus, although exact comparison studies are not available, existing data suggest that such studies, if performed using age, gender, and fitness-matched individuals, would find no correlation between protein intake and intrinsic renal disease. However, indirect effects may be observed in the form of renal stone formation.

Protein and Calcium Status

The calciuretic effect of high-protein diets is well established and has been demonstrated in both men (Allen et al., 1979; Linkswiler et al., 1974) and women (Hegsted and Linkswiler, 1981), although fewer studies have been done in women. Based on their review of 16 separate human studies, Kerstetter and Allen (1990) concluded that there is a linear relationship between dietary protein and urinary calcium such that for each 50-g increment of dietary protein, an extra 60 mg of urinary calcium is lost. This loss appears to be related to a direct effect of protein on renal function. An increase in glomerular filtration rate in response to high protein increases the filtered load of calcium. In addition, there is a decrease in fractional tubular reabsorption, which is thought to be related to the sulfur and acid load from protein (Zemel, 1988).

Body retention of calcium in response to high-protein diets is influenced by the dietary content of other nutrients, particularly calcium, phosphorus, and sodium. Thus, high intakes of various dietary sources of protein have differing effects on the magnitude of urinary calcium losses and calcium retention, depending on the minerals that they provide (Zemel, 1988). Spencer and coworkers (1988) failed to show increased calcium loss in response to high protein intake provided by red meat and other complex proteins, and attributed this difference in results to the phosphorus provided by meat. Although phosphorus is known to decrease urinary calcium losses, negative calcium retention in response to high protein intakes is not necessarily prevented by increasing the phosphorus content of the diet (Hegsted et al., 1981). In contrast to the results of Spencer and coworkers, the addition of meat to the diets of young men, which resulted in an increase in protein intake from 55 to 146 g and phosphorus from 890 to 1660 mg, led to increased urinary calcium losses and negative calcium balance (Schuette and Linkswiler, 1982). When the added protein was in the form of a mixture of meat and dairy products that provided calcium in addition to similar amounts of protein and phosphorus, an increase in urinary calcium also was observed, but calcium retention was positive.

Although it has been postulated that high protein intakes may represent a risk factor in osteoporosis because of their calciuretic effect, evidence to support or dispute such a relationship is limited. A recent study in young adults (seven men, eight women) showed that short-term intake of a high-protein (2.71 g/kg BW), high-calcium (1589 mg) diet had no effect on urinary pyridinium cross-link excretion, a sensitive indicator of bone resorption, compared to the effects of a low-protein diet (0.44 g/kg BW) with similar calcium content. A low-protein (0.49 g/kg BW), low-calcium (429 mg) diet, however, resulted in higher urinary pyridinium cross-link excretion, suggesting increased bone resorption (Shapses et al., 1995). Based on food frequency data and self-reports of bone fractures, a 12-year prospective study of participants in the Nurses' Health Study found that total dietary protein was associated with an increased risk of forearm fracture for women who consumed more than 95 g protein per day compared with those who consumed less than 68 g/d per day. An increase in risk of forearm fracture was also observed for animal protein but not vegetable protein. Women who consumed five or more servings of red meat per week also had a significantly increased risk of forearm fracture compared to women who ate red meat less than once a week. The incidence of hip fractures was not associated with protein intake (Feskanich et al., 1996).

The relatively high protein intakes of male soldiers both in garrison (98–132 g/d) and in field settings (105 g/d) would be expected to be associated with higher urinary calcium excretion than lower-protein diets. However, these protein intakes would not necessarily have a negative impact on calcium retention depending on other dietary factors such as the intake of phosphorus and calcium. Calcium intakes by soldiers in a variety of settings have been reported by Baker-Fulco (1995). Mean calcium intakes of male soldiers in field studies reached approximately 1000 mg/d (the dietary reference intake [DRI] for men; IOM, 1997) or more in five of the nine field studies reported, while calcium intake of men in garrison and at the U.S. Military Academy exceeded 1000 mg/d in all studies reported. In five of seven studies that included women, mean calcium intakes approached or exceeded 1000 mg, the DRI for women 19 to 50, while in the other two studies, intakes averaged around 750–800 mg (the MRDA for calcium).

Kerstetter and Allen (1990) suggested that calcium balance is close to equilibrium with daily protein intakes up to about 74 g and calcium intakes in the range of 500–1400 mg/d. Based on this statement, serious problems with calcium status in relation to protein intake would seem unlikely in military women. Although protein availability from operational rations is high, actual consumption is considerably lower. In addition, these diets are intended for short-term consumption, and although research is limited, some evidence suggests that regular weight-bearing physical activity contributes to bone strength. For these reasons, it would appear that operational rations would not likely be associated with a significantly increased risk of stress fracture or osteoporosis for women in the military.

Toxicity of Amino Acid Supplements

In 1994, the CMNR reviewed scientific information related to the use of selected supplements for enhancing performance (IOM, 1994) and concluded that the supplementation of certain amino acids at levels that were in the range of those found in a typical diet might have a positive impact on performance. The committee stated that the addition of amino acids at these levels is not likely to cause harm to healthy adults with short-term use.

The efficacy and safety of protein and amino acid supplements consumed by individuals in the hope of enhancing performance is revisited by Maher (see

Chapter 16). Maher points out that there is no credible scientific evidence to suggest that normal, healthy persons consuming diets adequate in protein would benefit nutritionally in any way from supplementation with any single amino acid. Further, he asserts that indiscriminate supplementation has the potential for real harm to people who eat less-than-ideal diets, because of the possible antinutritional (growth-inhibiting) effects of amino acid-imbalanced diets. In contrast, supplementation of diets containing poor-quality proteins with the amino acids that are limiting in the protein could be beneficial.

Maher also expresses concern regarding the consumption of amino acids at higher levels to achieve hypothetical pharmacological rather than nutritional benefits. He summarizes the findings of an expert panel (FASEB/LSRO, 1992) that reviewed the safety of consumption of amino acid supplements. Substantial potential exists for deleterious interactions of amino acids with a number of over-the-counter and prescription medicines. Maher points out that supplemental use of D-amino acids is especially risky since the D enantiomers not only have no nutritional value but also are likely to be more toxic at high doses (Friedman, 1991). According to Maher, lack of safety data regarding the consumption of high intakes of individual amino acids (D or L) suggests that recommendations should be conservative with regard to their use as supplements.

Maher further points out that the purity of individual amino acids is a very important consideration in their use as supplements. Experience with the consumption of L-tryptophan, which contained a suspected low-level contaminant, has shown that extreme harm can result (Hertzman et al., 1990).

High-Protein Diets, Amino Acid Supplements, and Pregnancy. The results of a controlled clinical trial in New York City that enrolled low-income, pregnant, African-American women suggest that the consumption of high-protein supplements during pregnancy may be detrimental to the fetus. Three dietary treatments were allocated randomly to evaluate fetal outcome: supplement, which consisted of two 8-ounce cans of a high-protein beverage; complement, which consisted of two 8-ounce cans of a balanced protein–energy beverage; and control, which consisted only of routine vitamin–mineral tablets (Rush et al., 1980). The protein content of the supplement was 8.5 g per 100 kcal, which provided 34 percent of calories compared with 1.9 g per 100 kcal, or 8 percent of calories, in the complement. With balanced protein–energy supplementation, gestational duration was increased, the proportion of low-birthweight infants was reduced, and mean birthweight was increased by 41 g (not statistically significant). With high-protein supplementation, a tendency toward increased incidence of very early premature births and associated neonatal deaths (at 20–35 weeks, not statistically significant) and significant growth retardation in the preterm infants were observed. Because the women in the supplement group who delivered prematurely consumed more of the supplement but fewer total calories, interpretation of the study findings is confounded; nevertheless, this historic study strongly warns against the use of high-protein supplements in pregnant women. It should be noted that there is no evidence that a high-protein diet from food is detrimental to pregnancy outcome.

There is no evidence that supplementation of the diet with a single amino acid during pregnancy would be of benefit nutritionally to normal, healthy individuals. On the contrary, supplementation of the diet with a single amino acid may be potentially dangerous to the developing fetus. Several studies in laboratory animals demonstrate antinutritional effects (that is, depressed growth and other adverse effects) associated with the intake of imbalanced amino acid diets. High doses of single amino acids given to rat dams elevated their plasma amino acid concentrations and resulted in offspring with lower birthweight, decreased brain weight, and altered behavior. Significant effects have been reported for such amino acids as leucine, isoleucine, valine, histidine, threonine, tryptophan, and tyrosine (Burns and Kacser, 1987; Frieder and Grimm, 1984; Funk et al., 1991; Huether, et al., 1992; Matsueda and Niiyama, 1982). The use of amino acid supplements in pregnant women, although not examined, might therefore also be expected to elevate maternal plasma amino acid levels and possibly lead to similar, negative effects in their offspring (as observed in rats). Based on the results of these studies, pregnant and lactating women might be at greater risk of adverse effects from ingestion of particular amino acids.

In summary, the results of human and animal studies show that consumption of single protein or amino acid supplements in excess of recommended intakes during pregnancy may have detrimental effects on fetal growth and development. Such effects must be considered in light of the increasing representation of women among deployed forces and the fact that, at any given time, approximately 10 percent of female active-duty personnel are pregnant.

REFERENCES

Albino, J.E., and R.B. Mateo. 1995. Nitric oxide. Pp. 99–123 in Amino Acid Metabolism and Therapy in Health and Nutritional Disease, L.A. Cynober, ed. Boca Raton, Fla.: CRC Press.

Allen, L.H., E.A. Oddoye, and S. Margen. 1979. Protein-induced hypercalciuria: A longer term study. Am. J. Clin. Nutr. 32:741–749.

AR (Army Regulation) 40-250. 1947. See U.S. Department of the Army, 1947.

Anderson, J.W., B.M. Johnstone, and M.E. Cooke-Newell. 1995. Meta-analysis of the effects of soy protein intake on serum lipids. N. Engl. J. Med. 333(5): 276–282.

Askew, E.W. 1989. Nutrition for a cold environment. Phys. Sportsmed. 17:77–89.

Askew, E.W., I. Munro, M.A. Sharp, S. Siegel, R. Popper, M.S. Rose, R.W. Hoyt, K. Reynolds, H.R. Lieberman, D. Engell, and C.P. Shaw. 1987. Nutritional status and physical and mental performance of soldiers consuming the Ration, Lightweight or the Meal, Ready-to-Eat military field ration during a 30 day field training exercise (RLW-

30). Technical Report No. T7-87. Natick, Mass.: U.S. Army Research Institute of Environmental Medicine.
Baker-Fulco, C.J. 1995. Overview of dietary intakes during military exercises. Pp. 121–149 in Not Eating Enough, Overcoming Underconsumption of Military Operational Rations, B.M. Marriott, ed. Institute of Medicine. Washington, D.C.: National Academy Press.
Baker-Fulco, C.J., J.C. Buchbinder, S.A. Torri, and E.W. Askew. 1992. Dietary Status of Marine Corps officer candidates. Fed. Am. Soc. Exp. Biol. J. [FASEB J] 6(4):A1682.
Baker-Fulco, C.J., S.A. Torri, J.E. Arsenault, and J.C. Buchbinder. 1994. Impact of menu changes designed to promote a training diet [abstract]. J. Am. Diet. Assn. 94:A9.
Banderet, L.E., and H.R. Lieberman. 1989. Treatment with tyrosine, a neurotransmitter precursor, reduces environmental stress in humans. Brain Res. Bull. 22:759–762.
Barbul, A., S.A. Lazarou, D.T. Efron, H.L. Wasserkrug, and G. Efron. 1990. Arginine enhances wound healing and lymphocyte immune responses in humans. Surgery 108(2):331–336.
Baumgartner, R.N., W.C. Chumlea, and A.F. Roche. 1989. Estimation of body composition from bioelectric impedance of body segments. Am. J. Clin. Nutr. 50(2):221–226.
Beisel, W.R. 1992. Metabolic responses of the host to infections. Pp. 1–13 in Textbook of Pediatric Infectious Diseases, Vol. I, 3rd ed., R.D. Feigin and J.D. Cherry, eds. Philadelphia: W.B. Saunders Co.
Berneis, K., R. Ninnis, J. Girard, B.M. Frey, and U. Keller. 1997. Effects of insulin-like growth factor I combined with growth hormone on glucocorticoid-induced whole-body protein catabolism in man. Clin. Endocrinol. Metab. 82:2528–2534.
Bhasin, S., T. Storer, N. Berman, C. Callegari, B. Clevenger, J. Phillips, T.J. Bunell, R. Tricker, A. Shirari, and R. Casaburi. 1996. The effects of supraphysiologic doses of testosterone on muscle size and strength in normal men. N. Engl. J. Med. 335:1–7.
Black, P.R., D.C. Brooks, P.Q. Bessey, R.R. Wolfe, and D.W. Wilmore. 1982. Mechanisms of insulin resistance following surgery. Ann. Surg. 196:420–435.
Blomstrand, E., P. Hassmén, B. Ekblom, and E.A. Newsholme. 1991. Administration of branched-chain amino acids during sustained exercise—Effects on performance and on plasma concentration of some amino acids. Eur. J. Appl. Physiol. 63:83–88.
Borchers, J., and G.E. Butterfield. 1992. The effect of meal composition on protein utilization following an exercise bout. Med. Sci. Sports. Exer. 24:S51.
Burns, J.E., and H. Kacser. 1987. Genetic effects on susceptibility to histidine induced teratogenesis in the mouse. Genet. Res. 50(2):147–153.
Buskirk, E.R. 1993. Energetics and climate with emphasis on heat: A historical perspective. Pp. 97–116 in Nutritional Needs in Hot Environments: Applications for Military Personnel in Field Operations. B.M. Marriott, ed. Institute of Medicine. Washington, D.C.: National Academy Press.
Buskirk, E.R. 1996. Exercise. Pp. 420–429, Chapter 41 in Present Knowledge in Nutrition, E.E. Ziegler and L.J. Filer, Jr., eds. Washington, D.C.: ILSI Press.
Butterfield, G.E. 1987. Whole-body protein utilization in humans. Med. Sci. Sports. Exerc. 19:S157–165.
Butterfield, G.E. 1996. Maintenance of body weight at high altitudes: In search of 500 kcal/day. Pp. 357–378 in Nutritional Needs in Cold and in High-Altitude Environments, B.M. Marriott and S.J. Carlson, eds. Institute of Medicine. Washington, D.C.: National Academy Press.
Butterfield, G.E., and D.H. Calloway. 1984. Physical activity improves protein utilization in young men. Br. J. Nutr. 11:171–184.
Butterfield, G.E., J. Gates, S. Fleming, G.A. Brooks, J.R. Sutton, and J.T. Reeves. 1992. Increased energy intake minimizes weight loss in men at high altitude. J. Appl. Physiol. 72:1741–1748.

Butterfield, G.E., J. Thompson, M.J. Rennie, R. Marcus, R.L. Hintz and A.R. Hoffman. 1997. Effect of rhGH and IGF-1 treatment on protein utilization in elderly women. Am. J. Physiol. 272:E94–E99.

Calloway, D.H. 1975. Nitrogen balance of men with marginal intakes of protein and energy. J. Nutr. 105:914–923.

Calloway, D.H., and H. Spector. 1954. Nitrogen balance as related to caloric and protein intake in active young men. Am. J. Clin. Nutr. 2:405–412.

Calloway, D.H., A.C.F. Odell, and S.J. Margen. 1971. Sweat and miscellaneous nitrogen losses in human balance studies. J. Nutr. 101:775–786.

Carlson, D.E., T.B. Dugan, J.C. Buchbinder, J.D. Allegretto, and D.D. Schnakenberg. 1987. Nutritional assessment of the Ft. Riley Non-commissioned Officer Academy dining facility. Technical Report T14-87. Natick, Mass.: U.S. Army Research Institute of Environmental Medicine.

Carraro, F., W.H. Hartl, C.A. Stuart, D.K. Layman, F. Jahoor, and R.R. Wolfe. 1990. Whole body and plasma protein synthesis in exercise and recovery in human subjects. J. Appl. Physiol. 258:E821–E831.

Castell, L.M., J.R. Poortmans, and E.A. Newsholme. 1996. Does glutamine have a role in reducing infection in animals. Eur. J. Appl. Physiol. 13:488–490.

Chaouloff, F. 1989. Physical exercise and brain monoamines: a review. Acta. Physiol. Scand. 137:1–113.

Chittenden, R.H. 1907. The Nutrition of Man. London: Heinemann.

Cleare, A.J., and A.J. Bond. 1994. Effects of alterations in plasma tryptophan levels on aggressive feelings. Arch. Gen. Psychiatry 51(12):1004–1005.

Cline, A.D., J.F. Patton, W.J. Tharion, S.R. Strowman, C.M. Champagne, J. Arsenault, K.L. Reynolds, J.P. Warber, C. Baker-Fulco, J. Rood, R.T. Tulley, and H.R. Lieberman. 1998. Assessment of the relationship between iron status, dietary intake, performance, and mood state of female Army officers in a basic training population. Technical Report No. T98-24. Natick, Mass.: U.S. Army Research Institute of Environmental Medicine.

Clugston, G., K.G. Dewey, C. Fjeld, J. Millward, P. Reeds, N.S. Scrimshaw, K. Tontisirin, J.C. Waterlow, and V.R. Young. 1996. Report of the working group on protein and amino acid requirements. Europ. J. Clin. Nutr. 50:S193–S195.

Collins, D.M., C.T. Rezzo, J.B. Kopp, P. Ruiz, T.M. Coffman, and P.E. Klotman. 1990. Chronic high protein feeding does not produce glomerulonephrosis or renal insufficiency in the normal rat. J. Am. Soc. Nephrol. 1:624.

Collipp, P.J., J. Thomas, V. Curti, R.K. Sharma, V.T. Maddaiah, S.E. Cohn. 1973. Body composition changes in children receiving human growth hormone. Metabolism 22:589–595.

Consolazio, C.F., and R. Shapiro. 1964. Energy requirements of men in extreme heat. Pp. 121–124 in Environmental Physiology and Psychology in Arid Conditions: Proceedings of the Lucknow Symposium. Liege, Belgium: United Nations, UNESCO.

Copeland, K.C., and K.S. Nair. 1994. Acute growth hormone effects on amino acid and lipid. J. Clin. Endocrinol. Metab. 78:1040–1047.

Cuneo, R.C., F. Salomon, C.M. Wiles, R. Hesp, P.H. Sonksen. 1991. Growth hormone treatment in growth hormone-deficient adults. II. Effects on exercise performance. J. Appl. Physiol. 70:695–700.

Cuthbertson, D. and H.N. Munro. 1937. A study of the effect of over-feeding on the protein metabolism of man. III. The protein-saving effect of carbohydrate and fat when superimposed on a diet adequate for maintenance. Biochem. J. 31:694–705.

Davis, J.M., S.P. Bailey, J.A. Woods, F.J. Galiano, M.T. Hamilton, and W.P. Bartoli. 1992. Effects of carbohydrate feedings on plasma free tryptophan and branched chain amino acids during prolonged cycling. Eur. J. Appl. Physiol. 65:513–519.

Deijen, J.B., and J.F. Orlebeke. 1994. Effect of tyrosine on cognitive function and blood pressure under stress. Brain Res. Bull. 33:319–323.

Delgado, P.L., D.S. Charney, L.H. Price, G.K. Aghajanian, H. Landis, and G.R. Heninger. 1990. Serotonin function and the mechanism of antidepressant action. Reversal of antidepressant-induced remission by rapid depletion of plasma tryptophan.. Arch. Gen. Psychiatry 47:411–418.

Duncan, A.M., R.O. Ball, and P.B. Pencharz. 1996. Lysine requirement of adult males is not affected by decreasing dietary protein intake. Am. J. Clin. Nutr. 64:718–725.

Edwards, J.S.A., E.W. Askew, N. King, C.S. Fulco, R.W. Hoyt, and J.P. DeLany. 1991. An assessment of the nutritional intake and energy expenditure of unacclimatized U.S. Army soldiers living and working at high altitude. Technical Report No. T10-91. Natick, Mass.: U.S. Army Research Institute of Environmental Medicine.

Elahi, D., M. McAloon-Dyke, N.K. Fukugawa, A.L. Sclater, G.A. Wong, R.P. Shannon, K.L. Minaker, J.M. Miles, A.H. Rubenstein, and C.J. Vandepol. 1993. Effects of recombinant human IGF-I on glucose and leucine kinetics in men. Am. J. Physiol. 265:E831–E838.

FASEB/LSRO. 1992. Safety of amino acids used as dietary supplements. Center for Food Safety and Applied Nutrition. FDA Contract No. 223-88-2124, Task No. 8.

FAO/WHO (Food and Agriculture Organization of the United Nations/World Health Organization). 1991. Protein Quality Evaluation. Report of a Joint FAO/WHO Expert Consultation. FAO Food and Nutrition Paper 51. Rome: FAO.

FAO/WHO/UNU (Food and Agriculture Organization of the United Nations/World Health Organization)/United Nations University). 1985. Energy and protein requirements. Report of a joint expert consultation. World Health Organization Technical Report Series No. 724. Geneva: World Health Organization.

Fernstrom, J.D. 1990. Aromatic amino acids and monoamine synthesis in the central nervous system: Influence of the diet. J. Nutr. Biochem. 1:508–517.

Fernstrom, J.D. 1994. Stress and monoamine neurons in the brain. Pp. 161–175 in Food Components to Enhance Performance, B.M. Marriott, ed. Institute of Medicine. Washington, D.C.: National Academy Press.

Fernstrom, J.D., and M.J. Hirsch. 1975. Rapid repletion of brain serotonin in malnourished, corn-fed rats following L-tryptophan injection. Life Sciences 17:455–464.

Fernstrom, J.D., and L.D. Lytle. 1976. Corn malnutrition, brain serotonin, and behavior. Nutr. Rev. 34:257–262.

Fernstrom, J.D., and R.J. Wurtman. 1971. Brain serotonin content: Physiological dependence on plasma tryptophan levels. Science 173:149–152.

Feskanich, D., W.C. Willett, M.J. Stampfer, and G.A. Colditz. 1996. Protein consumption and bone fractures in women. Am. J. Epidemiol. 143:472–479.

Fielding, R.A., C.N. Meredith, K.P. O'Reilly, W.R. Fontera, J.G. Cannon and N.J. Evans. 1991. Enhanced protein breakdown after eccentric exercise in young and older men. J. Appl. Physiol. 11:674–679.

Frieder, B., and V.E. Grimm. 1984. Prenatal monosodium glutamate (MSG) treatment given through the mother's diet causes behavioral deficits in rat offspring. Int. J. Neurosci. 23(2):117–126.

Friedl, K.E. 1997. Variability of fat and lean tissue loss during physical exertion with energy deficit. Pp. 431–450 in Physiology, Stress, and Malnutrition: Functional Correlates, Nutritional Intervention, J.M. Kinney and H.N. Tucker, eds. Philadelphia: Lippincott-Raven Publishers.

Friedman, M. 1991. Formation, nutritional value, and safety of d-amino acids. Pp. 447–482 in Nutritional and Toxicological Consequences of Food Processing. New York: Plenum Press.

Fryburg, D.A. 1994. Insulin-like growth factor I exerts growth hormone and insulin-like actions on human muscle protein metabolism. Am. J. Physiol. 267:E331–E336.

Fryburg, D.A. 1996. NG-monomethyl-L-arginine inhibits the blood flow but not the insulin-like response of forearm muscle to IGF- I: possible role of nitric oxide in muscle protein synthesis. J. Clin. Invest. 97:1319–1328.

Fryburg, D.A., and E.J. Barrett. 1993. Growth hormone acutely stimulates skeletal muscle but not whole-body protein synthesis in humans. Metabolism 42:1223–1227.

Fryburg, D.A., R.A. Gelfand, and E.J. Barrett. 1991. Growth hormone acutely stimulates forearm protein synthesis in normal subjects. Am. J. Physiol. 260:E499–E504

Fryburg, D.A., L.A. Jahn, S.A. Hill, D.M. Oliveras, and E.J. Barrett. 1995. Insulin and insulin-like growth factor I enhance human skeletal muscle protein anabolism during hyperaminoacidemia by different mechanisms. J. Clin. Invest. 96:1722–1729.

Fuller, M.F., and P.J. Garlick. 1994. Human amino acid requirements: Can the controversy be resolved? Ann. Rev. Nutr. 14:217–241.

Funk, D.N., B. Worthington-Roberts, and A. Fantel. 1991. Impact of supplemental lysine or tryptophan on pregnancy course and outcomes in rats. Nutr. Res. 11:501–512.

Gallagher, D., D. Belmonte, P. Deurenberg, Z. Wang, N. Krasnow, F.X. Pi-Sunyer, S.B. Heymsfield. 1998. Organ-tissue mass measurement allows modeling of REE and metabolically active tissue mass. Am. J. Physiol. 275:E249–E258.

Gontzea, I., P. Sutzescu, and S. Dumitrache. 1975. The influence of adaptation to physical effort on nitrogen balance in man. Nutrition Reports International 22:231–236.

Gore, D.C., D. Honeycutt, F. Jahoor, R.R. Wolfe, and D.N. Herndon. 1991. Effect of exogenous growth hormone on whole-body and isolated-limb protein kinetics in burned patients. Arch. Surg. 126:38–43.

Griffiths, R.D., C. Jones, and T.E.A. Palmer. 1997. Six-month outcome of critically ill patients given glutamine-supplemented parenteral nutrition. Nutrition 13(4):295–302.

Hartmann, E., and D. Greenwald. 1984. Tryptophan and human sleep: An analysis of 43 studies. Pp. 297–304 in Progress in Tryptophan and Serotonin Research, H.G. Schlossberger, W. Kochen, B. Linzen, and H. Steinhart, eds. Berlin: Walter de Gruyter.

Hegsted, M., and H.M. Linkswiler. 1981. Long-term effects of level of protein intake on calcium metabolism in young adult women. J. Nutr. 111:244–251.

Hegsted, M., S.A. Schuette, M.B. Zemel, and H.M. Linkswiler. 1981. Urinary calcium and calcium balance in young men as affected by level of protein and phosphorus intake. J. Nutr. 111:553–562.

Hertzman, P.A., W.L. Blevins, J. Mayer, B. Greenfield, M. Ting, and G.J. Gleich. 1990. Association of the eosinophilia–myalgia syndrome with the ingestion of tryptophan. N. Engl. J. Med. 322:869–873.

Heymsfield, S.B., D. Gallagher, M. Visser, C. Nuñez, and Z-M. Wang. 1995. Measurement of skeletal muscle: Laboratory and epidemiological methods. J. Gerontol. 50A:23–29.

Heymsfield, S.B., R. Ross, Z. Wang, D. Frager. 1997. Imaging Techniques of Body Composition: Advantages of Measurement and New Uses. Pp. 127-150 in Emerging Technologies for Nutrition Research, S.J. Carlson-Newberry and R.B. Costello, eds. Institute of Medicine. Washington, DC: National Academy Press.

Hirsch, E., W. Johnson, P. Dunne, C. Shaw, N. Hotson, W. Tharion, H. Lieberman, R. Hoyt, and D. Dacumos. In press. The effects of diet composition on food intake, food selection, and water balance in a hot environment. Technical Report. Natick, Mass.: Natick Research, Development and Engineering Center.

Hoyt, R.W., and A. Honig. 1996. Body fluid and energy metabolism at high altitude. Pp. 1277–1289 in Handbook of Physiology, Section 4: Environmental Physiology, C.M. Blatteis and M.J. Fregly, eds. New York: Oxford University Press for the American Physiological Society.

Huether, G., F. Thomke, and L. Adler. 1992. Administration of tryptophan-enriched diets to pregnant rats retards the development of the serotonergic system in their offspring. Brain Res. Dev. Brain Res. 68(2):175–181.

IOM (Institute of Medicine). 1992. A Nutritional Assessment of U.S. Army Ranger Training Class 11/91. March 23. Washington, D.C.

IOM. 1993b. Review of the Results of Nutritional Intervention, U.S. Army Ranger Training Class 11/92 (Ranger II), B.M. Marriott, ed. Washington, D.C.: National Academy Press.

IOM. 1994. Food Components to Enhance Performance, An Evaluation of Potential Peformance-Enhancing Food Components for Operational Rations, B.M. Marriott, ed. Washington, D.C.: National Academy Press.

IOM. 1995. Not Eating Enough, Overcoming Underconsumption of Military Operational Rations, B.M. Marriott, ed. Washington, D.C.: National Academy Press.

IOM. 1996. Nutritional Needs in Cold and in High-Altitude Environments, Applications for Military Personnel in Field Operations, B.M. Marriott and S.J. Carlson, eds. Washington, D.C.: National Academy Press.

IOM. 1997. Dietary Reference Intakes: Calcium, Phosphorus, Magnesium, Vitamin D and Fluoride. Washington D.C.: National Academy Press.

IOM. 1998. Assessing Readiness in Military Women: The Relationship of Body Composition, Nutrition, and Health. Washington, D.C.: National Academy Press.

IOM. 1999. Military Strategies for Sustainment of Nutrition and Immune Function in the Field. Washington, D.C.: National Academy Press.

Jacobi, C.A., J. Ordemann, F. Wenger, K. Zuckerman, H.D. Volk, and J.M. Muller. 1997. The influence of glutamine substitution in postoperative parenteral nutrition on immunologic function. First results of a prospective randomized trial (abstract). Shock 7(S):605.

Jacobs, B.L. and C.A. Fornel. 1993. 5-Hydroxytryptamine and motor control: a hypothesis. Trends in Neurosciences. 16:346–352.

Jones, P.J.H., and I.K.K. Lee. 1996. Macronutrient requirements for work in cold environments. Pp. 189–202 in Nutritional Needs in Cold and in High-Altitude Environments: Applications for Military Personnel in Field Operations, B.M. Marriott and S.J. Carlson, eds. Institute of Medicine. Washington, D.C.: National Academy Press.

Jorgensen, J.O.L., L. Thuesen, T. Ingemann-Hansen, S.A. Pedersen, J. Jorgensen, N.E. Skakkebaek, and J.S. Christiansen. 1989. Beneficial effects of growth hormone treatment in GH deficient adults. Lancet 1:1221–1225.

Katz, A., S. Broberg, K. Sahlin, and J. Wahren. 1986. Muscle ammonia and amino acid metabolism during dynamic exercise in man. Clin. Physiol. 6:365–379.

Kerr, G.R., E.S. Lee, M.M. Lan, R.J. Lorimor, E. Randall, R.N. Forthofer, M.A. Davis, and S.M. Magnetti. 1982. Relationships between dietary and biochemical measures of nutritional status in NHANES I data. Am. J. Clin. Nutr. 35:294–308.

Kerstetter, J.E., and L.H. Allen. 1990. Dietary protein increases urinary calcium. J. Nutr. 120:134–136.

King, N., K.E. Fridlund, and E.W. Askew. 1993. Nutrition issues of military women. J. Am. Coll. Nutr. 12:344–348.

Kretsch, M.J., P.M. Conforti, and H.E. Sauberlich. 1986. Nutrient intake evaluation of male and female cadets at the United States Military Academy, West Point, New York, Report No. 218. Presidio of San Francisco, Calif. Letterman Army Institute of Research.

Kishi, K., S. Miyatani, and G. Inoue. 1978. Requirements and utilization of egg protein by Japanese young men with marginal intakes of energy. J. Nutr. 108:658–669.
King, N., J.E. Arsenault, S.H. Mutter, C.M. Champagne, T.C. Murphy, K.A. Westphal, and E.W. Askew. 1994. Nutritional intake of female soldiers during the U.S. Army basic combat training. Technical Report No. T94-17. Natick, Mass.: U.S. Army Research Institute of Environmental Medicine.
Kinney, J.M., and D.H. Elwyn. 1995. Amino acid metabolism in health and nutritional disease. Pp. 1–12 in Amino Acid Metabolism in Health and Nutritional Disease, L.A. Cynober, ed. Boca Raton, Fla.: CRC Press.
Klicka, M.V., D.E. Sherman, N. King, K.E. Friedl, and E.W. Askew. 1993. Nutritional assessment of cadets at the U.S. Military Academy: Part 2. Assessment of nutritional intake. Technical Report T94-1. Natick, Mass.: U.S. Army Research Institute of Environmental Medicine.
Kreider, R.B., V. Miriel, amd E. Bertum. 1993. Amino acid supplementation and exercise performance—Analysis of the proposed ergogenic value. Sports Med. 16:190–209.
Kurzer, M.S. and D.H. Calloway. 1986. Effects of energy deprivation on sex hormone patterns in healthy menstruating women. Am. J. Physiol. 251:E483–E488.
LeBlanc, J.A. 1996. Cold exposure, appetite, and energy balance. Pp. 203–214 in Nutritional Needs in Cold and in High-Altitude Environments: Applications for Military Personnel in Field Operations, B.M. Marriott and S.J. Carlson, eds. Institute of Medicine. Washington, D.C.: National Academy Press.
Lehnert, H.R., D.K. Reinstein, and R.J. Wurtman. 1984a. Tyrosine reverses the depletion of brain norepinephrine and the behavioral deficits caused by tail-shock stress in rats. Pp. 81–91 in Stress: The Role of Catecholamines and Other Neurotransmitters, E. Usdin and R. Kvetnansky, eds. New York: Gordon and Beach.
Lemon, P.R., M.A. Tarnopolsky, J.D. MacDougall, and S.A. Atkinson. 1992. Protein requirements and muscle mass/strength changes during intensive training in novice bodybuilders. J. Appl. Physiol. 73:767–775.
Lieberman, H.R. 1994. Tyrosine and stress: Human and animal studies. Pp. 277–299 in Food Components to Enhance Performance, An Evaluation of Potential Performance—Enhancing Food Components for Operational Rations, B.M. Marriott, ed. Washington, D.C.: National Academy Press.
Lieberman, H.R. and B. Shukitt-Hale. 1996. Food components and other treatments that may enhance performance at high altitude and in the cold. Pp. 453–465 in Nutritional Needs in Cold and in High Altitude Environments, B. Marriott and S. Newberry, eds. Washington, D.C.: National Academy Press.
Lieberman, H.R., S. Corkin, B.J. Spring, J.H. Growdin, and R.J. Wurtman. 1983. Mood, performance, and pain sensitivity: Changes induced by food constituents. J. Psychiatr. Res. 17(2):135–145.
Lieberman, H.R., S. Corkin, B.J. Spring, R.J. Wurtman, and J.H. Growdon. 1985. The effects of dietary neurotransmitter precursors on human behavior. Am. J. Clin. Nutr. 42:366–370.
Lieberman, S.A., G.E. Butterfield, D. Harrison, and A.R. Hoffman. 1994. Anabolic effects of recombinant insulin-like growth factor-I in cachectic patients with the acquired immunodeficiency syndrome. J. Clin. Endocrinol. Metab. 78:404–410.
Linkswiler, H.M., C.L. Joyce, and R. Anand. 1974. Calcium retention of young adult males as affected by level of protein and of calcium intake. Proc. N.Y. Acad. Sci. 36:333–340.
Longenecker, J.B. 1961. Relationship between plasma amino acids and clinical chemistry of dogs. Pp. 469–485 in Progress in Meeting Protein Needs of Infants and Pre-school Children. Publ. 843. Washington, D.C.: National Academy of Sciences.
Longenecker, J.B. 1963. Utilization of dietary protein. Pp. 113–144, Chapter 2, in Newer

Methods of Nutritional Biochemistry, A.A. Albanese, ed. New York: Academic Press.
Longenecker, J.B., and N.L. Hause. 1959. Relationship between plasma amino acids and composition of the ingested protein. Arch. Biochem. Biophys. 84:46–60.
Longenecker, J.B., and N.L. Hause. 1961. Relationship between plasma amino acids and composition of the ingested protein. II. A shortened procedure to determine plasma amino acid (PAA) ratios. Am. J. Clin. Nutr. 9:356–363.
Lukaski, H.C., J. Mendez, E.R. Buskirk, and S.H. Cohn. 1981. Relationship between endogenous 3-methylhistidine excretion and body composition. Am. J. Physiol. 240:E302–E307.
Marchini, J.S., J. Cortiella, T. Hiramatsu, T.E. Chapman, and V.R. Young. 1993. Requirements for indispensable amino acids in adult humans: Longer term amino acid kinetic study with support for the adequacy of the Massachusetts Institute of Technology amino acid requirement pattern. Am. J. Clin. Nutr. 58:670–683.
Matsueda, S., and Y. Niiyama. 1982. The effects of excess amino acids on maintenance of pregnancy and fetal growth in rats. J. Nutr. Sci. Vitaminol. (Tokyo). 28:557–573.
McLarney, M.J., P.L. Pellett, and V.R. Young. 1996. Pattern of amino acid requirements in humans: An interspecies comparison using published amino acid requirements recommendations. J. Nutr. 126:1871–1882.
Meredith, C.N., M.J. Zackin, W.R. Frontera, and W.J. Evans. 1989. Dietary protein requirements and body protein metabolism in endurance-trained men. J. Appl. Physiol. 66:2850–2856.
Messing, R.B., and L.D. Lytle. 1977. Serotonin-containing neurons: their possible role in pain and analgesia. Pain 4:1–21.
Millward, D.J. 1994. Can we define indispensable amino acid requirements and assess protein quality in adults? J. Nutr. 124:1509S–1516S.
Millward, D.J., and J.P. Rivers. 1988. The nutritional role of indispensable amino acids and the metabolic basis for their requirements. Eur. J. Clin. Nutr. 42:367–393.
Millward, D.J., and J.P. Rivers. 1989. The need for indispensable amino acids: The concept of the anabolic drive. Diab. Metab. Rev. 5(2):191–211.
Millward, D.J., and J.C. Waterlow. 1996. Letter to the editor. Eur. J. Clin. Nutr. 50:832–833.
Millward, D.J., J.L. Bowtell, P. Pacy, and M.J. Rennie. 1994. Physical activity, protein metabolism and protein requirements. Proc. Nutr. Soc. 53(1):223–240.
Mitchell, H.H., and M. Edman. 1949. Nutrition and Resistance to Climatic Stress, with Reference to Man. Chicago, Ill.: Quartermaster Food and Container Institute for the Armed Forces.
Mitchell, H.H., and M. Edman. 1951. Nutrition and Resistance to Climatic Stress, with Particular Reference to Man. Springfield, Ill.: Charles C. Thomas.
Motil, K.J., C.M. Montandon, M. Thotathuchery, and C. Garza. 1990. Dietary protein and nitrogen balance in lactating and nonlactating women. Am. J. Clin. Nutr. 51:378–384.
Motil, K.J., T.A. Davis, C.M. Montandon, W.W. Wong, and P.D. Klein. 1996. Whole-body protein turnover in the fed state is reduced in response to dietary protein restriction in lactating women. Am. J. Clin. Nutr. 64:32–39.
Mulligan, K., and G.E. Butterfield. 1990. Discrepancies between energy intake and expenditure in physically active women. Br. J. Nutr. 64(1):23–36.
Munro, H.N., and M.C. Crim. 1994. Protein and amino acids. In Modern Nutrition in Health and Disease, M.E. Shils, J.A. Olson, and M. Shike, eds. Philadelphia: Lea and Febiger.
Neri, D.F., D. Wiegmann, R.R. Stanny, S.A. Shappell, A. McCardie, and D.L. McKay. 1995. The effects of tyrosine on cognitive performance during extended wakefulness. Aviat. Space Environ. Med. 66:313–319.

Nindl, B.C., K.E. Friedl, P.N. Frykman, L.J. Marchitelli, R.L. Shippee, and J.F. Patton. 1997. Physical performance and metabolic recovery among lean, healthy men following a prolonged energy deficit. Int. J. Sports Med. 18:1–8.
NRC (National Research Council). 1941. Recommended Dietary Allowances. Food and Nutrition Board. Washington, D.C.: National Academy Press.
NRC. 1989. Recommended Dietary Allowances, 10th ed. Institute of Medicine. Washington, D.C.: National Academy Press.
Nuñez, C., D. Gallagher, and S.B. Heymsfield. 1995. Appendicular skeletal muscle mass: Measurement with single frequency bioimpedance analysis. FASEB J. 9(4):A1012.
Nuñez, C., D. Gallagher, M. Visser, F.X. Pi-Sunyer, Z. Wang, and S.B. Heymsfield. 1997. Bioimpedance analysis: Evaluation of leg-to-leg system based on pressure contact foot-pad electrodes. Med. Sci. Sports Exerc. 29:524–31.
O'Riordain, M., K.C. Fearon, J.A. Ross, P. Rogers, J.S. Falconer, D.C. Bartolo, O.J. Garden, and D.C. Carter. 1994. Glutamine-supplemented parenteral nutrition enhances T-lymphocyte response in surgical patients undergoing colorectal resection. Ann Surg. 220:212–221.
Owen, O.E., K.J. Smalley, D.A. D'Alessio, M.A. Mozzoli, E.K. Dawson. 1998. Protein, fat, and carbohydrate requirements during starvation: anaplerosis and cataplerosis. Am. J. Clin. Nutr. 68:12–34.
Pardridge, W.M. 1977. Regulation of amino acid availability to the brain. Pp. 141–190 in Nutrition and the Brain, Vol. 1, R.J. Wurtman and J.J. Wurtman, eds. New York: Raven Press.
Paul, G.L. 1989. Dietary protein requirements of physically active individuals. Sports Med. 8:154–176.
Pearlstone, D.B., R.F. Wolf, R.S. Berman, M. Burt, M.F. Brennan. 1994. Effect of systemic insulin on protein kinetics in postoperative cancer patients. Ann. Surg. Oncol. 1(4):321–332.
Phillips, S.M., S.A. Atkinson, M.A. Tarnopolsky, and J.D. MacDougal. 1993. Gender differences in leucine kinetics and nitrogen balance in endurance athletes. J. Appl. Physiol. 75:2134–2141.
Picou, D., and T. Taylor-Roberts. 1969. The measurement of total protein synthesis and catabolism and nitrogen turnover in infants in different nutritional states and receiving different amounts of dietary protein.. Clin. Sci. 36:283–296
Reeds, P.J., and P.R. Becket. 1996. Protein and amino acids. Pp. 67–86 in Present Knowledge in Nutrition, 7th ed., E.E. Ziegler and L.J. Filer, eds. Washington, D.C.: ILSI Press.
Rennie, M.J. 1996. Influence of exercise on protein and amino acid metabolism. Pp. 995–1035 in American Physiological Society Handbook of Physiology on Exercise, Chapter 12, Section 12, Control of Energy Metabolism During Exercise, R. L. Terjung, ed. Bethesda, Md.: American Physiological Society.
Robertson, W.G., P.J. Heyburn, M. Peacock, F.A. Hanes, and R. Swaminathan. 1979b. The effect of high animal protein intake on the risk of calcium-stone-formation in the urinary tract. Clin. Sci. 57:285–288.
Rooyackers, O., and K.S. Nair. 1997. Hormonal regulation of human muscle protein metabolism. Ann. Rev. Nutr. 17:457–485.
Rose, M.S. and D.E. Carlson. 1986. Effects of A Ration meals on body weight during sustained field operations. Technical Report T2-87. Natick, Mass.: U.S. Army Research Institute of Environmental Medicine.
Rose, M.S., P.C. Szlyk, R.P. Francesconi, L.S. Lester, L. Armstrong, W. Matthew, A.V. Cardello, R.D. Popper, I. Sils, G. Thomas, D. Schilling, and R. Whang. 1989. Effectiveness and acceptability of nutrient solutions in enhancing fluid intake in the

heat. Technical Report No. T10-89. Natick, Mass.: U.S. Army Research Institute of Environmental Medicine.
Rose, R.W., C.J. Baker, W. Wisnaskas, J.S.A. Edwards, and M.S. Rose. 1989. Dietary assessment of U.S. Army basic trainees at Fort Jackson, South Carolina. Technical Report No. T6-89. Natick, Mass.: U.S. Army Research Institute of Environmental Medicine.
Rowbottom, D.G., D. Keast, C. Goodman, and A.R. Morton 1995. The haematological, biochemical and immunological profile of athletes suffering from the overtraining syndrome. Eur. J. Appl. Physiol. 70:502–509.
Rush, D., Z. Stein, and M.A. Susser. 1980. A randomized controlled trial of prenatal nutritional supplementation in New York City. Pediatrics 65:683–697.
Sahlin, K., A. Katz, and S. Broberg. 1990. Tricarboxylic acid cycle intermediates in human muscle during prolonged exercise. Am. J. Physiol. 159:C834–C841.
Sahlin, K., L. Jorfeldt, and K.G. Henriksson. 1995. Tricarboxylic acid cycle intermediates during incremental exercise in healthy subjects and in patients with McArdle's disease. Clin. Sci. 19:687–693.
Sakurai, Y., A. Aarsland, D.N. Herndon, D. L. Chinkes, E. Pierre, T.T. Nguyen, B.W. Patterson, and R.R. Wolfe. 1995. Stimulation of muscle protein synthesis by long-term insulin infusion in severely burned patients. Ann. Surg. 222(3):283–294.
Schuette, S.A., and H.M. Linkswiler. 1982. Effects on Ca and P metabolism in humans by adding meat, meat plus milk, or purified proteins plus Ca and P to a low protein diet. J. Nutr. 112:338–349.
Schwartz, R.S. 1995. Trophic factor supplementation: effect on the age-associated changes in body composition. J. Gerontol. A. Biol. Sci. Med. Sci. 50:151–156.
Seltzer, S., D. Dewart, R. L. Pollack, and E. Jackson. 1983. The effects of dietary tryptophan on chronic maxillofacial pain and experimental pain tolerance. J. Psychiat. Res. 17(2):181–186.
Shapses, S.A., S.P. Robins, E.I. Schwartz, and H. Chowdhury. 1995. Short-term changes in calcium but not protein intake alter the rate of bone resorption in healthy subjects as assessed by urinary pyridinium cross-link excretion. J. Nutr. 125:2814–2821.
Sharp, T., S.R. Bramwell, and D.G. Grahame-Smith. 1992. Effect of acute administration of L-tryptophan on the release of 5-HT in rat hippocampus in relation to serotoninergic neuronal activity: An *in vivo* microdialysis study. Life Sci. 50:1215–1223.
Shukitt-Hale, B., M.J. Stillman, and H.R. Lieberman. 1996. Tyrosine administration prevents hypoxia-induced decrements in learning and memory. Physiol. Behav. 59:867–871.
Shurtleff, D., J.R. Thomas, S.T. Ahlers, and J. Schrot. 1993. Tyrosine ameliorates a cold-induced delayed matching-to-sample performance decrement in rats. Psychopharmacol. 112:228–232.
Shurtleff, D., J.R. Thomas, J. Schrot, K. Kowalski, and R. Harford. 1994. Tyrosine reverses a cold-induced working memory deficit in humans. Pharmacol. Biochem. Behav. 47(4):935–941.
Souba, W.W., and D.W. Wilmore. 1994. Diet and nutrition in the care of the patient with surgery, trauma, and sepsis. Pp. 1207–1240 in Modern Nutrition in Health and Disease, 8th e., M.E. Shils, J.A. Olson, and M. Shike, eds. Philadelphia: Lea and Febiger.
Spencer, H., L. Kramer, and D. Osis. 1988. Do protein and phosphorus cause calcium loss? J. Nutr. 118:657–660.
Stein, T.P., R.W. Hoyt, M.O. Toole, M.J. Leskiw, and M.D. Schluter. 1989. Protein and energy metabolism during prolonged exercise in trained athletes. Int. J. Sports Med. 10:311–316.
Stone, E. A. 1975. Stress and catecholamines. Pp. 31–71 in Catecholamines and Behavior, A.J. Freidhoff, ed. New York: Plenum Press.

Stroud, M.A., A.A. Jackson, and J.C. Waterlow. 1996. Protein turnover rates of two human subjects during an unassisted crossing of Antarctica. Br. J. Nutr. 16:165–174.
Stucky, W.P., and A.E. Harper. 1962. Effects of altering indispensable to dispensable amino acids in diets for rats. J. Nutr. 78:278–286.
Szeto, E.G., D.E. Carlson, T.B. Dugan, and J.C. Buchbinder. 1987. A comparison of nutrient intakes between a Ft. Riley contractor-operated and a Ft. Lewis military-operated garrison dining facility. Technical Report No. T2-88. Natick, Mass.: U.S. Army Research Institute of Environmental Medicine.
Szeto, E.G., T.B. Dugan, and J.A. Gallo. 1988. Assessment of habitual diners' nutrient intakes in a military-operated garrison dining facility, Ft. Devens I. Technical Report No. T3-89. Natick, Mass.: U.S. Army Research Institute of Environmental Medicine.
Szeto, E.G., J.A. Gallo, and K.W. Samonds. 1989. Passive nutrition intervention in a military-operated garrison dining facility, Ft. Devens II. Technical Report No. T7-89. Natick, Mass: U.S. Army Research Institute of Environmental Medicine.
Tarnopolsky, M.A., J.D. Mac Dougal, and S.A. Atkinson. 1988. Influence of protein intake and training status on nitrogen balance and lean body mass. J. Appl. Physiol. 64:187–193.
Tarnopolsky, M.A., P.W.R. Lemon, J.D. MacDougall, and J.A. Atkinson. 1990a. Effect of body building exercise on protein requirements. Can. J. Sport Sci. 15:225–226.
Tarnopolsky, L.J., J.D. MacDougall, S.A. Atkinson, M.A. Tarnopolsky, and J.R. Sutton. 1990b. Gender differences in substrate for endurance exercise. J. Appl. Physiol. 68:302–308.
Tarnopolsky, M.A., S.A. Atkinson, S.M. Phillips, and J.D. Mac Dougal. 1995. Carbohydrate loading and metabolism during exercise in men and women. J. Appl. Physiol. 78:1360–1368.
Thomas, C.D., K.E. Friedl, M.Z. Mays, S.H. Mutter, and R.J. Moore. 1995. Nutrient intakes and nutritional status of soldiers consuming the Meal, Ready-to-Eat (MRE XII) during a 30-day field training exercise. Technical Report T95-6. Natick, Mass.: U.S. Army Research Institute of Environmental Medicine.
Tipton, K.D., and R.R. Wolfe. 1998. Exercise-induced changes in protein metabolism. Acta Physiol. Scand. 162(3): 377–387.
Tobin, J., and D. Spector. 1986. Dietary protein has no effect on future creatinine clearance. Gerontologist 25:59A.
Todd, K.S., G.E. Butterfield, and D.H. Calloway. 1984. Nitrogen balance in men with adequate and deficient energy intake at three levels of work. J. Nutr. 114:2107–2118.
Tschope, W., and E. Ritz. 1985. Sulfur-containing amino acids are the major determinant of urinary calcium. Mineral Electrolyte Metab. 11:137–139.
USACDEC/USARIEM (U.S. Army Combat Developments and Experimentation Center and U.S. Army Research Institute of Environmental Medicine). 1986. Combat Field Feeding System-Force Development Test and Experimentation (CFFS-FDTE) Technical Report CDEC-TR-85-006A. Vol. 1, Basic Report; vol. 2, Appendix A; vol. 3, Appendixes B through L. Fort Ord, Calif.: U.S. Army Combat Developments and Experimentation Center.
U.S. Department of the Army. 1947. Army Regulation 40-250. Nutrition. Washington, D.C.
Van der Hulst, R.R., B.K. van Kreel, M.F. von Meyenfeldt, R.J. Brummer, J.W. Arends, N.E. Deutz, and P.B. Soeters. 1993. Glutamine and the preservation of gut integrity. Lancet 341(8857):1363–1365.
Walser, M. 1992. Dietary proteins and their relationship to kidney disease. Pp. 168–178 in Dietary Proteins in Health and Disease, G.U. Liepa, ed. Champaign, Ill.: American Oil Chemists' Society.
Wang, Z., M. Visser, R. Ma, R. Baumgartner, D. Kotler, D. Gallagher, and S.B. Heymsfield.

1996. Skeletal muscle mass: Evaluation of neutron activation and dual-energy x-ray absorptiometry methods. J. Appl. Physiol. 80(3):824–831.

Wang, Z., P. Deurenberg, D.E. Matthews, and S.B. Heymsfield. 1998. Urinary 3-methylhistidine excretion: Association with total body skeletal muscle mass by computerized axial tomography. J. Parenter. Enteral Nutr. 22(2): 82–86.

Warber, J.P., F.M. Kramer, S.M. McGraw, L.L. Lesher, W. Johnson, and A.D. Cline. 1996. The Army Food and Nutrition Survey, 1995–97. Technical Report. Natick, Mass.: U.S. Army Research Institute of Environmental Medicine.

Walters, J.K., M. Davis, M.H. Sheard. 1979. Tryptophan-free diet: effects on the acoustic startle reflex in rats. Psychopharmacology (Berl) 62(2):103–109.

Waterlow, J.C. 1996. The requirements of adult man for indispensable amino acids. Eur. J. Clin. Nutr. 50:S151–176.

Welle, S., C. Thornton, M. Statt, and B. McHenry. 1996. Growth hormone increases muscle mass and strength but does not rejuvenate myofibrillar protein synthesis in healthy subjects over 60 years old. J. Clin. Endocrinol. Metab. 81:3239–3243.

Wilmore, D.W. 1991. Catabolic illness: Strategies for enhancing recovery. N. Engl. J. Med. 325(10):695–702.

Wilmore, D.W. 1997a. Glutamine saves lives! What does it mean? Nutrition 13(4):375–376.

Wolf, R.F., D.B. Pearlstone, E. Newman, M.J. Heslin, A. Gonenne, M.E. Burt, and M.F. Brennan. 1992. Growth hormone and insulin reverse net whole body and skeletal muscle protein catabolism in cancer patients. Ann. Surg. 216:280–288.

Wurtman, J.J., and J.D. Fernstrom. 1979. Free amino acid, protein and fat contents of breast milk from Guatemalan mothers consuming a corn-based diet. Early Human Development 3:67–77.

Wurtman, R.J., F. Hefti, and E. Melamed. 1981. Precursor control of neurotransmitter synthesis. Pharmacol. Rev. 32:315–335.

Yarasheski, K.E., J.J. Zachwieja, J.A. Campell, and D.M. Bier. 1995. Effect of growth hormone and resistance training on muscle growth and strength in older men. Am. J. Physiol. 268:E268–E276.

Young, V.R. 1987. McCollum Award Lecture: Kinetics of human amino acid metabolism: Nutritional implications and some lessons. Am. J. Clin. Nutr. 46:709–725.

Young, V.R. 1994. Adult amino acid requirement: The case for a major revision in current recommendations. J. Nutr. 124:1517S–1523S.

Young, V.R., and A. E. El-Khoury. 1995a. Can amino acid requirements for nutritional maintenance in adult humans be approximated from the amino acid composition of body mixed proteins? Proc. Natl. Acad. Sci. 921:300–304.

Young, V.R., and J.S. Marchini. 1990. Mechanisms and nutritional significance of metabolic responses to altered intakes of protein and amino acids, with reference to nutritional adaptation in humans. Am. J. Clin. Nutr. 51:270–289.

Young, V.R., D.M. Bier, and P.L. Pellet. 1989. A theoretical basis for increasing current estimates of the amino acid requirements in adult man with experimental support. Am. J. Clin. Nutr. 50:80–92.

Zawadzki, K.M., B.B. Yaspelkis, and J.L. Ivy. 1992. Carbohydrate–protein complex increases the rate of muscle glycogen storage after exercise. J. Appl. Physiol. 72:1854–1859.

Zello, G.A., P.B. Pencharz, and R.O. Ball. 1993. Dietary lysine requirement of young adult males determined by oxidation of 1-[1-^{13}C]phenylalanine. Am. J. Physiol. 264:E677–E685.

Zemel, M.B. 1988. Calcium utilization: Effect of varying level and source of dietary protein. Am. J. Clin Nutr. 48:880–883.

Ziegler, T.R., L.S. Young, K. Benfell, M. Scheltinga, K. Hortos, R. Bye, F.D. Morrow, D.O.

Jacobs, R.J. Smith, J.H. Antin, and D.W. Wilmore. 1992. Clinical and metabolic efficacy of glutamine-supplemented parenteral nutrition after bone marrow transplantation. A randomized, double-blind, controlled study. Ann. Intern. Med. 116(10):821–828.

Ziegler, T.R., R.L. Bye, R.L. Persinger, L.S. Young, J.H. Antin, and D.W. Wilmore. 1994. Glutamine-enriched parenteral nutrition increases circulating lymphocytes after bone marrow transplantation. J. Parenter. Enteral Nutr. 18:17S.

2

Responses to Questions, Conclusions, and Recommendations

In this chapter, responses are provided to the questions raised by the Army. These responses are based on the material presented in Chapter 1 and form the conclusions drawn by the committee on which its recommendations are based.

1. Do protein requirements increase with military operational stressors, including high workload with or without energy deficit? Are there gender differences in protein requirements in endurance exercise?

At the present time, controversy exists regarding the validity of recent estimations of protein and amino acid requirements (particularly the latter) for adults, a controversy that is based on methodological questions.

In addition, the evidence that high levels of physical activity increase protein requirements for individuals whose energy intake matches their output is equivocal. There is clear evidence that moderate physical activity increases the efficiency of protein utilization. However, strenuous endurance-type exercise has been shown to increase protein requirements above the recommended dietary allowance (RDA), but not the Military Recommended Dietary Allowance (MRDA). In contrast, resistance exercise does not appear to increase the requirement for maintenance of lean mass, although the protein intake that would be required for active individuals to increase tissue mass (1.2–1.5 g/kg BW/d) may be higher than that for sedentary individuals. There is also strong evidence that the efficiency of protein utilization is decreased (and the requirements increased) by a state of negative energy balance.

However, much of the research on the effects of physical activity on protein requirements and the effects of altered protein intakes on performance is difficult to interpret because of the time required for the body to adapt to changes in protein intake. One implication of this adaptation that is of concern for service personnel is that continuous excessive intake of protein may cause increased protein catabolism, resulting in greater risk when protein intake is reduced.

Systemic infection and serious injuries clearly increase protein requirements. However, data suggest that in patients recovering from burns or any major trauma, an increase in dietary protein intake does not permit the recovery of muscle mass to begin immediately, due to the acute-phase response, which is accompanied by changes in hormonal status. Longer term studies are therefore needed during recovery periods. Research on the effects of treatment with anabolic hormones, which stimulate protein synthesis or decrease protein breakdown, is ongoing.

Results of studies of protein requirements in hot, cold, and high-altitude environments suggest that these conditions do not increase protein requirements beyond currently recommended levels. In addition, because increases in protein intake also increase fluid requirements and sources of fluid for drinking are often limited during operations in extreme environments, previous reports of the Committee on Military Nutrition Research (CMNR) have cautioned against excessive protein intake under such circumstances.

The effects of combined stressors and other factors such as emotional stress on protein requirements have not been documented.

As emphasized in earlier IOM reports (IOM, 1992, 1995), the importance of adequate energy intake (sufficient to match output and avoid weight loss) and protein intake should be emphasized to soldiers as the primary means of maintaining lean tissue mass. Research is needed to resolve the controversy regarding the adult requirement for indispensable amino acids and to quantitate more

precisely the effect of energy deficit on protein and indispensable amino acid requirements.

Military researchers and physicians should pay careful attention to civilian research on the effects of treatment with anabolic hormones on recovery from burns and other injuries. Where appropriate, military-specific models should be developed.

2. What is the optimal protein content (and protein–energy ratio) for standard operational rations, and specifically, is the protein Military Recommended Dietary Allowance for operational rations (100 g/d for men and 80 g/d for women) appropriate? Is the protein MRDA for women appropriate during pregnancy and lactation?

Without more data on the functional implications of varying protein intakes, it is not possible to define with accuracy the optimal protein content of standard operational rations. However, based on currently available data, the use of the MRDA for operational rations is appropriate and provides a generous level of protein intake. The MRDA covers the protein requirements of pregnant and lactating women.

Current MRDAs for protein should be maintained. Provided that energy intake is adequate, no increase in MRDAs is necessary for pregnant or lactating women.

3. Is there evidence that supplementation with specific amino acids (AAs) or modification of dietary protein quality would optimize military performance, either cognitive or physical, during high workload, psychological stress, or energy deficit. What are the risks of amino acid supplements and high-protein diets?

At the present time, considerable debate surrounds the adult requirement for indispensable amino acids and thus high-quality proteins. Research fails to support the use of protein supplements to facilitate muscle building or improve physical performance under conditions of adequate energy and protein intake. In addition, research supporting the use of tyrosine supplements to enhance cognitive performance under field conditions is inconclusive. Supplemental glutamine and arginine have yet to show conclusively beneficial effects on immune function. The MRDA, if consumed, provides adequate protein and energy to sustain immune function under normal field conditions. Furthermore, with the exception of tryptophan, commercial preparations of which have been documented to cause specific toxic effects, there is a lack of safety data on the consumption of high levels of individual amino acids.

Some plant proteins such as those from soy and other legumes have an adequate balance of essential amino acids to meet the protein needs of military personnel. These plant foods may have the advantage of decreasing the risk of cardiovascular disease due to their content of soluble carbohydrates, their lower sodium and lower fat contents, and the presence of other as yet unidentified substances.

Current intakes of protein among military populations are high and show no apparent harmful effects, provided fluid intake is adequate. There is little evidence of increased health risks from a high intake of dietary protein; however, an amino acid imbalance may be created with the use of single amino acid or protein supplements. Although no data are available from groups similar in age and fitness characteristics to military personnel, a review of the information available shows that high protein intake is not associated with direct effects on renal dysfunction, although high-protein diets may indirectly stimulate renal stone formation and result in an increased renal workload because of the need to concentrate urine. High protein intake has been shown to increase urinary calcium loss, but there is no definitive evidence that the level of protein intake observed in Army women in field conditions represents a risk factor for osteoporosis.

Given adequate nutritional intake, soldiers should not use protein supplements for muscle building. Military researchers and physicians should pay careful attention to civilian research on the use of anabolic hormones to increase muscle or lean tissue mass.

Protein supplied in operational rations should be of high quality and digestibility. Energy intakes should be adequate, and sources of energy should be consumed within 2 hours of an intense bout of endurance exercise, to replace depleted muscle glycogen.

Soy food products are a healthful substitute for animal-based products; however individual products should be tested for their acceptability to soldiers.

Single amino acid supplements should not be used to modify cognitive performance, due to potential toxicity and insufficient evidence of efficacy.

The military should test the ability of supplemental glutamine and arginine to enhance the immune response and decrease rates of in-

fection under field conditions and in seriously injured hospitalized patients.

Given the high protein content of operational rations, adequate fluid intake should be emphasized, as recommended by the "Fluid Doctrine" (IOM, 1994).

REFERENCES

IOM (Institute of Medicine). 1992. A Nutritional Assessment of U.S. Army Ranger Training Class 11/91. March 23. Washington, D.C.

IOM. 1994. Food Components to Enhance Performance, An Evaluation of Potential Peformance-Enhancing Food Components for Operational Rations, B.M. Marriott, ed. Washington, D.C.: National Academy Press.

IOM. 1995. Not Eating Enough, Overcoming Underconsumption of Military Operational Rations, B.M. Marriott, ed. Washington, D.C.: National Academy Press.

II

Authored Papers and Workshop Discussions

The papers presented at the workshop comprise part II of this report. These chapters (3 through 16), have undergone limited editorial change, have not been reviewed by an outside group, and represent the views of the individual authors. Selected questions and the speakers' responses are included to provide the flavor of the workshop discussion.

Chapter 3 presents the rationale for the military's interest in protein and amino acids to maintain physical and cognitive performance with a brief overview of potential benefits. Historical data and current information from food intake surveys of military personnel in garrison and field situations are presented in Chapter 4. Male military personnel maintain high protein intakes from food, however, female personnel generally consume less energy and protein than the MRDA guidelines in field situations. A substantial number of military personnel consume protein powders and supplemental amino acids, with use among male personnel double that of female personnel.

Chapter 5 sets the stage for subsequent chapters with a general overview of the energy demands of protein metabolism, including the energy required for synthesis, regulation, and breakdown. Using the effect of aging as an example, Chapter 6 outlines some of the control mechanisms and hormones involved in the regulation of muscle mass and function. The maintenance of functioning muscle mass is a complex process involving the orchestration of the effects of anabolic and catabolic hormones, nutritional state, and supply of substrates to the site of protein synthesis together with physical activity and genetic factors.

The effect of protein intake on renal function and development of renal disease is discussed in Chapter 7. Concern about the adverse effects of high protein intake on renal function in healthy people, and in particular on the decline of renal function with age is ill advised. In fact, restricted protein intake in the elderly appears to be a cause of that decline. Restricted protein intake appears to be beneficial for those at risk of acute renal failure, as well as for individuals with nephrolithiasis.

Chapter 8 outlines changes that occur in whole body protein metabolism following infection and injury. Systemic infection and serious injury trigger rapid breakdown of skeletal muscle protein to supply amino acids for specific immune responses in the liver and immune tissues, leading to a marked translocation of nitrogen from muscle to viscera, and a loss of skeletal muscle mass.

Chapters 9 and 10 provide a dynamic point/counterpoint discussion between D.J. Millward and V.E. Young concerning the numerous difficulties involved in determining indispensable amino acid requirements (Millward) and the necessity and feasibility of using nitrogen balance studies and stable-isotope tracer techniques to make tentative estimates of these requirements (Young). Each author provides what they perceive to be the inherent difficulties.

The impact of exercise and contractile activity on muscle protein turnover and the potential for increased amino acid requirements is discussed in Chapter 11. Exercise stimulates protein catabolism, however, catabolism is actually increased by higher protein intakes. Catabolism during exercise is balanced by an increase in protein synthesis during the post-exercise period.

Chapter 12 provides a review and critique of techniques currently available for measuring skeletal muscle mass, and changes in muscle mass over time, while Chapter 13 describes the changes in protein metabolism that occur with severe injury and infection. Use of various hormones to reverse injury-induced muscle protein breakdown is also discussed.

The effect of individual amino acid supplementation on cognitive performance and brain function; and on metabolism and physical performance is presented in Chapter 14 and Chapter 15 respectively. Maintenance of appropriate plasma levels of tryptophan is essential for optimal brain function and cognitive performance, optimal levels of tyrosine appear to be less critical than that of tryptophan. Supplementation of branched-chain amino acids did not affect performance during endurance exercise, and in fact ingestion could lead to premature fatigue and loss of coordination under conditions where muscle glycogen stores have been depleted.

Finally, Chapter 16 reviews the safety and efficacy of dietary supplements of amino acids, and concludes that supplemental amino acids should be used for pharmacological rather than nutritional purposes and that current scientific literature does not support a safe upper limit for supplementation of any amino acid beyond that found in dietary protein.

3

Protein and Amino Acids: Physiological Optimization for Current and Future Military Operational Scenarios

LTC Karl E. Friedl[1]

INTRODUCTION

In conceptualizations of the battlefield of the future, the battle space involves three-dimensional swarms instead of conventional lines of attack; the tempo is faster; and war fighters are fewer in number, are more dispersed, and perform more functions. Increased technological complexity and lethality add to the already considerable stress on the individual. In this setting, individual lapses in judgment and less-than-perfect performance may be catastrophic. Furthermore, new tactics and equipment may be wasted if their requirements outstrip human capabilities. The protein content of operational rations as well as

---

[1] Karl E. Friedl, Army Operational Medicine Research Program, U.S. Army Medical Research and Materiel Command, Fort Detrick, MD 21702-5012.

specific amino acid supplements may provide a decisive advantage in such high-stress scenarios, including improved resistance to disease, preservation of muscular strength through maintenance of muscle tissue, and optimal cognitive performance even in the face of intense stressors.

One of the near-term requirements is to produce a limited-use operational ration that supports optimal cognitive and physical metabolic function while promoting utilization of the soldier's existing fat stores. This ration could be useful in short-duration, direct-action missions and in survival kits (Jones et al., 1993). Such a ration is not a new concept in military nutrition research; nearly 20 years ago, Consolazio and his colleagues (1979) asked:

> Would a planned ration of 600 kcal, if consumed, be more beneficial to the soldier than the remnants of a 3600 kcal ration, the majority of which was indiscriminately discarded because of its heavy weight?

Consolazio's studies reiterated previous findings that 100 g of glucose and some electrolytes were important constituents of such a limited-use ration just to ensure minimal function (Taylor et al., 1957). Within the past 3 years, carbohydrate drink and food bar supplements have been developed because of the clearly demonstrated benefits to military performance (Murphy et al., 1994). Current scientific advances may now allow us to consider the specific protein and amino acid content that could be sustaining and even enhancing. The purpose of this review by the Committee on Military Nutrition Research is to evaluate the state of knowledge from basic research and suggest promising research directions in protein and amino acid modulation of military performance in stressful operational scenarios.

The principal operational ration, the Meal, Ready-to-Eat (MRE), is highly fortified in protein (providing 2 g/kg body weight [BW]/d for the typical 75 kg male soldier). This ensures that the estimated daily requirement for protein is met even by Ranger students subsisting largely on MREs but offered less than a full daily ration. Thus, even with an average daily deficit of 1200 kcal/d for 8 weeks, Ranger students still average in excess of 100 g of protein/d (Moore et al., 1992). Although more sensitive markers of protein status such as circulating insulin-like growth factor-I and retinol-binding protein were markedly suppressed during periods of reduced food intake (Nindl et al., 1997), the soldiers developed no clinical, biochemical (e.g., serum protein and albumin), or gross physiological signs of protein deficiency (Martinez-Lopez et al., 1993). The Ranger students also lost a considerable amount of lean mass, although the proportion lost was inversely related to initial fat energy stores (Friedl et al., 1997). The question remains whether a higher protein diet or a carefully crafted supplement could provide any protection to lean mass in comparison with a group receiving the same total (and deficient) calories. A possible approach may come from consideration of factors altered in exercise, which have been suggested to reduce whole-body protein catabolism even in hypocaloric settings (Carraro et al., 1990).

Excessive levels of protein in operational rations may have undesirable effects. In a study of combat engineers engaged in a 30-d field-bridging exercise, men subsisting on the MRE consumed only 2,500 kcal/d during the study. However, even with wastage of at least one-third of the rations provided, they consumed components that provided 103 g protein per day, achieving 100 percent of the military Recommended Daily Allowance (Thomas et al., 1995). As further evidence of the high protein intake, 24-h urinary nitrogen excretion doubled by 30 days on the MRE diet. This level of intake was equivalent to that of energy-restricted Ranger students, except that for the engineers this was a voluntary restriction. High protein intake and positive balance may even produce premature satiety in field feeding. Tyrosine augments the anorectic action of sympathomimetic drugs in rats (Hull and Maher, 1990); the right mix in a high-protein food might similarly augment the sympathetic activation of soldiers working hard in the field and produce an anorectic effect. Although highly speculative, it is conceivable that this explains the inadequate energy intake observed in field studies with the MRE, where the average intake is 3000 kcal/d, and the average estimated energy requirement is 4000 kcal/d. High protein content of the rations also potentially increases calcium excretion, although studies feeding protein up to 2 g/kg BW indicate no calciuretic effects (Spencer and Kramer, 1986). Protein intake in excess of requirements increases water loss with increasing urea excretion, a factor that could be decisive in isolated desert environments. Conceivably, a low protein ration could be developed for limited use in dry environments to reduce the logistical burden of water transport requirements.

The potential brain effects of some of the amino acids that serve as neurotransmitter precursors, such as tyrosine and tryptophan, are intriguing for actions beyond appetite control. Manipulation of serotonin levels with tryptophan (Spring, 1984) might prove useful in the prevention of stress casualties, estimated to be as high as one in every four medical casualties in future conflicts. Tyrosine, acting as a precursor of catecholamines, may be useful in sustaining soldiers' performance in high-stress environments. Several human and animal studies suggest that 85–179 mg/kg of tyrosine can improve mental performance and reduce anxiety in stressful conditions and with inadequate rest (Lieberman, 1994; Neri et al., 1995). John Thomas and his colleagues at the Naval Medical Research Institute have reported preliminary data from studies with Marine sharpshooters on maneuvers in Alaska that indicated a trend toward restoring marksmanship performance degraded by cold and fatigue (Shurtleff et al., 1994). Improved field tests for assessment of military performance still need to be developed to better evaluate such ration benefits, as recommended following the first major workshop of the Committee on Military Nutrition Research (CMNR) (National Research Council, 1986).

In the study of Ranger students in stressful training, infection rates were notably elevated in association with derangements in indices of immune function (Kramer et al., 1997; Martinez-Lopez et al., 1993). Medical researchers were asked to determine if a simple fix such as a vitamin pill or other supplement could sustain the health of soldiers in stressful training without having to ease the rigors of the Ranger course; however, plasma biochemical measures indicated no vitamin or nutrient deficiencies (Moore et al., 1992). A similar model of stress-induced alteration in immune function indices was established by Colonel Fairbrother in soldiers during the 21-d Special Forces Assessment and Selection (SFAS) course (Fairbrother et al., 1995), although there was no significant incidence of infectious disease. This model has since been used for empirical test and evaluation of various "magic bullet" nutritional interventions. On the basis of the important role of glutamine in lymphocyte function, several researchers have suggested that this might sustain immune function in catabolic subjects (Ziegler et al., 1998). In fact, in the review of data from the 1991 Ranger study, the CMNR recommended that the effects of training and associated stressors on protein and amino acid (including glutamine) metabolism should be studied, particularly for the purpose of potential special supplements (CMNR, 1992). A second study of soldiers in the SFAS course tested such an intervention, comparing glutamine supplementation with an isonitrogenous control. The results demonstrated no effect on a wide variety of immune parameters (Shippee et al., 1995). This result does not exclude a benefit from glutamine supplementation but suggests that the solution may be more complex than simply adding back the amino acid (Ziegler et al., 1996). Glutamine may be beneficial in the same setting for other reasons. Presumably as a consequence of the semistarvation in the Ranger studies, a subclinical edema was noted that was reflected in an increased proportion of total body water in the lean mass. In studies with hospitalized catabolic patients, extracellular fluid excess was successfully attenuated with glutamine feeding (Scheltinga et al., 1991). Since we don't understand the mechanism or potential adaptive value of this disproportionate fluid retention in otherwise healthy subjects, interventions to prevent it may be premature.

The specific protein requirements of servicewomen may not differ from those of servicemen, but these requirements have been inadequately studied. At least two Defense Womens' Health Research Program grants center on this issue. In one, Vernon Young (Massachusetts Institute of Technology) is testing substrate utilization in subjects working intensively while on energy-deficient diets in a partial simulation of Army Ranger training. The key experiment in this study is a 21-d hypocaloric challenge to men and women, where the protein content of the diet is kept high (1.2 g/kg/d). This experiment will address gender differences in metabolic, physical, and mental performance in chronic hypocaloria. In another study, Anne Loucks (Ohio University) is attempting to define a threshold of energy deficiency that produces amenorrhea, having previously demonstrated an effect of inadequate protein intake on thyroid hormone and menstrual status in exercising women. The central thesis is that men are less susceptible to dietary disruption of luteinizing hormone pulsatility than are women. Conceivably, in future ration supplements, servicewomen could benefit from a high-protein or high-amino acid supplement that modifies endocrine responses to stress and hypocaloria.

In the distant future, regulation of urea formation and recycling of protein, using something like a biochemical hibernation trigger observed in bears (Nelson et al., 1975), may be important in manipulations of soldiers in stasis during long-term travel, during special surveillance missions, or after injury. Instead of increasing protein intake in an attempt to preserve lean mass, the strategy would be to prevent the loss of urea nitrogen. Some of these adaptations appear to be present already in endurance-trained athletes where intensive prolonged exercise does not shift the balance to one of net protein catabolism (Stein et al., 1989).

Future research on nutritional interventions to counter the effects of operational stressors may rely heavily on protein and amino acid components because of the important potential for these nutrients to modulate cognitive function and sustain performance.

REFERENCES

Carraro, F., W.H. Hartl, C.A. Stuart, D.K. Layman, F. Jahoor, and R.R. Wolfe. 1990. Whole body and plasma protein synthesis in exercise and recovery in human subjects. J. Appl. Physiol. 258:E821–E831.

Consolazio, C.F., H.L. Johnson, R.A. Nelson, R. Dowdy, and H.J. Krzywicki. 1979. The relationship of diet to the performance of the combat soldier: Minimal calorie intake during combat patrols in a hot humid environment (Panama). Technical Report 76. San Francisco, Calif.: Letterman Army Institute of Research.

Fairbrother, B., R.E. Shippee, T.R. Kramer, E.W. Askew, and M.Z. Mays. 1995. Nutritional and immunological assessment of soldiers during the Special Forces Assessment and Selection Course. Report USARIEM-T95-22, AD-A299 556. Natick, Mass.: Army Research Institute of Environmental Medicine.

Friedl, K.E. 1997. Variability of fat and lean tissue loss during physical exertion with energy deficit. Pp. 431–450 in Physiology, Stress, and Malnutrition: Functional Correlates, Nutritional Intervention. J.M. Kinney and H.N. Tucker eds. New York: Lippincott-Raven Publishers.

Hoffer, L.J., and R.A. Forse. 1990. Protein metabolic effects of a prolonged fast and hypocaloric refeeding. Am. J. Physiol. 258:E832–E840.

Hull, K.M., and T.L. Maher. 1990. L-tyrosine potentiates the anorexia induced by mixed-acting sympathomimetic drugs in hyperphagic rats. J. Pharmacol. Exp. Ther. 255:403–409.

IOM (Institute of Medicine).1992. A Nutritional Assessment of U.S. Army Ranger Training Class 11/91. March 23. Washington, D.C.

Jones, T.E., S.H. Mutter, J.M. Aylward, J.P. DeLany, and R.L. Stephens. 1993. Nutrition and hydration status of aircrew members consuming the Food Packet, Survival, General Purpose, Improved during a simulated survival scenario. Report USARIEM-T1-93. Natick, Mass.: U.S. Army Research Institute of Environmental Medicine.

Kramer, T.R., R.J. Moore, R.L. Shippee, K.E. Friedl, L.E. Martinez-Lopez, M.M. Chan, and E.W. Askew. 1997. Effects of food restriction in military training on T-lymphocyte responses. Int. J. Sports Med. 18(1):S84–S90.

Lieberman, H.R. 1994. Tyrosine and stress: Human and animal studies. Pp. 277–299 in Food Components to Enhance Performance. An Evaluation of Potential Performance— Enhancing Food Components for Operational Rations. Marriott B.M. ed. Institute of Medicine. Washington D.C.: National Academy Press.

Loucks A.B., and E.M. Heath. 1994. Induction of low-T3 syndrome in exercising women occurs at a threshold of energy availability. Am. J. Physiol. 266:R817–R823.

Martinez-Lopez, L.E., K.E. Friedl, R.J. Moore, and T.R. Kramer. 1993. A prospective epidemiological study of infection rates and injuries of Ranger students. Mil. Med. 158:433–437.

Moore, R.J., K.E. Friedl, T.R. Kramer, L.E. Martinez-Lopez, R.W. Hoyt, R.E. Tulley, J.P. DeLany, E.W. Askew, and J.A. Vogel. 1992. Changes in soldier nutritional status and immune function during the Ranger training course. Report USARIEM-T13-92, AD-A257 437. Natick, Mass.: Army Research Institute of Environmental Medicine.

Murphy, T.C., R.W. Hoyt, T.E. Jones, C.L. Gabaree, and E.W. Askew. 1994. Performance enhancing ration components program: Supplemental carbohydrate test. Report USARIEM-T95-2, AD-A288 560. Natick Mass.: Army Research Institute of Environmental Medicine.

Nelson. R.A., J.D. Jones, H.W. Wahner, D.B. McGill, and C.F. Code. 1975. Nitrogen metabolism in bears: Urea metabolism in summer starvation and in winter sleep and role of urinary bladder in water and nitrogen conservation. Mayo Clin. Proc. 50:141–146.

Neri, D.F., D. Wiegmann, R.R. Stanny, S.A. Shappell, A. McCardie, and D.L. McKay. 1995. The effects of tyrosine on cognitive performance during extended wakefulness. Aviat. Space Environ. Med. 66:313–319.

Nindl, B.C., K.E. Friedl, P.N. Frykman, L.J. Marchitelli, R.L. Shippee, and J.F. Patton. 1997. Physical performance and metabolic recovery among lean, healthy men following a prolonged energy deficit. Int. J. Sports Med. 18:1–8.

NRC (National Research Council). 1986. Cognitive Testing Methodology. Washington, DC: National Academy Press.

Scheltinga, M.R., L.S. Young, K. Benfell, R.L. Bye, T.R. Ziegler, A.A. Santos, J.H. Antin, P.R. Schloerb, and D.W. Wilmore. 1991. Glutamine-enriched intravenous feedings attenuate extracellular fluid expansion after a standard stress. Ann. Surg. 214:385–393.

Shippee, R.L., S. Wood, P. Anderson, T.R. Kramer, M. Neita, and K. Wolcott. 1995. Effects of glutamine supplementation on immunological responses of soldiers during the Special Forces Assessment and Selection Course [abstract]. FASEB J. 9:731.

Shurtleff, D., J.R. Thomas, J. Schrot, K. Kowalski, and R. Harford. 1994. Tyrosine reverses a cold-induced working memory deficit in humans. Pharmacol. Biochem. Behav. 47:935–941.

Spencer, H., and L. Kramer. 1986. The calcium requirements and factors causing calcium loss. Fed. Proc. . 45(12):2758–2762.

Spring, B. 1984. Recent research on the behavioral effects of tryptophan and carbohydrate. Nutr. Health 3:55–67.

Stein, T.P., R.W. Hoyt, M.O. Toole, M.J. Leskiw, and M.D. Schluter. 1989. Protein and energy metabolism during prolonged exercise in trained athletes. Int. J. Sports Med. 10:311–316.

Taylor, H.L., E.R. Buskirk, J. Brozek, J.T. Anderson, and F. Grande. 1957. Performance capacity and effects of caloric restriction with hard physical work on young men. J. Appl. Physiol. 10:421–429.

Thomas, C.D., K.E. Friedl, M.Z. Mays, S.H. Mutter, and R.J. Moore. 1995. Nutrient intakes and nutritional status of soldiers consuming the Meal, Ready-to-Eat (MRE XII) during a

30-day field training exercise. USARIEM Technical Report T95-6. Natick, Mass.: U.S. Army Research Institute of Environmental Medicine.

Ziegler, T.R., R.L. Bye, R.L. Persinger, L.S. Young, J.H. Antin, and D.W. Wilmore. 1998. Effects of glutamine supplementation on circulating lymphocytes after bone marrow transplantation: A pilot study. Am. J. Med. Sci. 315:4–10.

Ziegler, T.R., M.P. Mantell, J.C. Chow, J.H. Rombeau, and R.J. Smith. 1996. Gut adaptation and the insulin-like growth factor system: Regulation by glutamine and IGF-I administration. Am. J. Physiol. 271:G866–G875.

4

Overview of Garrison, Field, and Supplemental Protein Intake by U.S. Military Personnel

LTC (ret) Alana D. Cline[1] and John P. Warber

INTRODUCTION

Adequacy of nutrient intake by military personnel has been evaluated on a periodic basis since World War II; the responsibility for evaluating the nutritional status of military personnel and prescribing standards for operational rations has been that of the Army Surgeon General (U.S. Department of the Army, 1945; U.S. War Department, 1944). As new rations have been developed, their acceptability and effects on the health and performance of military personnel have been assessed. Subsequently, modifications have been recommended after evaluation of their nutritional impact.

The nutrient composition of operational rations is designed by food technologists at the U.S. Army Natick Research, Development and Engineering Center (NRDEC) to meet nutritional standards that are based on published

---

[1] Alana D. Cline, Military Nutrition and Biochemistry Division, U.S. Army Research Institute of Environmental Medicine, Natick, MA 01760-5007. Currently of Pennington Biomedical Research Center, Baton Rouge, LA 70808.

recommendations such as the Recommended Dietary Allowances (NRC, 1980) and those of other recognized bodies of nutrition scientists. Ideally, rations should be designed to meet not only the minimum nutritional requirements, but the requirements for optimal nutrition. In practice, however, the requirements for optimal nutrition have yet to be determined, and may differ among individuals. During World War II, recommendations of the first Food and Nutrition Board (FNB) (NRC, 1941, 1945) were used as a metric for determining adequacy of the rations. Since 1947, standards have been specifically established for the military, based on the FNB Recommended Dietary Allowances (RDAs), and adapted for emerging physical requirements.

Food intake by military personnel has been monitored periodically over several decades to determine whether changes in consumption of various nutrients have occurred. Because of recent public interest in strength training and body building, which has been associated with increased intake of various protein and amino acid supplements by individuals who have the expectation that their muscle strength, size, and performance will improve, the Army has recently surveyed military personnel to determine the extent of supplement use.

What has not been determined is whether protein intake recommendations established during World War II and still used remain appropriate for military personnel today. A related question is what was the general range of protein intake by soldiers during that time, and how has it changed since those guidelines were initially established.

This chapter presents an overview of ration studies that were conducted during World War II and compares them with more recent studies on energy and protein intake and requirements of military personnel in garrison and operational settings. Reported consumption of amino acid and protein supplements will also be addressed, with frequency of consumption identified by gender, age, and military specialty.

PROTEIN AVAILABILITY FROM OPERATIONAL RATIONS

Operational rations have been divided into those prepared in field kitchens for groups of military personnel and those the individual soldier must carry, prepare, and consume. By design, they provide an excess of calories and protein, when possible, to allow for some food choice by the individual and still allow for adequate nutrient intake (Samuels et al., 1947). Former and current rations most widely used for tactical consumption are compared in Table 4-1.

The individual ration most widely used during World War II was the C Ration, which provided a combination of canned foods and packaged dehydrated or dried foods. Nutritional composition was 2,794 kcal and 121 g protein. To provide a ration with the greatest caloric density in the smallest weight and space, the K Ration was developed in 1941; it provided 2,842 kcal and 79 g protein.

TABLE 4-1 Energy and Protein Content of Operational Rations

Ration Type	Energy (kcal)	Protein (g)	% Energy
C Ration	2,794	121	17
K Ration	2,842	79	11
5-1	3,383	98	12
10-1	4,188	124	12
B Ration	4,300	140	13
MRE	3,819	136	14
UGR	3,973	142	14
Ration, Cold Weather	4,500	90	8
Go-to-War Ration	3,600	144	16

NOTE: 5-1, Five-in-One; 10-1, Ten-in-One; MRE, Meal, Ready-to-Eat; UGR, Unitized Group Ration.

The most widely used rations packaged for group feeding in World War II were the Five-in-One Ration (5-1), providing 3,383 kcal and 98 g protein, and the Ten-in-One Ration (10-1), providing 4,188 kcal and 124 g protein. These rations were designed to provide adequate food for 1 day's consumption by a unit of 5 or 10 men, respectively, and required minimal food preparation for small groups of individuals away from food preparation facilities. The B Ration was developed to provide foods in bulk for a minimum of 100 individuals. It comprised packages and cans of bulk foods not requiring refrigeration, but needing reconstitution or rehydration during meal preparation. A series of 10 days' menus were provided, complete with recipes. Kitchen-prepared A (perishable foods, needing refrigeration) or B Rations were those on which the military depended for regular daily feeding; they were required to be fully adequate for all nutrients, meeting standards set by the National Research Council to provide at least 3,000 kcal and 70 g protein.

Current operational rations include the MRE for individual consumption and the UGR for group feeding, each exceeding the operational ration standards of 3,600 kcal and 100 g protein (AR 40-25, 1985). In addition, vegetarian MRE meals have been recently added (2/case of 12) to provide choices for individuals not consuming meat products. The Ration, Cold Weather (4,500 kcal, 90 g protein) is used to sustain an individual during operations occurring under frigid conditions, and the Go-to-War Ration (3,900 kcal, 156 g protein) was designed for the early stages of mobilization until such time that the ration industry can meet deployment demands. Both rations are designed to provide short-term support, so they do not meet the full nutritional requirements for operational rations.

HISTORIC RATION INTAKE

Major ration studies from 1941 to 1946 utilized several methods of data collection for troops deployed worldwide to varying environmental conditions. Techniques used for data collection included both field tests and individual and group surveys conducted on site; most subjects were Army soldiers, with the exception of several Air Force units and flight crews.

Studies evaluating male soldiers during World War II reported mean intakes of energy in garrison ranging from 3,400 to 3,800 kcal with protein intake from 110 to 132 g; intakes during field training were similar: 3,200 to 4,100 kcal and 100 to 125 g protein, with percentage of energy from protein at 13 percent for both garrison and field (Table 4-2).

In one of the first rigorous ration tests during World War II (Johnson and Kark, 1946), the B Ration was tested (Camp Lee, Va.) on soldiers who were placed on a rigid activity program that would reflect the ordinarily increased energy expenditure of deployment. During the pre-experimental or control period, test subjects consumed dining hall A Rations and participated in a work schedule that resulted in an estimated energy expenditure of 3,100 kcal/d. Energy intake (3,800 kcal) was highest during the control phase of the study while protein intake was 116 g (Table 4-2). As the physical activity schedule

TABLE 4-2 Energy and Protein Intake by Male Soldiers: Selected Trials 1940s

Study	N	Ration Type	Energy (kcal)	Protein (g)	% Energy
Camp Lee, 1943 (G)*, (Johnson and Kark, 1946)	65	A	3,800	116	12.2
Camp Lee, 1943 (G), (Johnson and Kar, 1946)	65	B	3,600	132	14.7
Mess Survey, 1941–1943 (G), (Howe and Berryman, 1945)	†	A	3,790	125	13.2
Camp Carson, 1944 (F)‡, (Bean et al., 1944)	118	B	3,930	125	12.7
Camp Carson, 1944 (F), (Bean el al., 1944)	125	10-1	4,100	125	12.2
Pacific Islands, 1945 (G), (Bean et al., 1946)	50	A	3,400	110	12.9
Luzon, 1945 (F), (Bean et al., 1946)	50	C	3,200	100	12.5
Average (G)			3,648	121	13.3
Average (F)			3,743	117	12.5

* G = Garrison study.
‡ F = Field study.
† Calculated from average ration intake from 455 garrison messes.

became more intense (4,000 kcal) during the test period, energy intake actually decreased while protein intake increased. Although the B Ration offered during the test phase provided 3,800 kcal, soldiers consumed only 3,600 kcal, and 132 g protein and generally complained of hunger.

Additional food consumption studies conducted at U.S. Army training camps from 1941 to 1943 were reported by Howe and Berryman (1945). Surveys took place at 50 posts and 455 mess facilities using the difference between weights of food issued and food discarded or wasted. Although consumption of foods away from the mess was not recorded, it was estimated that 350 to 400 kcal/d were purchased at the canteen or were received in personal packages. Results again showed a daily consumption of approximately 3,800 kcal of energy and 125 g of protein.

In 1944, acceptability and adequacy of several field rations were tested at 2,700 m elevation in the Rocky Mountains with Army troops from Camp Carson during summer maneuvers. Daily energy intake ranged from 2,880 kcal for a small group consuming the K Ration to 4,100 kcal, with protein intake averaging 12 percent of energy (110–125 g). Over a study period of 55 days, measurements of physical fitness, nutritional status, biochemical indices, and rifle firing all improved, with no differences seen between ration types (Bean et al., 1944).

Surveys of the health, fitness, and nutrition of troops in the Pacific were conducted in 1945 to compare nutrient intake of noncombat garrison soldiers with those who had been in combat continuously for 4½ months. Average nutrient intake of garrison soldiers in the Pacific was similar to intake by soldiers in training camps in the United States, in part due to ample supplies of fresh and frozen foods. Troops living exclusively on packaged rations (C Rations) in Luzon had a daily caloric intake 200 to 300 kcal less than garrison troops, but protein intake remained similar among the soldiers at 13 percent of total energy consumed (Bean et al., 1946).

CURRENT RATION INTAKE

More recent (1984–1996) studies have reported intakes by men in garrison ranging from 2,773 to 3,173 kcal and 98 to 132 g protein, averaging 15.2 percent of energy from protein; field intakes have ranged from 2,009 to 3,050 kcal and 86 to 126 g protein, or 16 percent of energy from protein (Table 4-3). Intakes by women have also been reported, with intakes in garrison ranging from 1,832 to 2,592 kcal and 75 to 96 g protein, and intakes in field exercises ranging from 1,668 to 2,343 kcal and 68 to 82 g protein, with approximately 15 percent of energy from protein in both garrison and field assessments (Table 4-4). It should be noted that mean weight of male soldiers increased from 68.4 kg in the 1940s to a current mean weight of 78.9 kg; current mean weight of female soldiers is 63.6 kg. In data recently collected from the Army Food and

TABLE 4-3 Energy and Protein Intake by Male Soldiers: Recent Studies

Study	N	Ration Type	Energy (kcal)	Protein (g)	% Energy
Pohakuloa, 1984 (F)*, (Askew et al., 1986)		MRE	2,273	99	17.4
Ft. Riley, 1986 (G)†, (Szeto et al., 1987)	43	A	3,112	123	15.8
Ft. Lewis, 1986 (G), (Szeto et al., 1987)	31	A	3,173	125	15.8
Ft. Devens I, 1987 (G), (Szeto et al., 1988)	54	A	2,978	111	14.9
Ft. Devens II, 1988 (G), (Szeto et al., 1989)	52	A	3,165	132	16.6
Alaska, 1989 (F), (Edwards et al., 1989)	31	MRE	2,009*	91§	NA
Ft. Chaffee, 1991 (F), (Thomas et al., 1995)	34	MRE + pouch bread	2,462	103	16.7
Ft. Chaffee, 1991 (F), (Thomas et al., 1995)	32	A	2,911	126	17.3
Chocolate Mtn., 1994 (F), (Tharion et al., 1997)	31	UGR	2,631	105	15.9
Chocolate Mtn., 1994 (F), (Tharion et al., 1997)	32	UGR + Suppl‡	3,050	93	12.2
Ft. Polk, 1995 (G), (Cline et al., 1997)	38	A	3,003	103	13.7
Hunter Army Airfield, 1996 (G), (Champagne et al., 1997)	73	A	2,773	98	14.1
Hunter Army Airfield, 1996 (F), (Champagne et al., 1997)	31	MRE	2,439	86	14.1
Average G			3,034	115	15.2
Average F			2,587	104	16.1

NOTE: MRE, Meal Ready-to-Eat; UGR, Unitized Group Ration; NA, not available.

* F = Field study.
† G = Garrison study.
‡ Carbohydrate beverage supplement provided *ad libitum*.
§ Data reported as mean of days, not mean of individuals.

TABLE 4-4 Energy and Protein Intake, Female Soldiers

Study	N	Ration Type	Energy (kcal)	Protein (g)	% Energy
Ft. Lewis, 1987 (G)[*], (Szeto et al., 1987)	12	A	1,832	75	16.4
Ft. Jackson, 1988 (G)[†], (R.W. Rose et al., 1989)	40	A	2,467	96	15.6
Ft. Hood, 1988 (F)[‡], (M.S. Rose et al., 1989)	27	2B + 1MRE	2,343	82	14.0
Bolivia, 1990 (F), (Edwards et al., 1991)	13	2B + 1MRE	1,668	68	16.3
Ft. Jackson, 1993 (G)[†], (King et al., 1994)	49	A	2,592	82	12.7
Ft. Sam Houston, 1995 (G)[†], (Cline et al., 1998)	56	A	2,037	75	14.7
Camp Parks, 1996 (F), (Hirsch et al., in press)	19	MRE	2,161	82	15.2
Average G			2,232	82	14.9
Average F			2,057	77	15.0

NOTE: MRE, Meal Ready-to-Eat.

[*] G = Garrison study.
[†] These soldiers in basic training course.
[‡] F = Field study.

Nutrition Survey I (Warber et al., 1996) at 33 Army installations worldwide, over one-fourth (26%) of male respondents and nearly one-third (31%) of female respondents replied that they used field feeding as a way to lose weight.

Energy and Protein Intakes of Military Men

Infantry soldiers from Fort Shafter, Hawaii, participated in a study at Pohakuloa Training Area, Hawaii (2,160 m), to determine the adequacy of the Meal, Ready-to-Eat (MRE) during strenuous cross-country running under high-altitude field conditions (Askew et al., 1986). During the 10-d exercise, soldiers consumed less than 67 percent of calories recommended for energy balance and lost 3 percent of their body weight, 10 percent of their body fat, and experienced a 5 percent decline in maximal aerobic capacity. Protein intake was 99 percent of recommended intake. By comparison, infantry soldiers training in Alaska under winter conditions were assessed for adequacy of nutritional intake over a 10-d field training exercise. Although individuals were given 4 MREs/d, consumption again remained significantly lower than the Military Recommended Dietary Allowances (MRDAs) for energy and protein (Edwards et al., 1989).

Because of concerns that logistical problems with food supply may require individuals to subsist solely on the MRE for up to 30 days, a study was conducted at Fort Chaffee, Arkansas, to assess the ability of the MRE to meet soldiers' nutritional needs and maintain performance in a field environment for an extended period (Thomas et al., 1995). Participants were combat engineers on a regularly scheduled 30-d field training exercise; they were divided into one group eating three MREs and two pouch bread (190 kcal, 5 g protein each) per day and one group eating two A Rations and one MRE per day. Nearly one-third of the subjects indicated they wanted to lose weight during the training exercise, and this was reflected in a low mean energy intake of 2,462 kcal for the MRE group and 2,911 kcal for the A Ration group. Even with incomplete consumption of the MREs provided, soldiers in both groups obtained 100 percent of their MRDA for protein and demonstrated a positive nitrogen balance, which indicates that although caloric intake was low, performance and overall nutritional status were not impaired when soldiers consumed only MREs for 30 days.

A test was conducted with a Marine battery-sized field artillery unit in 1994 at Chocolate Mountain Desert Gunnery Range, California (Tharion et al., 1997). The primary purpose of the test was to assess the ability of the new Unitized Group Ration (UGR) to meet nutritional requirements of individuals working in a desert environment. The UGR used in this study was a combination of A, B, and T Rations, which provided hot meals in a group feeding setting for two meals per day, with the third meal a MRE. Either a carbohydrate or a placebo beverage was also provided *ad libitum* to two supplemental groups to assess effects of additional nutrient intake and hydration status. In the placebo group, mean energy intake was 73 percent of MRDA (2,631 kcal), while protein intake was 105 percent (105 g). Although protein intake appeared to be adequate, energy intake was well below the calculated energy expenditure of approximately 4,300 kcal/d. The carbohydrate beverage did increase mean daily energy consumption (3,050 kcal), but protein intake was lower (93 g). This study provides further evidence that even when troops in field training exercises are provided two hot meals per day and consume adequate protein, their caloric intake may be inadequate for energy needs.

Alternately, studies of nutrient intake in garrison dining facilities continue to show that soldiers are consuming adequate energy and protein. Reports on assessment of nutritional status of soldiers at Fort Riley, Kansas; Fort Lewis, Washington (Szeto et al., 1987); and Fort Devens, Massachusetts (Szeto et al., 1988, 1989), concluded that dining facility consumption provided adequate daily energy and protein intakes for individuals, even when some meals and snacks were being consumed elsewhere. By contrast, air defense artillery soldiers at Fort Polk, Louisiana (Cline et al., 1997) consumed only approximately half of their meals in the dining facility. Overall nutrient intake from all foods, including those consumed elsewhere, was adequate in energy and protein. Nutrient intakes were reported to be 3,091 kcal and 105 g protein, with 13.6 percent of total energy obtained from protein. Although mean energy intake was below the MRDA, protein intake was adequate.

A nutritional assessment of Army Rangers at Hunter Army Airfield in 1996 was the first of an Army Special Operations unit (Champagne et al., 1997). Rangers are routinely involved with extensive deployments and field training exercises. This study was designed to compare dietary intake of Rangers before deployment with that during a field training exercise and during recovery after return. Phase I (predeployment) assessed a group subsisting in a garrison dining facility for the 6 days immediately prior to deployment on the exercise, while Phase II (training) assessed the dietary intake of the same unit during the 6-d field training exercise while consuming MREs (Table 4-3). Phase III assessed postdeployment garrison intake of a subgroup ($n = 40$) of subjects for the 2 days immediately after return from the field; mean recovery intake was 2,965 kcal and 98 g protein (13.2% of total energy). Mean energy and protein intakes for all phases were lower than what have been reported on previous studies. As with the soldiers at Fort Polk, a substantial amount of food (33% of energy) was consumed away from the dining facility. This trend is similar to eating patterns reported in national nutrition monitoring studies of eating patterns of nonurban households, where more than one-third of total food dollars were spent on food away from home (Interagency Board, 1993).

Energy and Protein Intakes of Military Women

Until recently, few studies have assessed nutritional intake of military women during field exercises or deployment. Women who were members of an Army engineering group deployed to high altitude in Bolivia were assessed for nutrient intake while consuming a combination of two B Ration meals and one MRE per day. They were also given a high-carbohydrate supplemental pack to consume between meals to determine whether increased carbohydrate intake was preferable at high altitude. Energy and protein intake were well below MRDA; approximately half of the subjects indicated they had acute mountain sickness (AMS) symptoms, which most likely affected appetite (Edwards et al., 1991).

Two field studies that included women as subjects were completed with Army Reserve hospitals conducting their annual field training exercises and subjects consuming operational rations. At Fort Hood, Texas, two A Ration meals and one MRE per day were provided during an 8-d test period; individuals also had access to additional foods they brought or that could be purchased from a PX mobile kitchen or fast-food establishments in the vicinity of the field site. Although 26 percent of the subjects stated that they were attempting to lose weight during the training exercise, mean energy intake was 2,343 kcal and mean protein intake was 82 g, both of which exceeded the MRDA for women (M.S. Rose et al., 1989). Similar results were reported in an Army Reserve hospital unit training at Camp Parks, California (Hirsch et al., in press).

When comparing nutrient intakes between women in enlisted basic training at Fort Jackson, South Carolina (R.W. Rose et al., 1989; King et al., 1994) and women in officer basic training at Fort Sam Houston, Texas (Cline et al., 1998), a difference was observed in both energy and protein intakes. Meals for enlisted women are provided in a dining facility at scheduled times, with free access to a variety of food selections. Alternately, officers are given a monetary allowance to purchase food, but they live in temporary housing, such as motels or officers quarters, that has limited or no cooking facilities. This could account for lower energy and protein intake by the officers. No recent studies have been completed on nutrient intake of career military women in garrison.

DETERMINATION OF PROTEIN REQUIREMENTS FOR OPERATIONAL RATIONS

The first standard recommendations for nutrient requirements for military personnel were formulated by the FNB of the National Research Council, a project that was organized in 1940 in connection with the defense program (Samuels et al., 1947). Recommendations by the board addressed allowances needed to maintain optimal nutritional status. Of concern was the fact that rations designed for short-term use, not previously required to be nutritionally adequate, were being used over long periods of time. Maximum nutrient availability was emphasized in all but a few of the survival rations.

The standard originally established for protein was 70 g/3,000 kcal, for a reference man weighing 70 kg. This recommendation was equivalent to 9 percent of energy from protein, and 0.8 to 1.0 g/kg of body weight (kcal requirements of 3,000–3,600). After reviewing research on energy consumption and expenditure of soldiers during World War II, the Army raised energy requirements to 3,600 kcal for physically active personnel in temperate climates, with a protein requirement of 100 g (AR 40-250, 1947). The RDA (NRC, 1989) for protein is 58 to 63 g for males and 46 to 50 g for females in the age categories of military personnel; recommendations are lower because reference weights and activity levels used for calculations are less than those for the military population.

Several investigators have recommended that the RDA for protein be increased to 1.5 g/kg for endurance athletes and to 2.0 g/kg for strength athletes (Brotherhood, 1984; Williams, 1995; Lemon, 1996). Recent research has shown an increase in the weight of Army soldiers from 68 to 78 kg for men and 61 to 62 kg for women (Gordon et al., 1989) which would increase recommendations to 117 g of protein (1.5 g/kg) for operational rations if soldiers are involved in heavy work. If protein in rations is increased to accommodate recommendations of 1.5g/kg for endurance activity in men, it would contribute 13% of food energy at the 3,600 kcal level. Food consumption data on the U.S. population from NHANES III (Third National Health and Nutrition Examination Survey) indicate that 14 to 16 percent of the total food energy intake is derived from protein (Interagency Board, 1995). This proportion remains similar for both sexes.

Energy requirements should be evaluated with full consideration of the fuel necessary to balance energy expenditure. For example, energy expenditure during field operations for Special Forces students has been reported at levels up to 6,000 kcal/d (Shippee et al., 1994).

SUPPLEMENT INTAKE

In data from the Army Food and Nutrition Survey I (Warber el al., 1996) at 33 Army installations worldwide, supplementation with amino acid (AA) products or protein powders (PP) was reported by military members, with usage differing by age and military job specialty (Table 4-5). Individuals less than 30 years old reported the highest use by age, with approximately one-third using both types of products. When classified by military job specialty, individuals in combat arms reported the highest usage, followed by combat service support and combat support. Percentages of men using AA (33% vs. 16%) or PP (27% vs. 11%) are at least double that of women, and individuals required to eat in military dining facilities report a higher percentage of usage (AA = 35%, PP = 29%) than those who receive food allowances to eat elsewhere (AA = 30%, PP = 24%).

Many amino acid and protein powder supplements have become available for purchase in military commissaries and exchanges, as well as in fitness centers on military installations. Newly established "nutrition store" franchises are also opening on numerous bases, offering products with a myriad of claims for benefit to performance. An expanding selection of fitness magazines is also available for purchase, providing advertisements as well as feature articles on claims of the benefits of the protein and amino acid products. Thus, military personnel have increased their consumption of these products because of their desire to improve physical performance and the belief that they will receive some benefit from using them on a regular basis.

Considerable variability persists, however. During a nutrient intake study at Fort Polk, Louisiana, in 1995 (Cline et al., 1997), only 4 percent of young male soldiers in an Air Defense Artillery company replied that they were using protein supplements. These responses contrasted with a 44 percent consumption rate among Army Rangers participating in a similar study at Hunter Army Airfield, Georgia, in 1996 (Unpublished data, J. P. Warber, USARIEM, Natick, Mass, 1997).

TABLE 4-5 Percentage of Military Personnel Reporting Use of Protein Supplements

Group (N)	Amino Acids (%)	Protein Powders (%)*
Men (1,941)	33	27
Women (291)	16	11
Age < 30 yr (1,238)	36	30
Age > 30 yr (1,040)	26	19
Combat Arms† (784)	38	32
Combat Service Support‡ (949)	28	22
Combat Support§ (536)	27	20
No Separate Rations" (545)	35	29
Separate Rations# (1,546)	30	24

SOURCE: Data from Warber et al., 1996.

* Some respondents reported consumption of both amino acids and protein powders.

† Infantry, Armor, Field Artillery, Air Defense, Special Forces.

‡ Ordnance, Quartermaster, Transportation, Adjutant, Chaplain, Finance, Judge Advocate General, Inspector General, Medical Department.

§ Engineer, Chemical, Military Intelligence, Military Police, Signal, Aviation, Civil Affairs.

" Meals provided in military dining facility.

Salary provides for purchase of foods of choice.

In a recent study on women entering the Army for basic training, only 2 of 105 replied that they had consumed supplemental protein products prior to entry for training (Cline and Pusateri, 1996). Women who have been on active duty for a longer period of time have reported a much higher rate of consumption (Table 4-5), leading one to question whether they have been influenced by the performance and fitness attitudes of their male counterparts with whom they exercise.

AUTHORS' CONCLUSIONS AND RECOMMENDATIONS

Male military personnel maintain high protein intakes from food consumption in garrison as well as during field operations. Females, however, generally consume less energy and protein than MRDA guidelines require during field exercises where access to foods is limited to operational rations.

Protein supplements are being used by a substantial number of military personnel, although no documented benefits from their use have been reported. This practice has been encouraged by easy access to products for purchase on military installations and a very active informal information network among military personnel that indicates perceived benefits of these products.

What has not been addressed in detail is whether a change has taken place in the food contribution of protein. Is the highest proportion of protein consumed provided by meats and dairy products, or have other food groups gradually replaced these in their contribution of protein to the diet? Dietary protein quality and digestibility of proteins in specific foods consumed need further investigation.

REFERENCES

AR (Army Regulation) 40-250. 1947. *See* U.S. Department of the Army. 1947. *See* U.S. Department of the Army.

AR (Army Regulation) 40-25. 1985. *See* U.S. Departments of the Army, the Navy, and the Air Force.

Askew, E.W., J.R. Claybaugh, S.A. Cucinell, A.J. Young, and E.G. Szeto. 1986. Nutrient intakes and work performance of soldiers during seven days of exercise at 7,200 ft. altitude consuming the meal, ready-to-eat ration. Technical Report No. T3-87. Natick, Mass.: U.S. Army Research Institute of Environmental Medicine.

Bean, WB., J.B. Youmans, W.F. Ashe, N. Nelson, D.M. Bell, L.M. Richardson, C.E. French, C.R. Henderson, R.E. Johnson, G.M. Ashmore, K.N. Halverson, and J. Wright. 1944. Project No. 30: Test of Acceptability and Adequacy of U.S. Army C, K, and 10-in-1, and Canadian Army Mess Tin Rations. Fort Knox, Ky.: Armored Medical Research Laboratory.

Bean, W.B., C.R. Henderson, R.E. Johnson, and L.M. Richardson. 1946. Nutrition Survey in Pacific Theater of Operations. Report to the Surgeon General. Bull. U.S. Army Med. Dept. 5(6):697.

Brotherhood, J.R. 1984. Nutrition and sports performance. Sports Med. 1:350–389.

Champagne, C.M., J.P. Warber, and H.R. Allen. 1997. Dietary intake of U.S. Army Rangers: Estimation of nutrient intake before and after field training. Presented at the XVI International Congress of Nutrition, Montreal.

Cline, A.D., and A.E. Pusateri. 1996. Comparisons of iron status, physical activity, and nutritional intake of women entering Army officer and enlisted basic training. Proceedings of 21st National Nutrient Databank Conference, Baton Rouge, La.

Cline, A.D., J.P. Warber, and C.M. Champagne. 1997. Assessment of meal consumption behaviors by physically active young men dining in a cafeteria setting. Presented at Experimental Biology Annual Meeting, New Orleans, LA. FASEB J; 11:A184.

Cline, A.D., J.F. Patton, W.J. Tharion, S.R. Strowman, C.M. Champagne, J. Arsenault, K.L. Reynolds, J.P. Warber, C. Baker-Fulco, J. Rood, R.T. Tulley, and H.R. Lieberman. 1998. Assessment of the relationship between iron status, dietary intake, performance, and mood state of female Army officers in a basic training population. Technical Report No. T98-24. Natick, Mass.: U.S. Army Research Institute of Environmental Medicine.

Edwards, J.S.A., E.W. Askew, N. King, C.S. Fulco, R.W. Hoyt, and J.P. DeLany. 1991. An assessment of the nutritional intake and energy expenditure of unacclimatized U.S. Army soldiers living and working at high altitude. Technical Report No. T10-91. Natick, Mass.: U.S. Army Research Institute of Environmental Medicine.

Edwards, J.S.A., D.E. Roberts, T.E. Morgan, and L.S. Lester. 1989. An evaluation of the nutritional intake and acceptability of the Meal, Ready-To-Eat consumed with and without a supplemental pack in a cold environment. Technical Report No. T18-89. Natick, Mass.: U.S. Army Research Institute of Environmental Medicine.

Gordon, C.C., T. Churchill, C.E. Clauser, B. Bradtmiller, J.T. McConville, I. Tebbetts, and R.A. Walker. 1989. 1988 Anthropometric survey of U.S. Army personnel: Methods and summary statistics. Technical Report No. TR-89/044. Natick, Mass.: U.S. Army Natick Research, Development and Engineering Center.

Hirsch, E., W. Johnson, P. Dunne, C. Shaw, N. Hotson, W. Tharion, H. Lieberman, R. Hoyt, and D. Dacumos. In Press. The effects of diet composition on food intake, food selection, and water balance in a hot environment. Technical Report. Natick, Mass.: Natick Research, Development and Engineering Center.

Howe, P.E., and G.E. Berryman. 1945. Average food consumption in the training camps of the United States Army. Am. J. Physiol. 144:588–594.

Interagency Board (Interagency Board for Nutrition Monitoring and Related Research). 1993. Chartbook I: Selected Findings From the National Nutrition Monitoring and Related Research Program. DHHS Publication No. (PHS) 93-1255-2. Hyattsville, Md.: Public Health Service.

Interagency Board (Interagency Board for Nutrition Monitoring and Related Research). 1995. Third Report on Nutrition Monitoring in the United States, vol. 2. Life Sciences Research Office, Federation of American Societies for Experimental Biology. Washington, D.C.: U.S. Government Printing Office.

Johnson, R.E., and R.M. Kark. 1946. Feeding problems in man as related to environment: An analysis of United States and Canadian Army ration trials and surveys, 1941–1946. Research Report. Chicago, Ill.: Quartermaster Food and Container Institute for the Armed Forces, Research and Development Branch, Office of the Quartermaster General.

King, N., J.E. Arsenault, S.H. Mutter, C.M. Champagne, T.C. Murphy, K.A. Westphal, and E.W. Askew. 1994. Nutritional intake of female soldiers during the U.S. Army basic combat training. Technical Report No. T94-17. Natick, Mass.: U.S. Army Research Institute of Environmental Medicine.

Lemon, P.W. 1996. Is increased dietary protein necessary or beneficial for individuals with a physically active lifestyle? Nutr. Rev. 54:169-175.

NRC (National Research Council). 1941. Recommended Dietary Allowances. Washington, D.C.: National Academy Press.

NRC. 1945. Recommended Dietary Allowances. Washington, D.C.: National Academy Press.

NRC. 1980. Recommended Dietary Allowances, 9th ed. Washington, D.C.: National Academy Press.

NRC. 1989. Recommended Dietary Allowances, 10th ed. Washington, D.C.: National Academy Press.

Rose, R.W., C.J. Baker, W. Wisnaskas, J.S.A. Edwards, and M.S. Rose. 1989. Dietary assessment of U.S. Army basic trainees at Fort Jackson, South Carolina. Technical Report No. T6-89. Natick, Mass.: U.S. Army Research Institute of Environmental Medicine.

Rose, M.S., P.C. Szlyk, R.P. Francesconi, L.S. Lester, L. Armstrong, W. Matthew, A.V. Cardello, R.D. Popper, I. Sils, G. Thomas, D. Schilling, and R. Whang. 1989. Effectiveness and acceptability of nutrient solutions in enhancing fluid intake in the heat. Technical Report No. T10-89. Natick, Mass.: U.S. Army Research Institute of Environmental Medicine.

Samuels, J.P., R.P. McDevitt, M.C. Bollman, W. Maclinn, L.M. Richardson, and L.G. Voss. 1947. Ration Development: A Report of Wartime Problems in Subsistence Research and Development, vol. 12. Chicago, Ill.: Quartermaster Food and Container Institute for the Armed Forces, Research and Development Branch, Office of the Quartermaster General.

Shippee, R., E.W. Askew, M. Mays, B. Fairbrother, K. Friedl, J. Vogel, R. Hoyt, L. Marchitelli, B. Nindl, P. Frykman, L. Martinez-Lopez, E. Bernton, M. Kramer, R. Tulley, J. Rood, J. DeLany, D. Jezior, and J. Arsenault. 1994. Nutritional and immunological assessment of Ranger students with increased caloric intake. Technical Report No. T95-5. Natick, Mass.: U.S. Army Research Institute of Environmental Medicine.

Szeto, E.G., D.E. Carlson, T.B. Dugan, and J.C. Buchbinder. 1987. A comparison of nutrient intakes between a Ft. Riley contractor-operated and a Ft. Lewis military-operated garrison dining facility. Technical Report No. T2-88. Natick, Mass.: U.S. Army Research Institute of Environmental Medicine.

Szeto, E.G., T.B. Dugan, and J.A. Gallo. 1988. Assessment of habitual diners' nutrient intakes in a military-operated garrison dining facility. Ft. Devens I. Technical Report No. T3-89. Natick, Mass.: U.S. Army Research Institute of Environmental Medicine.

Szeto, E.G., J.A. Gallo, and K.W. Samonds. 1989. Passive nutrition intervention in a military-operated garrison dining facility. Ft. Devens II. Technical Report No. T7-89. Natick, Mass: U.S. Army Research Institute of Environmental Medicine.

Tharion, W.J., A.D. Cline, N. Hotson, W. Johnson, P. Niro, C.J. Baker-Fulco, S. McGraw, R.L. Shippee, T.M. Skibinski, R.W. Hoyt, J.P. Delany, R.E. Tulley, J. Rood, W.R. Santee, S.H.M. Boquist, M. Bordic, M. Kramer, S.H. Slade, and H.R. Lieberman. 1997. Nutritional challenges for field feeding in a desert environment: Use of the Unitized Group Ration (UGR) and a supplemental carbohydrate beverage. Technical Report No. T97-9. Natick, Mass.: U.S. Army Research Institute of Environmental Medicine.

Thomas, C.D., K.E. Friedl, M.Z. Mays, S.H. Mutter, R.J. Moore, D.A. Jezior, C.J. Baker-Fulco, L.J. Marchitelli, R.T. Tulley, and E.W. Askew. 1995. Nutrient intakes and nutritional status of soldiers consuming the Meal, Ready-To-Eat (MRE XIII) during a 30-day field training exercise. Technical Report No. T95-6. Natick, Mass.: U.S. Army Research Institute of Environmental Medicine.

U.S. Department of the Army. 1947. Army Regulation 40-250. "Nutrition" Washington, D.C.

U.S. Departments of the Army, the Navy, and the Air Force. 1985. Army Regulation 40-25/Naval Command Medical Instruction 10110.1/Air Force Regulation 160-95. Nutritional Allowances, Standards and Education. May 15. Washington, D.C.

U.S. War Department. 1944. Nutrition. War Department Circular No. 98. Washington, D.C.

Warber, J.P., F.M. Kramer, S.M. McGraw, L.L. Lesher, W. Johnson, and A.D. Cline. 1996. The Army Food and Nutrition Survey, 1995-97. Technical Report. Natick, Mass.: U.S. Army Research Institute of Environmental Medicine.

Williams, C. 1995. Macronutrients and performance. J. Sports Sci. 13 Spec No: S1-10.

DISCUSSION

PATRICK DUNNE: The Savannah study subjects, in part, were by design a test population who consumed a diet that was purposefully lower in protein but higher in carbohydrate. I think we need to look at the trade-offs here because we got them to achieve well above the military 400 grams of carbohydrate intake, and it was designed to be a hot-weather study. So that was a designed diet, not a free choice; it was rather selective. What we really want to do is look at impact and overall turnover of water and hydration as part of that study. Jim hopefully has some data for us on that.

I think that one of the drivers that leads the ration developers and the logistics community to look at protein is not just performance but it is actually the cost of the ration. So there are some major trade-offs in overall metabolism. If you want to trade protein for carbohydrate, maybe that is good, but not protein for fat. That was our design.

ALANA CLINE: The data that I used for that study actually were from the control group that did not have the higher carbohydrate intake. The protein intake was even lower with the test group. So I went ahead and just used the control group for that and it is still a little bit lower.

DOUGLAS WILMORE: Thank you for the presentation. Do you have information on the source of the dietary protein over the period of 30 years or so?

ALANA CLINE: I do not have the historical data on the source of the protein. Some of our more recent studies are showing that there is a higher intake now of protein from non-meat sources. That is something that we would want to consider because there is a difference in the quality of the protein that is being consumed. That is something that we do need to look at more closely.

DOUGLAS WILMORE: So we have no idea about protein efficiency?

ALANA CLINE: Not for the historical data.

ROBERT NESHEIM: One last question.

ROBERT WOLFE: On that slide showing voluntary intake of supplements you had two columns, amino acids and protein, and one was like 36 percent and the other was 30. Does that mean that 36 percent were taking amino acids and a separate 30 percent were taking protein.

ALANA CLINE: Yes. There were two separate questions.

ROBERT WOLFE: So that was the total of the two things they were taking in terms of dietary supplements?

ALANA CLINE: No, I am sorry. It was either one or the other. So it would be about 30 percent. They could be taking both. We were not able to really clarify that with the questionnaire that we had.

5

The Energy Costs of Protein Metabolism: Lean and Mean on Uncle Sam's Team

Dennis M. Bier[1]

ORIGIN OF THE WORD "PROTEIN"

How protein got its name is an interesting story (Hartley, 1951). The traditional credit for coining the word "protein" goes to the Dutch chemist Gerardus Johannes Mulder, who in an article published in *the Bulletin des Sciences Physiques et Naturelles en Neerlande* on July 30, 1838, stated (in French) that this material was the essential general principle of all of the constituents of the animal body and defined it by the Greek word "proteus," which he translated in Latin to "primarius," that is, the primary constituent of the body (Hartley, 1951). The interesting part of the story is that Mulder appears to have taken the term directly from Swedish chemist Jac Berzelius, who, on July 10, sent Mulder a letter in which Berzelius suggested the name "protein."

---

[1] Dennis M. Bier, Children's Nutrition Research Center, Baylor College of Medicine, Houston, TX 77030-2600.

Besides the apparent use of the new term without proper attribution was the situation of a Dutch chemist, writing in a Dutch journal, defining a new word in French that was derived from Greek, and then qualifying its meaning in Latin! Further, considering that this episode took place 160 years ago, not only was the mail a lot faster, but also the publication time was substantially shorter than it is today, since the entire story took place in a period of about 3 weeks!

Admittedly, Berzelius and Mulder were right: Protein is the essential general principle of the constituents of the animal body. Thus, one might briefly summarize the physiological roles of protein in metabolism as "responsible for just about everything." But, this is neither particularly helpful nor informative.

FUNCTIONS OF BODY PROTEINS AND AMINO ACIDS

Table 5-1 lists a variety of the functions of body proteins and amino acids. From the practical standpoint of integrative human physiology helpful to the military, the major roles of body proteins are (1) those relating to protein synthesis and protein breakdown in the context of maintaining lean body mass, (2) efficient operation of regulatory proteins required for conduct and optimization of body functions, and (3) the energy costs of the above, including the cost of oxidation and excretion of protein metabolites resulting from the metabolic reactions constituting these events. The net physiological results that are useful to military personnel include increased strength, improved endurance, optimization of "fight or flight" reactions (preferably the former), efficient blood coagulation and wound healing, enhanced immunological functions with improved disease resistance, and peak mental alertness and memory.

What is often forgotten in this context is that protein turnover is an energy-requiring process. Further, the energy cost of protein metabolism is higher than conventionally estimated. These observations are important because one of the principal lessons learned from prior Committee on Military Nutrition Research reports, such as the Ranger studies, is that biological energy demands often far exceed dietary energy intakes in combat field circumstances. Thus, while strength and endurance might immediately be identified with protein metabolism, the energy costs of (1) the protein enzymatic reactions involved in fight or flight reactions, (2) the synthesis of host defense proteins, or (3) memory storage might not be so readily apparent.

PROTEIN METABOLISM AS AN ENERGY-REQUIRING PROCESS

In 1989, Waterlow and Millward calculated the daily energy cost of protein turnover as approximately 18 kJ (4.3 kcal)/kg body weight, or about 20 percent of the basal metabolic rate. This estimate, made from the best information available at the time, tallied the energy costs of protein breakdown and regulation as "probably negligible" and did not include the energy costs of posttranslational modifications. Today, however, it is known that these costs are not negligible. Using another approach relating protein (ATP) utilization to total ATP synthesis, Reeds et al. (1985) came to roughly similar conclusions but, after addressing the potential additional costs of all the cellular regulatory reactions, estimated that these control reactions and other, unaccounted-for, costs would increase the true energy cost of protein metabolism significantly.

Young and Yu (1996) have listed some of these energy-dependent processes (Table 5-2). In addition to the costs of protein synthesis, there is an energy-dependent cost of protein degradation via the ubiquitin pathway.

TABLE 5-1 Some Functions of Amino Acids and Their Products

Function	Example
Substrates for protein synthesis	Those amino acids for which there is a codon
Regulators of protein turnover	Leucine, arginine?
Regulators of enzyme activity	Arginine and N-acetylglutamate synthetase Phenylalanine and phenylalanine hydroxylase activation
Precursor of signal transducer	Arginine and nitric oxide
Methylation reactions	Methionine
Neurotransmitter	Tryptophan, glutamate
Ion fluxes	Taurine, glutamate
Physiologic molecular precursors	Arginine, glutamine, purines
Nitrogen transporters	Alanine, glutamine
Circulating transporters	Ceruloplasmin; apolipoproteins Vitamin and hormone binding proteins
Messengers/Signals	Insulin, growth factors
Movement	Actin, kinesin
Immunity	Antibodies, interleukins
Growth, differentiation, gene expression	Growth factors, transcription factors

SOURCE: Adapted from Young and Yu, 1996.

TABLE 5-2 Some Energy-Dependent Processes Associated with Protein Metabolism

1. Protein Turnover

Formation of initiation complexes; peptide bond synthesis

Protein breakdown via ubiquitin-proteosome pathway

2. RNA Turnover

Ribosomal RNA; transfer RNA; pre-messenger RNA splicing; and messenger RNA

3. Amino acid transport

4. Regulatory processes

Reversible phosphorylation, GTP-GDP exchange proteins,

Ion pumps and channels, second messengers

5. Nitrogen metabolism

Glutamine/glutamate cycle

Glucose/alanine cycle

Urea synthesis

SOURCE: Adapted from Young and Yu, 1996.

Similarly, there are a large number of energy-dependent processes involved in the regulation of protein metabolism, as expanded on below, and there are other assorted costs attached to the body's nitrogen metabolism cycles. The latter cycles are addressed by others in this volume.

ENERGY COSTS OF PROTEIN SYNTHESIS AND REGULATION

Of first concern are the energy costs of cell replication. Before the cost of protein synthesis itself can be calculated, one must recognize that the proteolytic degradation of cyclins regulates the movement from one phase of the cell cycle to the next through energy-dependent ubiquitin pathways (King et al., 1996; Martin-Castellanos and Moreno, 1997). Of second concern is that mitotic events, including spindle microtubular growth, and spindle elongation during anaphase require hydrolysis of GTP or ATP.

Protein synthesis itself has a higher cost than previously estimated. While transcription of the amino acid mRNA codon requires six ATP per amino acid and activation to amino acyl-tRNA requires another two ATP, there is disagreement about the total energy cost of mRNA translation. In addition to the one ATP per peptide required for capping the 5-prime end of the peptide, at least one GTP per peptide bond is required for initiation, two GTP per bond for elongation, and one GTP per peptide for termination. However, recent evidence suggests that one additional molecule of GTP is required for chain elongation (Schimmel, 1993), and hydrolysis of an additional GTP might be required during initiation and/or at the termination step as well. If hydrolysis of these additional high-energy bonds is proven correct, the net energy cost of protein synthesis alone will increase significantly from estimates made a decade ago.

In addition, there are a variety of other costs that are difficult to estimate. All of the additional sequences that are involved in, for example, synthesizing "pre-proteins" and "pre-pro-proteins" and the costs of alternate splicing are not easy to quantify. In a sense, synthesizing and then removing these unused peptide sequences is wasted energy unless some as-yet-unknown energy advantage is discovered for this process. Similarly, the cost of synthesizing nonessential amino acids that are required for protein synthesis and the costs of posttranslational modifications are not known with certainty.

Further, the folding (Hartl, 1996) and the movement of the synthesized proteins to their sites of action (Rothman and Wieland, 1996) are highly energy-dependent processes. ATP-dependent mechanisms are required for polypeptide chain folding by heat-shock protein 70 and the chaperonin families of molecular chaperones (Hartl, 1996). Translocation across the membrane of the rough endoplasmic reticulum is an energy-dependent process, as is each transport step to the cis, medial, and trans Golgi compartments (Rothman and Wieland, 1996).

TABLE 5-3 Protein Targeting Costs

Nonsecretory Proteins	Secretory Proteins
Importation of mitochondrial proteins	Translocation across the rough endoplasmic reticulum membrane
Transit peptide receptor binding- peptide unfolding;	Transport to cis, medial, and trans Golgi
Cytosolic competence factor binding & dissociation	Lysosomal and secretory sorting vesicles
Matrix/stromal peptidase processing	Acidification of secretary organelles
Importation of peroxisomal proteins	Phosphorylation of receptors/hands
Importation of nuclear proteins	Protein folding or "proofreading"
	Activation of cytoskeleton motors

TABLE 5-4 Energy-Dependent Processes in the Turnover of Regulatory Proteins

Reversible phosphorylation

Signal transduction
    Protein kinases

Second messengers
    Phosphatidylinositol kinases

Folding and translocation
    Chaperone proteins

Table 5-3 shows some of the protein targeting costs for secretory and nonsecretory proteins. The net actual costs of these events to the whole organism are difficult to estimate with any accuracy or precision, but they are likely to be significant.

In addressing the potential energy costs of regulatory proteins (Table 5-4), one is struck by the fact that, although these proteins do not represent much in the way of mass compared with structural proteins, their turnover rates are very high. Therefore, they may represent a significant energy drain. It is now known, for example, that an immense number of processes are controlled by reversible enzymatic phosphorylation/dephosphorylation reactions. This central mechanism of regulatory control occurs in every cell at an unimaginable number of times per minute. What the actual net energy cost of these regulatory events is to the whole body is only speculative, but likely significant.

Signal transduction processes, including those mediated by the more than 100 known members of the protein kinase family, and the energy-dependent costs of the second messenger families, for example the phosphatidyl and inositol kinases, are additional energy costs of regulatory protein metabolism. Furthermore, the energy costs of posttranslational modifications such as protein glycosylation are additional. The addition of each nucleotide sugar costs the hydrolysis of one uridine triphosphate bond. Approximately 10 percent of proteins are glycosylated. Each protein averages between 2 and 5 sugar chains, and each sugar chain averages about 12 sugars. Precisely how to tally this cost in the overall sum of the total daily energy cost of protein turnover, is not known, but the cost is potentially large.

ENERGY COSTS OF PROTEIN BREAKDOWN

Finally, it is important to address the potential energy costs of protein breakdown. There are two principal routes of protein degradation (Figure 5-1) (Mitch and Goldberg, 1996; Young and Yu, 1996). Most cytosolic, nuclear, and

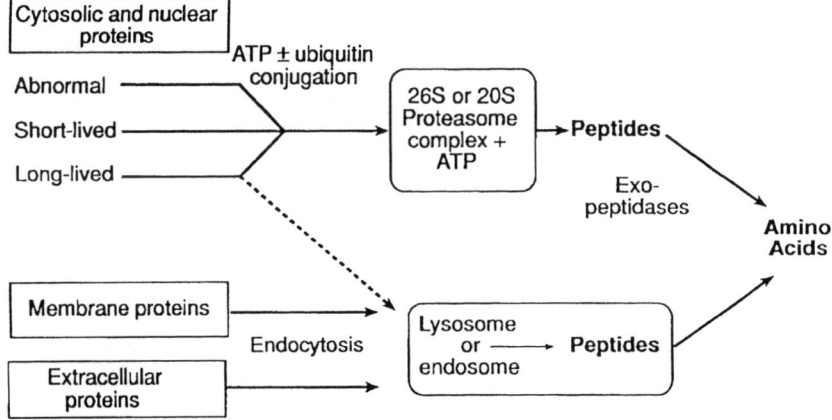

FIGURE 5-1 Major Pathways of Protein Degradation. SOURCE: Young, V.R. and Y.M. Yu, 1996. Protein and amino acid metabolism. Pp. 159–200 in Nutrition and Metabolism in the Surgical Patient, 2nd Edition, Josef E. Fischer, ed. Boston: Little, Brown and Company.

mitochondrial proteins are degraded through energy-dependent pathways requiring hydrolysis of ATP (Gottesman et al., 1997). Extracellular and membrane proteins are degraded through the lysosomal pathway. Although the latter route is not an energy-dependent process in itself, energy is required for proton movement into and maintenance of the acid environment within the lysosome. In Table 5-5, Mitch and Goldberg (1996) have summarized some of the conditions that alter muscle protein degradation through the energy-dependent ubiquitin-proteosome pathway. It is clear from this table that the

TABLE 5-5 Conditions That Alter Muscle Protein Degradation Through the Ubiquitin-Proteosome Pathway

Rat Models	Humans
Increased Protein Degradation	
Fasting	Eating disorders
Metabolic acidosis	Renal tubular defects
Renal failure	Uremia
Diabetes mellitus	Diabetes mellitus
Thermal injury	Burns
Endotoxin, bacteria	Sepsis
Tumors	Cancer cachexia
Glucocorticoids	Cushing's Syndrome
Decreased Protein Degradation	
Deficient dietary protein	Malnutrition
Hypothyroidism	Hypothyroidism

SOURCE: Adapted from Mitch and Goldberg, 1996

pathophysiological regulation of this pathway might affect energy demands in a wide variety of clinical circumstances.

This thesis is not merely an academic one. Price et al. (1996) demonstrated that the accelerated proteolysis observed in insulinopenic, streptozocin-diabetic rats was the consequence of activation of the energy-dependent ubiquitin-proteosome pathway. Similarly, Tiao et al. (1997) have demonstrated an increase in the ubiquitin and 20S proteosome subunit HC3 mRNA levels in rectus abdominis muscle obtained from septic patients.

CONCLUSION

The energy-dependent processes of maintaining the turnover of body proteins, including synthesis, folding, targeting, regulatory processes, and protein breakdown, have an overall cost to body energy homeostasis that is significantly higher than previously appreciated. The work of Tiao et al. (1997) confirms the human implications of animal studies like those of Price et al. (1996), and strongly suggests that circumstances that increase energy-dependent proteolytic mechanisms have practical clinical consequences for whole body energy homeostasis. It is not known yet how to approach a precise quantification of these costs at the whole body level. Nonetheless, investigators should give more thought to this question; there are important clinical circumstances where the energy costs of protein metabolism are likely to result in disproportionately increased energy demands. The high physical exertion levels of military personnel in the field may be one of these circumstances.

REFERENCES

Gottesman, S., M.R. Maurizi, and S. Wickner. 1997. Regulatory subunits of energy-dependent proteases. Cell 91: 435–438.
Hartl, F. U. 1966. Molecular chaperones in cellular protein folding. Nature 381: 571–580.
Hartley, H. 1951. Origin of the word "protein." Nature 168:244.
King, R.W., R.J. Deshaies, J.-M. Peters, and M.W. Kirschner. 1996. How proteolysis drives the cell cycle. Science: 274: 1652–1659.
Martin-Castellanos, C., and S. Moreno. 1997. Recent advances on cyclins, CDKs and CDK inhibitors. Trends in Cell Biol. 7: 95–98.
Mitch, W.E., and A.L. Goldberg. 1996. Mechanisms of muscle wasting. N. Engl. J. Med. 335: 1897–1905.
Price, S.R., J.L. Bailey, X. Wang, C. Jurkovitz, B.K. England, X. Ding, L. S. Phillips, and W.E. Mitch. 1996. Muscle wasting in insulinopenic rats results from activation of the ATP-dependent, ubiquitin-proteosome proteolytic pathway by a mechanism including gene transcription. J. Clin. Invest. 98: 1703–1708.
Reeds, P.J., M.F. Fuller, and B.A. Nicholson. 1985. Metabolic basis of energy expenditure with particular reference to protein. Pp. 46–47 in Substrate and Energy Metabolism in Man, J. S. Garrow and D. Halliday, eds. London: John Libby.
Rothman, J. E., and F. T. Wieland. 1996. Protein sorting by transport vesicles. Science 272: 227–234.

Schimmel, P. 1993. GTP hydrolysis in protein synthesis: two for tu? Science 259: 1264–1265.

Tiao, G., S. Hobler, J.J. Wang, T.A. Meyer, F.A. Luchette, J.E. Fischer, and P-0. Hasselgren. 1997. Sepsis is associated with increased mRNAs of ubiquitin-proteasome proteolytic pathway in human skeletal muscle. J. Clin. Invest. 99:163–168.

Waterlow, J. C., and Millward, D. J. 10. 1989. Energy cost of turnover of protein and other cellular constituents. Pp. 277–282 in Energy Transformations in Cells and Organisms, W. Wieser and E. Gnaiger, eds. Stuttgart: Georg Thieme Verlag.

Young, V.R. and Y.M. Yu. 1996. Protein and amino acid metabolism. Pp. 159–200 in Nutrition and Metabolism in the Surgical Patient, 2nd Edition, Josef E. Fischer, ed. Boston: Little, Brown and Company.

DISCUSSION

DR. NESHEIM: Questions?

DR. MILLWARD: If I can just speak briefly in defense of John Waterloow's original interpretation.

DR. BIER: By the way, do not take this as a criticism of John. He worked with the data available.

DR. MILLWARD: I would just like to make the comment that one of the difficulties with these sorts of arguments is definition of what you are talking about when you are talking about protein turnover. You made the initial statement that proteins are involved in everything that goes on in the cell. You talked about a wide range of protein-related functions that are energy dependent. There is no question about that. Therefore, it is possible to subsume ultimately all energy expenditure under the heading of protein-related functions if you are not careful in the way in which you have defined the process that you are talking about.

DR. BIER: I agree with that. That is, in the end, since you cannot live without the proteins required for life, then essentially all energy generation is dependent on protein turnover. I think that is a very good point, Joe, and you are correct in the sense of defining the limits of the argument.

I guess what I was trying to do was to move us from what I would call the traditional nutritional thinking on this issue, that is, that the costs are related to the net synthesis and oxidation of protein, the protein turnover of macromolecules, and the maintenance of lean mass, and move us to appreciate further that there are additional costs that many physiologists have not spent a lot of time thinking about.

DR. LIEBERMAN: Under conditions of relatively mild undernutrition, similar to what we see in your study, do you think there might be some better markers to tell us that protein balance is not where we would like it to be?

DR. BIER: I am sure there are, since use of all available probes of a very complicated system should give you more information. But is there any individual marker? I do not know. I think that some of the new things, like ubiquitin mRNA and the like, may give us an indication.

DR. LIEBERMAN: It would be nicer if we had something that was more fine grained.

LTC FRIEDL: I would not dismiss the turnover rate. We see that in undernutrition. That is a good example of where all these processes can be reduced. What are the consequences in the big picture? How much leeway do we have in adjusting the rate of protein turnover.

DR. BIER: Well, I am not sure you can turn it down in the circumstance of active physical exertion, for example. That is, during the period of exercise itself, protein turnover may "turn down, " but there is a period where you have to then "turn it up". I do not think in the context of continued strenuous activity you can significantly reduce the overall cost.

LTC FRIEDL: Maybe there is a compromise between the demands of physical activity and deficient energy intake. Undernutrition seems to favor the process of turning down the overall turnover.

DR. BIER: Yes, that is in a circumstance where the person is generally lying in bed or voluntarily reducing exertion overall.

LTC FRIEDL: It is not a suboptimal performance.

DR. BIER: Michael Rennie is going to talk about physical exercise later, and I will defer this to him.

DR. NESHEIM: Yes.

DR. HOYT: Colonel Friedl, Jim Delaney, and I did a study of Norwegian Rangers who did not eat or sleep for a week, and we looked at their body composition with dual X-ray Absorptiometry and their energy expenditure with doubly-labeled water. You can look at the contribution of fat and carbohydrate and protein to that negative energy balance. Would you expect anything other than a 4 kcal/g contribution from the protein loss reflecting the savings associated with not supporting some of the synthetic costs?

DR. BIER: That is a net protein loss. The actual cost is higher. But I take Joe's argument that, in the end, one can attribute all the energy cost somehow or other to protein. I guess, then, this goes back to the issue of whether we can change the efficiency or improve the net energy balance cost by adding protein in that sort of circumstance? I am not sure that I know the answer.

DR. NESHEIM: Thank you. We need to move on. This is a very good stage setting for a lot of discussion that will come along. Now, we want to move to a discussion of the regulation of muscle mass and functions by Dr. Nair from the Mayo Clinic

6

Regulation of Muscle Mass and Function: Effects of Aging and Hormones

Niels Moller and K. Sreekumaran Nair[1]

INTRODUCTION

Comprising close to 50 percent of total body weight in lean individuals, skeletal (or striated) muscle constitutes the largest single component of the body and serves as the major repository of protein (close to 50 percent of total body protein) and free amino acids in the body. Besides its locomotive functions, skeletal muscle is also an important metabolic organ. Metabolic functions are crucial not only for locomotion but also for maintaining homeostasis of substrates in the circulation and providing amino acids for various body functions. Mitochondria in skeletal muscle convert energy from nutrients into adenosine triphosphate (ATP) by oxidative phosphorylation. For mechanical functions involving mainly locomotion, this chemical energy (ATP) is further

---

[1] K. Sreekumaran Nair, Endocrine Research Unit, Mayo Clinic and Foundation, Rochester, MN 55905.
This work was supported by Public Health Service grants RO1 DK41973, AG09531 and General Clinical Research Center grants RR109 and RR00585.

converted into mechanical energy by the enzymatic (ATPase) action of myosin and actin in the sarcomere (Figure 6-1). In this process, a major proportion of the energy loss inevitably appears as heat. Under resting conditions, muscle accounts for 20 to 30 percent of total resting energy expenditure, the variability of which to a large extent is determined by differences in muscle metabolism (Zurlo et al., 1990). Under conditions of cold exposure and shivering thermogenesis, the function of muscle as a "heater" for the body and the resultant energy loss become still more conspicuous. In addition, skeletal muscle supplies amino acids for synthesis of proteins in other tissues (crucial during wound healing), for the immune functions, and for gluconeogenesis (alanine and glutamine) under catabolic conditions. Skeletal muscle also oxidizes glucose and fatty acids and stores large amounts of glycogen postprandially.

To fulfill these requirements optimally, muscle tissue function relies on a variety of factors as illustrated in Figure 6-1. First, the contractile myofibrillar apparatus consisting of actin and myosin filaments must exist in sufficient quality and quantity, and nervous activation of the contractile elements must take place in a controlled fashion. Second, the quality and quantity of mitochondrial proteins (mostly enzymes) must be sufficiently maintained to ensure efficient ATP production. Also critical, that the metabolic demands of the muscle cell be met to ensure a continuous ATP production (i.e., supplies of fuel substrates such as glucose and fatty acids and oxygen must be adequate) and that the metabolic by-products such as carbon dioxide are removed efficiently and in a timely manner. The quality of all proteins, both structural and metabolically active (such as myosin, actin, and mitochondrial enzymes), are maintained by a continuous remodeling process involving protein breakdown and protein

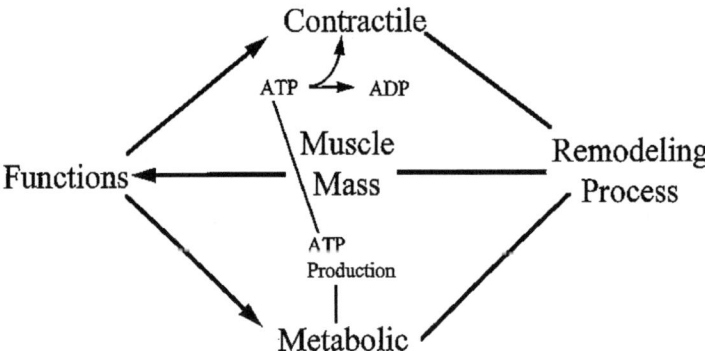

FIGURE 6-1 Muscle mass and quality are determined by a remodeling process. Muscle mass is one determinant of muscle function, which is also dependent on the quality of distinct muscle proteins. Muscle functions include both contractile and metabolic aspects. Two metabolic processes critical for contractility are mitochondrial ATP production and subsequent hydrolysis of ATP in the sarcomere (by myosin and actin) to release chemical energy for contractile function.

synthesis. The tissue concentration of a specific protein is determined by the balance between protein breakdown and protein synthesis. Nutrient supply to the muscle tissue (for ATP production) and the removal of metabolic by-products (e.g., carbon dioxide) are dependent on uninterrupted and dynamic circulatory systems. All of these individual processes are controlled by regulatory mechanisms, which include circulating and local levels of hormones and substrates, which in turn are influenced by the physiological state of the individual in terms of age, gender, nutritional status, exercise, and chronic or acute illness. Using aging as an example, this brief review outlines some of the control mechanisms and other biological factors involved in the regulation of muscle mass and function.

SARCOPENIA OF AGING

Aging can be described as a model in which many of the regulatory mechanisms are disrupted, resulting in functional disabilities involving both locomotive and metabolic aspects.

In spite of vigorous attempts by individuals to avert the physical impact of age, all population-based studies show a relentless loss of muscle mass and strength with aging. In addition, increased muscle fatigability and decline in endurance capacity substantially retard the functional capabilities of the elderly population. Some evidence indicates that this phenomenon is not solely due to a loss of muscle quantity, but also to an impairment of muscle quality (Reed et al., 1991; Rooyackers et al., 1996) (Figures 6-2 and 6-3). The combination of loss

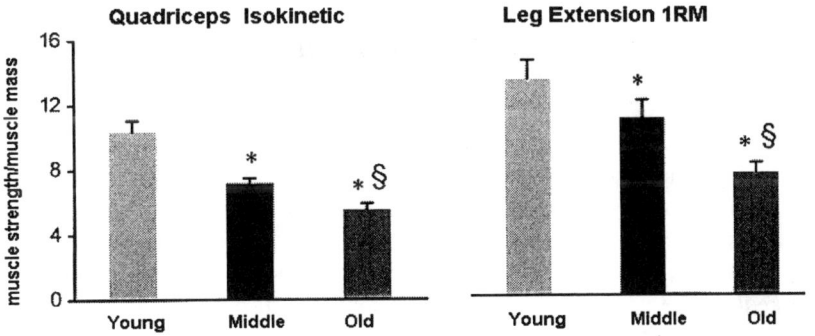

FIGURE 6-2 Muscle strength is not solely determined by muscle mass. With advancing age, a continuous loss of muscle efficiency occurs, indicating that muscle quality is declining. This study normalized muscle strength, quadriceps isokinetic strength, and leg extension for regional muscle mass (measured by Dual Photon X-ray) and showed a progressive decline with aging ($P < 0.05$–0.01). * Significant difference ($P < 0.01$) from young age group; § Significant difference ($P < 0.05$) from middle age group. SOURCE: Adapted from Balagopal et al. (1997).

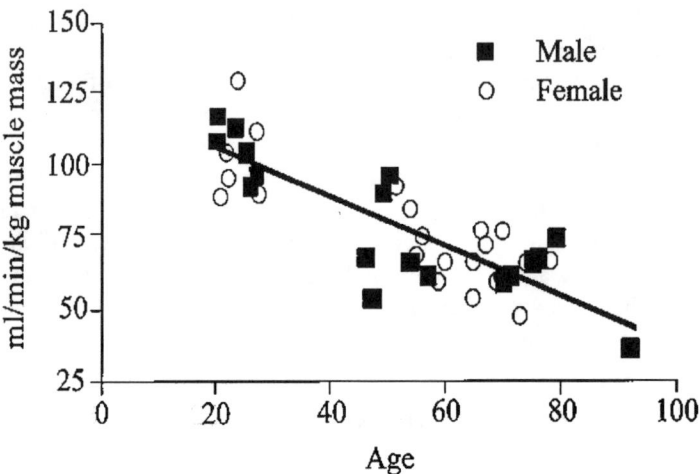

FIGURE 6-3 VO_{2max} declines with aging after normalizing for muscle mass, indicating that the decline in endurance capacity depends not only on muscle mass but also on other factors. SOURCE: Rooyackers, O.E., D.B. Adey, P.A. Ades, and K.S. Nair. 1996. Effect of age on in vivo rates of mitochondrial protein synthesis in human skeletal muscle. Proc. Natl. Acad. Sci. USA 93:15364–15369. Copyright 1996 National Academy of Sciences, U.S.A.

of muscle mass and increased muscle weakness and fatigability, which results in substantial impairment of muscle function, has been coined *sarcopenia of aging* and may contribute substantially to morbidity of the elderly by restricting physical activity, increasing the risk of falls and fractures, and causing changes in body metabolism and composition, which results in increased incidence of noninsulin-dependent diabetes mellitus.

It has been reported that in elderly in comparison with young subjects, there is a decline in the synthesis rate of mixed muscle protein—both total and myofibrillar proteins (Welle et al., 1993; Yarasheski et al., 1993). Interestingly, a recent study demonstrated that not only did synthesis of muscle mitochondrial proteins (pivotal to oxidative phosphorylation and ATP generation) decrease in the elderly, but also that this 40-percent decrease in mitochondrial protein synthesis occurred as early as middle age (average age 52 years) (Rooyackers et al., 1996) (Figure 6-4). The decline in mitochondrial protein synthesis was markedly more pronounced than the concomitant 10- to 15-percent decline in synthesis rates of mixed muscle proteins (Rooyackers et al., 1996). These changes were also associated with a decline in cytochrome-c-oxidase activity and endurance capacity (Rooyackers et al., 1996) (Figure 6-5). It is possible that the decline in mitochondrial protein synthesis may cause the impairment of endurance capacity and the more pronounced muscle fatigability in the aging population. In addition, robust ATP production is crucial for synthesis of other muscle proteins. A general decline in synthesis rates of several muscle proteins

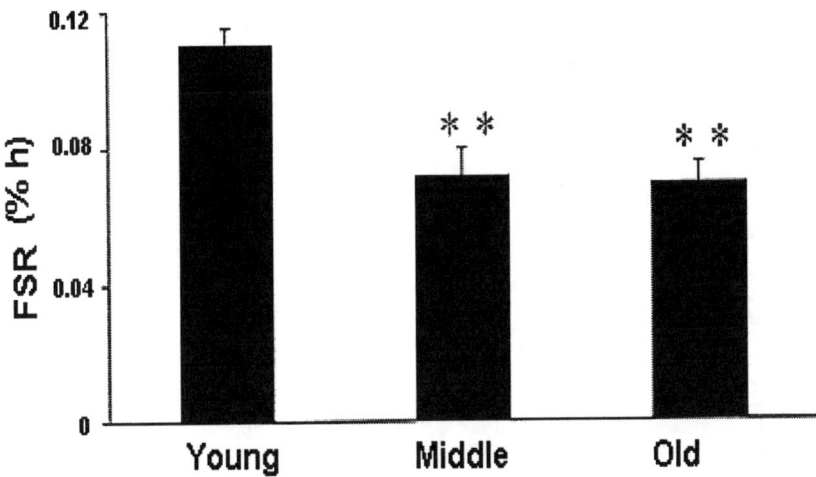

FIGURE 6-4 A decline in fractional muscle mitochondrial protein synthesis occurred with age. Approximately a 40 percent decline occurred by middle age ($P < 0.01$), but there was no further decline with advancing age. **Indicates significant difference from young age. SOURCE: Rooyackers, O.E., D.B. Adey, P.A. Ades, and K.S. Nair. 1996. Effect of age on in vivo rates of mitochondrial protein synthesis in human skeletal muscle. Proc. Natl. Acad. Sci. USA 93:15364–15369. Copyright 1996 National Academy of Sciences, U.S.A.

(such as myosin heavy chain and mitochondrial proteins) occurs with age (Balagopal et al., 1997; Rooyackers et al., 1996; Welle et al., 1993; Yarasheski et al., 1993), perhaps reflecting the inability of mitochondria to produce sufficient ATP. Furthermore, recent studies have demonstrated that synthesis rates of myosin heavy chain, a major myofibrillar protein involved in hydrolysis of ATP and conversion of chemical energy from ATP to mechanical energy, also decline by middle age (Balagopal et al., 1997). These results suggest that the aging process selectively affects the ATP-generating machinery of muscle and imply that any intervention should seek to restrict this loss of mitochondrial capacity. In addition, the reduced synthesis rate of myosin heavy chain is compatible with the notion that the ability to maintain adequate muscle protein quality declines with age, thereby potentially compromising the efficiency of the locomotive apparatus to extract mechanical energy from fuel stores. As discussed below, it is likely that age associated decrements in circulating levels of anabolic hormones, such as growth hormone (GH), insulin-like growth factor I (IGF-I), sex steroids, and fading effectiveness of insulin are all involved in the involution of muscle that occurs with aging.

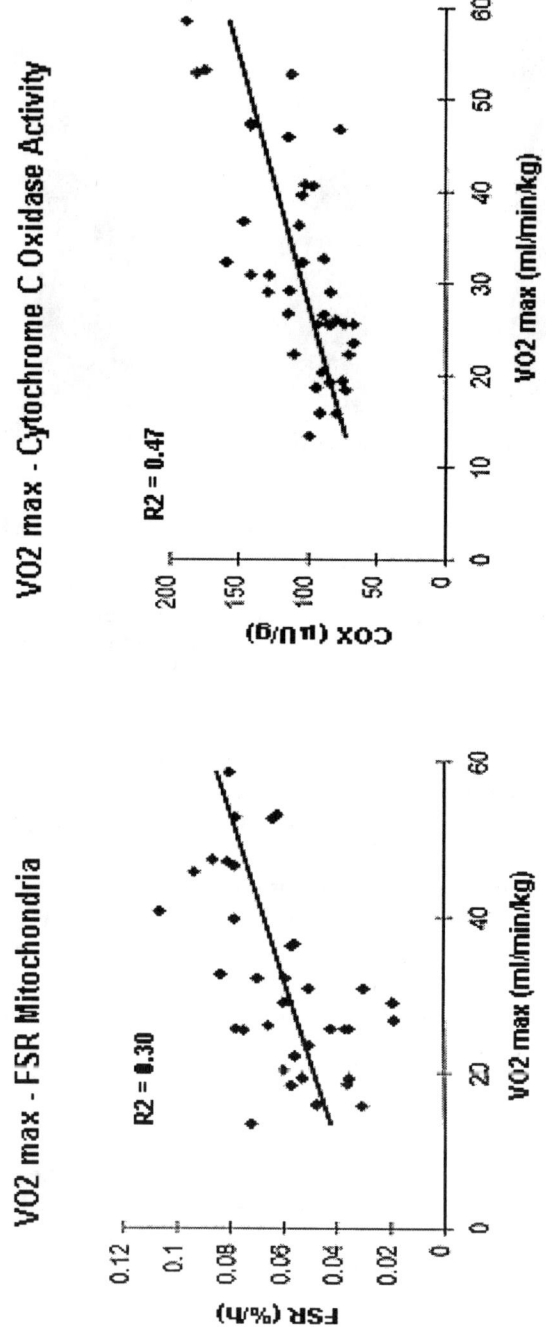

FIGURE 6-5 Relationship between maximal aerobic capacity per kilogram muscle mass and the fractional synthesis rate (FSR) of mitochondrial proteins and the activity of cytochrome *c* oxidase (COX) in muscle biopsies. Data (derived from Rooyackers et al. [1996]) indicate that cytochrome oxidase activity and fractional muscle protein rate are significantly correlated to VO_{2max} ($P < 0.01$).

HORMONAL EFFECTS ON MUSCLE PROTEIN

Insulin

Insulin is an anabolic hormone. After a meal, the ensuing insulin surge suppresses lipolysis and lipid oxidation and stimulates glucose storage and oxidation in skeletal muscle. The effects of insulin on muscle protein metabolism are less certain. Based on the clinical observation that insulin-deprived diabetic patients become cachectic and sarcopenic, it is, however, undisputed that insulin action is fundamental to preservation of functioning muscle mass. This clinical observation was convincingly confirmed when it was shown that insulin treatment increases nitrogen content in diabetic patients and that perfusion of the forearm with insulin in humans decreased amino acid release (Pozefsky et al., 1969). It remained to be resolved, however, whether this anticatabolic effect of insulin was due to stimulation of protein synthesis or inhibition of protein breakdown.

Studies performed in Type I diabetic patients showed that insulin replacement inhibits both protein breakdown and synthesis at the whole body level, but protein conservation occurs because the magnitude of insulin's effect on protein breakdown is more pronounced than on protein synthesis (Nair et al., 1995). It was also observed that insulin replacement has no effect on synthesis rate of muscle protein, indicating that the main effect of insulin on protein synthesis occurs in tissues other than muscle (Nair et al., 1995) (Figure 6-6). In that study, which employed sampling of the femoral artery, femoral vein, and hepatic vein and administration of amino acid tracers in insulin-deprived patients with diabetes, it was demonstrated that insulin replacement inhibited protein breakdown in the leg, with no effect on protein synthesis, while it inhibited protein breakdown and synthesis in the splanchnic bed, indicating that insulin's anticatabolic effect is largely due to its inhibition of muscle protein breakdown (Nair et al., 1995). This concept is supported (Gelfand et al., 1987) by the finding of one study that local forearm perfusion with insulin inhibited protein degradation without any effect on protein synthesis. However, the bulk of *in vitro* experiments suggest that insulin stimulates protein synthesis (Kimball and Jefferson, 1988), and a recent human in vivo study reported, based on data from arteriovenous differences combined with a muscle biopsy, that insulin augmented protein synthesis in the perfused leg (Biolo et al., 1995).

The reason for these discrepancies is not clear, but it may relate to the possibility that the response to insulin depends on the prevailing concentration of substrate (amino acids), in particular intracellularly, in the immediate precursor pool for protein synthesis (amino acyl tRNA). In addition, hyperinsulinemia is normally prompted by meals, and most human studies were performed in the postabsorptive state because of theoretical problems relating to isotope modeling in a nonsteady state. In the postprandial state, especially

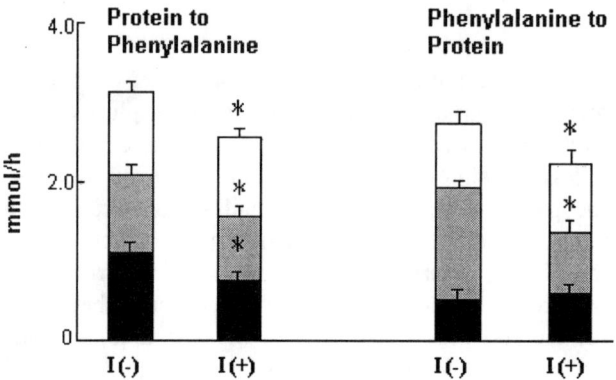

FIGURE 6-6 Phenylalanine kinetics in insulin-dependent diabetic subjects in the insulin deprived [I(−)] and insulin treated [I(+)] state. Splanchnic region ( ) accounted for a similar proportion of approximately 30 percent of whole body breakdown on both occasions, whereas the contribution of skeletal muscle (■) decreased from 37 to 30 percent with insulin treatment. The skeletal muscle component of whole body protein synthesis increased from 19 to 27 percent during insulin treatment, concomitant with a decrease of splanchnic protein synthesis from 50 to 34 percent. The entire decrease in whole body protein synthesis during insulin treatment is due to changes in the splanchnic bed. This study clearly demonstrated that insulin replacement in the diabetic patients achieves protein conservation mainly by decreasing muscle protein breakdown. SOURCE: Data derived from Nair et al. (1995).

after a mixed meal, amino acid concentrations increase, whereas insulin administration in the postabsorptive state decreases amino acid concentration. It is therefore crucial to determine whether insulin's effect on muscle protein synthesis is qualitatively different in the postabsorptive and postprandial states. Thus, the precise mechanism of insulin's anticatabolic effect, in particular on skeletal muscle, remains to be clearly defined.

Another pertinent question is whether insulin differentially affects the individual muscle proteins. Novel techniques that allow assessment of fractional synthesis and breakdown rates of specific proteins, such as mitochondrial proteins, myosin heavy chain, and myosin isoforms, have been, or are being, developed and will soon serve to clarify these issues.

Growth Hormone and IGF-I

Growth hormone (GH) has widespread metabolic actions, including a catabolic lipolytic effect in adipose tissue (Moller et al., 1995). In most other tissues, GH acts anabolically, in part by increasing concentrations of IGF-I in the circulation and locally. It is well established that patients with GH deficiency have a decreased muscle mass and impaired muscle function (Jorgensen et al., 1989) and that both abnormalities tend to normalize during treatment. When high doses of GH are given for a week to normal volunteers, whole body protein synthesis increases (Horber and Haymond, 1990), but this acute increase seems to occur mostly in tissues other than skeletal muscle (Copeland and Nair, 1994). Studies assessing the specific effects of GH on muscle protein synthesis and breakdown have yielded conflicting results (Butterfield et al., 1997; Copeland and Nair, 1994; Fryburg et al., 1991; Welle et al., 1996; Yarasheski et al., 1995), although infusion of GH directly into an artery has been reported to enhance muscle protein synthesis (Fryburg et al., 1991). Recently, one study was able to detect a 50-percent increase in muscle protein synthesis after 1 week of GH treatment in elderly women (Butterfield et al., 1997). Other studies, however, failed to show any effect of GH on muscle protein synthesis in elderly subjects in comparison with placebo (Welle et al., 1996) or in young or elderly subjects undergoing a resistance training program (Yarasheski et al., 1995). It is intriguing, however, that muscle mass and strength increased on GH administration (Welle et al., 1996) in these subjects without any effect on myofibrillar protein synthesis. It is possible that GH-induced accretion of protein occurs during the postprandial state since all muscle protein synthesis measurements were performed in the postabsorptive state. It is also important to include measurements of synthesis of specific proteins, which may clarify the discrepancies among many previous studies.

When considering the effects of GH on muscle metabolism, it is important to remember that GH exposure invariably leads to increments in the levels of insulin, IGF-I, and in general, free fatty acids and that all of these compounds have an independent protein anabolic impact.

IGF-I has potent anabolic properties. Dissection of the metabolic effects of IGF-I is entangled by the facts that (1) in the circulation only a minute percentage of IGF-I is free, the remainder being bound to numerous distinct binding proteins with specific kinetic characteristics, and (2) many of the actions of IGF-I may be executed in an autocrine or paracrine manner in body compartments not readily accessible for investigations (LeRoith, 1997).

Studies in which IGF-I has been perfused locally indicate that IGF-I preferentially stimulates protein synthesis in muscle (Fryburg, 1994), whereas systemically administered IGF-I only stimulates muscle protein synthesis when additional amino acids are supplied (Fryburg et al., 1995). The roles of circulating versus local IGF-I and of free versus bound fractions, as well as the importance of each individual binding protein, remain to be determined.

Sex Steroids

Little is known about the possible effects of estrogen and progesterone on muscle metabolism, and it is uncertain whether the age-related decline in muscle mass in postmenopausal women is associated with the loss of ovarian function.

Testosterone has anabolic actions on muscle. Many years ago, testosterone replacement in adult hypogonadal men was shown to enhance nitrogen retention (Kenyon et al., 1940). A recent study concluded that supraphysiological doses of testosterone increased muscle mass and strength in normal men, in particular when combined with a muscle training program (Bhasin et al., 1996). It has also been shown that 6 months of testosterone replacement in five hypogonadal men led to a 20 percent increase in muscle mass and a 50 to 60 percent increase in mixed muscle protein synthesis (Brodsky et al., 1996). There was a nonsignificant increase in the synthesis rate of myosin heavy-chain, a component of the contractile apparatus (Brodsky et al., 1996) (Figure 6-7). Administration of testosterone has also been reported to enhance muscle protein synthesis in elderly men (Urban et al., 1995). Although it has been clearly

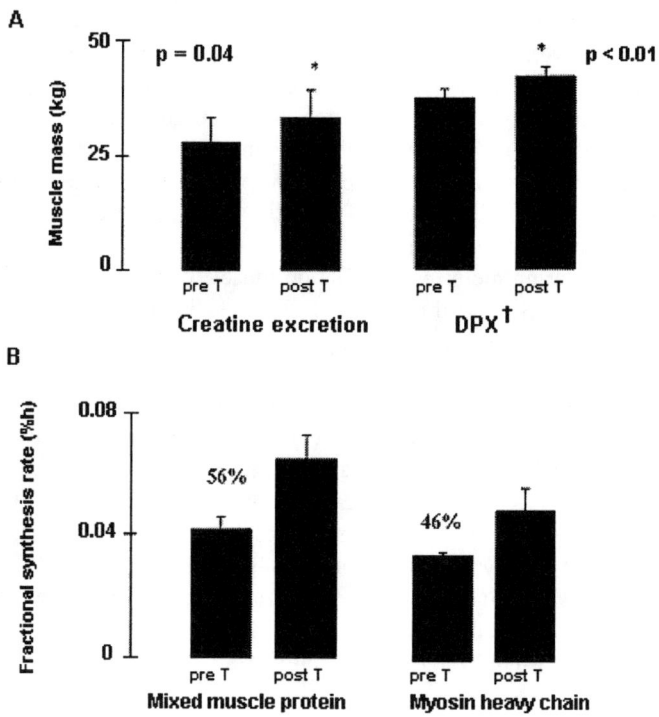

FIGURE 6-7 Six months of testosterone replacement in hypogonadal men increased muscle mass ($P < 0.05$) and decreased fat mass; an associated increase in fractional synthesis of mixed muscle protein ($P < 0.05$) and myosin heavy chain ($P < 0.09$) is likely to explain the increase in muscle mass. † Dual Photon X-ray. SOURCE: Data derived from Brodsky et al. (1996).

established that levels of biologically available testosterone fall with aging, the effects of testosterone on age-related sarcopenia and other problems associated with aging remain to be clearly defined. It is, however, also unclear whether testosterone replacement in elderly men may increase the incidence of prostatic cancer and/or other problems such as lipid disorders.

Catabolic Hormones
(Glucagon, Cortisol, Epinephrine, and Thyroid Hormones)

Glucagon, a catabolic hormone, increases the net loss of protein from the body in the postprandial state due both to an acceleration of amino acid disposal and to inhibition of protein synthesis (Charlton et al., 1996). There is no evidence that glucagon has any direct effect on muscle; rather, it seems likely that the primary effect is a stimulation of hepatic ureagenesis and consumption of glucogenic amino acids for glucose production leading to decreased circulating levels of amino acids. Glucagon has been shown to increase leucine oxidation during insulin deficiency (Nair et al., 1987), but this effect seems to be modest, in particular when insulin is replaced (Hartl et al., 1990).

In patients with Cushing's syndrome due to glucocorticoid excess, muscle wasting is a symptom. On the whole body level, it has been observed that 1 week of prednisone treatment increased protein breakdown and oxidation (Horber and Haymond, 1990). Whether this overall effect is secondary to specific events in muscle or caused by alterations in, for instance, hepatic amino acid metabolism is unclear at present.

Although catecholamines are generally regarded as being catabolic hormones, there is no solid evidence to support the concept that catecholamines (epi- and norepinephrine) have any negative impact on protein balances. As catabolic hormones, they do, however, cause a hypermetabolic state, which in turn may increase overall breakdown of protein.

The clinical observation that both hypo- and hyperthyroid patients have muscle weakness suggests that thyroid hormones may have a diphasic effect on muscle mass and function with an interposed "euthyroid" optimum. Arteriovenous studies demonstrated that hyperthyroidism is associated with a net increase of muscle protein breakdown, although no changes were observed in hypothyroid patients (Morrison et al., 1988). Direct measurement of muscle protein synthesis is required to determine the effect of thyroxine on muscle protein synthesis.

In addition to the hormones mentioned above, it is possible that cytokines and, probably, prostaglandins may have important actions on muscle protein metabolism. It should be noted that many of these agents have diverse effects when given systemically. For instance, administration of tumor necrosis factor-α to humans leads to a full-blown pyrogenic response with fever, malaise, and release of all major stress hormones together with hyperglycemia and signs of augmented lipolysis and ketogenesis (Starnes et al., 1988).

SUBSTRATES AND NUTRITION

Prevailing levels of substrates in the circulation or, more precisely, in the microenvironment have important implications for muscle function and muscle mass. An adequate fuel supply is crucial to maintain tissue protein turnover and to preserve appropriate tissue balances. There is mounting evidence that the abundance of fuel substrates, *per se*, including amino acids is a significant regulator of protein metabolism in muscle. After a meal, concentrations of glucose, fatty acids, and amino acids rise together with pancreatic secretion of insulin. Under these conditions, protein accretion in muscle is increased due to both enhanced synthesis and inhibited protein breakdown (Tessari et al., 1996). A number of studies have shown that elevation of plasma amino acid concentrations stimulates protein synthesis and inhibits breakdown (Nair et al., 1992; Svanberg et al., 1996; Tessari et al., 1996). Glucose and fatty acids have protein-sparing effects (Tessari et al., 1986), but to what extent these effects are independent of insulin is currently unresolved. In addition, ketone bodies (specifically 3-hydroxy-butyrate, a by-product of fatty acid metabolism) have been shown to stimulate the synthesis rate of muscle protein (Nair et al., 1988).

It thus appears that under the free-living conditions of everyday life, hormones and substrates act in mutual support to restrict protein loss and preserve muscle mass and function.

AUTHORS' CONCLUSIONS

The maintenance of functioning muscle mass is a complex process that involves orchestration of the effects of anabolic and catabolic hormones, nutritional state, and supply of substrates to the site of protein synthesis together with physical activity. As illustrated in Figure 6-8, hormones play a key role, although the precise level of their action and the interaction of hormones with other factors such as substrate supply remains to be clearly defined. It is hoped that newly developed techniques that allow assessment of the contribution of each of these factors to the fractional synthesis of specific muscle proteins will substantially improve our understanding of this field. Regulation of breakdown of specific muscle proteins is beyond current investigative capabilities because of the paucity of technology to quantify breakdown of specific muscle proteins. It is possible, however, to measure the regulation of the enzyme systems responsible for protein breakdown. Further studies that integrate the effects of hormones and substrates on muscle protein turnover with the effects of genetic factors are necessary to fully understand the regulation of muscle mass and functions.

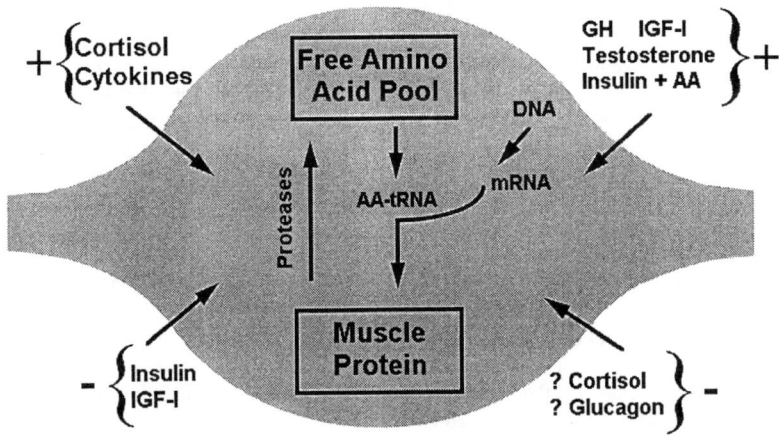

FIGURE 6-8 The quality and quantity of protein in a muscle cell are regulated by an intricate interplay between hormonal, nutritional, genetic, nervous, and physical factors, some of which are outlined here. The effects of any one of these factors on the specific processes of synthesis and breakdown of individual muscle proteins remain to be clearly defined.

REFERENCES

Balagopal, P., O.E. Rooyackers, D.B. Adey, P.A. Ades, and K.S. Nair. 1997. Effects of aging on in vivo synthesis of skeletal muscle myosin heavy-chain and sarcoplasmic protein in humans. Am. J. Physiol. 273:E790-E800

Bhasin, S., T. Storer, N. Berman, C. Callegari, B. Clevenger, J. Phillips, T.J. Bunell, R. Tricker, A. Shirari, and R. Casaburi. 1996. The effects of supraphysiologic doses of testosterone on muscle size and strength in normal men. N. Engl. J. Med. 335:1–7.

Biolo, G., R.Y.D. Fleming, and R.R. Wolfe. 1995a. Physiologic hyperinsulinemia stimulates protein synthesis and enhances transport of selected amino acids in human skeletal muscle. J. Clin. Invest. 95:811–819.

Brodsky, I.G., P. Balagopal, and K.S. Nair. 1996. Effects of testosterone replacement on muscle mass and muscle protein synthesis in hypogonadal men. J. Clin. Endocrinol. Metab. 81:3469–3475.

Butterfield, G.E., J. Thompson, M.J. Rennie, R. Marcus, R.L. Hintz and A.R. Hoffman. 1997. Effect of rhGH and IGF-I treatment on protein utilization in elderly women. Am. J. Physiol. 272:E94–E99.

Charlton, M.R., D.B. Adey, and K.S. Nair. 1996. Evidence for a catabolic role of glucagon during an amino acid load. J. Clin. Invest. 98:90–99.

Copeland, K.C., and K.S. Nair. 1994. Acute growth hormone effects on amino acid and lipid. J. Clin. Endocrinol. Metab. 78:1040–1047.

Fryburg, D.A. 1994. Insulin-like growth factor I exerts growth hormone and insulin-like actions on human muscle protein metabolism. Am. J. Physiol. 267:E331–E336.

Fryburg, D.A., R.A. Gelfand, and E.J. Barrett. 1991. Growth hormone acutely stimulates forearm protein synthesis in normal subjects. Am. J. Physiol. 260:E499–E504 metabolism.

Fryburg, D.A., L.A. Jahn, S.A. Hill, D.M. Oliveras, and E.J. Barrett. 1995. Insulin and insulin-like growth factor I enhance human skeletal muscle protein anabolism during hyperaminoacidemia by different mechanisms. J. Clin. Invest. 96:1722–1729.

Gelfand, R.A., and E.J. Barrett. 1987. Effect of physiologic hyperinsulinemia on skeletal muscle protein synthesis and breakdown in man. J. Clin. Invest. 80:1–6.

Hartl, W.H., H. Miyoshi, F. Jahoor, S. Klein, D. Elahi, and R.R. Wolfe. 1990. Bradykinin attenuates glucagon-induced leucine oxidation in humans. Am. J. Physiol. 259:E239–E245.

Horber, F.F., and M.W. Haymond. 1990. Human growth hormone prevents the protein catabolic side effects of prednisone in humans. J. Clin. Invest. 86:265–272.

Jorgensen, J.O.L., L. Thuesen, T. Ingemann-Hansen, S.A. Pedersen, J. Jorgensen, N.E. Skakkebaek, and J.S. Christiansen. 1989. Beneficial effects of growth hormone treatment in GH deficient adults. Lancet 1:1221–1225.

Kenyon, A.T., K. Knowlton, I. Sandiford, F.C. Koch, and G. Lotwin. 1940. A comparative study of the metabolic effects of testosterone proprionate in normal men and women and in eunuchoidism. Endocrinology 26:26–45.

Kimball, S.R., and L.S. Jefferson. 1988. Cellular mechanisms involved in the action of insulin on protein synthesis. Diabetes Metab. Rev. 4:773–787.

LeRoith, D. 1997. Insulin-like growth factors. N. Engl. J. Med. 336:633–640.

Moller, N., J.O.L. Jorgensen, J. Moller, P. Ovesen, O. Schmitz, J.S. Christiansen, and H. Orskov. 1995. Metabolic effects of growth hormone in humans. Metabolism 44(suppl.):33–36.

Morrison, W.L., J.N.A. Gibson, R.T. Jung, and M.J. Rennie. 1988. Skeletal muscle and whole body protein turnover in thyroid disease. Eur. J. Clin. Invest. 18:62–68.

Nair, K.S., G.C. Ford, K. Ekberg, E. Fernquist-Forbes, and J. Wahren. 1995. Protein dynamics in whole body and in splanchnic and leg tissues in type 1 diabetic patients. J. Clin. Invest. 95:2926–2937.

Nair, K.S., D. Halliday, D.E. Matthews, and S.L. Welle. 1987. Hyperglucagonemia during insulin deficiency accelerates protein catabolism. Am. J. Physiol. 253:E208–E213.

Nair, K.S., R.G. Schwartz, and S. Welle. 1992. Leucine as a regulator of whole body and skeletal muscle protein metabolism in humans. Am. J. Physiol. 263:E928–E934.

Nair, K.S., S. Welle, D. Halliday, and R.G. Campell. 1988. Effect of 3-hydroxybutyrate on whole-body leucine kinetics and fractional mixed skeletal muscle protein synthesis in humans. J. Clin. Invest. 82:198–205.

Pozefsky, T., P. Felig, J.D. Tobin, J.S. Soeldner, and G.F. Cahill. 1969. Amino acid balance across tissue of forearm in postabsorptive man. Effect of insulin at two dose levels. J. Clin. Invest. 48:2273–2282.

Reed, R.L., L. Pearlmutter, K. Yochum, E.E. Meredith, and A.D. Mooradian. 1991. The relationship between muscle mass and muscle strength in the elderly. J. Am. Geriatr. Soc. 39:555–561.

Rooyackers, O.E., D.B. Adey, P.A. Ades, and K.S. Nair. 1996. Effect of age on in vivo rates of mitochondrial protein synthesis in human skeletal muscle. Proc. Natl. Acad. Sci. 93:15364–15369.

Starnes, H.F., R.S. Warren, M. Jeevanandam, J.L. Gabrilove, W. Larchian, H.F. Oettgen, and M.F. Brennan. 1988. Tumor necrosis factor and the acute metabolic response to tissue injury in man. J. Clin. Invest. 82:1321–1325.

Svanberg, E., A.C. Moller-Loswick, D.E. Mathews, U. Korner, M. Andersson, and K. Lundholm. 1996. Effects of amino acids on synthesis and degradation of skeletal muscle proteins in humans. Am. J. Physiol. 271:E718–E724.

Tessari, P., S.L. Nissen, J. Miles, and M.W. Haymond. 1986. Inverse relationship of leucine flux and oxidation to free fatty acid availability in vivo. J. Clin. Invest. 77:575–581.

Tessari, P., M. Zanetti, R. Barazonni, M. Vettore, and F. Michielan. 1996. Mechanisms of postprandial protein accretion in human skeletal muscle. Insight from leucine and phenylalanine forearm kinetics. J. Clin. Invest. 98:1361–1372.

Urban, R.J., Y.H. Bodenburg, C. Gilkison, J. Foxworth, R.R. Wolfe, and A. Ferrado. 1995. Testosterone administration to healthy elderly men increases muscle strength and protein synthesis. Am. J. Physiol. 296:E820–E826.

Welle, S., C. Thornton, R. Jozefowicz, and M. Statt. 1993. Myofibrillar protein synthesis in young and old men. Am. J. Physiol. 264:E693–E698.

Welle, S., C. Thornton, M. Statt, and B. McHenry. 1996. Growth hormone increases muscle mass and strength but does not rejuvenate myofibrillar protein synthesis in healthy subjects over 60 years old. J. Clin. Endocrinol. Metab. 81:3239–3243.

Yarasheski, K.E., J.J. Zachwieja, and D.M. Bier. 1993. Acute effects of resistance exercise on muscle protein synthesis rate in young and elderly men and women. Am. J. Physiol. 265:E210–E214.

Yarasheski, K.E., J.J. Zachwieja, J.A. Campell, and D.M. Bier. 1995. Effect of growth hormone and resistance training on muscle growth and strength in older men. Am. J. Physiol. 268:E268–E276.

Zurlo, F., K. Larson, C. Bogardus, and E. Ravussin. 1990. Skeletal muscle is a major determinant of resting energy expenditure. J. Clin. Invest. 86:1423–1427.

DISCUSSION

ROBERT NESHEIM: Any questions?

JEFF ZACHWIEJA: What is the dose of testosterone that you used in the hypogonadal man?

K. SREEKUMARAN NAIR: Just a replacement dose. We administered, I believe, 3mg/kg of testosterone every two weeks.

JEFF ZACHWIEJA: You got a similar increase in muscle protein synthesis in men who are not hypogonadal?

K. SREEKUMARAN NAIR: No. Only in hypogonadal men. We did not administer testosterone to normal subjects. In the study that the UCLA group did (Bhasin et al., 1996), they administered testosterone suprophysiological doses and found substantial increase in muscle mass in men who are not hypogonadal. We administered replacement doses. We did not look at the dose effect or what is the effect of testosterone in healthy men on muscle protein synthesis. The UCLA group gave suprophysiological doses and measured the effect on muscle mass and strength.

JEFF ZACHWIEJA: Was it 600 milligrams?

K. SREEKUMARAN NAIR: That is right.

JEFF ZACHWIEJA: Was there any functional improvement in the muscle of those men?

K. SREEKUMARAN NAIR: In this group we did not measure their muscle strength. It was an outpatient study where the patients were administered testosterone for that study. The UCLA group showed increase of muscle strength on supraphysiological testosterone administration.

GAIL BUTTERFIELD: In terms of the military's interest, do you find the kinds of changes that you observed in mitochondrial protein and myosin heavy chain, with a decrease in energy intake or a decrease in protein intake?

K. SREEKUMARAN NAIR: We do not have any data on that. It is only the first study we have done in the aging population. But certainly, from the military point of view, I am sure that you may have the endurance capacity and also the muscle strength decline with aging. It is an open question what impact the changes in nutrition has on this age related changes.

ROBERT NESHEIM: Last question.

PATRICK DUNNE: With the military situation, it is important to consider the interactions of exercise and dietary carbohydrate intake and their impact on hormones. It may be desirable to design diets that are higher or lower in carbohydrate or protein for their effects on muscle function, but it is necessary to consider whether people will eat those diets. That is a factor that must be included in this discussion.

ROBERT NESHEIM. Thank you very much, Dr. Nair. I guess that one ought to stay young if one is going to be in the military. I am kidding.

7

Effects of Protein Intake on Renal Function and on the Development of Renal Disease

Mackenzie Walser[1]

INTRODUCTION

One of the major functions of the kidney is the elimination of the products of protein metabolism. It is not surprising, therefore, that protein intake exerts many diverse effects on the kidney. Most attention has been directed toward the role of protein intake in chronic renal failure. For at least a century, it has been known that reducing protein intake will ameliorate the symptoms of chronic renal failure, provided the reduction is not severe enough to induce protein deficiency. In recent years, much effort has been directed toward defining the optimum protein intake for patients with chronic renal failure, with the aim of slowing the rate of progression toward the end stage. Despite many reports, it is still not firmly established that protein restriction slows disease progression, although the bulk of available evidence supports this conclusion (for example, Fouque et al., 1992; Levey et al., 1996). In this review, however, attention will be limited to the effects of protein intake on the initiation of renal disease.

---

[1] Mackenzie Walser, The Johns Hopkins School of Medicine, Baltimore, MD 21205.

There are three ways in which protein intake may play a role in the development of renal disease: promotion of nephrolithiasis, enhancement of the morbidity of acute renal failure, and acceleration of glomerulosclerosis.

NEPHROLITHIASIS

During recent decades, nephrolithiasis has been increasing in frequency in developed countries in association with improved nutrition (Goldfarb, 1988; Robertson et al., 1979a). In Japan, for example, the incidence of renal stone disease has increased threefold since the period preceding World War II (Iguchi et al., 1990). Estimates of the prevalence of nephrolithiasis in the general population range from 1 percent to 15 percent (Johnson et al., 1979). Approximately 12 percent of the U.S. population will have a kidney stone at some time (Johnson et al., 1979; Sierakowski et al., 1978). One contributory factor is rising intake of protein-rich foods, which may promote urolithiasis (uric acid stone formation) in several ways (Robertson et al., 1979b).

Calcium Excretion and Calcium-Containing Stones

Calcium Intake

Calcific nephrolithiasis occurs when the activity product^2 of free calcium times free oxalate or phosphate exceeds a certain level, which is in turn modified by crystallization inhibitory factors. Hence, a reduction in urinary calcium excretion might be expected to reduce the occurrence of calcific nephrolithiasis. Indeed, restriction of calcium intake has been a mainstay of dietary advice for the prevention of recurrent nephrolithiasis. However, as pointed out by Goldfarb (1988), this approach is inappropriate and even potentially dangerous for several reasons: first, negative calcium balance may ensue because gastrointestinal calcium losses continue, with resultant osteoporosis during long-term application (Lalau et al., 1992); second, retrospective data indicate that the incidence of stone recurrence is *negatively* correlated with spontaneous calcium intake, probably because ingested calcium precipitates oxalate in the gut (Curhan et al., 1993, Lemann et al., 1996). Whatever the explanation, it is clear that attempts to reduce renal clearance of calcium will be more fruitful than reducing calcium intake. In this respect the high calcium intake typically associated with a high protein intake may be *protective* against calcific nephrolithiasis.

---

[2] In physical chemistry, an ideal concentration for which the law of mass action will apply perfectly. The ratio of the activity to the true concentration is the activity coefficient.

Determinants of Calcium Clearance

As first reported by this laboratory (Walser, 1961a), calcium clearance in the dog is approximately equal to sodium clearance during a variety of experimental maneuvers (Figure 7-1), including the administration of diuretics (with the exception of thiazides) (Walser and Trounce, 1961). Many workers have confirmed these relationships in humans. Thus, a high sodium intake augments and a low sodium intake diminishes calcium clearance. A second

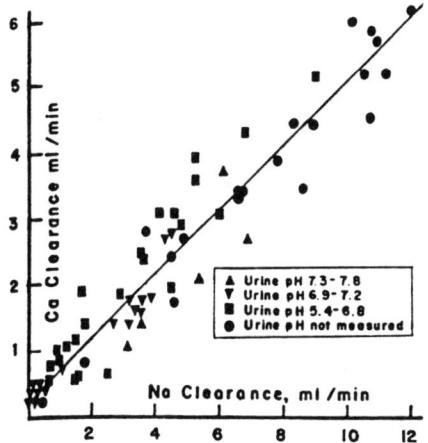

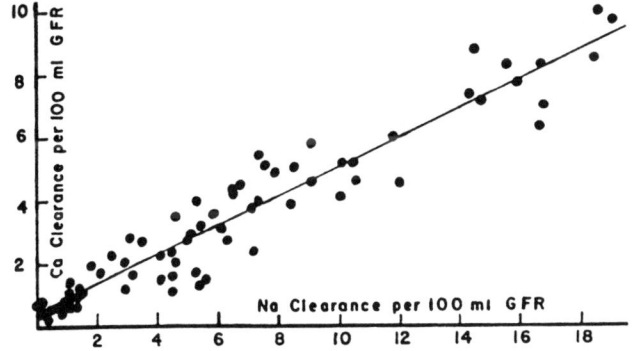

FIGURE 7-1 Calcium clearance in the dog as a function of sodium clearance during saline and osmotic diuresis, expressed as ml/min (above) and as ml/100 ml of glomerular filtration rate (below). Alkaline, neutral and acidic urines, indicated by different symbols, yield the same results. The correlations coefficients, r, are 0.95 (above) and 0.93 (below). A small but statistically significant ($p < 0.01$) intercept is present. SOURCE: Reprinted by permission from Walser (1961a).

determinant of calcium clearance is sulfate excretion. Walser and Browder (1959) reported that infusion of sodium sulfate in dogs caused a pronounced increase in calcium clearance (Figure 7-2)—greater than the attendant increase in sodium clearance. This study showed that ion-pair formation between calcium and sulfate accounted for this effect, since the clearance of free calcium ions remained approximately equal to the clearance of sodium (Figure 7-3) (Walser, 1961b). This was the first demonstration of the biological role of ion pairs, which are electrostatic complexes between oppositely charged ions.

Subsequent work has confirmed that sodium excretion and sulfate excretion are the most important determinants of calcium excretion. For example, Tschope and Ritz (1985) measured 24-hour urinary excretion rates of calcium, sodium, and sulfate (as well as other constituents) in normal individuals and found that the ratio of calcium to creatinine was closely correlated with the ratio of sulfate to creatinine ($r = 0.61, p < 0.001$), and also correlated with the ratio of sodium to creatinine ($r = 0.48, p < 0.01$). When multiple regression analysis was

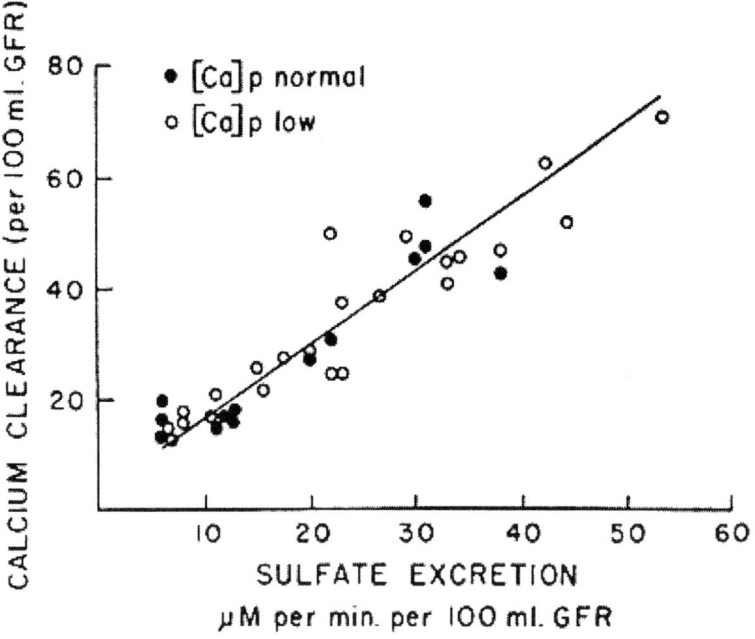

FIGURE 7-2 Calcium clearance per 100 ml of glomerular filtration rate as a function of sulfate excretion is dogs infused with sodium sulfate. SOURCE: Reproduced from The Journal of Clinical Investigation, 1959, 38:1404–1411.

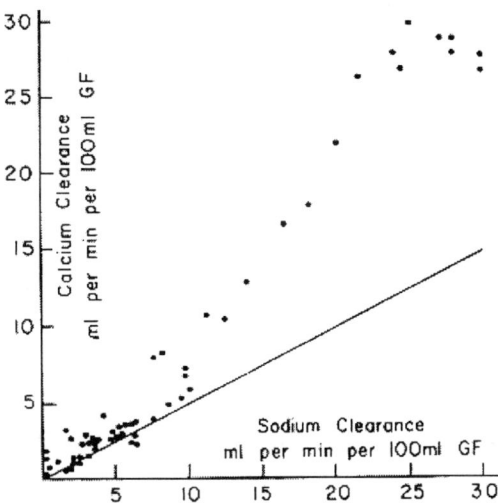

FIGURE 7-3 Calcium clearance as a function of sodium clearance in dogs infused with sodium sulfate. The diagonal line represents the relationship seen during saline or mannitol diuresis, as depicted in Figure 7-1. When calcium clearance is calculated as free calcium ion clearance, the relationship becomes the same as the diagonal line; the excess represents the calcium sulfate ion-pair complex. SOURCE: Reprinted by permission from Walser (1961b).

performed, calcium:creatinine remained highly correlated with sulfate:creatinine ($r = 0.76$, $p < 0.001$). Similar findings were reported by Puche et al. (1987). Tschope and Ritz (1985) also administered methionine (40 mmol/d for 5 days) to nine normal individuals and noted that calcium excretion increased significantly in conjunction with the increase in sulfate excretion and fall in urine pH (Table 7-1). Forty mmol sulfur is the amount present as cystine and methionine in 1 pound of sirloin.

Urinary acidification also increases calcium clearance (Lemann et al., 1967), independently of sodium excretion. Thus, part of the effect of sulfate loads on calcium excretion is attributable to the attendant decrease in urine pH. Lutz (1984) found that two thirds of the increment in calcium excretion that resulted from an increase in protein intake in normal subjects was eliminated if sufficient sodium bicarbonate was concurrently administered to prevent any fall in urine pH.

Potassium supplementation may prevent the calciuretic action of a protein load (Kaneko et al., 1990). Lemann et al. (1991) have demonstrated that potassium administration reduces and potassium deprivation increases urinary calcium, possibly because of associated changes in sodium excretion or in calcitriol synthesis.

TABLE 7-1 Effect of Methionine Ingestion (40 mmol/day) in Normal Subjects; Average Urinary Exretion During Five Control Days and Five Experimental Days

Urinary Constituent	Before	During	p
Calcium, mg/day	189 ± 94	276 ± 128	< 0.01
Magnesium, mg/day	122 ± 26	124 ± 24	n.s.
Sulfate, mg/day	2,382 ± 778	6,177 ± 893	< 0.01
Urea, g/day	23 ± 4	23 ± 3	n.s.
Citrate, mg/day	880 ± 231	327 ± 168	< 0.06
PH	5.57 – 6.70	5.02 – 6.37	<0.01

SOURCE: Reprinted by permission from Tschope and Ritz, 1985.

Contrary to the assertion of Trilok and Draper (1989), sulfur amino acids do not account entirely for the increase in net acid production associated with protein oxidation: phosphorylated amino acids and dibasic amino acids also contribute to endogenous acid production (Walser, 1986). This may be why other workers have found that sulfur amino acid content of dietary protein cannot account entirely for its calciuretic effect (Schuette et al., 1981; Zemel et al., 1981).

Patterns of Urinary Excretion in Stone Formers

In stone formers, urinary excretion of calcium and oxalate is often higher than in normal subjects, even though their diets may not differ in weekly intake of protein or calcium (Iguchi et al., 1990; Martini et al., 1993; Rao et al., 1982). Others have found, on the contrary, that stone formers consume more animal protein than controls (Iguchi et al., 1989; Robertson et al., 1979b; Trinchieri et al., 1991). Higher salt intake also may be a factor (Jaeger et al., 1988). The calciuric effect of dietary protein may be exaggerated in stone formers (Goldfarb, 1988; Jungers et al., 1993). Hypercalciuric patients tend to eat more protein at dinner and on weekends (Iguchi et al., 1989; Martini et al., 1993). Oxalate excretion after ingesting oxalate and the increase in calcium oxalate supersaturation of the urine are greater than in control subjects, provided that both groups are ingesting a high protein diet (Urivetzky et al., 1987).

Animal protein causes more calciuria than vegetable protein but, unlike the latter, does not increase urinary oxalate (Marangella et al., 1989). As is well known, animal protein contains a greater proportion of sulfur amino acids and therefore leads to a greater increment in urinary sulfate and net acid excretion.

Thus, it appears that sulfate excretion (and its effect on acid excretion) and sodium excretion, which are in turn functions of sulfur-amino acid intake and salt intake, respectively, are the principal determinants of calcium excretion by the kidney. Intake of calcium itself is of lesser importance. Phosphorus intake is also a factor: if phosphate intake increases at the same time as protein intake, less calciuria results (Hegsted et al., 1981; Linkswiler et al., 1981). Several other studies (Puche and Feldman, 1992; Puche et al., 1987; Schneider and Menden, 1988; Singh et al., 1993) have reached the same conclusions, although in one older study (Block et al., 1980), sulfur amino acid ingestion had no effect on calcium excretion in normal subjects.

It would seem to follow that dietary treatment of calcific nephrolithiasis should include restriction of the intake of salt and protein, particularly animal protein.

Uric Acid Stones

Acidification of the urine, which results from sulfur intake, increases the likelihood of uric acid stone formation. Furthermore, careful study of the ion products in the urine by Breslau et al. (1988), who compared subjects receiving a vegetable protein diet with subjects consuming an animal protein diet, found that the latter increased the tendency of the urine to form uric acid stones, but not calcium oxalate or calcium phosphate stones. Fellström et al. (1983, 1984) also reported that a high animal-protein diet increased urinary supersaturation with respect to uric acid and ammonium urate but not calcium oxalate. Thus although high animal protein diets may promote osteoporosis when consumed for decades, they are more likely to lead to uric acid stones than calcific nephrolithiasis. Nevertheless, uricosuria is associated with calcium oxalate urolithiasis (Coe, 1978), and uricosuria is increased by high protein intake (Robertson et al., 1979c).

Prophylactic Treatment of Nephrolithiasis

Allopurinol administration reduces the incidence of calcific nephrolithiasis in subjects who have hyperuricosuria but are not hypercalciuric (Ettinger et al., 1985). Protein restriction augments citrate excretion in hypercalciuric stone formers and reduces stone incidence but may not affect uric acid excretion (Goldfarb, 1988). Sodium restriction reduces urinary calcium in hypercalciuric subjects (Muldowney et al., 1982). As Churchill (1987) pointed out a decade ago, there are few scientifically adequate trials of any form of treatment in preventing recurrence of urinary stones. This remains true today.

ACUTE RENAL FAILURE

Andrews and Bates (1986) reported that the response of rat kidney to a 45-minute period of ischemia was dramatically affected by prior protein intake. For example, 93 percent of rats fed a high-protein diet for the preceding 2 weeks died of renal ischemia within 3 days, compared with 12 percent of rats fed a low-protein diet. One hundred percent of rats fed a zero protein diet survived, and most of them exhibited normal serum creatinine levels by the fourth day after ischemia.

Instituting protein restriction after a period of ischemia provided no protection, nor did switching from a zero-protein diet to a high-protein diet immediately after the ischemic insult cause any increase in mortality. Approximately 1 week of prior protein restriction was required to produce this remarkable effect.

The authors suggested that patients who are to be subjected to surgical procedures with a high risk of acute renal failure might benefit from protein restriction in the pre-operative period.

It is not clear how these findings might be applicable to soldiers, since their risk of acute renal failure, even in combat, is low. It is clear, however, that at least with regard to the risk of morbidity from acute renal failure, dietary protein restriction is more likely to confer benefit than to increase risk.

PROTEIN INTAKE AND THE DEVELOPMENT OF GLOMERULOSCLEROSIS

Glomerular filtration rate (GFR) declines with age for people over 40 in approximately two-thirds of persons without frank kidney disease (Lindeman et al., 1985), especially if they are hypertensive (Lindeman et al., 1984); in the remaining one third, GFR remains constant or increases with age (Lindeman, 1990). According to Fliser et al. (1997), about two-thirds of elderly subjects who are not in heart failure and not on diuretics (even if hypertensive) have GFRs within the range of younger subjects (Figure 7-4). Kidneys examined at autopsy of persons dying of causes other than kidney disease show a progressive, though extremely variable, increase in the fraction of glomeruli that are sclerotic (Figure 7-5) (Kaplan et al., 1975). Circulatory impairment may be another factor, since there may be localized areas of reduced flow within the kidneys of older subjects (Friedman et al., 1972).

In rats, the development of chronic renal failure with age is nearly universal (Coleman et al., 1977; Hayashida et al., 1986). The predominant lesion is glomerulosclerosis. Because this process was apparently attenuated by protein restriction (see review by Masoro and Yu, 1989), Anderson and Brenner (1987) suggested that the high protein intake of Western societies plays a central role in

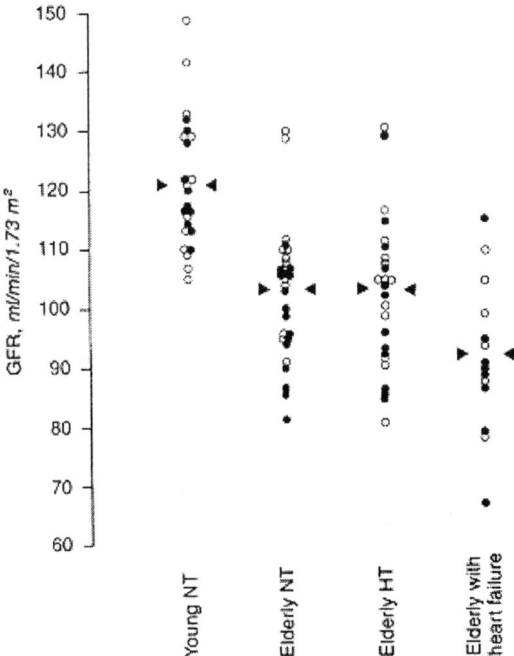

FIGURE 7-4 Glomerular filtration rate, measured as inulin clearance, in four groups of subjects. Young NT: healthy subjects 26 ± 3 years of age with normal blood pressure. Elderly NT: healthy subjects 68 ± 7 years of age with normal blood pressure. Elderly HT: subjects 70 ± 6 years of age with hypertension but not receiving diuretics. Elderly with heart failure: subjects 69 ± 6 years of age with heart failure. Symbols: open circles, men; solid circles, women. SOURCE: Used with permission from Kidney International, 1997, 51:1196–1204.

the decline of renal function with age and that restriction of dietary protein might prevent this decline.

This recommendation cannot be supported for a number of reasons as pointed out in a recent review (Walser, 1992). First, caloric restriction is more effective in rats than is protein restriction in retarding the age-associated decline in renal function (Tapp et al., 1989). Furthermore, caloric restriction without protein restriction markedly retarded the progression of glomerulosclerosis. Rats prefer rations containing higher proportions of protein, and the earlier studies indicating that protein restriction retarded renal damage failed to monitor food intake.

Second, protein restriction tends to lower GFR rather than increase it. Lew and Bosch (1991) recorded the dependence of creatinine clearance on spontaneous protein intake in subjects aged 22 to 50 years and in subjects aged

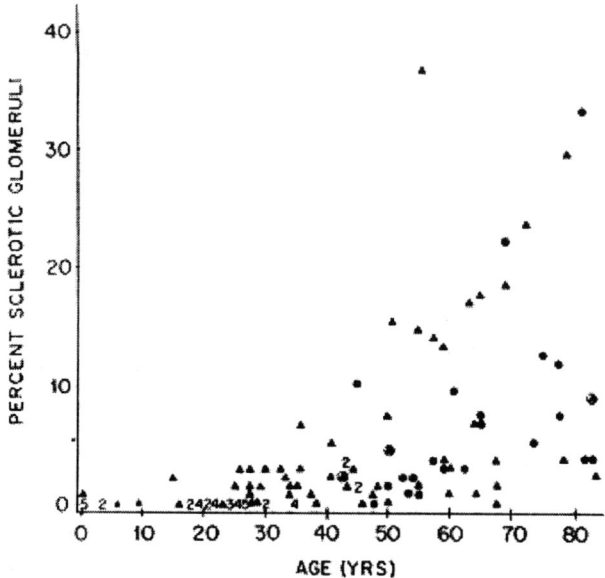

FIGURE 7-5 Percent sclerotic glomeruli in kidneys from 122 autopsied patients without known renal disease. SOURCE: Reprinted by permission from Kaplan et al. (1975).

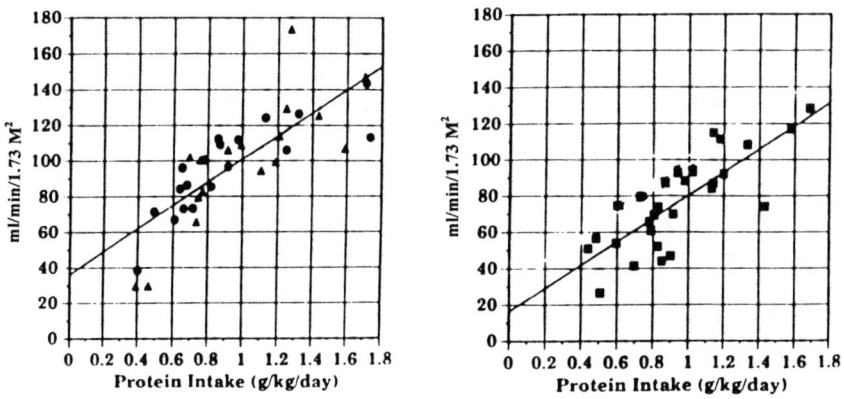

FIGURE 7-6 Dependence of creatinine clearance, in ml/min/1.73 M^2, on protein intake in healthy subjects aged 22–50 years (left) and 55–88 years (right). The slopes are not different, but the intercepts (43 ml/min and 31 ml/min, respectively) differ significantly. SOURCE: S.Q. Lew and J.P. Bosch, 1991, Effect of diet on creatinine clearance and excretion in young and elderly healthy subjects and in patients with renal disease, J. Am. Soc. Nephrol., 2:856–865.

55 to 88 years (Figure 7-6). In both groups, there was a similar pronounced dependence of clearance on protein intake; in the older subjects, a lower intercept was seen, reflecting the effect of age on GFR. If protein intake is not controlled, the apparent decline in GFR with age will be exaggerated, because older subjects consume less protein than younger subjects (Kerr et al., 1982).

Third, Tobin and Spector (1986) measured creatinine clearance in 198 normal men on two occasions, 10 to 18 years apart, and correlated the decline in clearance during this interval with protein intake: no relationship was detected. They found no evidence that a high intake of protein caused a progressive reduction in renal function.

Fourth, high-protein feeding (60%) for 2 years in rats had no effect on the percentage of sclerotic glomeruli (Collins et al., 1990).

Fifth, the progressive decline in renal function seen in rats after partial nephrectomy (Brenner et al., 1982) may be unique to this species. In dogs with 75 percent nephrectomy, GFR does not decline progressively with time for the ensuing 4 years, whether protein intake is high or low (Bovée, 1991). In baboons followed for 5 years after subtotal nephrectomy and on either 8 percent protein or 25 percent protein diets, renal failure did not occur; GFR, measured as inulin clearance, increased sharply in baboons fed 25 percent protein, and this difference tended to disappear with time (curiously, creatinine clearances did not decrease with time in either group). Even in baboons fed 8 percent protein, a slow decline in GFR with time was seen (5% per year). Proteinuria did not differ between the two groups and did not progress (Bourgoignie et al., 1994). In human kidney donors, mild proteinuria (but not albuminuria) is often seen, and the incidence of hypertension may be increased, but progressive renal failure rarely if ever occurs, and there is no correlation between protein intake and proteinuria (Anderson et al., 1985; Hakim et al., 1984). Such individuals are not generally advised to restrict their protein intake (Rocher and Swartz, 1987), although some clinicians have recommended that these individuals do so until the question is settled (Bay and Hebert, 1987). Likewise, protein restriction is not advised for persons with a solitary normal kidney (Mitch, 1989).

From these observations, it is clear that protein restriction does not prevent the decline in renal function with age and, in fact, is the major cause of that decline. A better way to prevent the decline would be to increase protein intake. Indeed, protein malnutrition in the elderly is a far more prevalent problem (Morgan et al., 1986; Rudman et al., 1989) than is nitrogen retention as a consequence of an age-related decline in renal function.

AUTHOR'S CONCLUSIONS AND RECOMMENDATIONS

Concerns about an adverse effect of high protein intake on renal function, and in particular on its decline with age, appear to be ill advised. Putting aside the potential adverse effects of protein intake on nephrolithiasis or on morbidity of acute renal failure, detailed earlier, there is no reason to restrict protein intake in healthy individuals in order to protect the kidney. Nevertheless, there is also no evidence that a more restricted protein intake has any harmful long-term effects on the kidney.

REFERENCES

Anderson, C.F., J.A. Velosa, P.P. Frohnert, V.E. Torres, K.P. Offord, J.P. Vogel, J.V. Donadio, Jr., and D.M. Wilson. 1985. The risks of unilateral nephrectomy: Status of kidney donors 10 to 20 years postoperatively. Mayo Clin. Proc. 60:367–374.

Anderson, S., and B.M. Brenner. 1987. The aging kidney: structure, function, mechanisms, and therapeutic implications. J. Am. Geriatr. Soc. 35:590–593.

Andrews, P.M., and S.B. Bates. 1986. Dietary protein prior to renal ischemia dramatically affects postischemic kidney damage. Kidney Int. 30:299–303.

Bay, W.H., and L.A. Hebert. 1987. Kidney donors and protein intake [letter to the editor]. Ann. Intern. Med. 107:427.

Block, G.D., R.J. Wood, and L.H. Allen. 1980. A comparison of the effects of feeding sulfur amino acids and protein on urine calcium in man. Am. J. Clin. Nutr. 33:2128–2136.

Bourgoignie, J.J., G. Gavellas, S.G. Sabnis, and T.T. Antonovych. 1994. Effect of protein diets on the renal function of baboons (*Papio hamadryas*) with remnant kidneys: A 5-year follow-up. Am. J. Kidney Dis. 23:199–204.

Bovée, K.C. 1991. Influence of dietary protein on renal function in dogs. J. Nutr. 121:S128–S139.

Brenner, B.M., T.W. Meyer, and T.W. Hostetter. 1982. Dietary protein intake and the progressive nature of kidney disease: the role of hemodynamically moderated glomerular injury in the pathogenesis of progressive glomerular sclerosis in aging, renal ablation and intrinsic renal disease. New Engl. J. Med. 307:652–659.

Breslau, N.A., L. Brinkley, K.D. Hill, and C.Y.C. Pak. 1988. Relationship of animal protein-rich diet to kidney stone formation and calcium metabolism. J. Clin. Endocrinol. Metab. 66:140–146.

Churchill, D.N. 1987. Medical treatment to prevent recurrent calcium urolithiasis. Miner. Electrolyte Metab. 13:294–304.

Coe, F.L. 1978. Calcium-uric acid nephrolithiasis. Arch. Intern. Med. 138:1090–1093.

Coleman, G.L., W. Barthold, G.W. Osbaldiston, S.J. Foster, and A.M. Jonas. 1977. Pathological changes during aging in barrier-reared Fischer 344 male rats. J. Gerontol. 32:258–278.

Collins, D.M., C.T. Rezzo, J.B. Kopp, P. Ruiz, T.M. Coffman, and P.E. Klotman. 1990. Chronic high protein feeding does not produce glomerulosclerosis or renal insufficiency in the normal rat. J. Am. Soc. Nephrol. 1:624.

Curhan, G.C., W.C. Willett, E.B. Rimm, and M.J. Stampfer. 1993. A prospective study of dietary calcium and other nutrients and the risk of symptomatic kidney stones. New Engl. J. Med. 328:833–838.

Ettinger, B., J.T. Citron, A. Tang, and B. Livermore. 1985. Prophylaxis of calcium oxalate stones; clinical trials of allopurinol, magnesium hydroxide, and chlorthalidone. Pp. 549–552 in Urolithiasis and Related Clinical Research, P.O. Schwille, L.H. Smith, W.G. Robertson, and W. Vahlensieck, eds. New York: Plenum.

Fellström, B., B.G. Danielson, B. Karlström, H. Lithell, S. Ljunghall, and B. Vessby. 1983. Dietary animal protein and urinary supersaturation in renal stone formers. Proc. Eur. Dial. Transplant. Assoc. 20:411–416.

Fellström, B., B.G. Danielson, B. Karlström, H. Lithell, S.L. Ljunghall, B. Vessby, and L. Wide. 1984. Effects of high intake of dietary animal protein on mineral metabolism and urinary supersaturation of calcium oxalate in renal stone formers. Brit. J. Urol. 56:263–269.

Fliser, D., E. Franek, M. Joest, S. Block, E. Mutschler, and E. Ritz. 1997. Renal function in the elderly: Impact of hypertension and cardiac function. Kidney Int. 51:1196–1204.

Fouque, D., M. Laville, J.P. Boissel, R. Chifflet, M. Labeeuw, and P.Y. Zech. 1992. Controlled low protein diets and chronic renal insufficiency: Meta-analysis. Br. Med. J. 304:216–220.

Friedman, S.A., A.E. Raizner, H. Rosen, N.A. Solomon, and W. Sy. 1972. Functional defects in the aging kidney. Ann. Intern. Med. 76:41–45.

Goldfarb, S. 1988. Dietary factors in the pathogenesis and prophylaxis of calcium nephrolithiasis. Kidney Int. 34:544–555.

Hakim, R.M., R.C. Goldszer, and B.M. Brenner. 1984. Hypertension and proteinuria: Long-term sequelae of uninephrectomy in humans. Kidney Int. 25:930–936.

Hayashida, M., B.P. Yu, E.J. Masoro, K. Iwasaki, and T. Ikeda. 1986. An electron microscopic examination of age-related changes in the rat kidney: The influence of diet. Exp. Gerontol. 21:535–553.

Hegsted, M., S.A. Schuette, M.B. Zemel, and H.M. Linkswiler. 1981. Urinary calcium and calcium balance in young men as affected by level of protein and phosphorus intake. J. Nutr. 111:553–562.

Iguchi, M., T. Umekawa, Y. Ishikawa, Y. Katayama, M. Kodama, M. Takada, Y. Katoh, K. Kataoka, K. Kohri, and T. Kurita. 1989. [Dietary habits of Japanese stone formers and clinical effects of prophylactic dietary treatment]. Hinyokika Kiyo 35:2115–2128.

Iguchi, M., T. Umekawa, Y. Ishikawa, Y. Katayama, M. Kodama, M. Takada, Y. Katoh, K. Kataoka, K. Kohri, and T. Kurita. 1990. Dietary intake and habits of Japanese renal stone patients. J. Urol. 143:1093–1095.

Jaeger, P., L. Portmann, J.M. Ginalski, and P. Burckhardt. 1988. [So-called "renal" idiopathic hypercalciuria most often has a dietary origin]. Schweiz. Med. Wochenschr. 118:15–17.

Johnson, C.M., D.M. Wilson, W.M. O'Fallon, R.S. Malek, and L.T. Kurland. 1979. Renal stone epidemiology: A 25-year study in Rochester, Minnesota. Kidney Int. 16:624–631.

Jungers, P., M. Daudon, C. Hennequin, and B. Lacour. 1993. [Correlations between protein and sodium intake and calciuria in calcium lithiasis]. Nephrologie 14:287–290.

Kaneko, K, U. Masaki, M. Aikyo, K. Yabuki, A. Haga, C. Matoba, H. Sasaki, and G. Koike. 1990. Urinary calcium and calcium balance in young women affected by high protein diet of soy protein isolate and adding sulfur-containing amino acids and/or potassium. J. Nutr. Sci. Vitaminol. (Tokyo) 36:105–116.

Kaplan, C., S. Pasternack, H. Shah, and G. Gallo. 1975. Age-related incidence of sclerotic glomeruli in human kidneys. Am. J. Pathol. 80:227–234.

Kerr, G.R., E. S. Lee, M.-K.M. Lam, R.J. Lorimor, E. Randall, R.N. Forthofer, M.A. Davis, and S.M. Magnetti. 1982. Relationships between dietary and biochemical measures of nutritional status in HANES I data. Am. J. Clin. Nutr. 35:294–308.

Lalau, J.D., J.M. Achard, P. Bataille, C. Bergot, I. Jans, B. Boudailliez, J. Petit, G. Henon, P.F. Westeel, N. El Esper, M.A. Laval-Jeantet, R. Bouillon, J.L. Sebert, and A. Fournier. 1992. [Vertebral density in hypercalciuric lithiasis: relationship with calcium and protein consumption and vitamin D metabolism]. Ann. Med. Interne 143:293–298.

Lemann, J. Jr., J.R. Litzow, and E.J. Lennon. 1967. Studies of the mechanism by which chronic metabolic acidosis augments urinary calcium excretion in man. J. Clin. Invest. 46:1318–1328.

Lemann, J., Jr., J.A. Pleuss, R.W. Gray, and R.G. Hoffmann. 1991. Potassium administration reduces and potassium deprivation increases urinary calcium excretion in healthy adults. Kidney Int. 39:973–983.

Lemann, J., Jr., J.A. Pleuss, E.M. Worcester, L. Hornick, D. Schrab, and R.G. Hoffmann. 1996. Urinary oxalate excretion increases with body size and decreases with increasing dietary calcium intake among healthy adults. Kidney Int. 49:200–208.

Levey, A.S., S. Adler, A.W. Caggiula, B.K. England, T. Greene, L.G. Hunsicker, J.W. Kusek, N.L. Rogers, and P.E. Teschan for the Modification of Diet in Renal Disease Study Group. 1996. Effects of dietary protein restriction on the progression of advanced renal disease in the Modification of Diet in Renal Disease Study. Am. J. Kidney Dis. 27:652–663.

Lew, S.Q., and J.P. Bosch. 1991. Effect of diet on creatinine clearance and excretion in young and elderly healthy subjects and in patients with renal disease. J. Am. Soc. Nephrol. 2:856–865.

Lindeman, R.D. 1990. Overview: Renal physiology and pathophysiology of aging. Am J. Kidney Dis. 16:276–282.

Lindeman, R.D., J. Tobin, and N.W. Shock. 1984. Association between blood pressure and the decline in renal function with age. Kidney Int. 26:861–268.

Lindeman, R.D., J. Tobin, and N.W. Shock. 1985. Longitudinal studies on the rate of decline in renal function with age. J. Am. Geriatr. Soc. 33:278–285.

Linkswiler, H.M., M.B. Zemel, M. Hegsted, and S. Schuette. 1981. Protein-induced hypercalciuria. Fed. Proc. 40:2429–2433.

Lutz, J. 1984. Calcium balance and acid-base status of women as affected by increased protein intake and by sodium bicarbonate ingestion. Am. J. Clin. Nutr. 39:281–288.

Marangella, M., O. Bianco, C.V. Martini, M. Petrarulo, C. Vitale, and F. Linari. 1989. Effect of animal and vegetable protein intake on oxalate excretion in idiopathic calcium stone formers. Br. J. Urol. 63:348–351.

Martini, L.A., I.P. Heilberg, L. Cuppari,. F.A. Medeiros, S.A. Draibe, H. Ajzen, and N. Schor. 1993. Dietary habits of calcium stone formers. Braz. J. Med. Biol. Res. 26:805–812.

Masoro, E.J., and B.P. Yu. 1989. Diet and nephropathy [editorial]. Lab. Invest. 60:165–167.

Mitch, W.E. 1989. Dietary restrictions for a single normal kidney. Pediatr. Nephrol. 3:129.

Morgan, D.B., H.M. Newton, C.J. Schorah, M.A. Jewitt, M.R. Hancock, and R.P. Hullin. 1986. Abnormal indices of nutrition in the elderly: A study of different clinical groups. Age Ageing 15:65–76.

Muldowney, F.P., R. Freaney, and M.F. Moloney. 1982. Importance of dietary sodium in the hypercalciuria syndrome. Kidney Int. 22:292–296.

Puche, R.C., and S. Feldman. 1992. Relative importance of urinary sulfate and net acid excretion as determinants of calciuria in normal subjects. Medicina (B Aires) 52:220–224.

Puche, R.C., A.F. Carlomagno, A. Gonzalez, and A. Sanchez. 1987. A correlation and path coefficient analysis of components of calciuria in normal subjects and idiopathic stone formers. Bone Miner. 2:405–411.

Rao, P.N., V. Prendiville, A. Buxton, D.G. Moss, and N.J. Blacklock. 1982. Dietary management of urinary risk factors in renal stone formers. Br. J. Urol. 54:578–583.

Robertson, W.G., M. Peacock, and A. Hodgkinson. 1979a. Dietary changes and the incidence of urinary calculi in the U.K. between 1958 and 1976. J. Chronic Dis. 32:469–476.

Robertson, W.G., P.J. Heyburn, M. Peacock, F.A. Hanes, and R. Swaminathan. 1979b. The effect of high animal protein intake on the risk of calcium-stone-formation in the urinary tract. Clin. Sci. 57:285–288.

Robertson, W.G., M. Peacock, P.J. Heyburn, F.A. Hanes, A. Rutherford, E. Clementson, R. Swaminthan, and P.B. Clark. 1979c. Should recurrent calcium oxalate stone formers become vegetarians? Brit J. Urol. 51:427–431.

Rocher, L.L., and R.D. Swartz. 1987. Kidney donors and protein intake [letter to the editor]. Ann. Intern. Med. 107:427.

Rudman, D., D.E. Mattson, A.G. Feller, R. Cotter, and R.C. Johnson. 1989. Fasting plasma amino acids in elderly men. Am. J. Clin. Nutr. 49:559–566.

Schneider, W., and E. Menden. 1988. [The effect of long-term increased protein administration on mineral metabolism and kidney function in the rat. I. Renal and enteral metabolism of calcium, magnesium, phosphorus, sulfate and acid]. Z. Ernährungwiss. 27:170–185.

Schuette, S.A., M. Hegsted, M.B. Zemel, and H.M. Linkswiler. 1981. Renal acid, urinary cyclic AMP, and hydroxyproline excretion as affected by the level of protein, sulfur amino acid, and phosphorus intake. J. Nutr. 111:2106–2116.

Sierakowski, R., B. Finlayson, R.R. Landes, C.D. Finlayson, and N. Sierakowski. 1978. The frequency of urolithiasis in hospital discharge diagnoses in the United States. Invest. Urol. 15:438–441.

Singh, P.P., F. Hussain, R.C. Gupta, A.K. Pendse, R. Kiran, and R. Ghosh. 1993. Effect of dietary methionine and inorganic sulfate with and without calcium supplementation on urinary calcium excretion of guinea pigs (*Cavia porcellus*). Indian J. Exp. Biol. 31:96–97.

Tapp, D.C., W.G. Wortham, J.F. Addison, D.N. Hammonds, J.L. Barnes, and M.A. Venkatachalam. 1989. Food restriction retards body growth and prevents end-stage renal pathology in remnant kidney of rats regardless of protein intake. Lab. Invest. 60:184–195.

Tobin, J, and D. Spector. 1986. Dietary protein has no effect on future creatinine clearance. Gerontologist 26(SI):59A.

Trilok, G., and H.H. Draper. 1989. Sources of protein-induced endogenous acid production and excretion by human adults. Calcif. Tissue Int. 44:335–338.

Trinchieri, A., A. Mandressi, P. Luongo, and E. Pisani. 1991. The influence of diet on urinary risk factors for stones in healthy subjects and idiopathic renal calcium stone formers. Brit. J. Urol. 67:230–236.

Tschope, W., and E. Ritz. 1985. Sulfur-containing amino acids are the major determinant of urinary calcium. Mineral Electrolyte Metab. 11:137–139.

Urivetzky, M., J. Motola, S. Braverman, and A.D. Smith. 1987. Dietary protein levels affect the excretion of oxalate and calcium in patients with absorptive hypercalciuria type II. J. Urol. 137:690–692.

Walser, M. 1961a. Calcium clearance as a function of sodium clearance in the dog. Am. J. Physiol. 200:1099–1104.

Walser, M. 1961b. Ion association. VII. Dependence of calciuresis on natriuresis during sulfate infusion. Am. J. Physiol. 201:769–773.

Walser, M. 1986. The roles of urea production, ammonium excretion, and amino acid oxidation in acid-base balance. Am. J. Physiol. 250:F181–F188.

Walser, M. 1992. Dietary proteins and their relationship to kidney disease. Pp. 168–178 in Dietary Proteins in Health and Disease, G.U. Liepa, ed. Champaign: American Oil Chemists' Society.

Walser, M., and A.A. Browder. 1959. Ion association. III. The effect of sulfate infusion on calcium excretion. J. Clin. Invest. 38:1404–1411.

Walser, M., and J.R. Trounce. 1961. The effect of diuresis and diuretics upon the renal tubular transport of alkaline earth cations. Biochem. Pharmacol. 8:157.

Zemel, M.B., S.A. Schuette, M. Hegsted, and H.M. Linkswiler. 1981. Role of sulfur-containing amino acids in protein-induced hypercalciuria in man. J. Nutr. 111:545–552.

DISCUSSION

JOAN CONWAY: Throughout your presentation this morning, you mentioned high and low protein intake. Can you quantitate those intakes so we have some ability to refer that to what we have been told military members consume?

MACKENZIE WALSER: Each study I reported to you would have a different answer to that question. I am not sure how you want me to answer that question because, as I say, no two studies are alike.

JOAN CONWAY: Well, if you say high and you say low, what would those be?

MACKENZIE WALSER: I think one gram per kilogram is pretty high. Some people would say that is not, but I think it is pretty high, and 0.6 is pretty low.

JOAN CONWAY: Okay. Thank you.

HARRIS LIEBERMAN: One gram per kilogram is much lower than what most people consume.

MACKENZIE WALSER: No. That is a myth. I know we are going to get into an argument about that. You are right about young medical students, but you are not right about older folks and women.

HARRIS LIEBERMAN: I am also right about soldiers.

MACKENZIE WALSER: Oh, that may well be. I have to re-orient my thinking in that regard.

STEVEN HEYMSFIELD: There were studies of protein restriction in people with chronic renal failure I believe. Is that right?

MACKENZIE WALSER: Yes.

STEVEN HEYMSFIELD: What ever happened with those studies?

MACKENZIE WALSER: Well, that work is going on and on. It looks like it is never going to end. I think that the bottom line at the moment is that protein restriction probably does slow progression. It certainly reduces symptoms, there is no question. That has been known for a hundred years. I cannot believe there are still people—telling their patients to go ahead and eat what they wish. That is a mistake. There is no question about that.

ROBERT NESHEIM: Dr. Millward.

D. JOE MILLWARD: Is there any information from the epidemiology of disease patterns in those groups of people who traditionally have had very high-protein diets, such as the Inuit Indians?

MACKENZIE WALSER: Is there any information about the incidence of renal disease?

D. JOE MILLWARD: Yes, they have low incidence of some chronic cardiovascular disease.

MACKENZIE WALSER: They do?

D. JOE MILLWARD: Yes, and, hence, all of the omega-three fatty acids are elevated. Traditionally, they have what is probably the highest protein intake of any group on the planet.

MACKENZIE WALSER: Yes. I have heard that. In fact, I learned somewhere that the reason that the Eskimos eat blubber is because if they eat just whale meat they get meat intoxication. They get up to about 400 grams a day of protein. That is why they eat some blubber to attenuate the protein intake. Have you read that?

D. JOE MILLWARD: Who knows why they do that?

MACKENZIE WALSER: You do not think they do so to prevent ammonia intoxication?

D. JOE MILLWARD: There are substantial long-term studies of their health.

MACKENZIE WALSER: I do not know the answer to your question. That is the straight answer.

ROBERT NESHEIM: Johanna?

JOHANNA DWYER: This was wonderful. With respect to the Brenner Hypothesis (Brenner et al, 1982) and later work that suggested maybe there was an energy deficit in the subjects on a low protein diet, how strong is the evidence?

MACKENZIE WALSER: Yes. Well, the progressive renal damage, according to Tap et al. (1989), is attributable to calories and not protein. That is right.

JOHANNA DWYER: Yes. If it is attributable to calories and not protein, but the reality of many patients with renal disease is that their energy intakes are also extremely low, do you think the Brenner hypothesis still has relevance to the human situation?

MACKENZIE WALSER: I think that it certainly must have relevance. I should add that those observations stimulated a great deal of research in this field without which the research never would have been done. Exactly where it fits in I am not sure.

For example, people born with one kidney or a small kidney, people with renal hypogenesis, do occasionally develop renal failure. So I am sure that there are instances in which renal failure results from partial nephrectomy or absence thereof and that probably could be ameliorated by protein restriction.

ROBERT NESHEIM: Thank you. We need to move on. It is time for a break here.

8

Infection and Injury: Effects on Whole Body Protein Metabolism

Douglas W. Wilmore[1]

Injury and infection elicit a rather stereotypic metabolic response characterized by hypermetabolism, accelerated gluconeogenesis, increased fat oxidation, and negative nitrogen balance (Wilmore, 1997). These responses are particularly evident in young, previously healthy, well-nourished individuals, such as members of the military, who have a well-developed lean body mass and no associated diseases or dysfunctional organs. Because body protein represents an indispensable structural and functional element of the body, the mechanisms that regulate the accelerated proteolysis during these catabolic states have received much attention. This is because significant erosion of lean body mass results in immunosuppression, poor wound healing, decreased strength and activity, and prolonged convalescence. Methods of attenuating the catabolic response or enhancing protein anabolism are now being evaluated in

---

[1] Douglas W. Wilmore, Department of Surgery, Harvard Medical School, Brigham and Women's Hospital, Boston, MA 02115.

patients in an effort to shorten convalescent recovery and decrease the length of hospital stay (Wilmore, 1991).

This chapter outlines some of the changes that occur in whole body protein metabolism following infection and injury. Evidence will be presented that these catabolic states result in marked translocation of protein from the carcass to visceral tissues, resulting in a net loss of skeletal muscle mass. Such loss results in decreased strength and activity.

BACKGROUND AND GENERAL RESPONSE CHARACTERISTICS

Over 200 years ago, John Hunter, the British surgeon and biologist, recognized that accidental injury initiated a series of responses in the host that presumably aided tissue repair and general recovery (Hunter, 1794). In the late 1800s, the concept of nitrogen balance was established by Voit and his German colleagues working in this area of physiological biochemistry (Munro, 1964). They noted that nitrogen balance could be related to the body's protein economy and was affected positively by the increased ingestion of both energy and protein and affected negatively by sepsis-induced fever. These concepts were confirmed and extended in the early 1900s, when Coleman and DuBois (1915) studied both energy and protein balance in a group of patients with typhoid fever. Using both direct calorimetric techniques and whole body balance methodology, they studied patients during the acute and convalescent phases of their illness. The investigators described the increased net loss of nitrogen that was associated with the febrile episodes of typhoid fever; the negative nitrogen balance could not be offset by the ingestion of large quantities of energy at the moderate level of protein intake that was utilized.

In 1932, Cuthbertson (1932) described a group of patients who had either undergone orthopedic surgical procedures on their lower extremities or who had sustained long bone fractures. He noted that these injuries to the long bone were associated with an enhanced loss of nitrogen from the body. This negative nitrogen balance was maximal during the first week postinjury and persisted for up to 1 month. Cuthbertson divided the metabolic response to injury into two phases: an early, acute "ebb" or shock phase and a latter "flow" or hyperdynamic phase. During the ebb phase, there was decreased metabolic activity, which was followed within 12 to 48 hours by the flow phase, a hypermetabolic state in which metabolic rate, temperature, and urinary nitrogen excretion all increased. Others have since contributed to the field of postinjury metabolism, with classic contributions being made by John Howard, Oliver Cope, Francis Moore, John Kinney, and many others (Wilmore, 1997).

During the early 1900s, researchers realized that infection was related to increased loss of protein from the body, but it was not until the 1960s that the systematic study of the metabolic response to infection was undertaken and published. William Beisel and colleagues (1967), working with the Army to understand the impact of infectious disease on the host, studied normal individuals who were infected under experimental conditions with tularemia, sandfly fever, or Q fever. The individuals received a fixed diet of known composition, and after metabolic baseline was established, the infectious agent was administered. A well-defined sequence of events occurred over time: phagocyte function increased; serum amino acids, zinc, and iron concentrations fell; glucocorticoids and growth hormone concentrations increased in the blood stream, and acute-phase protein levels were also altered. The patients subsequently became febrile, retained salt and water, but excreted more nitrogen in the form of urea. The negative nitrogen balance was related to the decreased dietary intake of energy and protein and also to the fever-related increases in metabolism. In a separate experiment, exogenous glucocorticoids were administered to noninfected controls. Although this mimicked in part the adrenal cortical response to infection, this perturbation did not simulate the negative nitrogen and mineral balance observed in the febrile volunteers. Likewise, when noninfected normals consumed a caloric and protein intake similar to that ingested by the subjects who had experimental tularemia, the negative nitrogen balance was slight (18 g cumulative), demonstrating only about one-third of the losses observed in the infected patients (52 g cumulative). Because of the extent of the protein loss, it often took weeks for the infected individuals to replete their lean body mass.

From these and related studies, some general conclusions can be made concerning the whole body protein catabolic response following injury or infection:

- The increased nitrogen loss occurs via the urine. Urea represents the major component of the nitrogen lost and contributes 85 to 90 percent to the negative nitrogen balance.
- The response pattern generally shows dose-response characteristics; that is, the greater the infection or injury, the more extensive the nitrogen loss (Table 8-1).
- The response follows a time course, with nitrogen excretion increasing in the first several days, peaking for several days or weeks, and then gradually returning to equilibrium as the inflammation resolves, food intake increases, and spontaneous physical activity occurs (Figure 8-1).
- More nitrogen is lost from a well-nourished individual than from a depleted patient following a comparable catabolic insult. Thus, a relationship exists between the size of the lean body mass and the extent of protein catabolism, and this may account for variation in nitrogen balance that occurs with age and gender. Both older individuals and women are thought to excrete less nitrogen following a standard stimulus than do their appropriate controls.

TABLE 8-1 Estimates of Nitrogen Loss Following Catabolic Illness (First 10 Days, *Ad Lib* Feedings)

Precipitating Factor	Cumulative Nitrogen Loss (g)
Injury	
Major burn	170
Multiple injury	150
Peritonitis	136
Simple fracture	115
Major operation	50
Minor operation	24
Infection	
Typhoid fever (untreated)	116
Pneumonia (untreated)	59
Tularemia (treated)	52
Q fever (treated)	40
Sandfly fever (untreated)	16

SOURCE: Adapted in part from Wannemacher, 1975.

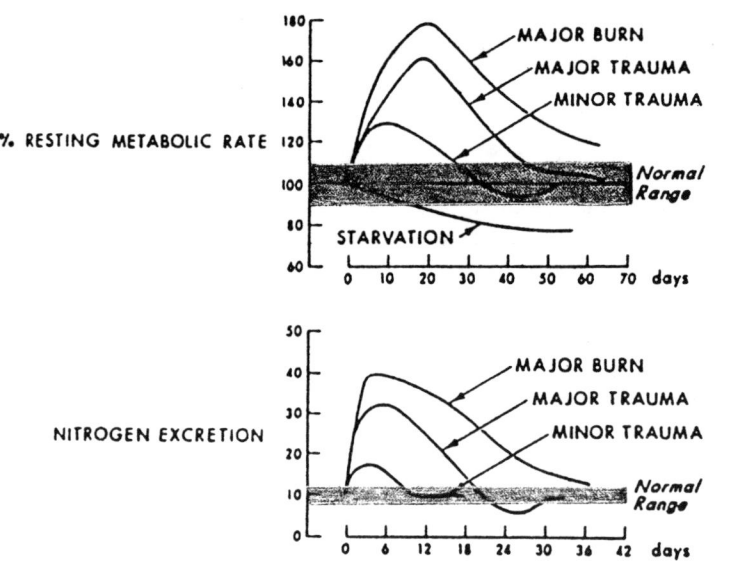

FIGURE 8-1 Hypermetabolism and nitrogen excretion are closely related, show a dose-response relationship, and resolve with time. Patients received 12 g of dietary nitrogen per day. SOURCE: Adapted from Kinney, 1966.

TABLE 8-2 Alterations in Rates of Protein Synthesis and Catabolism That May Affect Hospitalized Patients

Subject Characteristics	Synthesis	Catabolism
Normal-starvation	↓	0
Normal-fed, bedrest	↓	0
Elective surgical procedure	↓	0
Injury/sepsis-IV dextrose	↑↑	↑↑↑
Injury/sepsis-fed	↑↑↑	↑↑↑

NOTE: ↓, decrease; ↑, increase; 0, no change.
SOURCE: Adapted from Wilmore, 1977.

Negative nitrogen balance represents the net loss of nitrogen from the body. The nitrogen lost reflects a balance between protein synthesis and protein breakdown, which are simultaneously occurring in all tissues of the body. A variety of stable isotopic studies have attempted to measure these absolute rates. A summary of these studies is shown in Table 8-2. Injury and infection result in the increased whole body flux of amino acids throughout the body. Catabolic rates outstrip rates of protein synthesis in the partially fed state; feeding enhances protein synthesis and attenuates the negative nitrogen balance that occurs. Despite feeding, protein flux rates remain high in these catabolic states, and these rates do not return to normal until the inflammation is resolved.

THE TRANSLOCATION OF NITROGEN FROM CARCASS TO VISCERA

Although the protein-catabolic response may appear to be a generalized effect of net protein breakdown throughout the body, this is not the case. Skeletal muscle is known to become catabolic; arterial-venous measurements of amino acid concentration across noninjured extremities of injured patients have documented the accelerated release of nitrogen that occurs from skeletal muscle during the flow phase (Aulick and Wilmore, 1979). However, catheterization of vessels supplying and draining visceral organs in similar patients has demonstrated enhanced uptake of amino acids (Wilmore et al., 1980), and these rates of uptake generally match the rates of peripheral amino acid release. That translocation of nitrogen occurs from the carcass to visceral organs is also supported by the following phenomena:

- In addition to the nitrogen lost in the urine, there is increased excretion of potassium and phosphorous. All of these elements are lost in proportion to their concentrations found in skeletal muscle.

- Following a catabolic event, creatinuria occurs. In addition, there is increased excretion of 3-methyl histidine. Both of these substances are found predominantly in skeletal muscle.
- A common, clinical observation is that wasting of skeletal muscle occurs during catabolic states, and this is accompanied by muscle weakness.

A characteristic pattern of amino acids released from catabolic muscle has been described: alanine and glutamine constitute 50 to 70 percent of the amino acid nitrogen released (Muhlbacher et al., 1984). This represents accelerated *de novo* synthesis of these amino acids; hydrolysis and analysis of skeletal muscle protein reveals that both alanine and glutamine contribute less than 10 percent to the overall amino acid residues. Translocation of these amino acids to visceral tissues presumably occurs for a purpose; both amino acids serve as important glucose precursors and support the enhanced rate of gluconeogenesis. In the liver, nitrogen residue from this reaction is converted to other nonessential nitrogen compounds or converted to urea and excreted in the urine. Glutamine is also extracted by the kidneys, where it contributes ammonia to combine with hydrogen ions to form ammonium (NH_4^+), which is excreted in the urine. This is an important component of acid-base homeostasis, for the loss of every H+ in the urine is accompanied by the secretion of a bicarbonate ion into the bloodstream. Glutamine is also taken up by enterocytes, lymphocytes, and macrophages, where it serves as a primary fuel source for these tissues, supporting proliferation and signaling or supporting other important responses.

Thus, skeletal muscle serves as an important source of substrate during catabolic states to support vital structure and functions of the visceral tissue necessary for host survival. However, the increased loss of skeletal muscle protein occurs with a cost—the price paid is loss of muscle mass and the disturbance of normal muscle function. These alterations in composition of organs in catabolic states have been observed in both animal models (Artuson, 1961) and human patients following injury or infection (Bararc-Nieto et al., 1978) (Figure 8-2).

REGULATORS OF THE TRANSLOCATION OF PROTEIN

Various factors have been associated with the increased skeletal muscle proteolysis and translocation of nitrogen from the carcass to visceral organs. *Anorexia* commonly accompanies infection, and the diminished food intake causes hormonal changes, such as a fall in insulin and insulin-like growth factor-1, which decrease skeletal muscle protein synthesis. Infection and injury are also associated with a variety of constitutional symptoms including pain, fever, myalgia, headache, and fatigue. As activity decreases, or bed rest is imposed, the dynamics of net skeletal muscle protein synthesis are altered, enhancing net skeletal muscle proteolysis.

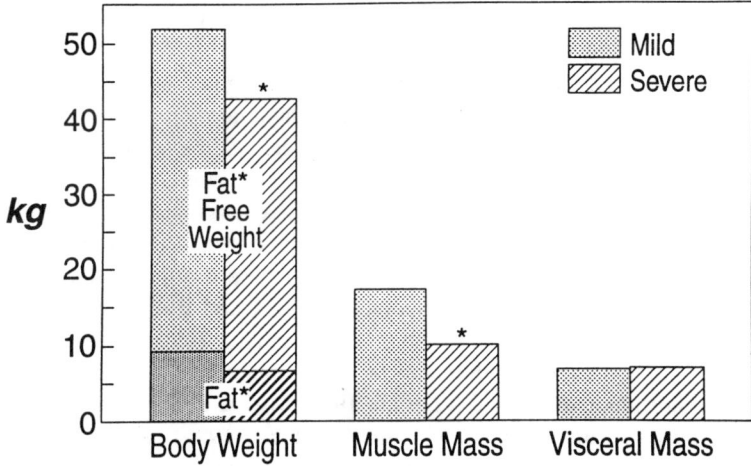

FIGURE 8-2 Body composition in patients with mild and severe chronic infection. Note the loss of fat-free tissue and muscle with severe infection, with maintenance of visceral tissues. *, $P < 0.05$. SOURCE: Adapted from M. Barac-Nieto et al., 1978.

Other factors that control intermediate metabolism are known to be associated with the altered protein metabolism that occurs during injury and infection. A characteristic *hormonal response* pattern has been observed during catabolism. Early in the time course, insulin levels are low, after which the blood levels generally rise to normal or supranormal levels. However, the counterregulatory hormones glucagon, glucocorticoids, and catecholemines are all elevated, and they generally remain so throughout the period of catabolism.

To evaluate the role of this altered hormonal environment, Bessey and associates (1984) and Watters and colleagues (1986) infused these catabolic hormones into normal volunteers with or without the addition of a pyrogen. Hormonal infusion caused many, but not all, of the responses observed following infection and injury. Specifically, when glucagon, glucocorticoids, and catecholemines were infused to achieve blood levels similar to those that occur in catabolic states, negative nitrogen balance, hypermetabolism, and accelerated gluconeogenesis were observed (Bessey et al., 1984). The administration of a pyrogen added fever and activation of many components of acute-phase response to this simple complex (Watters et al., 1986).

In subsequent studies, Bessey and Lowe (1993) used somatostatin to block insulin elaboration, as occurs during the early phases of injury or infection. Under these conditions, the triple hormone infusion resulted in much greater nitrogen losses, reminiscent of excretion rates observed in burn or seriously injured patients. Thus, the hormonal environment plays a major role in determining body protein balance. Hormonal changes during critical illness may explain much of the negative nitrogen balance observed, especially if food intake is provided by enteral or parenteral support.

Alterations in the hormonal environment are triggered by yet another set of signals, those of the cytokines and other pro-inflammatory mediators. Cytokines are glycoproteins synthesized by inflammatory cells and elsewhere in the body in response to inflammation and other noxious stimuli. Although these mediators primarily signal other cells in the local environment, they may be produced in abundance and can, on occasion, be detected in the blood stream (Cannon et al., 1990). Those cytokines central to the pro-inflammatory response are interleukin (IL-1), tumor necrosis factor (TNF), and IL-8 (a potent chemoattractant). These and other pro-inflammatory mediators (e.g., complement, fatty acid metabolites, vascular endothelial factors) initiate both cellular (Hill et al., 1996) and hormonal changes (Hill et al., 1996; Michie et al., 1988) that induce skeletal muscle proteolysis (Mitch and Goldberg, 1996), alter carbohydrate metabolism, and stimulate hypermetabolism, all of which are components of the metabolic responses observed following injury and infection (Ling et al., 1997).

These pro-inflammatory cytokines are balanced by other endogenous factors. The elaboration of glucocorticoids is a result of cytokines signaling the central nervous system to initiate a pituitary-adrenal cortical response. Glucocorticoid elaboration is an important mechanism that attenuates cytokine effects and thus modulates catabolism (Santos et al., 1993). Other anti-inflammatory cytokines are elaborated (IL-4, IL-10, and IL-13), and these substances dampen or attenuate the inflammatory responses and enhance tissue viability and/or promote repair. Nutritional elements also serve as important antioxidants to diminish the effect of free radicals, which are also generated during the inflammatory process. Of particular importance during bacterial infections are vitamins A, C, and E and the minerals zinc and selenium (Grimble, 1993). Glutathione is the most important intracellular antioxidant, and the provision of the amino acids cystine (as N-acetyl-cystine or methionine) and glutamine is necessary to support the ongoing synthesis of this tripeptide. Other nutrients, such as omega-3 fatty acids and arginine may also be beneficial to ameliorate inflammation.

CONCLUSIONS

The mediators of the protein catabolic response are decreased food intake, reduced exercise, and a set of hormonal and cytokine signals that favor protein breakdown. Only now are researchers learning how to modify some of these factors in order to maintain the beneficial aspects of the protein catabolic response while attenuating those elements that appear to be deleterious to the host.

REFERENCES

Artuson, G. 1961. Pathophysiological aspects of the burn syndrome. Acta Chirg. Scand. 274:7–64.
Aulick, L.H., and D.W. Wilmore. 1979. Increased peripheral amino acid release following burn injury. Surgery 30:196–197.
Bararc-Nieto, M., G.B. Spurr, H. Lotero, M.G. Maksud. 1978. Body composition in chronic under nutrition. Am. J. Clin. Nut. 31:23–41.
Beisel, W.R., W.D. Sawyer, E.D. Ryll, and D. Crozier. 1967. Metabolic effects of intracellular infection in man. Ann. Int. Med. 67:744–779.
Bessey, P.Q., and K.A. Lowe. 1993. Early hormonal changes effect the catabolic response to trauma. Ann. Surg. 218:476–489.
Bessey, P.Q., J.M. Watters, T.T. Aoki, and D.W. Wilmore. 1984. Combined hormonal infusions simulate the metabolic response to injury. Ann. Surg. 200:264–281.
Cannon, J.G., R.G. Tompkins, J.A. Gelfand, H.R. Michie, G.G. Stanford, J.W.M. van der Meer, S. Endres, G. Lonnemann, J. Corsetti, B. Chernow, D.W. Wilmore, S.M. Wolff, J.F. Burke, and C.A. Dinarello. 1990. Circulating interleukin-1 and tumor necrosis factor in septic shock and experimental endotoxin fever. J. Infect. Dis. 161:79–84.
Coleman, W., and E.F. DuBois. 1915. Clinical calorimetry. VII calorimetric observations on the metabolism of typhoid patients with and without food. Ann. Int. Med. 15:887–938.
Cuthbertson, D.P. 1932. Observations on disturbance of metabolism produced by injury to the limbs. Quart. J. Med. 25:233–246.
Grimble, R.F. 1993. The maintenance of antioxidant defenses during inflammation. Pp. 347–366 in Metabolic Support of the Critically Ill Patient, D.W. Wilmore and Y.A. Carpentier, eds. New York: Springer-Verlag.
Hill, A.G., L. Jacobson, J. Gonzalez, J. Rounds, J.A. Majzoub, and D.W. Wilmore. 1996. Chronic central nervous system exposure to interleukin-1β causes catabolism in the rat. Am. J. Physiol. 271:R1142–R1148.
Hunter, J. 1794. A Treatise on the Blood Inflammation and Gunshot Wounds. London.
Kinney, J.M. 1966. Energy deficits in acute illness and injury. P. 174 in Proceedings of a Conference on Energy Metabolism and Body Fuel Utilization. A.P. Morgan, ed. Cambridge: Harvard University Press.
Ling, P.R., J.H. Schwartz, and B.R. Bistrian. 1997. Mechanisms of host wasting induced by administration of cytokines in rats. Am. J. Physiol. 272:E333–E339.
Michie, H.R., K.R. Manogue, D.R. Spriggs, A. Revhaug, S. O'Dwyer, C.A. Dinarello, A. Cerami, S.M. Wolff, and D.W. Wilmore. 1988. Detection of circulating tumor necrosis factor after endotoxin administration. N. Engl. J. Med. 318:1481–1486.
Mitch, W.E., and A.L. Goldberg. 1996. Mechanisms of muscle wasting. N. Engl. J. Med. 335:1897–1905.
Muhlbacher, F., C.R. Kapadia, M.F. Colpys, R.J. Smith, and D.W. Wilmore. 1984. Effects of glucocorticoids on glutamine metabolism in skeletal muscle. Am. J. Physiol. 247:E75–E83.
Munro, H.N. 1964b. Historical introduction: The origin and growth of our present concept of protein metabolism. Pp. 1-29 in Mammalian Protein Metabolism, H.N. Munro and J.B. Allison, ed. New York: Academic Press.
Santos, A.A., M.R. Scheltinga, E. Lynch, E.F. Brown, P. Lawton, E. Chambers, J. Browning, and C.A. Dinarello. 1993. Receptor antagonist is not attenuated by glucocorticoids after endotoxemia. Arch. Surg. 128(2):138–143.
Wannamacher, R.W., Jr. 1975. Protein metabolism (applied biochemistry). P. 133 in Total Parenteral Nutrition: Premises and Promises, H. Ghadimi, ed. New York: John Wiley.

Watters, J.M., P.Q. Bessey, C.A. Dinarello, S.M. Wolf, and D.W. Wilmore. 1986. Both inflammatory and endocrine mediators stimulate host responses to sepsis. Arch. Surg. 121:179–190.

Wilmore, D.W., C.W. Goodwin, L.H. Aulick, M.C. Powanda, A.D. Mason, Jr., and B.A. Pruitt, Jr. 1980. Effect of injury and infection on visceral metabolism and circulation. Ann. Surg. 192:491–504.

Wilmore, D.W. 1991. Catabolic illness: Strategies for enhancing recovery. N. Engl. J. Med. 325(10):695–702.

Wilmore, D.W. 1997b. Homeostasis: Bodily changes in trauma and surgery. Pp. 55–67 in Textbook of Surgery, 15th ed., D.C. Sabiston, ed. Philadelphia: W.B. Saunders.

Wilmore, D.W. 1977. The Metabolic Management of the Critically Ill. New York: Plenum Medical.

DISCUSSION

ROBERT NESHEIM: Thank you, Doug, for that very excellent and interesting presentation. It is open for questions.

JOHANNA DWYER: Just a quick question, Doug, on the dietary aspects. You mentioned a lot of antioxidants, some of which are nutrients. What about BHA and BHT? I remember that those are very potent antioxidants.

DOUGLAS WILMORE: I think there are a whole variety of antioxidants that can be utilized, Johanna. The reports on those are really a mixed bag simply because it is probably going to be difficult to see effects with single antioxidants. There is a very nice study, for example, of high-doses of vitamin C in long-distance runners who over-train, for example. But it is rare to find reports of single nutrients alone working as antioxidants that have major types of effects.

ROBERT NESHEIM: Wanda.

WANDA CHENOWETH: You have talked about antioxidants. What about pro-oxidants for example?

DOUGLAS WILMORE: For example?

WANDA CHENOWETH: Iron?

DOUGLAS WILMORE: Iron? Well, it is a good point. Bill may be able to make some comments about this. The classic teaching is that iron is the single nutrient that should not be administered during infection and injury. If it is fixed in food and supplemented, in general, in a very controlled situation, you would take it out. You would not give it. This is simply because iron is redistributed after infection and your iron level in your blood stream becomes very low because it really is a very important substance for bacteria and some viruses to use for replication. So we would take it out or not give it. This, again, gets back to the problem of having fixed rations and all of the components in a ration, and the assumption that a ration that is appropriate for well people is appropriate for sick people, which is a concept that I do not embrace.

ROBERT NESHEIM: Dr. Millward.

D. JOE MILLWARD: I was just going to make the point in respect to BHT. I think that one of the interesting things about those sorts of antioxidants is that intakes of them are extremely high and we generally disregard them as important components of our food because they are there as additives. But, actually, we now take in large quantities of these compounds. It may well be that they are in fact a major component of our antioxidant defense mechanism. I think that we should be thinking about them to a much greater extent than we do.

ROBERT NESHEIM: John Vanderveen.

JOHN VANDERVEEN: Doug, the question is whether, on a long-term basis or a short-term basis, we can build up body stores that are going to help us in the case of an infection or injury event, or is it something that is best handled in terms of a short preliminary type of preparative activity?

DOUGLAS WILMORE: That is a very strategic question from the standpoint of preparing people to go into maneuvers or go to do a task. I guess that one way we could start thinking about it, and this may not be the only way to think about it, would be to think in terms of water soluble and fat soluble substances. Because we can pretty well store many of the fat-soluble substances, but we would probably have to provide many of the water-soluble substances. When we talk about genetic predisposition or phenotype changing, it may well be that omega-three fats could be fed over the period of about three to four weeks, if you will, and we could change responses in just that short a period of time.

Vitamin E loading, in adequate doses, for example, takes about four weeks. And there are nice clinical data now from heart biopsies of patients undergoing cardiac surgery where various doses and time intervals have been used for vitamin E loading as an antioxidant.

The University of Toronto group has reported that you need about four weeks before you can get adequate vitamin E loading in cardiac muscle. So that is one way to think about it.

There is another way to think about it too and that is—it is not all that bad for many of these people to have a surfeit of adipose tissue which is their energy supply and an adequate skeletal muscle mass, because history has really told us that you can draw on those stores and that will enhance your survival as long as you are within some sort of a functional zone. So there may be some strategies that we can think about that will have to do with loading or preparing people. Mackenzie Walser and others may recall the interest people had with using odd-chain fats. Odd-chain fats would then provide you with gluconeogenic precursors. If you were starving for a long period of time, these fats would be drawn from your adipose tissue stores, but if you were provided with gluconeogenic precursors and you would not have to break down muscle as fast. So there have been talks about that in the past.

ROBERT NESHEIM: Dr. Walser.

MACKENZIE WALSER: Is the mechanism of that known? Does it have etiological value?

DOUGLAS WILMORE: Presumably it is there to protect circulating blood volume and oxygen delivery capacity. The mechanisms are through a lot of normal pathways that conserve salt and water.

ROBERT NESHEIM: Dr. Lieberman.

HARRIS LIEBERMAN: Doug, do we know anything about patterns of change in amino acids as a consequence of either acute or chronic illness, particularly those that are important to the brain, such as tryptophan or tyrosine?

DOUGLAS WILMORE: In general, all of the indispensable amino acids are lowered, I mean, as a sort of a general rule without knowing the specific pattern for studies. In general, all essential amino acids are low. The dispensable amino acids are generally normal or slightly increased. Joe, are there other thoughts? (No response.)

ROBERT NESHEIM: Steve Heymsfield, and then we will need to go on. We have a discussion period following the next presentation so we can cover many of these things then. Steve.

STEVEN HEYMSFIELD: You mentioned genetics as being one factor. Do gender or age affect the metabolic response to injury in any way?

DOUGLAS WILMORE: They must, but the age issue is exceedingly difficult to study because you introduce progressive organ dysfunction or change; you bring in other disease processes and things like that.

But when I was in the Army Burn Unit in the 1970s, we were able to sort surviving from nonsurviving patients by just their response to a cool environment. A cool environment, incidentally, was about this temperature (e.g. the temperature of an air conditioned room). It is one of the issues of why hospitals should not be cold. The older patients responded more rapidly to cold exposure than the younger individuals.

MICHAEL RENNIE: Can I just say something about that metabolic response? We have just done some studies on very young children, neonates. In fact, as far as we can tell, they do not show a metabolic response to injury. So that appears to be a developmentally-acquired response.

DOUGLAS WILMORE: And that is compatible with the fact that babies can actually undergo complete hormonal blockade with heart surgery and actually do better in terms of their overall response when compared with non blockaded patients.

ROBERT NESHEIM: Thank you very much. That was very interesting.

Protein and Amino Acids, 1999
Pp. 169–216. Washington, D.C.
National Academy Press

9

Inherent Difficulties in Defining Amino Acid Requirements

D. Joe Millward[1]

INTRODUCTION

To address the role of protein and amino acids in performance, this chapter is based on the premise that it is an inherently difficult problem to define the dietary requirements of human adults for indispensable amino acids and to assess the nutritional value (protein quality) of different food protein sources to provide for those needs. There are three major reasons for this difficulty. The first is adaptation, that is, a variable metabolic demand for amino acids set by the habitual intake. Thus, the extent to which any intake appears to be adequate will depend on the completeness of adaptation to that intake. The second is methodology, with no entirely satisfactory practical nitrogen or amino acid balance method or other measure of dietary adequacy. The third is lack of quantifiable, unequivocal metabolic indicators of adequacy that can validate

[1] D. Joe Millward, Centre for Nutrition and Food Safety, School of Biological Sciences, University of Surrey, Guildford GU2 5XH United Kingdom.

balance measurements. The questions posed in this review cannot currently be answered because of the absence of studies of outcome in terms of physical performance in long-term, controlled feeding trials.

BACKGROUND TO THE CURRENT CONTROVERSY

In 1985, the Food and Agriculture Organization (FAO) report on protein and energy requirements was published (FAO/WHO/UNU, 1985), a feature of which was the recommendation that protein quality should be evaluated by the PDCAAS method (protein-digestibility corrected amino acid score), making use of age-specific amino acid scoring patterns. Because the indispensable amino acid (IAA) requirement values used to calculate the scoring patterns fell markedly with age, from over 50 percent of total protein requirement in infants to only 16 percent in adults (see Table 9–1), the quality of any protein would now be assessed as higher when used for adults than for children. Furthermore the low requirement level of IAA in adults meant that all natural diets and food proteins would be adequate. Thus, apart from digestibility, protein quality ceased to be an issue in the nutrition of adults.

Considerable disquiet arose about the 1985 report. Young (1986) argued that the adult IAA requirement values were seriously flawed because of the way Rose (1957) conducted his nitrogen (N) balance studies (mainly excess energy and no account for miscellaneous N losses). Millward and Rivers (1988) reviewed the subject, paying particular attention to the adaptive changes in amino acid oxidation that can occur and that will influence requirement values. They argued that the marked fall with age in the requirement values was mainly a reflection of the methodologies used in their assessment. Thus, the infant values were largely patterned on the composition of breast milk, while the adult values, measured in balance studies with excess nonessential nitrogen and low levels of indispensable amino acid, would have identified minimum requirement values. They concluded that IAA requirements are complex, include an adaptive component, and can only be defined under specific artificial conditions that would allow definition of a minimum value and that "current estimates of adult requirements may be close to this level." To identify which IAA might be rate limiting for the obligatory N losses (ONL), they calculated the obligatory oxidative amino acid losses (OOL) as estimates of the losses of tissue IAAs that would give rise to the ONL, as discussed in detail below.

Young et al. (1989) then published a paper entitled "A Theoretical Basis for Increasing Current Estimates of the Amino Acid Requirements in Adult Man with Experimental Support." This paper reproduced the table of OOL values from Millward and Rivers (1988). After making some small adjustments in lysine, threonine, and valine values derived from their stable isotope studies and increasing all values assuming a 70 percent efficiency of utilization, Young and colleagues proposed that this pattern, the "MIT" (Massachusetts Institute of

TABLE 9-1 Protein and Indispensable Amino Acid Requirements (mg/kg/d) and Obligatory Indispensable Amino Acid Oxidative Losses*

	Infants (3–4 mo)	Children (2 y)	Schoolboys (10–12 y)	Adults	Obligatory oxidative losses†	
					Value	Requirement multiple
Protein						
Growth	625	187	106	—		
Maintenance	750	738	681	600		
Total	1,375	925	787	600	338	0.56
Amino acids						
Histidine	28	(20)	(20)	8–12	11.5	1.05
Isoleucine	70	31	30	10	16.2	1.62
Leucine	161	73	45	14	27.4	1.96
Lysine	103	64	60	12	30.1	2.51
TSA	58	27	27	13	13.5	1.01
TAA	125	69	27	14	27	1.93
Threonine	87	37	35	7	15.5	2.21
Tryptophan	17	12.5	4	3.5	4.0	1.14
Valine	93	38	33	10	16.9	1.69
Total	742	372	281	94	162	1.72
% Protein requirement	54	40	36	16	48	

NOTE: TSA, total sulfur amino acids; TAA, total aromatic amino acids.

*Protein requirements are mean values

†These are the rates of oxidative loss of IAA's predicted to occur based on the assumption that the ONL (54 mg N/kg/d) derive from the oxidation of amino acids liberated from body protein (amino acid composition as beef), and the composition of the free amino acid pool does not change.

SOURCE: Adapted from FAO/WHO/UNU, 1985.

Technology) scoring pattern, should be used as the basis for protein quality evaluation in adults and in children given the similarity between their pattern and that of the FAO preschool child pattern.

In 1989, FAO/WHO convened a meeting to consider protein quality evaluation and to endorse the PDCAAS method recommended by FAO in 1985. However the report (FAO/WHO, 1991) rejected both the 1985 adult and older school child IAA requirement values as flawed, was unable to identify any other appropriate adult scoring pattern, and proposed that the scoring pattern for the preschool child be utilized for older children and for adults as a strict interim measure. It was argued that (a) the preschool child data were reliable, (b) in the absence of any other data, some pattern was needed for older children and adults, and (c) the slow growth of children compared with adults means that a major change in the requirement pattern with age was unlikely.

Although Young and colleagues broadly agreed with this conclusion in that the MIT and preschool patterns were similar, Millward (1994) argued that the report was flawed. In fact, the data that formed the basis of the preschool child pattern had never been published and were not available for scrutiny except for some "typical" data for lysine published in a book review (Pineda et al., 1981). The data, which were derived from study of preschool children who had recovered from protein energy malnutrition (PEM), show that the N balances were so large in the children studied that they would have been exhibiting catch-up growth as far as lean tissue was concerned (growth rates and N retentions of 3 times the expected values). This growth would markedly increase the need for indispensable amino acids compared with that of normal preschool children, older children, and especially adults. Millward (1994) also argued against acceptance of the MIT scoring pattern on the grounds that amino acid requirements for maintenance cannot be predicted from the amino acid composition of body proteins.

Fuller and Garlick (1994) have reviewed the controversy, and while they did not endorse the MIT pattern, they did conclude that the FAO values were likely to be underestimates, having failed to include miscellaneous N losses in the original balance studies. They reported adjusted higher values, taking into account estimated miscellaneous N losses. They also raised concern that the tracer studies may also suffer from an underestimate of losses and that neither N nor tracer studies are inherently better than the other. These issues are considered in more detail below. Waterlow (1996) made a detailed analysis of the tracer studies and came to conclusions similar to Fuller and Garlick, that is, they did not accept the theoretical basis of the MIT pattern but recognized that the ^{13}C leucine studies do point to a higher leucine requirement than does the FAO value. The issue was considered at an international meeting of an expert group in London in 1994. However, contrary to what was published (Clugston et al., 1996), the MIT pattern was not endorsed at this meeting since, as subsequently reported by Millward and Waterlow (1996), the published statement "a large majority of the group accepted as an interim operational pattern that [that was] proposed by Young et al.," had in fact emerged during postmeeting editing.

The current views of Young and colleagues are described in the chapter following this one. What follows is an account of this author's current understanding of the debate. Much of the argument made here has been reported previously in publications by the author and in correspondence relating to publications by Young and colleagues (Millward and Rivers, 1988, 1989; Millward, 1990, 1991, 1992, 1993, 1994; Millward et al., 1989, 1990; Millward and Pacy, 1995).

METABOLIC BACKGROUND

One cause of potential confusion within this debate is lack of consistency in terminology. To avoid that here, protein and amino acid requirements will be discussed in terms of metabolic demand, dietary requirement, and dietary allowances. Metabolic demand (MD) is determined by the nature and extent of those metabolic pathways that consume amino acids. The dietary requirement is the amount of protein and/or its constituent amino acids that must be supplied in the diet to satisfy the metabolic demand, usually greater than the MD because of less-than-perfect protein utilization. Dietary Reference Values (U.K. terminology) or Recommended Dietary Allowances (U.S. terminology) are a range of intakes derived from estimates of individual requirements that are designed to meet the dietary requirements of the population and that take into account the variability among individuals in that dietary requirement. This chapter focuses on the MD and dietary requirement.

Obligatory Metabolic Demand

The MD for dietary protein is to provide amino acid precursors for the synthesis of tissue proteins and a range of nonprotein products. Although most proteins are in a dynamic state of constant turnover, little metabolic demand for amino acids is generated by this avenue because of amino acid recycling. "Wear and tear" as a driver of MD is not an appropriate biological analogy. Only net protein synthesis contributes to MD. This growth aspect of MD is straightforward in that the qualitative nature of MD is determined by the amino acid pattern of tissue protein deposited. This pattern is usually assumed to be influenced only by the changes in body composition occurring during growth and is not generally assumed to vary with the diet within cells and tissues. However, as discussed by Fuller and Garlick (1994), some evidence exists for changes in amino acid content of tissues during growth on amino acid-limiting diets.

Nonprotein products of amino acid precursors include nucleic acids, and a range of smaller molecules such as creatine, taurine, glutathione, hormones (e.g., catecholamines and thyroxine), neurotransmitters (serotonin, dopamine), and nitric oxide, a key regulator of blood flow and other physiological processes.

In human nutrition, growth occurs very slowly after the first few months of life. Net protein synthesis contributes a small and decreasing component of MD during pre- and immediate post-adolescence. In the adult, it comprises only that associated with continuing growth of skin and hair and the synthesis of those gastric secretions (e.g., threonine-rich mucus glycoproteins) that pass into the colon to be utilized for bacterial metabolism. Thus, apart from these small components, humans are normally at nitrogen or amino acid equilibrium, with MD reflecting mainly nonprotein pathways of amino acid metabolism and catabolism associated with maintenance of normal function and composition. In the traditional nutritional terminology of human growth and maintenance, growth needs are low at all ages after early infancy with maintenance dominating the MD. The task, then, is to define the amounts and amino acid pattern of the maintenance requirement.

Obligatory Metabolic Demands and Obligatory Oxidative Losses

The diverse obligatory maintenance MDs for amino acids represent an important, but small, intrinsic part of MD, the magnitude of which is the main subject of current debate. Table 9-1 shows the requirement values for IAAs and for protein as reported by the FAO (FAO/WHO/UNU, 1985). The feature of these values that has been at the heart of the controversy is that the IAA requirement as a proportion of the protein requirement falls markedly with age, from 54 percent in infants to 16 percent in adults. Millward and Rivers (1988) argued that some information could be obtained from the magnitude of the ONL. In subjects fed a protein-free but otherwise nutritionally adequate diet, N losses fall to a stable and reproducible low level after 7 to 14 days. Subjects lose body protein at a constant daily rate, about 54 mg/kg/day, which is equivalent to 0.34 g of protein/kg/day (FAO/WHO/UNU, 1985). These ONLs are assumed to represent nitrogen end products of amino acids derived from body protein and utilized for the obligatory metabolic demand (OMD) that is tacitly assumed to be the same in subjects consuming a protein-free diet as in subjects consuming a normal diet. The ONL is a function of body weight and, when normalized to "metabolic body size" ($kg^{0.75}$), varies little with age (FAO/WHO/UNU, 1985). Millward and Rivers (1988) reported a simple calculation of the OOLs. These are the oxidation rates of amino acids that give rise to the ONL and they are equal to the amounts of amino acids in the protein equivalent of the ONL. It was assumed that tissue protein composition could be approximated by that of muscle and the values for beef muscle listed by FAO (FAO/WHO/UNU, 1985) were used. This was only a first approximation calculation, since some of the ONL occurs as protein *per se* (skin, hair, some fecal nitrogen, and secretions), and protein is lost from several tissues in addition to skeletal muscle. They made the assumption that of the individual amino acids that comprise the obligatory MD (which, on a protein-free diet, is in effect fueled by tissue protein), one amino acid would be rate limiting, with the highest ratio of obligatory MD to OOL. All other amino acids with a lower ratio would be present in excess in the OOL but would be nevertheless oxidized because they could not be returned on their own to the tissue protein pool. They argued that if protein turnover is tightly regulated, allowing just enough of the rate-limiting amino acid to be withdrawn from tissue protein to provide for its MD, the OOL of this amino acid should be a reasonable guide to its requirement. For all others, the values for the OOL should be greater than the maintenance requirements.

Use of the Obligative Oxidative Loss Pattern to Predict a Requirement Pattern

The actual listed values of the OOL were adopted by Young et al. (1989) as the basis of the MIT scoring pattern, something that was contrary to what Millward and Rivers (1988) intended. In effect, this defined a maintenance pattern with the same composition as tissue protein, a novel assumption given the widespread assumption of different amino acid patterns for maintenance and growth.

Millward and Rivers (1988) assumed that the pattern of the obligatory MD is different from that of tissue protein, so that there would be a rate-limiting amino acid that "drives" the ONL. The identification of this driver can be done by reference to an actual requirement pattern. Thus, they compared the values of OOL with the 1985 FAO/WHO/UNU requirement values such that if the FAO values were accurate, the values would be similar for one amino acid. In fact, while the OOL for most amino acids was greater than the FAO requirement, with lysine, threonine, leucine, and the aromatic amino acids being particularly in excess (2–2.5 times the requirement), the OOL of the total sulfur amino acids (TSAs) was quite close to the FAO requirement values; this latter observation showed that the TSAs are rate determining for the mobilization of tissue protein to provide for obligatory MD. The possibility that the TSAs are the rate-limiting amino acids that drive the ONL in humans was attractive on the basis of animal studies. Providing the rate-limiting amino acid to the protein-free diet fed during measurement of ONL should result in a fall in N excretion to the rate determined by the demand for the second limiting amino acid. Studies in dogs (Allison et al., 1947) and rats (Yoshida, 1983) have shown that supplementation by S amino acids reduces N excretion.

Animal Data for the Pattern of the Obligatory Metabolic Demand

Animal data clearly indicate that the amino acid pattern of the obligatory maintenance MD differs from that of tissue protein, which necessarily represents the pattern for the growth requirement. However, this view is not accepted by Young (e.g., Young and El-Khoury, 1995), who, having used the OOL pattern to derive the MIT pattern, assumed maintenance requirements to be broadly similar to tissue protein.

Before reviewing the animal data, their relevance to human nutrition needs to be addressed. Young and El-Khoury (1995) have discussed the relevance of the high-quality nitrogen balance data obtained in the young pig (Fuller et al., 1989). They argued that any comparison of human requirement with that of the young pig is invalid because (a) relative amounts of maintenance and growth vary too much between the species (maintenance = < 10 percent total [growth + maintenance] requirement in the young pig and < 5 percent total in the rat), and (b) the efficiency of dietary protein utilization in the young pig at maintenance is much higher (100 percent) compared with human values (assumed to be 70 percent). In fact, the first point is irrelevant when human-animal comparisons are limited to discussions of either maintenance or growth needs specifically, and the second point is irrelevant when what is considered is MD rather than dietary requirement.

Young also argues that data from adult pigs (boars) support the proposition that the pattern of tissue protein is similar to the maintenance pattern. Yet he acknowledges the data to be poor and at variance with most other animal data.

Finally, Young and El-Khoury (1995) argue that maintenance patterns derived from rapidly growing animals held at maintenance by food restriction are unphysiological and, consequently, may be an unreliable guide to the human obligatory IAA MD. To date there has yet to be a claim that human IAA MD can be accurately predicted from animal values, only that consideration of the animal data as a whole, including both growing and adult data, may provide useful general information. There appear to be few major differences between mammalian species with regard to the fundamentals of amino acid and protein metabolism. With obvious exceptions (e.g., arginine requirements for growing cats and growing and adult dogs, a taurine requirement for the kittens, and a high-maintenance amino acid requirement in avian species for feather growth), interspecies comparisons appear to be legitimate since robust animal data should provide general principles about the nature of human needs.

Two kinds of studies are pertinent. The first is deletion studies, in which individual amino acids are removed from the diet and the extent of the negative balance is monitored. If the maintenance requirement patterns corresponded exactly to the patterns of tissue protein, then there should be a similar negative balance on removal of each IAA. If not, then negative balance will occur in proportion to the ratio of obligatory MD to tissue content of each amino acid.

Only one report exists for the adult rat (Said and Hegsted, 1970), a high-quality study based on measured changes in body water. Gahl et al. (1991) reported N balance data for young rats, while Fuller et al. (1989) studied 41-kg pigs, with N balance; data from the latter study are widely recognized as the most robust data for the pig.

Table 9–2 shows the relative losses normalized for the response to a protein-free diet. The first, second, and third limiting amino acids are threonine, TSA, and isoleucine for the growing rat; threonine, isoleucine, and tryptophan for the adult rat; and TSA, threonine, and tryptophan for the pig. The most highly conserved, least-limiting amino acids are lysine and leucine in the rat and all three branched-chain amino acids (BCAAs) and lysine in the pig.

Supplementation with each limiting amino acid allows the slope of the balance curve to be established and the consequent requirement values to be determined; these are shown in Table 9-3 compared with carcass protein content. For ease of comparison, these patterns have been normalized for threonine. Leucine and lysine are the two most abundant amino acids in carcass proteins and in the growth requirement patterns for both rat and pig; in the maintenance requirement patterns, the most abundant amino acids are threonine and TSA in the pig; threonine, isoleucine, valine, and TSA in both adult and growing rats.

TABLE 9-2 Responses (Negative Balance) to Deletion of Individual Indispensable Amino Acids or a Protein-Free Diet

Amino acid	Growing pig*	Growing rat†	Adult rat‡
Histidine	nd	0.31	0.55
Isoleucine	0.24	0.94	0.90
Leucine	0.21	0.38	0.36
Lysine	0.33	0.25	0.30
TSA	0.93	1.09	0.51
TAA	0.34	0.06	0.47
Threonine	0.68	1.19	0.92
Tryptophan	0.53	0.75	0.80
Valine	0.23	0.88	0.71
All (protein free)	1.00	1.00	1.00

NOTE: TAA, total aromatic amino acids; TSA, total sulfur amino acid; nd, not determined.

* Fuller et al. (1989).
† Gahl et al. (1991).
‡ Said and Hegsted (1970).

TABLE 9-3 Amino Acid Composition and Requirement Patterns

	Growing pig*			Growing rat†			Adult rat‡	
	Body	G	M	Body	G	M	G	M
Histidine	0.74	—	0.65	0.34	0.14	0.41	0.48	—
Isoleucine	0.92	0.92	0.30	0.81	1.07	0.56	1.08	1.03
Leucine	1.88	1.66	0.43	1.67	1.61	0.38	1.35	0.94
Lysine	1.87	1.45	0.68	1.58	1.23	0.33	1.76	0.74
TSA	0.74	0.76	0.92	0.87	1.40	0.63	0.98	0.97
TAA	1.89	1.80	0.70	1.67	1.25	0.20	1.41	1.15
Threonine	1.00	1.00	1.00	1.00	1.00	1.00	1.00	1.00
Tryptophan	nd	0.26	0.21	0.26	0.16	0.06	0.22	0.22
Valine	1.25	1.12	0.38	1.03	1.27	0.66	1.10	1.03

NOTE: G, growth; M, maintenance; TSA, total sulfur amino acids; TAA, total aromatic amino acids; nd, not determined.

* Fuller et al. (1969).
† Benevenga et al. (1994).
‡ Said and Hegsted (1970).

The major implication of these animal data is that there are marked differences between the MD for maintenance and for growth. It is clear that in the growing rat and pig and the adult rat, leucine and lysine exhibit the biggest difference between growth and maintenance patterns, these two amino acids being most abundant for growth and among the least abundant for maintenance. The practical consequence of this, as pointed out by Hegsted (1973), is that the balance-intake curve is extremely shallow for leucine and lysine both in the submaintenance and growth range. This means that small differences in balance result in large differences in maintenance intakes so that measurement of a requirement value for maintenance is very difficult and depends on the exact criterion for adequacy. The several early reports of rats maintaining body weight for 6-month periods on very low lysine diets (e.g., zein [Osborne and Mendel, 1916] or even lysine-free diets [Bender, 1961]) are probably explained by coprophagy, given the clear evidence of a metabolic need for lysine in terms of the rapid onset of symptoms on a lysine-free diet in humans (Rose, 1957). However, no evidence exists for anything other than a low metabolic need for this amino acid.

A second type of study which is pertinant is the work of Yoshida (1983) who has done most to explore the concept that rate-limiting amino acids at maintenance differ from those that rate-limit growth. Having established that in adult rats fed a protein-free diet, the most rate-limiting amino acids were threonine and TSA, he also showed that in adult rats fed limiting amounts of rice or wheat diets, the limiting amino acids were threonine and the sulfur amino acids. When these two amino acids were added to the cereal diets, they restored nitrogen balance and transformed body weight loss to growth (See Figure 9-1). This may explain why attempts to show in human adult supplementation trials that lysine is the limiting amino acid in wheat were so disappointing (Scrimshaw et al., 1973).

Although the nature of the relative metabolic need for individual amino acids is by no means clear, Fuller's work with the pig points to ileal amino acid losses as a partial explanation, accounting for some 60 percent of pig amino acid maintenance requirements (Wang and Fuller, 1989). Table 9-4 compares ileal losses of the pig and humans. These data show that in each case, threonine is the largest component, and while the patterns differ somewhat, the absolute values are much lower in humans than in the pig. Thus, despite discussion of the pig as an inappropriate model for humans, to the extent that ileal losses comprise a component of obligatory MD, these data point to a lower MD in humans than in the pig.

To summarize research on obligatory maintenance MD, a consistent and extensive body of animal data shows the maintenance pattern to differ from the growth pattern, with lower levels of lysine and leucine in the maintenance pattern. As a result, the rate-limiting amino acids in dietary proteins for maintenance may differ from those for growth, the example being that lysine limits wheat for growth but not for maintenance.

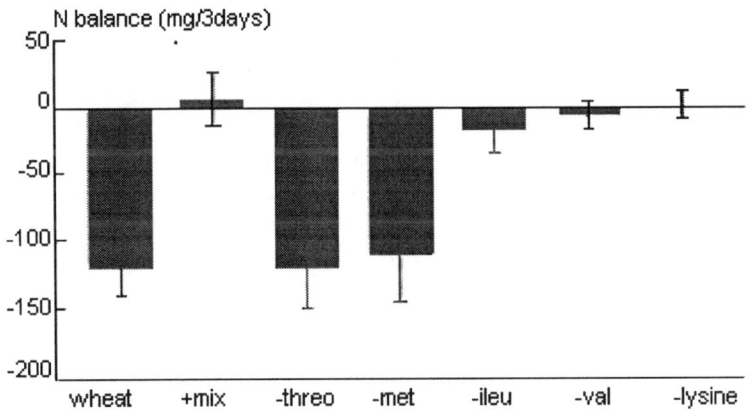

FIGURE 9-1 Adult rats were fed limiting amounts of a wheat diet that did not allow balance and were supplemented with a mixture of indispensable amino acids that did allow balance. By removing individual amino acids, the limiting amino acids were identified as threonine and the sulfur amino acids. Lysine was not needed to improve balance. NOTE: threo, threonine; met, methionine; ileu, isoleucine; val, valine; N, nitrogen; IAA, indispensable amino acids. SOURCE: Adapted from Yoshida (1983).

TABLE 9-4 Ileal Indispensable Amino Acid Losses of the Pig and Human

	Pig* 40 kg mg/kg/d	Human adult† 66 kg mg/kg/d
Isoleucine	11.3	1.7
Leucine	25	3.2
Lysine	15.3	3.9
TSA	18.3	1.8
TAA	28.5	3.9
Threonine	32.3	4.2
Valine	18.3	2.9

NOTE: TSA, total sulfur amino acid; TAA, total aromatic amino acid.
* Wang and Fuller (1989).
† Fuller et al. (1994).

The Adaptive Component of the Metabolic Demand

It may appear a simple task to identify and quantify the maintenance MD for each amino acid in terms of the various metabolic pathways involved. However, difficulty arises in distinguishing between obligatory intrinsic, functionally important demands and those that serve a purpose, but can and do vary according to circumstances, that is, an adaptive component. It is this adaptive component that brings complexity to the maintenance MD.

The ONL at 54 mg N or 0.34 g protein/kg/day, is only 50 percent of current estimates of the protein requirement (0.6 g/kg/d), and the nature of this additional need (the difference between 0.34 and 0.6 g protein) has in the past been difficult to account for. Usually, it has been attributed to an inefficiency of utilization, although why proteins such as those in milk, eggs, or meat were not utilized more efficiently was always puzzling. It is much easier to understand the inefficiency of utilization as representing an adaptive component of MD.

When subjects are fed a protein-free diet, their urinary N losses initially reflect their normal dietary protein intake and then falls over 7 to 14 days to reach a low stable output level (see FAO/WHO, 1973). That is, an additional loss of body N occurs on a daily basis for some time before equilibrium is reached at the lower level. This additional daily N loss demonstrates the existence of this adaptive component of the MD. Traditionally, this has been defined as "the labile protein reserves," which imply that metabolically it was a pool of protein that varied in size with the dietary protein intake. In the rat, liver and visceral protein content does vary directly with dietary protein intake (see Munro, 1964) in support of the labile protein reserve concept. However, no such protein pool has ever been identified in humans, and more importantly, recent detailed metabolic studies of balance regulation have pointed to an alternative explanation.

Thus, studies of the distribution of urinary N losses between the postabsorptive and postprandial periods, in subjects adapted to a wide range of protein intakes (Price et al., 1994), showed that not only did postprandial losses of N increase with increasing intakes as would be expected, but so did postabsorptive losses. Price and colleagues utilized both N and ^{13}C leucine balance methods to assess the magnitude of the total losses in the postabsorptive and postprandial state in normal adults with protein intakes ranging from a very low to a high level (0.36 g/kg–2.3 g/kg) for 2 weeks. Comparisons of the rate of ^{13}C leucine oxidation and total N excretion (corrected for acute changes in the body urea pool) indicated the two measures were proportional over the entire range of intakes, although leucine oxidation underestimated N excretion by about 25 percent (see below). The results calculated from postabsorptive ^{13}C-1 leucine oxidation are shown in Figure 9-2. The N equivalent of postabsorptive leucine loss is shown calculated as a daily rate and divided between the estimated obligatory MD (54 mg N/kg), with the remainder representing the adaptive MD. Even for the lowest intake, total losses (assumed to be tissue protein and association amino acids) were in excess of the obligatory MD, which indicates that subjects may not have fully adapted to this low intake in the relatively short (2-week) period of the study. However, it is clear that with higher intakes, the adaptive component of the MD varied up to more than twice the obligatory MD.

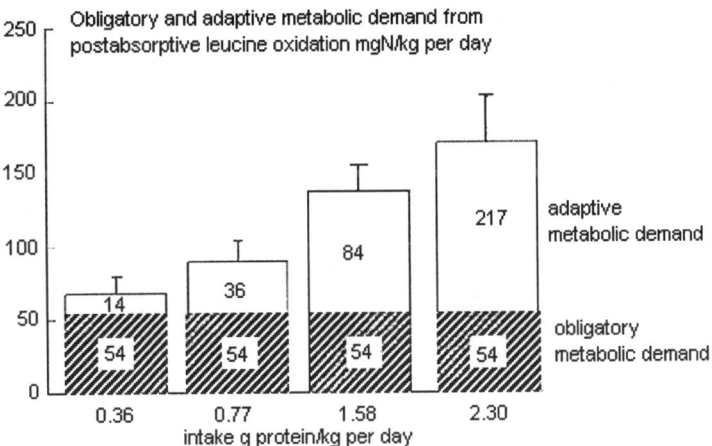

FIGURE 9-2 The magnitude of the obligatory and adaptive metabolic demand in normal adults. Values calculated from postabsorptive leucine oxidation rates measured in subjects who were fed the indicated protein concentrations for 2 weeks. SOURCE: Adapted from Price et al. (1994).

In other studies of the transition from a high-to-lower protein diet (Quevedo et al., 1994), it was shown that the expected period of negative balance involved a similar lag in reduction of these N losses in both postabsorptive and postprandial states (see Figure 9-3). As a result, there was an overall marked negative balance, and even at the end of 2 weeks, equilibrium had not been reached. Thus, these studies are inconsistent with the concept of labile protein reserves. Instead, they imply that the transitional losses during adaptation to a lower intake reflect the time taken for catabolic pathways of amino acid metabolism to adapt from a level set to deal with one level of protein intake to that required for a lower level.

The implications of these adaptive responses are that the increasing catabolic losses with increasing protein intake represent an actual MD generated in response to the presence of protein in the diet at levels in excess of minimal metabolic needs.

Since amino acids can only be stored as protein, the capacity for which is limited in the body, excess dietary amino acids are oxidized and converted to glucose or fat. As argued by Millward and Rivers (1988), many amino acids, especially the branched-chain, aromatic, and sulfur amino acids, represent a potentially toxic challenge to the organism and are maintained at very low levels in the tissue-free amino acid pools. Thus, after a meal, these amino acids are rapidly removed by oxidative catabolism if not deposited in protein by high-

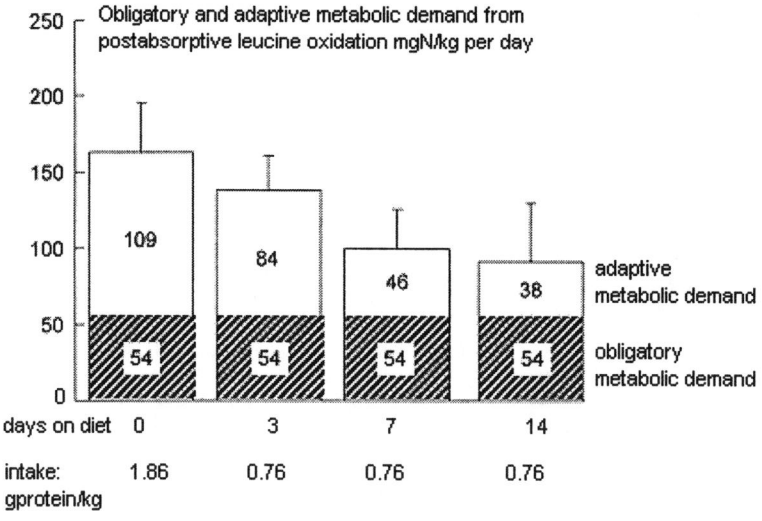

FIGURE 9-3 Rate of change in the adaptive metabolic demand during transition from a high- to lower-protein diet. Because of the slow change in the adaptive metabolic demand after the reduction in protein intake, there was a negative nitrogen balance throughout the 14 days after the diet change. SOURCE: Adapted from Quevedo et al. (1994).

capacity, sensitively regulated pathways. Furthermore, the capacity and activity of these pathways adapt to match the protein levels in the diet to ensure rapid postprandial disposal. Most importantly, because this adaptation of oxidative catabolism to a change of protein intake is relatively slow, the extent of postprandial oxidative catabolism reflects mainly habitual rather than actual protein intake in meals. In effect, the habitual level of protein in the diet creates a level of oxidative amino acid catabolism sufficient to avoid accumulation of toxic concentrations of certain free amino acids, and this becomes part of the MD. Apparently, during slow growth or at weight maintenance, in order to be able to rapidly dispose of dietary protein in excess of minimal needs, pathways of oxidative amino acid catabolism and particular catabolic enzymes are in effect primed to operate at the appropriate rate set by habitual protein intakes. This rate continues regardless of the actual acute intake, utilizing tissue protein if the dietary level falls or during the postabsorptive state, for as long as it takes to adapt to the lower level of intake.

The overall metabolic scheme describing these various aspects of the MD is shown in Figure 9-4. The metabolic fate of dietary protein is shown as providing amino acids into the tissue free amino acid pool, which is in a state of dynamic

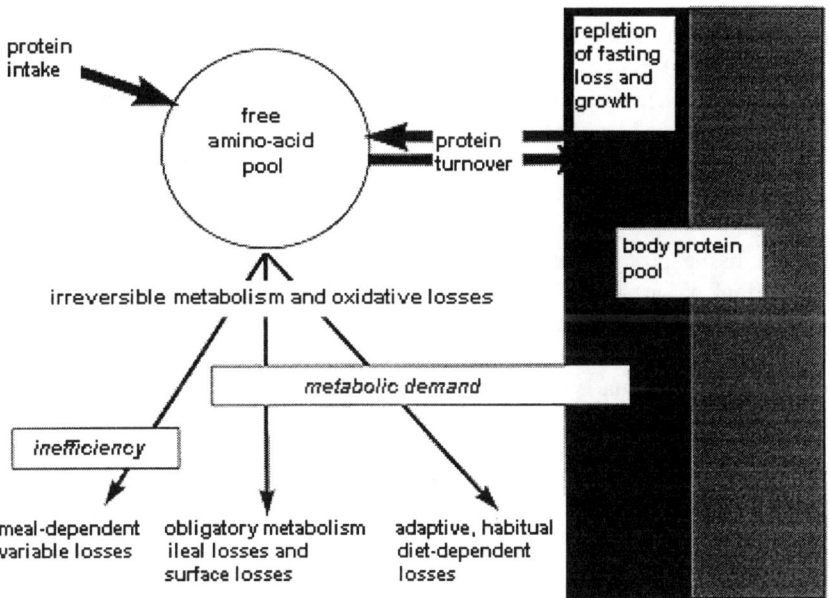

FIGURE 9-4 Scheme to describe the metabolic fate of amino acids in relation to the metabolic demand.

exchange with the protein pool through protein turnover. The MD includes both net protein synthesis (i.e., the repletion of postabsorptive losses and any growth) and irreversible amino acid metabolism and oxidative metabolism. The latter component is shown divided into three parts: (1) the metabolic demands for nonprotein products that eventually give rise to an obligatory N loss, (2) adaptive oxidative amino acid catabolism, which is also part of the MD, and (3) any additional oxidative amino acid catabolism due to an inefficiency of protein utilization. Within this metabolic framework the determination of protein requirements is a problem of assessing both intrinsic, fixed MDs and the adaptive oxidative catabolism, which together define the MD. Assessment of this latter component accounts for the major practical and conceptual difficulties in applying N balance techniques to the study of protein requirements.

Diurnal Cycling: A Qualitative Influence on Metabolic Demand?

There is an important but complex implication of diurnal feeding and fasting for the IAA pattern of the MD. Clearly, from the above discussion, with increasing dietary protein intake, there will be an increasing MD generated by the adaptive oxidative losses. With both adaptive and obligatory MD occurring continuously, overall daily balance is achieved within a complex diurnal cycle of postprandial repletion of tissue proteins mobilized to provide for the postabsorptive demand. Furthermore, the amplitude of this cycle increases with increasing habitual level of protein intake (Price et al., 1994). The key question is: To what extent does this diurnal cycle of body protein influence the IAA composition of the adaptive MD? Does it mean that regardless of the amino acid composition of the intake that induces the adaptive MD, the adaptive MD will have a tissue protein pattern of amino acids to enable postprandial protein deposition? Young and El-Khoury (1995) have assumed this to be the case and a partial justification for their assertion that the maintenance requirement pattern is like that of tissue protein.

In fact, the actual amount of amino acids needed in the diet to provide for this adaptive requirement cannot be predicted for three reasons. First, the amount of actual postabsorptive tissue protein loss depends on the extent of the true postabsorptive state. Price et al. (1994) utilized 12 + 12-hour diurnal cycles in their studies, but the actual amount will vary with the pattern of meal feeding, and individuals consuming both an early breakfast and late supper may spend less than 12 hours in a true postabsorptive state and mobilize less tissue protein. In subjects in the postprandial state, the diet will directly provide for the losses without inducing tissue protein loss.

Second, when true postabsorptive losses of tissue protein do occur it does not follow that all IAAs liberated from the net tissue proteolysis are oxidized. Although increases in the concentrations of the BC, aromatic, and sulfur amino acids liberated from tissue protein are minimized by increasing their oxidation, this is unlikely to be the case for lysine and threonine. These two amino acids differ from most other IAAs with higher Km^2 values for their main oxidative catabolic enzymes, (18 and 52 mM, respectively), larger pool sizes, (about 1 mM) and with less evidence of fine control of their catabolic pathways. The difference in the handling of lysine and leucine may well be important in allowing conservation of these amino acids and recycling from postabsorptive losses for postprandial gains.

Evidence for conservation of lysine and threonine comes from muscle biopsy studies after feeding subjects protein (albumin) or a protein-free diet (Bergstrom et al., 1990). These investigators biopsied human muscle at 3 and 7 hours after a meal of 50 g of albumin. They showed that for leucine and lysine, although the intakes were the same and removal of these two amino acids into protein will be at the same rate (since their concentrations in protein are similar), the increase in the concentration of lysine was twice that of leucine, and the same was true for threonine in comparison with valine. By 7 hours, the concentration of all the BCAAs, methionine, and the aromatic amino acids had fallen below the baseline, but for threonine and lysine, there was still an excess of amino acid over the baseline value.

Some indication of how much of these free pools of lysine and threonine might be available to supplement dietary protein that was inadequate in lysine and threonine was indicated by biopsy studies after feeding subjects a protein-free meal. Subsequently, protein deposition would have been entirely dependent on the free amino acid pools (Bergstrom et al., 1990). The fall in the free lysine and threonine pools in muscle were sufficient to have enabled about 250 mg protein deposition per kg body weight. Given that for an adult consuming 0.75 g/kg/d this author's studies indicate that about 50 percent of this is deposited during feeding, (i.e., 380 mg), the free pool of lysine and threonine could in theory contribute two-thirds of the subjects' needs if the dietary source was inadequate. If this is the case, then the organism in overall balance may be less sensitive to poor dietary quality than would be expected in terms of lysine and threonine needs for transient protein deposition.

Evidence for this lack of sensitivity to protein quality is the very small difference in the efficiency of postprandial protein utilization of wheat protein in normal adults compared with that of milk (see Millward et al., 1996). Studies with either frequent small meals (Fereday et al., 1994) or a larger single meal (Fereday et al., 1997) indicate that about 80 percent of dietary wheat protein is deposited in the tissues even though the lysine and threonine content of the wheat is insufficient to allow this. Recycling of the free lysine and threonine liberated in the postabsorptive state allows efficient postprandial protein utilization.

---

[2] The Km (the Michaelis-Menton constant) is the concentration of the substrate at which half the maximum velocity of an enzyme-catalyzed reaction is achieved.

Third, evidence supports that the amplitude of diurnal cycling and the consequent need for postprandial protein deposition is adaptive according to the IAA composition of the diet. Thus, Marchini et al. (1993) showed that postabsorptive leucine balance was 7.8, 9.3, and 12.8 µmol/kg/h for subjects on isonitrogenous diets fed for 3 weeks with FAO patterns (very low IAAs), MIT patterns (intermediate IAAs), or egg patterns (high IAAs) of amino acids, respectively. That is, postabsorptive losses with the FAO pattern fell to < 60 percent of those induced by the egg diet even though the overall level of total amino acid intake was the same. This is an important adaptive response. More recent 24-hour ^{13}C leucine balances in subjects fed purified amino acid diets that varied only in terms of leucine content (14, 38, and 89 mg/kg/d; El-Khoury et al., 1994a, b) also indicated adaptive reduction in postabsorptive losses, with postabsorptive leucine oxidation rates equivalent to 34.7, 20.9, and 15 mg leucine/kg/12 h. Thus, these results point to substantial adaptive reductions in the amplitude of diurnal cycling and in the consequent metabolic demand for IAAs in response to reductions in dietary protein quality. Therefore, the qualitative influence of diurnal cycling on the metabolic demand cannot be predicted.

Summary of the Metabolic Demand for Indispensable Amino Acids

In normal adults, the MD for IAAs includes the following:

• an obligatory component, which in most cases is less than that contained in 0.33 g tissue protein and has a pattern that cannot be predicted from first principles. However, on the basis of animal data, it is likely to contain considerably less lysine and leucine than in the tissue protein pattern.

• an adaptive component, the amount and composition of which is variable according to the nature and feeding pattern of the habitual dietary protein intake, with a particular adaptive mechanism allowing conservation of lysine and threonine when their intakes are low.

A scheme showing these components of the MD throughout the diurnal cycle is shown in Figure 9-5.

FAO REQUIREMENT VALUES AND N BALANCE STUDIES

The FAO requirement values were derived from N balance studies so that any consideration of these values requires consideration of the N balance technique.

Nitrogen balance studies were initiated in the mid-nineteenth century by Carl Voit, and they have been central to the definition of protein and amino acid

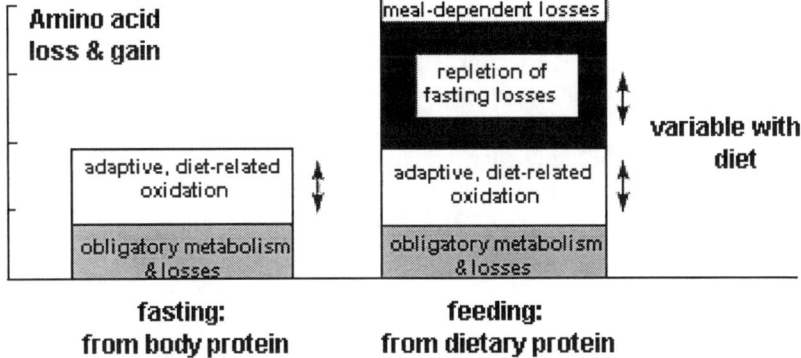

FIGURE 9-5 During fasting, body protein provides for both obligatory and adaptive demands. During feeding, the diet provides for obligatory and adaptive demands, for repletion of tissue protein lost during fasting, and for losses due to inefficiency of utilization. The adaptive components vary markedly with dietary intake, as shown in Figure 9-2, and change only slowly with intake (see Figure 9-3).

requirements ever since. Notwithstanding the simple aim of nitrogen balance studies (that is, definition of the relationship between intake and all losses—urinary, fecal, and surface), the methodology has sustained considerable criticism. For example, Young and colleagues have argued that because of "conceptual limitations" and poor reproducibility, N balance studies in adults in relation to assessment of amino acid requirements are of questionable value (Marchini et al., 1993; Young, 1986). Some of the difficulties reflect the practical problems that arise in performing N balance studies.

Measurements of N balance (i.e., N intake minus all N losses) are relatively imprecise, with balance being a small value compared with the much larger values of N intake and N excretion. This results in considerable error in the prediction of balance (see Forbes, 1973; Wallace, 1959). Hegsted (1976) commented that intake tends to be overestimated because of spillage and incomplete recovery of uneaten food, while excreta are often less than completely recovered. This has variously been attributed to denitrification by bacteria in the colon with loss of N gas (Costa et al., 1968); to an underestimate of urea losses from skin and expired ammonia (Calloway et al., 1971) with losses of nitrogen as gaseous ammonia rarely measured; to nitrate in food and urine, which is not measured by the Kjeldahl technique (Kurzer and Calloway, 1981); and possibly to endogenous NO production via NO synthase (Anggard, 1994). NO is oxidized to nitrate, which is excreted in urine, or diffuses into the colon, where it is reduced to ammonia and nitrite (MacFarlane and Cummings,

1991). Nitrate excretion and fecal N-nitroso compound levels increase in response to increased protein, nitrate, and meat intakes (Bingham et al., 1996; Rowland et al., 1991). Furthermore, although fecal losses might be assumed to be predictable according to knowledge of dietary protein digestibility, the major determinant of fecal N output is biomass arising from bacterial fermentation in the colon (Stephen and Cummings, 1979). It is dependent on dietary nonstarch polysaccharide (NSP) and other fermentable carbohydrate intake (Cummings et al., 1992). These factors result in typical errors estimated in the range of -11 to $+11$ mg/kg/d (see Millward and Roberts, 1996), more than the entire miscellaneous losses (8 mg N/kg).

A problem in very short-term studies is how to account for the changing size of the body urea pool. In subjects fed a high-protein diet, losses from the body urea N pool in a 12-hour postabsorptive period accounted for on average 34 percent of the observed N balance with individual values as high as 66 percent, with gains of similar magnitude in the postprandial state (Price et al., 1994). Without correcting for these changes the distribution of N losses between the postabsorptive and postprandial phase of the diurnal cycle would be markedly in error, with an underestimate of postprandial and an overestimate of postabsorbtive losses.

Another difficulty stems from the analysis of the balance data and estimation of a requirement value, given the nonlinearity of the balance curve. In multi-level feeding trials in adults which start at low intakes, losses rise to match intakes when body protein reaches its maximum regulated level. Thus, the intake balance curve is, in theory, curvilinear. Balance increases with intake asymptotically toward zero or (as is often observed) some positive value, as losses increase to eventually match intake. This means that there is no simple term to define the overall shape of the balance curve and allow prediction of the requirement (as the zero balance intake). In fact, as is well documented in protein requirement balance studies, prediction of a zero balance intake intercept from a few balance points by linear regression will result in requirement values that vary according to where the intake values lie on the balance curve (see Millward and Roberts, 1996). Thus, studies conducted at low intakes will underestimate requirements, while studies conducted with supramaintenance intakes will overestimate requirements. The logic of this is that (a) reliable balance studies are those conducted with intakes very close to the actual requirement and (b) studies with intakes based on preconceived requirement values will tend to confirm such preconceptions.

Finally, the conceptual limitations raised by Young relate to the problem of evaluating whether zero balance, when obtained, signifies an adequate metabolic state (i.e., appropriate rates of protein turnover and lack of any cellular pathology). Young discussed attainment of balance in terms of adaptation (metabolic changes with no adverse consequences) and accommodation, metabolic changes which include some adverse response such as a loss of body tissue or reduction of protein turnover rates. Clearly, such a response would indicate a lack of adequacy. In this author's view, minimum requirements need to be defined both in terms of balance and body nitrogen, the latter being a level of body protein judged to be appropriate for the subject's height and frame size. This raises a major design difficulty in balance studies. Losses of body N and achievement of balance at a lower body N level must be differentiated from any transient losses during an adaptation period that are gradually replaced, thus restoring appropriate body protein stores. Many of the balance studies reported have involved this initial low-protein period and the restoration stage has generally been ignored, with the test diets fed to subjects who would be depleted. This will result in very high efficiency of utilization and a probable overestimation of the requirement. Experimentally, however, the problem becomes more difficult since studies need to be long enough not only for balance to be achieved at the lowered level, but also for repletion of losses induced during the adaptation to occur. This will be most difficult to measure.

A final issue is raised by Fuller and Garlick (1994), who pointed to animal data and indicated that nitrogen equilibrium may not necessarily mean amino acid equilibrium. They cited the depletion of the tissue peptides, carnosine (β-alanyl histidine) and anserine (β-alanyl-l-methyl histidine), and the depletion of hemoglobin, a protein very rich in histidine, when histidine-free diets are fed as was shown in the original balance studies of Rose and others. They also pointed to animal experiments that showed changes in the amino acid composition of the body induced by feeding amino acid-deficient diets; for example, pigs fed low-lysine diets have less lysine, phenylalanine, and tyrosine, and more glycine and arginine in whole body protein than those fed an adequate diet.

Are There Robust N Balance Data that Support the FAO Requirement Values?

Whether the FAO requirement levels maintain N balance in normal adults has been examined in N balance studies. Weller et al. (1971) tested in one group the pattern that Rose fed or in another group the pattern with a 30 percent addition of all essential amino acids together with nonessential N to give a total level of 7 g N. These diets did not maintain N balance, but an egg pattern with at least 2 times the Rose values did maintain balance. Note that in these studies, no histidine was fed, and the overall level of N at 7 g (0.6 g protein/kg) is low compared with usual intakes.

In contrast, Young and colleagues (Marchini et al., 1993) tested the 1985 FAO pattern (with histidine), fed for 3 weeks at a higher total N level (9.8 g N, 0.9 g protein/kg), against the MIT and an egg pattern and showed that all three diets maintained N balance; as shown in Figure 9-6, there were no significant differences among FAO, MIT, or egg diets. Young rejected these data as evidence for the validity of the FAO requirement values. Since the balance studies were carefully done, with due consideration given to miscellaneous losses, rejection implies nonacceptance of N balance in adults under any circumstances, which in this author's view is an extreme and unjustifiable position.

Biological Value of Plant Proteins in Human N Balance Trials

The extent to which a very low IAA requirement for maintenance is valid can be examined in terms of whether differences in the apparent quality of plant proteins are observed in N balance trials. Such trials have been reported in children and in adults. In children, responses need to be considered in relation to the expected rates of growth and the relative magnitude of growth and maintenance demands. In fact, tissue growth represents the major part of human requirements only in the first few months of life, 60 percent of metabolic demand at 1 month, 20 percent at 12 months, 10 percent at 2 years, and lower subsequently (Dewey et al., 1996). When rapid growth is occurring, differences in dietary protein quality are observed. Graham and colleagues studied the nutritive value of various unsupplemented plant protein sources (wheat, maize, potato, rice, beans, and sorghum) in terms of both digestibility and biological value relative to casein in young children recovering from malnutrition (Graham et al.; 1979; Maclean et al., 1981). They observed that sorghum and wheat were poorly utilized, which is consistent with a lysine limitation. However, the relatively small increases in lysine and other IAAs provided by rice, maize, and improved maize strains increased the biological value (BV) to between 80 and

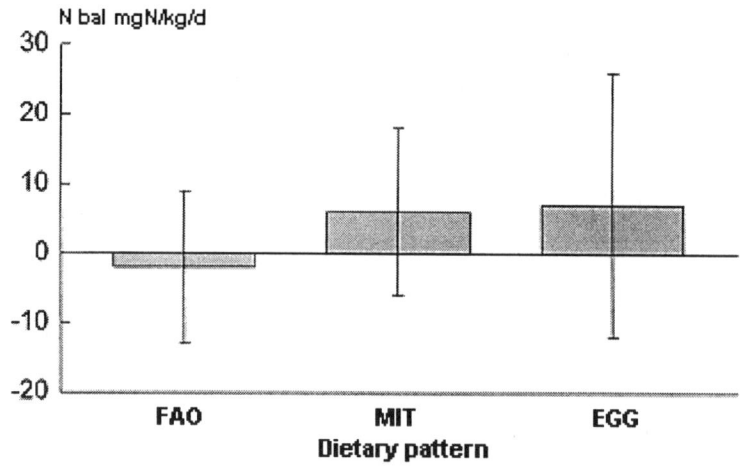

FIGURE 9-6 Nitrogen balances observed in young adults fed diets containing the FAO pattern, the MIT pattern, and an egg pattern. Diets contained 9.8 g/N, 0.9 g protein/kg, and subjects were fed for 3 weeks. NOTE: FAO, Food and Agricultural Organization; MIT, Massachusetts Institute of Technology. SOURCE: Adapted from Marchini et al. (1993).

95 percent that of casein, with potato protein actually higher than casein. Furthermore, in a more recent study (Graham et al., 1990), they showed that in children growing at 2 to 3 g/kg/d after the most rapid phase of catch-up growth has passed, the genetically improved maize was as effective as casein in supporting both weight and height growth. This suggests that in these relatively rapidly growing children, their growth needs do influence their amino acid requirement, and the maintenance component of their needs, which is still substantial, includes a lower level of IAAs compared with their growth needs. This is supported by the observation that lysine supplementation of Indian children fed a wheat-based diet had no effect on weight gain or N balance (Reddy, 1971). In other preschool children fed either wheat- or rice-based diets, there was lower height growth with wheat compared with rice, which improved with lysine supplementation. However, the rice-based diet, which would have had a low BV in rat growth trials, allowed both weight and height growth at the fiftieth centile of the National Center for Health Statistics (NCHS) standards (Begum et al., 1970).

A large number of adult trials of protein quality have been conducted with a variable range of findings. Although differences among protein sources have been reported in N balance studies in young adults (e.g., relative biological values of 0.66 for wheat compared with beef [Young et al., 1975]), when the calculated BVs of a range of plant and animal proteins measured in separate studies are examined together (see Millward et al., 1989), the differences between wheat and other proteins and mixed diets are much less apparent. This is because of the lack of reproducibility between studies with the same protein. Within individual studies, interindividual variability is very marked, with biological values often associated with coefficients of variation (CVs) of 15 to 20 percent (e.g., Young et al., 1973), and even 50 percent (Young et al., 1984). In a study of the effect of lysine supplementation of a wheat gluten-based diet, (Scrimshaw et al., 1973), which is widely quoted as indicating protein quality effects in adults, the magnitude of the response (a 2.9–7.7 percent fall in urea excretion) was such that with the CV of mean urea excretions ranging from 11 to 36 percent, the response was not statistically significant. The combination of within-study variability and poor reproducibility between trials means that statistical analysis of the data is nearly impossible.

However, there have been several long-term trials examining the effects of a wheat-based diet on both weight and maintenance of N balance. In the Minnesota bread study (Bolourchi et al., 1968), young adults were fed a "bread" diet for 50 days. The bread contained 90 to 95 percent protein from wheat flour, which supplied 67 g protein/d, 0.95 g/kg, and 18 mg/kg lysine. The result was a negative balance during the first 10 days, followed by a positive balance of 1 g N/d, sufficient to cover miscellaneous losses. Body weight was stable, and fitness, measured in terms of pulse rate with exercise, improved.

In the North Carolina wheat study (Edwards et al., 1971), responses were measured over 60 days in young men who were fed bread-based diets supplying

46 g protein [35 g from wheat (76%) and 11 g from other plant sources]. Body weights were maintained (± 4 pounds) as was N balance (at +1 g N/d), and this was not influenced by replacing 20 percent wheat by beans, rice, or peanuts.

Clark et al. (1962) fed 56 g protein (9 g N), of which 45 percent was from wheat flour and corn, together with amino acids, with lysine supplying 600, 700 or 950 mg/d (14 mg/kg). There was a progressive increase in N retention for 30 days, and at 700 and 950 mg/d balance of lysine, was maintained when 0.5 g N is included for miscellaneous losses.

Taken together, these N balance data do point to a relatively low MD for IAAs in the human maintenance requirement. Regardless of one's view about the validity of N balance, weight and fitness maintenance data from the Minnesota bread study strongly support the possibility that the MD for IAA's is very low.

Adjustment of the Food and Agriculture Organization Requirement Values

Although many of these potential sources of error in the N balance methodology preclude correction of published data, the error induced by not including surface losses, as in the N balance experiments of Rose, Leverton, and others, can be corrected. Hegsted (1963) performed a meta-analysis on published data, excluding the Rose data, which he considered poor (see Hegsted, 1973), and performing a regression analysis to calculate the requirement for zero balance and suggested that an adjustment could be made for unmeasured losses. Clearly, the magnitude of the additional losses requires careful consideration. Calloway et al. (1971) made comprehensive measurements of dermal and other N losses in adult subjects, reporting greatly increased dermal losses with exercise-induced sweating and with protein intake. They estimated a potential total loss of about 0.5 g/d; a value essentially adopted by FAO/UNU/WHO (1985) (8 mg/kg/d). Fuller and Garlick argued that since the loss is variable with protein intake and exercise, such an adjustment may be excessive in the balance experiments reviewed by Hegsted. They proposed a lower value estimated by Calloway et al. (1971) for sedentary subjects on moderate nitrogen intakes: 0.30 g N/d, 4 to 5 mg N/kg. The impact of this adjustment is shown in Table 9–5; the total requirement values were doubled for most amino acids. Nevertheless, when these values are compared with the pattern obtained using OOL, the former (values recalculated for sedentary individuals on moderate N intakes) are lower than the latter, with the exception of the TSAs, which are similar to the OOL as would be expected from the calculations. Lysine in particular is half the concentration in the OOL pattern. Given the care with which Hegsted selected and analyzed the published data to arrive at his values in Table 9-5, and given that none of the studies involved excessive energy intakes, since the Rose values were excluded, the recalculated values in Table 9-5 are probably the best estimates of the minimum requirement for indispensable amino acids available.

Stable Isotope Estimates of Requirement Values

Isotopic studies of amino acid oxidation particularly based on ^{13}C were proposed as an alternative to N balance by Young (1986). Initially, such isotopic studies involved an examination of the relationship between intake and some response, such as protein synthesis or amino acid oxidation in order to define an endpoint (e.g., Meredith et al., 1982). Subsequently, daily balances were calculated from measured rates of amino acid oxidation over a range of intakes. Balances were reported for leucine (Meguid et al., 1986a), lysine (Meredith et al., 1986), threonine (Zhao et al., 1986), and valine (Meguid et al., 1986b). Subsequently, a large number of ^{13}C leucine and some valine balances have

TABLE 9-5 FAO Requirement Values Recalculated to Allow for Miscellaneous Losses

	FAO* mg/d	mg/kg (70)§	Recalculated† mg/d	mg/kg(70)	OOL‡
Histidine	—	—	—	—	11.5
Ileucine	550	7.9	1,083	15.5	16.2
Leucine	730	10.4	1,580	22.6	27.4
Lysine	545	7.8	1,118	16.0	30.1
TSA	350	5.0	960	13.7	13.4
TAA	350	5.0	1,184	16.9	27.0
Threonine	375	5.4	942	13.5	15.5
Tryptophan	168	2.4	222	3.2	4.0
Valine	622	8.9	811	11.6	16.9
Total	3,690		7,898		162
% protein	10.5		22		48

NOTE: FAO, Food and Agricultural Organization; OOL, Obligatory oxidative loss; TAA, total aromatic amino acid; TSA, total sulfur amino acid.

* Values from Hegsted (1963) calculated for zero balance.

† With the exception of lysine, TSA and TAA, all values computed from regression equations reported by Hegsted (1963) with addition of 0.3gN losses.

Lysine is recalculated from Jones et al. (1958) with original balance data expressed per kg bodyweight, adjusted for 5 mgN/kg per day additional losses and with zero balance calculated from regression of balance (mgN/kg per d) on $\log_e$ intake (mgN/kg per d) with all reported balance periods included in the regression.

‡ Pattern in an amount of tissue protein equal to the obligatory nitrogen loss (54 mg N/kg).

§ Based on a 70-kg man.

TSA is the OOL adjusted for utilization as described by Young et al. (1989).

TAA is from Tolbert and Watts (1963).

been reported, culminating with the most recent reports in which leucine oxidation and balance have been monitored for entire 24-hour periods (El-Khoury et al., 1994a,b, 1995). Taken together, this represents a large and complex body of experimental work. Reviewing the work is difficult since the experimental design and calculation methods have changed over time as potential problems have emerged. For example, the way in which leucine oxidation has been calculated has changed from an approach that by attempting to compensate for the tracer would have underestimated the true oxidation by 5 to 15 percent (see Millward, 1993) to calculation of actual rate in current reports (El-Khoury et al., 1994b). The current 24-hour studies are state of the art although, as discussed below, some problems persist. However, as discussed elsewhere (Millward and Rivers, 1988), the initial leucine, lysine, valine, and threonine studies, which have been used as experimental support for the MIT scoring pattern, are deeply flawed. Oxidation rates are underestimated due to use of the plasma amino acid labeling as opposed to intracellular amino acid labeling as a precursor, with no account taken for losses of ^{13}C from threonine by routes other than CO_2, and with balance calculated from a single fed value and an assumed (nonmeasured) postabsorptive value. In essence, ^{13}C tracer balances can be applied in one of two ways: as measures of the intake-balance relationship for the particular tracer under examination (as with the leucine, lysine, valine, and threonine studies) or, alternatively, the balance of one amino acid such as leucine can be used as an alternative to N balance. Thus, for an indispensable amino acid like leucine, which has a small, highly regulated free pool, if the intake and oxidative loss are measured accurately, then the net accretion or loss of that amino acid can be assumed to be protein. This was the approach used by Marchini et al. (1993) to compare the FAO, MIT, and egg IAA patterns.

Many of the kinetic assumptions and isotope-related problems associated with this approach are well known (Bier et al., 1985) and were reviewed by Millward and Rivers (1988) and by Millward et al. (1991). The main problems are listed in Table 9-6. This list of problems is by no means exhaustive; it has been separated into problems that are ultimately soluble and those that are currently less tractable. Problem 1 is unavoidable with current methodology. Although the 24-hour studies (El-Khoury et al., 1994a, b) represent heroic experiments, they represent a short period in the context of balance studies, in general, with subjects necessarily restricted resulting in unknown consequences for protein balance. Day-to-day variation is large for N balance studies, some of which may reflect the operation of the regulator of body composition. The extent of day-to-day variation in leucine balance is currently unknown. El-Khoury et al. (1994a, b) reported daily balances in subjects on a high leucine intake (89 mg/d) ranging from −3.58 to +5.83 mg leucine/kg (0.76 ± 2.99 mg/kg/d) and falling to −13.4 ± 5.2 mg/kg/d on an intake of 14 mg/kg/d. This would indicate less inherent variability than N balance.

TABLE 9-6 Potential Problems Relating to ^{13}C Isotopic Measurements of Amino Acid Oxidation Rates and Amino Acid Balance

A. Practical problems with potential solutions?

1. Need for subject restriction and limitation to short periods of study.

2. The extent of isotope retention as bicarbonate.

3. Measurement of CO_2 production rates.

4. Variation in background tracer enrichment.

5. Quantitative excretion of labeled carbon as CO_2.

6. Oxidation rates as a tracer for overall N excretion influenced by mismatch of food and body tissue amino acid composition.

7. Underestimation of losses due to non-CO_2 routes.

8. Excessive positive balances in some studies.

B. Model problems: Less tractable and possibly insoluble.

9. Amount of tracer excessive and may influence balance.

10. True precursor amino acid enrichment of measured value.

Problems 2 to 4 are solvable by appropriate methodologies. Problem 5 can probably be ignored for leucine and most carboxyl-labeled amino acids, but possibly not for threonine, which exhibits unrealistic positive balances (Zhao et al., 1986), most likely because of substantial fixation as glycine (Ballevre et al., 1990). Problem 6 is important when oxidation rates are measured to calculate overall amino acid oxidation and N excretion. Thus, during the fed state with milk-based food, for example, the higher leucine content in milk (10%) compared with tissue protein (8%) means that excess leucine oxidation during feeding will occur, and unless corrected for, will result in an underestimate of overall amino acid balance during feeding (see Price et al., 1994). Problem 7 is important but usually ignored. Any route of leucine loss as the entire amino acid (free or protein bound in skin, hair, urinary and fecal peptides, or fecal bacterial protein) will not be measured. Although the magnitude of this loss is unknown, it could represent the equivalent of several mgs of N/kg. Problem 8 is observed in several reported studies. Indeed, in the studies of Marchini et al. (1993), the leucine balances ranged from −100 µmol/kg/d on the FAO diet to +180 µmol/kg/d on the egg diet. These values were equivalent to changes in body weight over the 3 weeks of the study of −1.5 kg to +2.5 kg. Whether such changes in body weight occurred is not mentioned by the authors.

As for problem 9, the fact that in stable isotope studies, the tracer is not massless but may be infused at significant rates compared with either dietary intakes or measured oxidation rates is a serious problem. In some circumstances, the tracer infusion rate can be of the same magnitude as the oxidation rate that it is purporting to trace. For example in a valine infusion (Pelletier et al., 1991a), the tracer infusion rate was 80 percent of the oxidation rate being studied. In the lysine studies reported by Meredith et al. (1986) on the lowest intakes of lysine, the infusion rate was 72 percent of the reported oxidation rate, equal to 3 times the rate of lysine intake from the diet at the lowest dietary intake studied. In other leucine infusion studies (Pelletier et al., 1991b), 8.2 μmol/kg/h of leucine was infused, equivalent over 12 hours to an intake of 13 mg/kg leucine in individuals fed 13 mg/d of leucine. Clearly, the way in which the infusion intake is assumed to influence the system under study and the consequent way it is treated in calculating oxidation rates and balance can markedly influence the results.

This problem was discussed by Fuller and Garlick (1994), who considered several issues. One was treatment of the tracer in 24-hour balances extrapolated from measurements made during a few hours in subjects in the fed and fasted states. In the initial studies, only the labeled amino acid intake given during the actual period of infusion was included (e.g., 6, 42–44), which would have underestimated balance by 9 to 13 mg. A second issue was that the tracer represents an additional level of intake above the dietary level to which the subjects have adapted, so that what is studied is not what has been previously fed. In the 24-hour studies (El-Khoury et al., 1994a,b), the diet was adjusted on the infusion day by removing an amount of leucine equal to that infused (about 10 mg). Although this means an overall (24-hour) leucine intake that is the same as during the adaptation period, the diet supplied less leucine than the nominal intake during the feeding period because of the tracer infused in the postabsorptive period. In fact at the lowest dietary level of 14 mg/kg/d, this reduced the actual intake during the feeding period (10 hours) to about 8 mg (food + tracer), the other 6 mg (tracer) being supplied during the postabsorptive 14-hour period. The likely effect of this would be to reduce the capacity for net postprandial protein synthesis through leucine limitation and to increase oxidation and negative balance in the postabsorptive state, with an overall leucine balance that is more negative than otherwise. El-Khoury et al. (1994) did show a gradual increase in leucine oxidation during the fasting period, as the excess was oxidized, and a fall during feeding, because of the high efficiency of utilization on the very low-leucine diet. Attempting to study the limits of adaptation of leucine oxidation and balance with a nutritionally significant amount of the tracer amino acid is highly unsatisfactory. Fuller and Garlick (1994) concluded that this problem is likely to give rise to an appreciable overestimate of the amino acid requirement.

Finally, Problem 10 represents a "model" problem resulting from an inadequate understanding of the kinetic model under analysis and especially the uncertainty about the true precursor isotopic abundance compared with what is measured. Currently, only leucine and the other branched-chain amino acids allow any solution in terms of their keto acids in plasma, which derive from actual intracellular precursor pools. Even in these cases, the extent of any error remains unknown.

Price et al. (1994) attempted to test the assumptions empirically by comparing rates of leucine oxidation with measured N excretion rates in subjects in the fed and fasted states who were fed protein in amounts ranging from 0.33 to 2.2 g/kg/day. Leucine oxidation rates varied in proportion to N excretion over the entire range of N excretion rates, although the amounts of leucine oxidation were lower than expected. The ratio of predicted versus measured N excretion for the entire group was 0.79 (standard deviation 0.23, n = 38). In other words, leucine oxidation resulted in a 27 percent overestimate of balance compared with N.

In summary, ^{13}C leucine oxidation studies, while not entirely free from potential problems, do appear to allow reasonable estimates of whole body protein balance when applied with care, particularly when intakes of leucine are not especially low. The 24-hour ^{13}C leucine balances currently employed by Young and colleagues involve far fewer assumptions and uncertainties compared with the original tracer studies used to validate the MIT scoring pattern, which can, in this author's view, be generally discounted. Nevertheless at low leucine intakes, the tracer problem remains a serious one. Uncertainty still remains as to whether the apparent inability of 14 mg/kg/d to maintain balance is due to excessive postabsorptive oxidation of the tracer.

Toronto "Break Point" Studies

The Toronto Break Point studies are important because they represent the only stable isotope data on phenylalanine (Zello et al., 1990) and the only lysine study (Zello et al., 1992) apart from the MIT data. Furthermore, the lysine study involved a method that does not depend on balance, that is, the indicator amino acid oxidation method.

As shown in studies of animals fed amino acid mixtures, when the test amino acid intake falls below the requirement for adequate postprandial net protein synthesis, oxidation of a labeled indicator amino acid increases. The lysine study (Zello et al., 1992), measured ^{13}C phenylalanine oxidation in subjects fed diets with fixed phenylalanine intakes but varying lysine intakes. The break point or increase in phenylalanine oxidation indicated the lysine intake that is inadequate. The phenylalanine study (Zello et al., 1990) involved measurement of postprandial phenylalanine oxidation on diets with decreasing amounts of phenylalanine.

The important problem with these studies is their design and interpretation. Phenylalanine oxidation is calculated from the isotopic enrichment of plasma phenylalanine, not tyrosine, and in fact, the investigators give excess tyrosine in the diets. The authors argued that phenylalanine is oxidized from a pool that does not equilibrate with tyrosine so that plasma tyrosine enrichment, which was very low compared with that of phenylalanine, was not the relevant precursor enrichment. If their assumption is not correct, then their data are uninterpretable because (a) the $^{13}CO_2$ will reflect the oxidation rates of both phenylalanine and tyrosine and (b) the excess tyrosine will dominate the oxidation of phenylalanine and blunt any usefulness the protocol might have in assessing phenylalanine oxidation.

It is difficult to accept the assumptions of Zello's studies in light of previous studies by Thompson et al. (1989), who developed Clark and Bier's original study (1982). Tyrosine labeling would be expected to be low since the labeled tyrosine formed by phenylalanine hydroxylation would be diluted by the tyrosine flux and would not be expected to achieve plateau enrichment within the 4 hour infusion without priming. Since these studies were measuring phenylalanine oxidation, the true shape of the oxidation curve in relation to the lysine intake is not known. In this author's view, neither study can be interpreted in a meaningful way in relation to either lysine or phenylalanine requirements. Therefore, there are no reliable kinetic data for a minimum lysine requirement.

Functional Indicators of Adequacy of Intakes Allowing Zero Balance

Protein Turnover

One proposed advantage of tracer studies is that they allow validation of balance in terms of measurement of protein turnover. In this context, Young and Marchini (1990) reported marked reductions in whole body protein synthesis in subjects fed low levels of specific IAAs. In fact, it is likely that these data are artifactual.

Most would agree that if reductions in protein turnover occurred in response to reduced protein or amino acid intakes, it would be a cause for concern. In fact, although the animal data for reductions in protein turnover with low-protein diets are quite clear (e.g., Jepson et al., 1988; Waterlow et al., 1978), evidence for this in humans is generally scant even in chronically undernourished Indian laborers (Soares et al., 1991). Considerations of the significance of changes in protein synthesis need to be made in an appropriate framework. In the context of the diurnal cycling model of balance, the increasing amplitude of gains and losses with increasing protein intake, does involve changes in protein synthesis (although changes in proteolysis are the main mediator of cycling). As a result, postprandial synthesis is higher for someone on a high- rather than low-protein intake, although postabsorptive synthesis rates are not changed (Pacy et al., 1994). Thus the changes in protein synthesis are only related to the increasing amplitude of diurnal cycling. Since there is no obvious physiological benefit from any particular amplitude of diurnal cycling as long as overall balance is achieved, there is no obvious reason to be concerned if dietary protein or indispensable amino acid levels result in lower amplitudes of diurnal cycling and lower rates of postprandial protein synthesis. In fact, Pacy et al. (1994) reported that the mean daily rate of protein synthesis, S, was higher on very high protein intakes, (S=4.55 ± 0.5 g/kg/d), compared with 2 weeks on a very low protein diet of 0.36 g/kg/d (S=4.09 ± 0.63 g/kg/d), although $P > 0.05$, but even this was explainable by the fact that overall daily leucine balances became positive at 0.82 g protein/kg/d. Thus turnover *per se* (i.e., replacement) was not increased, and it was concluded that any influence of protein intake on protein turnover was below the detection limits of current methods. In general, previous results provided little support for a nutritional sensitivity of whole body protein turnover within the range of intakes likely to be consumed.

Marchini et al. (1993) reported no significant changes in protein synthesis in the postabsorptive or postprandial state on diets with the FAO, MIT, or egg amino acid patterns. The only kinetic responses commented upon were a reduced rate of postabsorptive proteolysis after 3 weeks on the FAO diet and a lack of a feeding-induced inhibition. Although this explains the lack of postprandial gain and overall negative balance, it appears that no changes in turnover occurred in these studies.

The one exceptional body of data is the large reduction in protein synthesis observed in subjects fed very low levels of individual amino acids in the MIT studies (Young and Marchini, 1990). However, Pacy and coworkers (1994) have suggested that since such responses are only observed when the tracer is the same as the limiting amino acid, and not observed in response to an identical dietary protocol when turnover is measured by a nonlimiting labeled amino acid (e.g., Zello et al., 1992), a methodological problem associated with either compartmentation or the nonsteady state may exist.

On this basis, it appears that the data suggesting that protein synthesis falls at inadequate intakes of individual amino acids are artifactual. Thus, with balanced amino acid mixtures and whole proteins, whole body protein synthesis changes only minimally at intakes below those capable of supporting overall balance. Thus, protein turnover is not a useful indicator of nutritional adequacy that can be used in the debate.

Other Metabolic Influences of Amino Acids

A minimal requirement for protein and amino acids may be identifiable with sufficient time allowed for adaptation. This value is likely to be much lower than the amounts provided by natural diets that also provide sufficient energy and other nutrients. Thus, the magnitude of this minimal requirement becomes to some extent only an issue of scientific curiosity. It might be considered that the requirement for balance is less important than the functional consequences of particular levels of protein in the diet. In other words, the key issue is that of an optimal requirement, which not only allows balance but also supports optimal body function such as optimal immune function, minimal long-term deterioration of renal function, and in the present context, physical performance, especially under stress.

Although the issue of an optimal protein requirement is generally recognized as important, there is little qualitative or quantitative information. In growing animals, dietary protein intakes in excess of those associated with maximum efficiency of protein utilization have been shown to increase rates of bone growth through hormonal mechanisms involving insulin, thyroid hormones, and insulin-like growth factor-I (IGF-I). This has led to the concept of an "anabolic drive" of dietary protein exerted by protein intakes in excess of minimal needs (see Millward and Rivers, 1989). However, the extent to which this is important in human growth has yet to be explored.

The importance of dietary protein for the maintenance of the immune system has also been suggested, with dietary protein influencing the human response to infection through several mechanisms (see Reeds and Becket, 1996). One mechanism is the maintenance of gut barrier function, through provision of threonine, cysteine, and other amino acids involved in synthesis of mucus glycoproteins. Another mechanism is maintenance of general immuno-competence through provision of specific amino acids for synthesis of cellular proteins of the immune system and support of the hepatic acute-phase protein response. In particular, glutathione, a key free radical scavenger synthesized from glutamate (glutamine), glycine, and cysteine is depleted in hepatic and intestinal mucosa in protein-restricted animals, in the erythrocytes of infants suffering from kwashiorkor. Also infants with low birth weight have low circulating levels of glutathione. The reduced levels of glutathione in rats are restored by providing cysteine in the diet. The influence of protein intake on glutamine levels is also a possible factor influencing immune function.

Glutamine is strongly concentrated in skeletal muscle, appears to play a specific role in maintaining function of rapidly proliferating cells such as lymphocytes and mucosal enterocytes, and regulates muscle protein turnover. Under conditions of infection and trauma, muscle concentrations of glutamine fall, and this may limit the provision of glutamine for the immune system or to the splanchnic bed, where it appears to play a specific role in maintaining glutathione synthesis during trauma. Taurine, a beta-amino sulfonic acid derived from cysteine, appears to be an effective scavenger of peroxidation products (particularly those containing the oxychloride groups) and acts as a neuromodulatory agent. Taurine is specifically concentrated in skeletal musculature and in the central nervous system. Creatine is a compound that is crucial for energy flow within skeletal muscle and is synthesized from glycine, arginine, and a suitable source of methyl groups. Creatine is concentrated in skeletal muscle and brain and is lost from muscle during infection and trauma, with a deterioration of muscle contractile function. Because glutamine, creatine, and taurine are maintained at substantial concentrations in the free amino acid fraction of milk, provision of all three compounds is believed to be important in supporting postnatal development. Finally, provision of arginine is important in relation to the immune system. This may relate to its role as precursor for synthesis of nitric oxide, a key regulator of a variety of physiological processes that include the regulation of hepatic protein metabolism, macrophages' killer function, and interactions between macrophages and lymphocyte adhesion and activation. However, as yet the quantitative relationship between intakes of any of these individual amino acids and function has not been quantified.

As for physical activity, it is often assumed that increasing protein intakes is of benefit (e.g., Lemon, 1996), but much of the data indicating apparent increased protein requirements of athletes are based on arguably misleading N balance studies. In this author's view, there is no unequivocal evidence and some data suggesting potential disadvantages (see Millward et al., 1994). However, as is often the case with the nutrition literature, the key experiments are described in the classic texts written decades ago. Chittenden took a group of soldiers and elite Yale University athletes and persuaded them to reduce their protein intake by 50 percent over 5 months, mainly by switching to a vegetarian diet of about 0.75 g protein/kg/d. Extensive measures of strength were made. Over the 5 months, soldiers' strength increased by > 85 percent (largely due to a training effect), but even the athletes' strength increased on average by 35 percent (see Table 9-7), coupled with a fall in perceived fatigue. Chittenden

TABLE 9-7 Strength of Male Athletes Before and After a 5-Month Period of a Reduced-Protein Diet[*]

Athlete	Total Strength[†]		
	Initial	Final	Increase (%)
1	4,913	5,722	16
2	6,016	9,472	57
3	5,993	8,165	36
4	2,154	3,983	85
5	4,584	5,917	29
6	4,548	5,917	30
7	5,728	7,135	23
8	5,351	6,833	28
		Mean	38

[*] Exclusion of most dietary meat and reduction of protein intake by "more than 50%" to about 55 g protein/d over 5 months.

[†] Sum of 15 distinct tests of measured strength and work performed in various exercises (Chittenden, 1907).

concluded that his experiments "afford reasonable proof of the beneficial effects of a lowered protein intake upon the muscular strength of man," and that "man can profitably maintain nitrogen equilibrium and body weight upon a much smaller amount of protein food than he is accustomed to consume (Chittenden, 1907, p. 207)." In the context of minimal and optimal protein requirements, Chittenden defined an optimal requirement of about 0.75 g/kg, the current safe allowance, with most of it coming from plant protein sources. This is significantly less than average intakes. Chittenden's views were controversial at the time and have been given less prominence than they deserve in the intervening years of continued controversy. However, the data speak for themselves.

AUTHOR'S CONCLUSIONS AND RECOMMENDATIONS

Defining amino acid requirements is inherently difficult even in the context of "normal" needs, let alone in the specific context of this workshop. Figure 9-7 summarizes what can and cannot be defined in relation to amino acid requirements and, especially, the problem of assessing nutritional adequacy of proteins by scoring. The scheme is drawn for lysine, but the principle applies generally. A requirement for rapid growth can be defined, and it may well be close to the FAO preschool values. A minimum obligatory metabolic demand

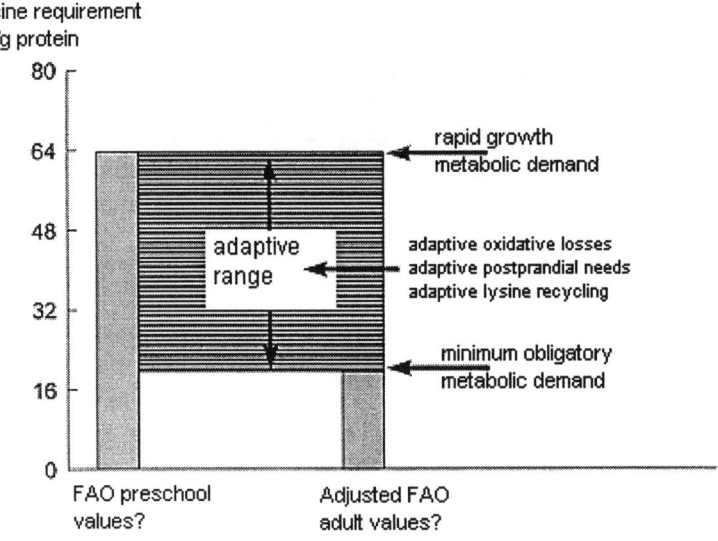

FIGURE 9-7 Amino acid requirements can be defined for rapid growth, which may be close to the FAO preschool values, and for a minimum obligatory metabolic demand, possibly represented by the adjusted values in Table 9-5. Assessment of the nutritional value of intakes in the adaptive range between these two extremes is problematic. NOTE: FAO, Food and Agricultural Organization.

can be defined, and this is low, possibly represented by the adjusted values in Table 9-5. The problem arises in assessing the nutritional value of intakes in the adaptive range between these two extremes. There are mechanisms allowing considerable adaptation of MD to match intakes in this adaptive range, but because it is not easy to predict whether complete adaptation will occur; one cannot be confident of the nutritional value of the protein. This means that the protein cannot be scored. Protein scoring as a means of predicting net protein utilization (NPU) in animals was adopted after demonstrating that it worked (more or less). That is, the score was closely correlated with experimentally determined NPU within a relatively simple animal model in which tissue growth dominated needs. Humans have a much more complex MD that cannot be defined as a fixed quantity. In the absence of suitable indicators of adequacy that allow judgments to be made about optimal requirements, which are independent of balance, a scoring pattern for assessment of nutritional value in the adaptive range cannot be defined and is unlikely to be useful. Although a scoring pattern based on the adjusted FAO values in Table 9-5 may represent an ultimate minimum nutritional value, actual adequacy could not be judged without long-term studies of the extent to which diets with intakes as low as this allowed balance, health, and fitness.

Thus, the key test of adequacy of either protein or amino acid intake must be the long-term response in terms of the specific function of interest. A specific question that this workshop addressed related to amino acid requirements is the following: Is there evidence that a different mix of amino acids would optimize military performance (cognitive function, long-term muscle mass preservation) during high workload, psychological stress, and/or energy deficit? Given the uncertainty about amino acid requirements for normal health, it is not surprising that this question cannot currently be answered. As far as gender issues are concerned, none of the investigations that provided the FAO values identified gender differences in normal requirements.

Regarding protein needs, what would be of interest is a controlled trial repeating Chittenden's measurements, with both protein level and quality as independent variables. Although N balance and stable isotope studies would be of great interest in such studies, the main outcome test of adequacy would be, simply, performance. There is no reason why such a trial should not be undertaken.

REFERENCES

Allison, J.B., J.A. Anderson, and R.D. Seeley. 1947. Some effects of methionine on the utilization of nitrogen in the adult dog J. Nutr. 33:361–370.

Anggard, E. 1994. Nitric oxide: mediator, murderer and medicine. Lancet 343:1199–1206.

Ballevre, O., A. Cadenhead, A.G. Calder, W.D. Rees, G.E. Lobley, M.F. Fuller, and P.J. Garlick. 1990. Quantitative partition of threonine oxidation in pigs: Effect of dietary threonine. Am. J. Physiol. 259:E483–E491.

Begum, A., A.N. Radhakrishnan, and S.M. Pereira. 1970. Effect of amino acid composition of cereal-based diets on growth of preschool children. Am. J. Clin. Nutr. 23:1175–1183.

Bender, A.E. 1961. Determination of the nutritive value of proteins by chemical analysis. Pp. 407–415 in Progress in Meeting Protein Needs of Infants and Preschool Children. Pub. 843. Washington, D.C.: National Academy of Sciences and National Research Council.

Benevenga, N.J., M.J. Gahl, T.D. Crenshaw, and M.D. Fink. 1994. Protein and amino acid requirements for maintenance and growth of laboratory rats. J. Nutr. 124:451–453.

Bergstrom, J, P. Furst, and E. Vinnars. 1990. Effect of a test meal with and without protein on muscle and plasma free amino acids. Clin. Sci. 79:331–337.

Bier, D.M., D.E. Matthews, and V.R. Young. 1985. Interpretation of amino acid kinetic studies in the context of whole body protein metabolism. Pp. 27–36 in Substrate and Energy Metabolism in Man, J.S. Garrow and D. Halliday, eds. London and Paris: John Libbey.

Bingham, S., B. Pignatelli, J. Pollock, A. Ellul, C. Mallaveille, G. Gross, S. Runswick, J.H. Cummings, and I.K. O'Neill. 1996. Does increased formation of endogenous N nitroso compounds in the human colon explain the association between red meat and colon cancer? Carcinogenesis 17:515–523.

Bolourchi, S., C.M. Friedeman, and O. Mickelson. 1968. Wheat Flour as a source of protein for adult human subjects. Am. J. Clin. Nutr 21:827–835.

Calloway, D.H., A.C.F. Odell, and S. Margen. 1971. Sweat and miscellaneous nitrogen losses in human balance studies. J. Nutr. 101:775–786.

Chittenden, R.H. 1907. The Nutrition of Man. London: Heinemann.

Clugston, G., K.G. Dewey, C. Fjeld, D.J. Millward, P. Reeds, N.S. Scrimshaw, K. Tontisirin, J.C. Waterlow, and V.R. Young. 1996. Report of the working party on protein and amino acid requirements. Eur. J. Clin. Nutr. Suppl. 1:S193–S195.

Clark, H.E., L.L. Reitz, T.S. Vacharotayan, and E.T. Mertz. 1962. Effect of certain factors on nitrogen retention and lysine requirements of adult human subjects. J. Nutr. 78:173–178.

Clarke, J.T.R., and D.M. Bier. 1982. The conversion of phenylalanine to tyrosine in man. Direct measurement by continuous intravenous tracer infusions of L-[$ring$-$^2H_5$] phenylalanine and L-[1-^{13}C] tyrosine in the postabsorptive state. Metabol. 31:999–1005.

Costa, G., L. Ullrich, F. Kantor, and J.F. Holland. 1968. Production of elemental nitrogen by certain mammals including man. Nature 281:546–551.

Cummings, J.H., S.A. Bingham, K.W. Heaton, and M.A. Eastwood. 1992. Quantitative estimates of bowel habit, bowel disease risk and the role of diet. Gastroenterol. 103:1783–1789.

Dewey, K.G., G. Beaton, C. Fjeld, B. Lonnerdal, and P. Reeds. 1996. Protein requirements of infants and children. Eur. J. Clin. Nutr. 50:(suppl. 1):S119–S147.

Edwards, C.H., L.K. Booker, C.H. Rumph, W.G. Wright, M.S. Ganapathy, and Ganapathy Seetha. 1971. Utilization of wheat by adult man: nitrogen metabolism, plasma amino acids and lipids. Am. J. Clin. Nutr. 24:181–193.

El-Khoury, A.E., N.K. Fukagawa, M. Sanchez, R.H. Tsay, R.E. Gleason, T.E. Chapman, and V.R. Young. 1994a. Validation of the tracer-balance concept with reference to leucine: 24-h intravenous tracer studies with L-[1-^{13}C]leucine and [15N-15N]urea. Am. J. Clin. Nutr. 59(5):1000–1011.

El-Khoury, A.E., N.K. Fukagawa, M. Sanchez, R.H. Tsay, R.E. Gleason, T.E. Chapman, and V.R. Young. 1994b. The 24-h pattern and rate of leucine oxidation, with particular reference to tracer estimates of leucine requirements in healthy adults. Am. J. Clin. Nutr. 59(5):1012–1020.

El-Khoury, A.E., M. Sanchez, N.K. Fukagawa, R.E. Gleason, R.H. Tsay, and V.R. Young. 1995a. The 24-h kinetics of leucine oxidation in healthy adults receiving a generous leucine intake via three discrete meals. Am. J. Clin. Nutr. 62(3):579–590.

FAO/WHO. 1973. Energy and protein requirements. Report of a joint FAO/WHO Ad Hoc expert committee. WHO Technical Report Series No. 522. Geneva: World Health Organization.

FAO/WHO. 1991. Protein quality evaluation. Report of a joint FAO/WHO expert consultation. Food and Agricultural Organization Food and Nutrition Paper 51. Geneva: Food and Agricultural Organization.

FAO/WHO/UNU. 1985. Energy and protein requirements. Report of a joint expert consultation. World Health Organization Technical Report Series No. 724. Geneva: World Health Organization.

Fereday, A., N. Gibson, M. Cox, D. Halliday, P.J. Pacy, and D.J. Millward. 1994. Postprandial utilization of wheat protein in normal adults. Proc. Nutr. Soc. 53:201a.

Fereday, A., N. Gibson, M. Cox, D. Halliday, P.J. Pacy, and D.J. Millward. 1997a. Postprandial protein utilization of wheat protein from a single meal in normal adults. Proc. Nutr. Soc. 56:80a.

Forbes, G.B. 1973. Another source of error in the metabolic balance method. Nutr. Rev. 31:297–300.

Fuller, M.F., and P.J. Garlick. 1994. Human amino acid requirements: Can the controversy be resolved? Annu. Rev. Nutr. 14:217–241.

Fuller, M.F., R. McWilliam, T.C. Wang, and L.R. Giles. 1989. The optimum dietary amino acid pattern for growing pigs; requirements for maintenance and for tissue protein accretion. Brit. J. Nutr. 62:255–267.

Fuller, M.F., A. Milne, C.I. Harris, T.M. Reid, and R. Keenan. 1994. Amino acid losses in ileostomy fluid on a protein-free diet. Am. J. Clin. Nutr. 59(1):70–73.

Gahl, M.J., M.D. Finke, T.D. Crenshaw, and N.J. Benevenga. 1991. Use of a four-parameter logistic equation to evaluate the response of growing rats to ten levels of each indispensable amino acid. J. Nutr. 121(11):1720–1729

Graham, G.G., E. Morales R.P. Placko, and W.C. Maclean. 1979. Nutritive value of brown and black beans for infants and small children. Am. J. Clin. Nutr 32:2362–2366.

Graham, G.G., J. Lembcke, and E. Morales. 1991. Quality-protein maize as the sole source of dietary protein and fat for rapidly growing young children. Pediatrics 85(1):85–91.

Hegsted, D.M. 1973. The amino acid requirements of rats and human beings. Pp. 275–292 in Proteins in Human Nutrition, J.W.G. Porter and B.A. Rolls, eds. London: Academic Press.

Hegsted, D.M. 1963. Variation in requirements of nutrients: Amino acids. Fed. Proc. 22:1424–1430.

Hegsted, D.M. 1976. Balance studies. J. Nutr. 106:307–311.

Jepson, M.M., P.C. Bates, and D.J. Millward. 1988. The role of insulin and thyroid hormones in the regulation of muscle growth and protein turnover in response to dietary protein. Brit. J. Nutr. 59:397–415.

Jones E.M., C.A. Bauman and M.S. Reynolds. 1956. Nitrogen balances in women maintained on various levels of Lysine. J. Nut. 60:549-559.

Kurzer, M.S., and D.H. Calloway. 1981. Nitrate and nitrogen metabolism in men. Am. J. Clin. Nutr. 34:1305–1313.

Lemon, P.W. 1996. Is increased dietary protein necessary or beneficial for individuals with a physically active lifestyle? Nutr. Rev. 54:S169–S175.

MacFarlane, G., and J.H. Cummings. 1991. The colonic flora, fermentation, and large bowel digestive function. Pp. 51–92 in The Large Intestine: Physiology, Pathophysiology and Disease, S. Phillips, J.H. Peniberton, and R.G. Shorter, eds. New York: Raven.

Maclean, W.C., G.L. De Romana, R.P., Placko, and G.G. Graham. 1981. Protein quality and digestibility of sorghum in pre school children: Balance studies and plasma free amino acids. J. Nutr. 111:1928–1936.

Marchini, J.S., J. Cortiella, T. Hiramatsu, T.E. Chapman, and V.R. Young. 1993. Requirements for indispensable amino acids in adult humans: Longer-term amino acid kinetic study with support for the adequacy of the Massachusetts Institute of Technology amino acid requirement pattern. Am. J. Clin. Nutr. 58:670–683.

Meguid, M.M., D.E. Matthews, D.M. Bier, C.N. Meredith, J.S. Soeldner, and V.R. Young 1986a. Leucine kinetics at graded leucine intakes in young men. Am. J. Clin. Nutr. 43:770–780.

Meguid, M.M., D.E. Matthews, D.M. Bier, C.N. Meredith, and V.R. Young. 1986b. Valine kinetics at graded valine intakes in young men. Am. J. Clin. Nutr. 43:781–786.

Meredith, C., D.M. Bier, M.M. Meguid, D.E. Matthews, Z. Wen, and V.R. Young. 1982. Whole body amino acid turnover with ^{13}C tracers: A new approach for estimation of human amino acid requirements. Pp. 42–59 in Clinical Nutrition '81, R.I.C. Wesdorp and P.B. Soeters, eds. Edinburgh and London: Churchill Livingstone.

Meredith, C.N., Z-M. Wen, D.M. Bier, D.E. Matthews, and V.R. Young. 1986. Lysine kinetics at graded lysine levels in young men. Am. J. Clin. Nutr. 43:787–794.

Millward, D.J. 1990. Amino acid requirements in adult man. Am. J. Clin. Nutr. 51:492–493.

Millward, D.J. 1992. The metabolic basis of the amino acid requirement. Pp. 31–57 in Protein-Energy Interactions I/D/E/C/G, N.W. Scrimshaw and B. Schurch, eds. Lausanne, Switzerland: Nestle Foundation.

Millward, D.J. 1993. Stable-isotope-tracer studies of amino acid balance and human indispensable amino acid requirements Am. J. Clin. Nutr. 57(1):81–86.

Millward, D.J. 1994. Can we define indispensable amino acid requirements and assess protein quality in adults? J. Nutr. 124:1509–1516.

Millward, D.J., and J.P.W. Rivers. 1988. The nutritional role of indispensible amino acids and the metabolic basis for their requirements. Eur. J. Clin. Nutr. 42:367–393.

Millward, D.J., and J.P. Rivers. 1989. The need for indispensable amino acids: The concept of the anabolic drive. Diab. Metab. Rev. 5(2):191–211.

Millward, D.J., and P.J. Pacy. 1995. Postprandial protein utilization and protein quality assessment in man. Clin. Sci. 88:597–606.

Millward, D.J., and J.C. Waterlow. 1996. Letter to the editor. Eur. J. Clin. Nutr. 50:832–833.

Millward, D.J., and S.R. Roberts. 1996. Protein requirement of older individuals. Nutr. Res. Rev. 9:67–88.

Millward, D.J., A.A. Jackson, G. Price, and J.P.W. Rivers. 1989. Human amino acid and protein requirements: Current dilemmas and uncertainties Nutr. Res. Revs. 2:109–132.

Millward, D.J., G.M. Price, P.J.H. Pacy, and D. Halliday. 1990. Maintenance protein requirements: the need for conceptual revaluation. Proc. Nutr. Soc. 49:473–487.

Millward, D.J., G. Price, P.J.H. Pacy, and D. Halliday. 1991. Whole body protein and amino acid turnover in man: What can we measure with confidence? Proc. Nutr. Soc. 50:197–216.

Millward, D.J., J.L. Bowtell, P. Pacy, and M.J. Rennie. 1994. Physical activity, protein metabolism and protein requirements. Proc. Nutr. Soc. 53(1):223–240.

Millward, D.J., A. Fereday, N. Gibson, and P.J. Pacy. 1996. Postprandial protein metabolism. Ballieres Clin. Endocrinol. Metabol. 10(4):533–549.

Munro, H.N. 1964a. General aspects of the regulation of protein metabolism by diet and hormones. Pp. 381-481 in Mammalian Protein Metabolism, Vol. 1., Munro H.N. and J.B. Allison, eds. New York: Academic Press.

Osborne, T.B., and L.B. Mendel. 1916. The amino-acid minimum for maintenance and growth, as exemplified by further experiments with lysine and tryptophan. J. Biol. Chem. 25:1–8.

Pacy, P.J., G.M. Price, D. Halliday, M.R. Quevedo, and D.J. Millward. 1994. Nitrogen homeostasis in man: The diurnal responses of protein synthesis and degradation and amino acid oxidation to diets with increasing protein intakes. Clin. Sci. 86(1):103–116.

Pelletier, V., L. Marks, D.A. Wagner, R.A. Hoerr, and V.R. Young. 1991a. Branched-chain amino acid interactions with reference to amino acid requirements in adult men: Valine metabolism at different leucine intakes. Am. J. Clin. Nutr. 54:395–401.

Pelletier, V., L. Marks, D.A. Wagner, R.A. Hoerr, and V.R. Young. 1991b. Branched-chain amino acid interactions with reference to amino acid requirements in adult men: Leucine metabolism at different valine and isoleucine intakes. Am. J. Clin. Nutr. 54:402–407.

Pineda, O., B. Torun, F.E. Viteri, and G. Arroyave. 1981. Protein quality in relation to estimates of essential amino acids requirements. Pp. 131-139 in Protein Quality in Humans: Assessment and *In Vitro* Estimation, C.E. Bodwell, ed. Westport, Conn.: AVI Publishing Company Inc.

Price, G.M., D. Halliday, P.J. Pacy, M.R. Quevedo, and D.J. Millward. 1994. Nitrogen homeostasis in man: 1. Influence of protein intake on the amplitude of diurnal cycling of body nitrogen. Clin. Sci. 86:91–102.

Quevedo, M.R., G.M. Price, D. Halliday, P.J. Pacy, and D.J. Millward. 1994. Nitrogen homeostasis in man: 3. Diurnal changes in nitrogen excretion, leucine oxidation and whole body leucine kinetics during a reduction from a high to a moderate protein intake. Clin. Sci. 86:185–193.

Reddy, V. 1971. Lysine supplementation of wheat and nitrogen retention in children. Am. J. Clin. Nutr. 24:1246–1249.

Reeds, P.J., and P.R. Becket. 1996. Protein and amino acids. Pp. 67–86 in Present Knowledge in Nutrition, 7th ed., E.E. Ziegler and L.J. Filer, eds. Washington, D.C. : ILSI Press.

Rose, W.C. 1957. The amino acid requirements of adult man. Nutr. Abstr. Rev. 27(3):631–647.

Rowland, I.R., T. Granli, O.C. Bockman, P.E. Key, and R.C. Massey. 1991. Endogenous N nitrosation in man assessed by measurement of apparent total N nitroso compounds in faeces. Carcinogenesis 12:1359–1401.

Said, A.K., and D.M. Hegsted. 1970. Response of adult rats to low dietary levels of essential amino acids. J. Nutr. 100:1363–1376.

Scrimshaw, N.S., Y. Taylor, and V.R. Young. 1973. Lysine supplementation of wheat gluten at adequate and restricted energy intakes in young men. Am. J. Clin. Nutr. 26:965–972.

Soares, M.J., L.S. Piers, P.S. Shetty, S. Robinson, A.A. Jackson, and J.C. Waterlow. 1991. Basal metabolic rate, body composition and whole body protein turnover in Indian men with differing nutritional status. Clin. Sci. Colch. 81(3):419–425.

Stephen, A.M., and J.H. Cummings. 1979. The influence of dietary fibre on fecal nitrogen excretion in man. Proc. Nutr. Soc. 38:141A.

Thompson, G.N., P.J.H. Pacy, H. Merritt, G.C. Ford, M.A. Read, K.N. Cheng, and D. Halliday. 1989. Rapid measurement of whole body and forearm protein turnover using a [^{2}H$_5$]phenylalanine model. Am. J. Physiol. 256:E631–E639.

Tolbert, B., and J.H. Watts. 1963. Phenylalanine requirements of women consuming a minimal tyrosine diet and the sparing effect of tyrosine on the phenylalanine requirement. J. Nut. 80, 111-117.

Wallace, W.M. 1959. Nitrogen content of the body and its relation to retention and loss of nitrogen. Fed. Proc. 18:1125–1130.

Wang, T.C., and M.F. Fuller. 1989. The optimum dietary amino acid pattern for growing pigs. 1. Experiments by amino acid deletion. Br. J. Nutr. 62(1):77–89.

Waterlow, J.C. 1996. The requirements of adult man for indispensable amino acids. Eur. J. Clin. Nutr. 50 (suppl. 1):S151–176.

Waterlow, J.C., P.J. Garlick, and D.J. Millward. 1978. Protein Turnover in Mammalian Tissues and the Whole Body. Amsterdam: North Holland Elsevier.

Weller, L.A., D.H. Calloway, S. Margen. 1971. Nitrogen balance of men fed amino acid mixtures based on Rose's requirements, egg white protein, and serum free amino acid patterns. J. Nutr. 101(11):1499-1507.

Yoshida, A. 1983. Specificity of amino acids for the nutritional evaluation of proteins. Pp. 163–182 in Proceedings of the International Association of Cereal Chemists Symposium on Amino Acid Composition and Biological Value of Cereal Proteins, R. Lasztity and M. Hidvegi, eds. Budapest: Akademiai Kiado.

Young, V.R. 1986. Nutritional balance studies: Indicators of human requirements or adaptive mechanisms. J. Nutr. 116:700–703.

Young, V.R., and A. E. El-Khoury. 1995a. Can amino acid requirements for nutritional maintenance in adult humans be approximated from the amino acid composition of body mixed proteins? Proc. Natl. Acad. Sci. USA 921:300–304.

Young, V.R., and J.S. Marchini. 1990. Mechanisms and nutritional significance of metabolic responses to altered intakes of protein and amino acids with reference to nutritional adaptation in humans Am. J. Clin. Nutr. 51:270–289.

Young, V.R., Y.S.M. Taylor, W.R. Rand, and N.S. Scrimshaw. 1973. Protein requirements of man: efficiency of egg protein utilization at maintenance and sub-maintenance levels in young men. J. Nutr. 103:1164–1174.

Young, V.R., L. Fajardo, E. Murray, W.M. Rand, and N.S. Scrimshaw. 1975. Protein requirements of man: Comparative nitrogen balance response within the submaintenance-to-maintenance range of intakes of wheat and beef proteins. J. Nutr. 105:534–544.

Young, V.R., M. Puig, E. Queiroz, N.S. Scrimshaw, and W.M. Rand. 1984. Evaluation of the protein quality of an isolated soy protein in young men: Relative nitrogen requirements and effect of methionine supplementation. Am. J. Clin. Nutr. 39:16–24.

Young, V.R., D.M. Bier, and P.L. Pellet. 1989. A theoretical basis for increasing current estimates of the amino acid requirements in adult man with experimental support. Am. J. Clin. Nutr. 50:80–92.

Zello, G.A., P.B. Pencharz, and R.O. Ball. 1990. Phenylalanine flux, oxidation and conversion to tyrosine in humans studied with ^{13}C phenylalanine. Am. J. Physiol. 259:E835–E843.

Zello, G.A., P.B. Pencharz, and R.O. Ball. 1992. Lysine requirement in young adult males. Am. J. Physiol. 264:E677–E685.

Zhao, X-H., Z-M. Wen, C.N. Meredith, D.E. Matthews, D.M. Bier, and V.R. Young. 1986. Threonine kinetics at graded threonine intakes in young men. Am. J. Clin. Nutr. 43:795–802.

DISCUSSION

ROBERT NESHEIM: Thank you. Are there questions? Yes, sir, John.

JOHN VANDERVEEN: I am interested in the reality of the current situation. At the present time, our troops are getting plenty of protein in amounts far beyond this whole discussion. But the issue of adaptation raises the issue of survival. If, indeed, higher levels of protein intake change our metabolism to the point where additional adaptation, if you will, occurs in our needs, are we putting soldiers at risk to survive starvation or low protein intake for a period of time on rations?

D. JOE MILLWARD: I think it is a very complicated question. I think that it is quite possible that we are putting soldiers at risk. Certainly, the view that Chittenden expressed was that high protein intakes represent a metabolic load that is hard to deal with.

Now, Mike Rennie will present data on physical activity and intakes and deal with this problem to a certain extent. But, in my view, the real point about the human response to protein intake is that we adapt our metabolic demand according to how much we eat. The point about any individuals who stuff themselves with very high-protein diets are two things. Firstly, as soon as they stop taking in that high protein diet, they lose lean tissue very, very rapidly. So the soldier who suddenly finds himself without food for a few days on a background of a high protein diet will lose more lean body mass than somebody who has been on a lower intake.

Secondly, the response to exercise, I think, can be shown to induce a bigger loss when one's background intake is higher. Now, there are things about infection and all of that which complicate the issue and that I do not think we have an answer to.

ROBERT NESHEIM: Yes?

G. RICHARD JANSEN: Joe, I guess, having been recruited to work on the ill-fated lysine program at Dupont 40 years ago, this is like dèja vu all over again.

I remember a 1960 conference in Chicago at the Federation Meetings discussing protein reserves. If you forget the turnover, it was the flux data. It is the same argument basically. But it seems to me that your position as you have published it is more agnostic than atheistic to Vernon's dogma here. And it seems to me you have not really addressed his argument that there is an advantage in terms of metabolic control to that increased oxidation. And, in fact, in your slide there you said, number one, that we basically cannot determine the indispensable requirements because we do not have enough data. However, you say protein scoring is invalid. It seems to me that those two statements cannot be made together. If we cannot define it, we cannot determine what is valid or invalid yet.

D. JOE MILLWARD: Okay. The point about protein scoring is that it is a procedure that was adopted for quality assessment in the rat because it was shown to work. In other words, scoring correlated with net protein utilization in the growing rat. The reason that it correlates is simply because for the growing rat the requirement is quite simple and straight-forward. It has to do with the requirement for tissue deposition.

What I am saying is that for adult humans, our requirement is more complicated because it is variable. We have this fixed requirement that may well be very low. And, in fact, FAO took the view that you use the minimum requirement value and you score protein with that. If you do that, then all proteins in the world are nutritionally adequate. Now, they are only adequate if you can actually adapt to them successfully. So for a population with a habitual protein intake that is relatively high, the question is what is adequate for that population. That is what is very difficult to define, because ultimately one has to test it against that group and against the extent to which they are adapted to their intakes.

If you add the problem that there may well be other components of intake that are more important than balance, like performance, like the immune response, like any of these things, and we talked about the anabolic drive on growth and development, then it becomes even more difficult to define nutritional adequacy against this single pattern of individual amino acids. That is the point that I am trying to make.

G. RICHARD JANSEN: I assumed that. But I think, at the same time, I do not think that you addressed at all Young's argument that there is an advantage in terms of metabolic flux and control to that increased protein turnover, which then would make the argument that he has made, that the pattern should be what it is and, therefore, it would be valid. I understand the rat gross data. I have run a lot of those experiments; but you are not addressing the argument he is making.

D. JOE MILLWARD: Well, I did address it partly. That argument is made basically in the Young and Marchini paper, which he quotes in his abstract, where they show that whole body protein synthesis rates plummet dramatically as the intake of essential amino acids plummet. I think that data are wrong, to be quite honest. I think they are artifactual.

There are other data where the same diet is fed, for example, with decreasing lysine intake, but the protein turnover is measured with another amino acid, namely phenylalanine, and there is no change in the phenylalanine flux over the whole range of intake.

So I think there is a technical problem with the observation that higher intakes are associated with higher rates of protein turnover. Because I do not think there are higher rates of protein turnover with higher intakes. I do not think they actually change.

SUSAN HUDSON: I have a question. We have argued about this for years. But is there any way to make the steady isotope labeling data more precise, that is, those you say are flawed? Is there any kind of information that could be gathered that would, say for leucine, make the data better?

D. JOE MILLWARD: I think that if Vernon were here, what he would say is that over the years, the studies have gotten better. There has been a better understanding of what was going on. The way the calculations have been done has been improved. The studies that Vernon is doing at the moment in India where he looking at lysine intakes and he is using leucine balances are probably as good as you can actually get. So we do await the outcome of those studies. The difficulty is that many of the data that have been used in the past to support the MIT scoring pattern, in my view, simply are unusable because they are seriously flawed. So we are still waiting for hard data that support a better requirement pattern. Now, Denny [Bier] was a co-author of those papers, so I am sure he will want to respond.

ROBERT NESHEIM: Denny?

DENNIS BIER: Well, first, I think that Joe has probably presented the clearest and the fairest view of this subject that I have heard in the last 10 years. I would like to see it come out as clearly as it was presented here today. I would agree with you on this whole business of requirements, that is, the adaptive component and how it affects those requirements. I think it is precisely in that slide you showed us of the obligatory losses. I think that, unfortunately, that is really what we would like to know. That is what the military would like to know in the circumstances that they are addressing. How do we get at that?

I think we are trying to address a problem about intake and requirements in the face of a body that is filled with amino acids, as you pointed out in your remark about the utilization and recycling. It becomes very difficult, I think, to solve the problems in that context when really what we are looking for is some change from the baseline.

In addition, the balance studies really do not tell us in any way, of course, what happened to what was in the body at that time and whether or not there has been an advantage or disadvantage to what it took to get the balance in the context of these sort of intakes.

So adding functional measurements, adding things like whole body nitrogens or whole body potassium, something that really lets us know over the long term what happened to the body proteins, would be very useful.

Now, to go back to the models themselves, my position on this, and I have stated it many times, is that the models are not very good. But the original data that went into some of these calculations (I am not going to talk about scoring patterns because I have no idea what they even mean), but, the original data that went into the calculations were based not on any complicated model but simply on the amount of C^{13} recovered outside of the body during the course of the tracer study. Admittedly, that has its own set of problems. But, if anything, most of those problems entail underestimation of loss rather than overestimation.

If you just take those numbers, they were those net losses that generated the initial part of this. So, admittedly, they were done at MIT with young, healthy adult men, adapted longterm, and using a different protein technique for a maximum of three weeks. I have no idea, frankly, whether that applies to Northern Thais who eat rice or adapt to a different type of protein for the rest of their lives.

K. SREEKUMARAN NAIR: I want to echo what was said. This is a complex problem. My concern is we are going to try to solve this problem by a very simplistic approach.

What we need is a long-term study on the effects of different proteins on performance changes. They are not easy to perform.

D. JOE MILLWARD: I think we have tried many times to erect very complicated models to understand amino acid and protein metabolism. I have always taken the view that, if you cannot actually measure the total amount of body protein that is there and whether it is going up or whether it is going down, any other information is basically irrelevant. You have got to start with good data about what is the state of the overall system, i.e. the lean body mass. And, while that is a simplistic approach, it is a necessary precondition to answering any of the other questions.

DENNIS BIER: But the balance method does not allow that—the only way to do that is with the total body nitrogen.

D. JOE MILLWARD: Exactly. I think that we need much more data on whole body nitrogen.

NANCY BUTTE: Joe, the committee has been asked to address a very straightforward question. Is the military RDA for protein, which is one hundred grams per men and 80 for women, appropriate? Given the current state of knowledge with the techniques we have, would you recommend that they even attempt to revise that right now?

D. JOE MILLWARD: Well, there is certainly, in my mind, no evidence to suggest that it is inadequate. The only issue is whether that level is too much.

NANCY BUTTE: And what would you advise them if you were asked to advise them at this time? Do you think the recommendations are too high?

D. JOE MILLWARD: Well, I think it is. To be quite honest, I do not think we have yet done the right studies. I mean, if you work on the basis of this very significant series of studies that were done over three or four years on Army recruits and young athletes, then you would have to conclude that the current requirements are too high and that soldiers may well perform better on a lower intake. Mike [Rennie] is going to present data this afternoon that suggest that for physical activity, there may well be benefits of lower protein intakes.

ROBERT WOLFE: I would just like to follow up on Dr. Jansen's point. I do not know that it can be so readily dismissed that a higher protein intake affects protein turnover, because even your own anecdotal comment, that if you eat a higher protein intake and then stop the protein intake, you will lose protein faster. The only way that could occur would be if you have a higher protein turnover; the higher protein intake caused the higher protein turnover. The only way you would lose it more rapidly would be if it were turning over more rapidly before you stopped. But, furthermore, there are some studies that have done what you have suggested and avoided the problem of using the same amino acid tracer. Because we performed a study a number of years ago, albeit not in normal subjects, but in severely burned patients in a crossover study in which they were given different levels of protein, and three different essential amino acids were used, and all three tracers showed parallel increases in whole body protein turnover when the protein intake was increased, although none indicated an improvement in balance.

So I think the fact that the same amino acid tracer as the test compound has been used with different levels of protein intake may not be valid, and that it is a jump to go from that observation to the conclusion that there is no effect of protein intake on protein turnover.

D. JOE MILLWARD: Let me say that it was not an anecdotal statement. We have published a series of studies where we used a multiple tracer approach. We used N^{15}, we used C^{13} leucine, we used deuterated phenylalanine to address the question of whether the level of protein intake affects the overall rate of protein turnover. The outcome of that experiment was that from an intake of .3 grams per kilogram, up to an intake of 2.5 grams per kilogram over three weeks, there was no effect on the overall replacement rate. However, those subjects who were in positive balance obviously had a higher rate of synthesis than proteolysis and those on the lower intake had a lower rate of synthesis than proteolysis. But, if you calculated rate of replacement as the lower of the two processes, that rate did not change over the entire range.

DENNIS BIER: Two comments, first on the last statement. If you change the protein intake, one way or another you at least have to change components of the protein turnover. That is, you have to deal with the protein that comes in. So the total protein turnover may not change, but how it is distributed has to be changed, because you have a new term in the equation that is introducing protein into the body. So the fact is that something has to happen, not to the total number [turnover] necessarily, but to how it is distributed. Our models do not really allow us to analyze that.

The second comment goes back to Chittenden's study. How did he control for the training effect on this?

D. JOE MILLWARD: Well, that was the basis of the problem with his initial study, because he did that study on army recruits. There was this enormous increase in strength. He concluded immediately that it was the weekly measurement over five months that trained them and that was responsible for the increase in strength. So he only controlled for it by recruiting elite athletes. He had the U.S. National Decathlon equivalent champion among his subjects. They were as highly trained as they could get. That was how he got the results that he did. But it is not a perfect study. But it is the only study that has ever seriously addressed the problem. Just because it was published ninety years ago, it is completely ignored.

DENNIS BIER: Well, you know, it is not on Medline!

JEFF ZACHWIEJA: There are some recent data that are germane to that issue. Others may help me out with this. I think that Bill Evans' group (when he was still at Tufts), recently published data that suggested that exercise training in older individuals, whether they had a dietary protein intake of .8 grams per kg or 1.6 grams per kg, essentially achieved the same level of muscle strength and I think those individuals who were on the lower protein intake were still in positive nitrogen balance, but not as great as those on the higher protein intake. But, in terms of performance, the increments in performance were very similar on those two types of diets. These results spoke to the issue of the body being more efficient or becoming more efficient on the lower protein intake to achieve essentially the same functional outcome.

ROBERT NESHEIM: I think we could have more discussion on this, and we will have more time later.

HARRIS LIEBERMAN: I just have one more comment in regard to that issue of one hundred g protein per day. There is a lot of confusion about that. The really critical issue for us is not the MRDA level of protein that is recommended but the standard for operational rations. Right now, the standard operational ration must include at least a hundred grams of protein. That typically is not fully consumed. If you lower that ration content, a hundred grams or 90 grams will not be consumed; substantially less than that will be consumed. So when we think about it and talk about it, please keep in mind that what really is critical for us is what we recommend to the ration developers at Natick, the minimum level of protein they can include in a combat ration.

JOHN VANDERVEEN: I have one last comment. The concern about ration design depends on how you calculate that 100 grams of protein as well. What protein quality factor do you use? Do you have a cut-off in what you consider protein? If the protein has an amino acid score that is such that you have a cut-off at 20 percent or 30 percent, or whatever, is that counted in? I think that those are issues.

I guess maybe the last question I would raise is whether there is any thinking as to how long the amino acids are available in the pool so that you can take advantage of a lower-quality protein at some later time in the day or at a different time?

D. JOE MILLWARD: Well, there are good data on that. There are good data on muscle biopsy studies, which I referred to in my talk, of the rate at which amino acids disappeared from the intracellular muscle pool after a meal. It is quite clear that, whereas leucine, all three BCAAs, the aromatic amino acids, and the sulfur amino acids disappear quite quickly, lysine and threonine stick around for a long time. Seven hours after an albumin meal, lysine and threonine levels are still elevated, whereas all of the branch chains have fallen below baseline levels. I think that what we know about the Kms of the catabolic pathways for lysine and threonine would indicate that the body does not turn up their oxidation quickly after a meal in the same way that it does for the BCAAs.

ROBERT NESHEIM: One more question.

PATRICK DUNNE: It is more of a request for background information. It is really our group that, over the years, has been doing the nutritional assessment of the ration contents. What we do not know is the amino acid composition of the diets. It is way too expensive to do that. We can make some guestimates like anyone else can from the databases. We would say that the majority of our protein intake is coming from meat. But that is switching as we look at newer concepts in our rations, where we have more vegetarian items. We are getting better at providing bread to people in the field and developing shelf-staple breads. That is a changing phenomenon and may be an area where we need to gather more data.

JOHN VANDERVEEN: So you are telling us that you take N values and multiply times the conversion factor and that is how you estimate ration protein content?

PATRICK DUNNE: I would say that is correct, with some minor corrections using different factors for dairy versus muscle proteins. You typically perform kjeldahl nitrogen assays and that is what you get. You get approximate values for N, and that forms the database. We are trying to reconcile that database with the nutrient content of the meal, ready to eat (MRE)-15 as best we can. That is the level of knowledge.

D. JOE MILLWARD: In fact, I think it is highly unlikely that protein quality is going to be an issue in the sort of rations that are provided unless they are really quite unusual. In other words, unless they are basically unsupplemented wheat. There have been a number of papers recently that have used rats to do protein quality assays of basically vegetarian diets, and these have shown that vegetable-type mixtures can have higher NPU values than those for meat and milk. So, unless these military diets were really very peculiar, then I do not think that the amino acid composition would become an issue.

PATRICK DUNNE: Right. However, in the field setting, we probably have the normal level of intake of protein, lower in dairy proteins than you would have in the garrisons just because of what is shelf stable. We do look at cross-linking effects and do not think it is a problem to really reduce the quality of protein during the normal storage period, which could be three-plus years.

We also have some data relative to the antioxidant issue. Until recently, when more commercial items were added to our rations, we were not feeding our troops a lot of BHA and BHT. We were feeding them lots of vitamin C and vitamin E. As we add more commercial items, you are going to see changes in macro- and micronutrients, something for the good.

ROBERT NESHEIM: Thank you very much. I think it has been a very interesting morning. It is time to break for lunch.

Protein and Amino Acids, 1999
Pp. 217–242 Washington, D.C.
National Academy Press

10

Amino Acid Flux and Requirements: Counterpoint; Tentative Estimates Are Feasible and Necessary

Vernon R. Young

INTRODUCTION

This chapter serves as a counterpoint to that by Millward (see Chapter 9), in which he largely evaluates my views and those of my colleagues concerning the requirements for nutritionally indispensable amino acids in healthy adults. His points are multiple and well articulated, although he presents little direct experimental data from his own laboratories that are directly relevant to the quantitative determination of adult human amino acid requirements or that are in contrast to our own findings. Nevertheless, I agree with a number of important elements in his argument; indeed, I accept some of his criticisms of our work, which is overviewed briefly below. However, other major components of his position lack strength and/or validity. Of course, he, too, has made a similar statement about some of my ideas, concepts, and experimental data, while attempting to reassess the quantitative needs for indispensable amino acids in adult humans and to assess their practical significance in human protein nutrition.

At the outset, I share his view that there are inherent difficulties in defining amino acid requirements. Indeed, Millward has provided an excellent backdrop against which I will make the case that the current international amino acid requirement values (FAO/WHO/UNU, 1985) for adults are "no longer acceptable or nutritionally relevant" (Clugston et al.1996). Further, I will argue that the tentative requirement values that we have proposed (Young 1991, 1992; Young and El-Khoury, 1995; Young and Pellett, 1990) should be used for evaluating the amino acid adequacy of diets and/or for planning diets intended to meet adequately the physiological needs of consumers in healthy populations. With respect to the issue of indispensable amino acid requirements and dietary protein quality, the most recent UK expert panel (Department of Health, 1991), of which Millward was a member, stated: "DRVs (dietary reference values) assume that the dietary protein pattern includes sufficient variety of different protein-containing foods, or sufficient high quality protein sources to provide for indispensable amino acid (IAA) requirements." This assessment or advice would not be at all necessary if the IAA requirement estimates proposed by FAO/WHO/UNU (1985) and assumed to be adequate by Millward were valid. It would seem worthwhile then, to make our case for a significant revision in the current UN amino acid requirement values, while addressing a number of Millward's points and before concluding with a few comments about dietary protein and amino acids in relation to performance and exercise.

As Millward reminds us, much of the argument he makes has been reported previously. Similarly, my colleagues and I have also published our assessments and views elsewhere (Young et al., 1989; Young and El-Khoury, 1995, 1996). Hence, this chapter concentrates on many, but not all, of the particular issues raised by him. It is agreed, as he states: "The task in hand then is to define the amounts and amino acid pattern of the maintenance requirement."

Proposed Massachusetts Institute of Technology Amino Acid Requirement Pattern

In 1989, following up and extending on an idea concerning obligatory nitrogen (N) losses that was originally proposed by Millward and Rivers (1989), we adopted the values for the estimated obligatory amino acid losses (OAALs) as a partial basis for deriving the Massachusetts Institute of Technology Amino Acid Requirement Pattern (MIT-AARP). We also applied some of our data from ^{13}C-amino acid tracer studies in deriving the AARP (Young et al., 1989). This pattern is summarized in Table 10-1, together with the proposed 1985 FAO/WHO/UNU patterns for preschool children and adults. Millward points out that the approach we adopted meant that we had defined the maintenance pattern

TABLE 10-1 The MIT Amino Acid Requirement Pattern (MIT-AARP) Compared with the 1985 FAO/WHO/UNU Patterns for Adults and Preschool Children

Amino Acid	MIT-AARP	1985 FAO/WHO/UNU	
		Preschool Children	Adults
Isoleucine	38	28	13
Leucine	65	66	19
Lysine	50	58	16
Methionine + Cystine	25	25	17
Phenylalanine + Tyrosine	65	63	19
Threonine	25	34	9
Tryptophan	10	11	5
Valine	35	35	13

NOTE: All values are mg amino acid per g protein (N × 6.25).

with the same amino acid composition of the mixed proteins in the body. He is correct except that the values for lysine, sulfur amino acids, and threonine, in particular, were derived more from the results of our tracer studies, which were limited in scope. This detail becomes important when we respond to the issues raised by Millward in his chapter.

Millward has three objections to the approach we followed in applying the OAAL. First, he criticizes the approach, because the OAAL for most amino acids were greater by 2 or more times than the FAO/WHO/UNU (1985) requirement values, with the exception of the total sulfur amino acids (methionine + cystine; SAA) for which the losses and FAO/WHO/UNU (1985) requirement values were quite close. He interprets this comparison by suggesting that the SAA "drives" the obligatory nitrogen loss. Further, this appears to be why Millward objects to our use of OAAL for purposes of deriving a tentative amino acid requirement pattern. However, the power of his argument depends, in the first instance, on an assumption of the validity and acceptance of the nutritional significance of the FAO/WHO/UNU amino acid requirement estimates. Because the latter can be legitimately questioned, his reasoning is problematical. Furthermore, it has been shown that the presence of a source of dietary cystine spares the methionine requirement, possibly by as much as 90 percent (Rose, 1957; Williams et al., 1974). Hence, in theory, the OAAL for methionine could be reduced to the equivalent of about one-quarter of the methionine loss under conditions of a high cystine intake. I am not aware, however, that dietary cystine, in contrast to methionine (Yoshida, 1986), reduces N excretion when experimental animals are given a protein-free diet. Furthermore, not only are short-term animal feeding studies of this kind difficult to interpret for their human nutritional significance, but similar experiments in adult humans are lacking. Thus, it is a matter of speculation whether the SAAs are rate determining for mobilization of whole body proteins under conditions of protein-free or amino acid-inadequate diets.

Second, in contrast to the position taken by Millward (1998, see Chapter 9), I conclude that there is a broad similarity between the amino acid pattern of mixed body proteins and the adult maintenance pattern. Indeed, the comparison that we made previously for the 1985 FAO/WHO/UNU preschool age pattern and body protein shows a close similarity (Young and El-Khoury, 1995). I recognize that this comparison may be complicated by the fact that the children in the studies that provide amino acid requirement values appear to have been retaining body protein at a much higher-than-normal rate (cf. FAO/WHO/UNU, 1985; Pineda et al., 1981). Thus, the dietary retention of indispensable amino acids may well have been more efficient than for a fully repleted child who was gaining body protein at a more usual rate. The latter rate accounts for a relatively small proportion of the total daily requirement (Young, 1991), but this does not necessarily mean that the amino acid requirement pattern would be that different.

Additionally, body protein maintenance in the adult involves depletion of body proteins during the fasting period of the day and their repletion during the fed period of the day. Thus, the pattern and composition of the retained amino acids must be that of mixed body proteins. The question then is whether this prandial gain would require, under steady state conditions of N balance and at intakes of N that met but did not significantly exceed the physiological need, an amino acid pattern (also at limiting and not excessive intakes) of IAA that is similar to that of body mixed proteins. Using leucine as an example, our tracer studies indicated that the leucine needed to achieve this balance, per unit of N intake, is in proportion to leucine in body proteins (El-Khoury et al., 1994a). Lysine metabolism shows a significant adaptive capacity when intakes are limiting. Thus, the question is whether the conclusion we have drawn for leucine also applies to lysine. Although we do not have the same extensive data for lysine, that we do for leucine, the fasting loss of lysine at a generous (77 mg/kg/day) lysine intake appears to be about 70 or 90 percent of that which we would predict from leucine oxidation, depending upon whether the route of ^{13}C-lysine administration was intravenous or oral (El-Khoury et al., in press). At a limiting, probably inadequate, intake of lysine (12mg/kg/d) the fasting state rate of oxidation appears to be about 50 to 110 percent of the rate of leucine loss at an equivalent low leucine intake (Unpublished data, A.E. El-Khoury and V.R. Young, 1998). Hence, there appears to be some conservation of the lysine liberated during the postabsorptive period that may later be used for retention of protein and replenishment of earlier losses during the prandial phase. However, the difference between the measured and predicted oxidative loss of lysine and its replenishment during the prandial phase does not seem to be profoundly different from that for leucine. On this basis, a general similarity between the requirement pattern of amino acids and that of the mixed proteins in the body (with some differences that may not be large due to the specific characteristics of metabolism of the individual amino acids) appears to be a reasonable hypothesis. For lysine, the difference between our proposed lysine requirement (expressed per unit of protein requirement) and the concentration of lysine in whole body mixed proteins amounts to about 30 percent, which is consistent with the lysine oxidation data mentioned above. This matter will be raised again below, in some detail, and with respect to the diurnal cycling of protein retention and loss.

Third, Millward concludes, from the animal data, that the maintenance pattern differs from that of growth, and criticizes our estimates for adult human amino acid requirements. I have said before that maintenance in a growing animal is a very different metabolic condition than that of nutritional maintenance in an adult and probably also that of a pre-school child. Further, the data for requirements for adult maintenance in different animal species are not only limited but also contradictory, as we both recognize (McLarney et al., 1996; Millward, 1998, see Chapter 9).

Millward's three positions are not supported adequately. A reasonable initial and tentative definition of the pattern and level of requirements for the IAA in adult humans can be made from estimates of OAALs (Young and El-Khoury, 1995).

Adaptive Aspects of Amino Acid Metabolism

Millward appropriately emphasizes that the adaptive component of amino acid metabolism brings complexity to the assessment of maintenance amino acid needs. The important practical question, however, is how long it takes the integrative aspects of whole body amino acid metabolism to adjust to a new level of amino acid, or total protein, intake so that a reasonable estimate of the steady state amino acid needs for that intake level and status of nutriture can be made by either N balance or tracer techniques. In our studies, we have used relatively short experimental diet periods of 6 to 7 days prior to the conduct of ^{13}C-amino acid oxidation studies (Young, 1994). Millward (Chapter 9) concludes that the adaptation of oxidative catabolism to a change in protein intake is relatively slow and that the extent of postprandial oxidative catabolism reflects mainly the habitual rather than the actual or present protein intake supplied in meals. The validity of this conclusion depends on whether the metabolic responses and adaptations to altered protein or amino acid intakes are indeed as slow as Millward concludes. He bases his argument on his own studies of N balance and leucine oxidation following the transition from a high- to a low-protein diet in healthy adults.

The following six points are relevant, therefore, to the question of whether a 6-day period of "adaptation" is long enough to permit a suitable derivation of nutritionally meaningful amino acid requirement values:

1. Based on the earlier N excretion and balance studies carried out in our laboratories (Rand et al., 1976, 1979, 1985) and on our evaluation of the data published by others (Rand et al., 1981) it is entirely reasonable to propose that a 6 day adjustment period is suitable for purposes of establishing the amino acid needs of initially well-nourished subjects. Further, relatively short-term

experimental diet studies provide the database used by the UN group to establish adult amino acid requirements for populations worldwide (FAO/WHO/UNU, 1985). Quevedo et al. (1994), from their study on the transition of N balance from a high (292 mg N/kg/day) to lower (124 mg N/kg/day) protein diet, suggested that it takes longer and possibly more than 2 weeks for a new N equilibrium to be reached. However, their nitrogen excretion results are confounded by the apparently low energy intake supplied by their experimental diet. My calculations suggest that the energy supply in that experiment approximated 34 Kcal/kg/day, which means a physical activity level (PAL; total energy expenditure expressed as a multiple of the basal metabolic rate) of about 1.4. This value is much lower than the expected value of approximately 1.8 and above for young healthy adults (Goran et al., 1993; Roberts et al., 1990) and 1.6 to 1.8 for free-living elderly men (Roberts et al., 1995). On this basis, it appears likely that energy intake was inadequate and this raises doubts about whether their nitrogen balance data can be used to help answer adequately the question of the time required to achieve a new steady state of whole body protein metabolism when amino acid/protein intakes are changed.

2. A study was carried out in our laboratories to determine the rates of leucine oxidation (Marchini et al., 1993) and phenylalanine hydroxylation (Marchini et al., 1994) at the end of 1 and 3 weeks after giving healthy subjects diets supplying adequate L-amino acid mixtures patterned after (a) the FAO/WHO/UNU (1985) adult amino acid requirements, (b) the MIT tentative requirement values, and (c) whole hen's egg proteins. The kinetics of whole body leucine and phenylalanine catabolism during the fast and fed states did not differ between the 1- and 3-week time periods, again supporting the premise that a 6-day diet adjustment period is suitable for exploring aromatic amino acid kinetics, balances, and requirements in healthy adults. In addition, Zello et al. (1990) found that phenylalanine flux and oxidation rates at test intakes of phenylalanine ranging from 5 to 60 mg/kg/day (in the presence of generous tyrosine) were not affected by giving subjects a diet supplying 4.2 or 14 mg of phenylalanine per kg per day for up to 9 days prior to the kinetic measurements.

3. It could be argued that even a 3-week dietary period might not be long enough to reach a full degree of adaptation. However, this becomes an exceedingly difficult issue to resolve both conceptually and with respect to the implications of adaptive change for the determination of functional state and nutritional requirements.

4. Equally important to ask in this context of adaptive oxidative losses is whether the amino acid requirements of individuals in populations in developing regions of the world, particularly where protein and/or dietary lysine are likely to be more limiting and lower than for the U.S. diet, are similar to or different from those for U.S. subjects. This question is raised despite the fact that the current international FAO/WHO/UNU (1985) amino acid requirement values, based largely on studies conducted mainly in young adult American subjects (Irwin and Hegsted, 1971) are recommended for application world-wide.

To date, no major research effort outside of North America has carried out ^{13}C-tracer studies of amino acid requirement comparable to those calculated at MIT or the University of Toronto. We have completed an initial study with our collaborators at St. John's Medical College, Bangalore, India (Kurpad et al., 1998), using the indicator amino acid oxidation technique (Zello et al., 1995) to assess the lysine requirement of healthy Indians whose long-term lysine intake appears to be about 60 percent of the level characteristic of the North American subjects. Results from this laboratory do not suggest that adequately nourished Indian subjects require a different lysine intake for maintenance as compared to MIT subjects. Further studies will need to be conducted to strengthen this initial but potentially very important conclusion.

5. There are few relevant data that can be used to predict whether the indispensable amino acid needs and the lysine requirement, in particular, are similar to or different among various population groups who have "adapted" to different dietary conditions. Studies of obligatory nitrogen losses in American (Calloway and Margen, 1971; Scrimshaw et al., 1972; Young and Scrimshaw, 1968), Chinese (Huang et al., 1972), Indian (Gopalan and Narasinga Rao, 1966), Nigerian (Atinmo et al., 1985; Nicol and Philips, 1976a) and Japanese men (Inoue et al., 1974) reveal that they are remarkably uniform (Bodwell et al., 1979). This implies similar obligatory amino acid losses (OAAL) and, by implication a probable similarity in the dietary requirements for indispensable amino acids (Young and El-Khoury, 1995), unless the efficiency of specific amino acid retention at *requirement intake* differs among apparently similar subjects in different population groups. Also, according to FAO/WHO/UNU (1985), nitrogen balance studies have not revealed any striking differences in estimates of total protein requirements in relation to body cell mass for well nourished subjects in different countries. The earlier studies by Nicol and Phillips (1976b), which might have suggested Nigerian men of low income utilize dietary protein more efficiently than, for example, U.S. students, are not appropriate to answer the question of whether requirements differ among populations. The N balance results in these Nigerian studies indicated that the subjects were depleted and were undergoing body protein repletion in response to the adequate diet given during the experiments. This confounds the nutritional interpretation of these studies and a later series carried out in Nigerian adult males (Atinmo et al., 1988) indicates that at maintenance nitrogen intakes, the efficiency of dietary protein utilization is the same as that for Caucasian and Asian subjects.

6. From our collaborative studies with Maroni and coworkers (Masud et al., 1994; Tom et al., 1995), it can be concluded that the postabsorptive rate of whole body leucine oxidation at the end of a 16 ± 2 month period during which non-acidotic chronic renal failure patients consumed a very low protein diet (0.28 g protein per kg per day) plus an amino acid-keto acid supplement was not different from that at the early phase of the low-protein intake. Furthermore, the leucine oxidation rate for these subjects was comparable to that reported in

healthy adults given either 14 or 38 mg leucine per kg per day for 6 days prior to the measurement of ^{13}C-leucine kinetics (El-Khoury et al., 1994b).

In summary, it is proposed that a 6-day period of dietary "lead-in" permits an appropriate adjustment, or adaptation, to variable intakes of specific IAAs. This hypothesis should now be further validated through metabolic studies of varying duration and, preferably, in population groups in different geographic regions of the world.

Diurnal Cycling and Amino Acid Requirements

Millward and coworkers (1996) have conducted interesting and important studies on the relationships between the postabsorptive N losses and prandial N gains and how the magnitude of this diurnal cycle of protein metabolism is affected by the habitual and prevailing intake dietary protein. Millward also points out in Chapter 9 that the key question is the extent to which this diurnal cycle ". . . influences the IAA composition of the adaptive metabolic demand" or, in other words, "the amounts and amino acid pattern of the maintenance requirement."

We accept, as concluded by Millward, that it does not necessarily follow that all of the IAAs liberated via tissue proteolysis are quantitatively oxidized. Therefore, a conservation of lysine (and perhaps threonine) may occur in the free amino acid pools, whereas there may be little conservation of other amino acids, such as leucine.

Regarding this metabolic issue, discussed above, evidence suggests some apparent conservation of lysine during the postabsorptive phase of amino acid metabolism. Millward uses the data of Bergstrom et al. (1990) to make his case for a significant contribution made by the free pool of lysine in muscle to the dietary retention of amino acids as body protein. Unfortunately, specific data on muscle free amino acid concentrations for the postabsorptive state are not given by Bergstrom et al. (1990). These data would be the most useful for evaluating the possibility of a postabsorptive conservation of lysine at different lysine intakes and in reference to estimations of the lysine requirement based on ^{13}C-tracer techniques.

We have, however, considered this postabsorptive retention of lysine in relation to the fasted:fed ratio of lysine oxidation at a generous intake and whether this is far lower than that for leucine. If so, this would indicate that a significantly greater proportion of the lysine released from protein breakdown during the fasting state is retained within the free lysine pool compared with that for leucine. Such a retention might occur, particularly in skeletal muscle and perhaps in other tissues and organs; the free lysine pool of muscle is relatively large and it responds to ingestion of protein-free and protein-containing meals (Bergstrom et al., 1990). Hence, there is a theoretical capacity to accommodate, or store, some free lysine that is liberated via proteolysis during the fasting period. However, for a daily balance condition to be achieved, this "retained" lysine must contribute, in effect, to the total dietary intake during the fed period. Thus, it would have to be removed from the free lysine pool either via oxidation or re-incorporation into protein during the fed phase. A key issue, therefore, is the nature and extent to which changes occur in the size of the free lysine pool with feeding and fasting, under varying conditions of lysine intake.

Unfortunately, few direct experimental data have been published on changes in the concentration of free lysine and of leucine in the "metabolic free amino acid pool." Bergstrom et al. (1990) showed that a protein-rich meal increased the free lysine and leucine concentrations in muscle by 30 percent and 60 percent, respectively, at 3 hours after beginning a meal. When a protein-free meal was given, lysine and leucine concentrations declined by about 30 to 40 percent. Although these data are limited, they help us to interpret our estimates of lysine oxidation that are derived from plasma ^{13}C-lysine values, in comparison with leucine oxidation-expected N losses (El-Khoury et al., in press). Thus, as summarized in Table 10-2, the measured rate of lysine oxidation during the 12-hour fast appeared to be close to 90 percent of that which would be predicted from leucine kinetics. This being so, the conclusion is that a relatively small retention of lysine causes a rise in the free lysine concentration in muscle tissue. The expected or predicted rise would fall well within the changes observed by Bergstrom et al. (1990) in subjects fed a protein meal. However, to maintain daily mass balance of lysine within the body free lysine pool, a fasting-state increase in muscle free lysine would have to be followed by an equivalent fall during the fed phase. This pattern of change is inconsistent with the data of Bergstrom et al. (1990), which again suggests that there might not be a significant retention of lysine arising from proteolysis in the free amino acid pool of muscle. Clearly, it would be worthwhile to obtain data on changes in the free amino acid concentrations in muscle tissue under the relevant dietary conditions of our ^{13}C-amino acid tracer studies. However, technical difficulties and ethical constraints make this an unlikely research activity in the near future.

TABLE 10-2 Comparison of Rates of ^{13}C-Lysine and ^{13}C-Leucine Oxidation (expressed as IPNL) in Fast and Fed States at Generous IAA Intakes

	Leucine	Lysine
12-h Fast	375	336 (90)
12-h Fed	625	567 (91)
24-h Day	1,000	903 (90)
Protein intake	1,000	1,000

NOTE: Values are mg protein for the relevant period. Values in parenthesis are percent of leucine value. IPNL, irreversible protein nitrogen loss. SOURCE: Adapted from El-Khoury et al., 1998

Nitrogen Balance and Estimations of IAA Requirements

We have argued that N balance studies in adults as an assessment of amino acid requirements are of questionable value (Young 1987; Young and Marchini, 1990; Young et al., 1989). However, one problem with N balance data is illustrated simply by reference to those of Fisher et al. (1969) from which an estimate of the lysine requirement of college women was attempted. As summarized in Table 10-3 and according to these N balance results, body N balance can be achieved at lysine intakes of between 0 and 50 mg/day. This level of intake is so much lower than (a) our minimal estimates of postabsorptive lysine oxidation losses (Young and El-Khoury, 1995) and (b) the ileal losses of lysine reported by Fuller et al. (1994), that it would seem to be an invalid determination of the lysine requirement. Clearly, this points out a serious limitation of the N balance technique. A second problem with N balances is that if they are not determined with accuracy, then the estimate of the requirement for a specific amino acid will be in error as illustrated, for example, by the data of Rose et al. (1955b) and summarized in Table 10-4. If the reported N balance data shown here are recalculated to account for unmeasured N losses that are assumed to be about 8 mg/kg/day (FAO/WHO/UNU, 1985), then following the interpretation of the N balance results as used by these investigators, it must be concluded that the phenylalanine requirement of this subject exceeded 40 mg/kg/d. This contrasts with the conclusion of Rose et al. (1955b) that the phenylalanine requirement for the subject was 11 mg/ kg/day.

A further problem that is related to the specific design of the N balance experiment, as well as the determination of body N balance *per se*, is illustrated in Figure 10-1. This figure depicts the sequence and length of a series of experimental diet periods providing different levels of lysine (Rose et al., 1955a). The lysine requirement was determined by these investigators to be 6.25 mg/kg/d (0.4 g lysine daily). However, the balance period that was used to derive this requirement immediately followed an earlier 5-day lysine-free diet period. This design would be expected to enhance the retention of lysine during the "requirement" phase of the study. For this reason, as well as the fact that only at

TABLE 10-3 An Illustration of the Limitation of Nitrogen Balance as a Basis for Estimation of Lysine Requirements in Five College Women

Time on Diet (days)	Lysine Intake (mg/d)	N balance (g/d)[1]
4	0	−0.67 ± 0.44
7	50	0.06 ± 0.08
5	0	−0.10 ± 0.15
5	50	0.43 ± 0.09

[1] Mean values ± SE for five subjects. SOURCE: Adapted from Fisher et al., 1969.

TABLE 10-4 Phenylalanine Requirement of an Adult Human (H.I.E.)

Period (days)	Balance (g N per day)		Phenylalanine intake	
	Reported	Recalculated	(g/d)	(mg/kg/d)
5	+0.70	+0.14	4.29	61.0
7	+0.36	-0.20	2.86	40.0
4	-1.28	-1.84	0.00	0.0
5	+0.22	-0.34	2.30	32.0
4	+0.32	-0.24	2.00	28.0
9	+0.27	-0.29	1.00	14.0
9	+0.12	-0.44	0.80	11.0
5	-0.37	-0.93	0.70	9.6

NOTE: The phenylanamine requirement was estimated by Rose et al. (1955b) to be 11 mg/kg/d and Body weight increased 2.3 kg at 53 kcal intake/kg/d. SOURCE: Adapted from Rose et al., 1955b.

the 1-g lysine intake level was N balance found to be positive when unmeasured losses of N were included in the determination of balance, it is reasonable to conclude that this requirement estimate of about 6 mg/kg is flawed.

In addition, even assuming that the design of the N balance study by Rose et al. (1955a) was appropriate, it is difficult to accept, on metabolic grounds, a lysine requirement of 6.25 mg/kg/d. The argument is as follows: if the plasma lysine flux is minimally about 70 to 100 µmol/kg/h (or 245 to 350 mg/kg/d), then at requirement intakes, the lysine oxidation would only amount to 1.7 to 2.6 percent of flux. At a lysine intake of 15 mg/kg/d, or a level that Millward

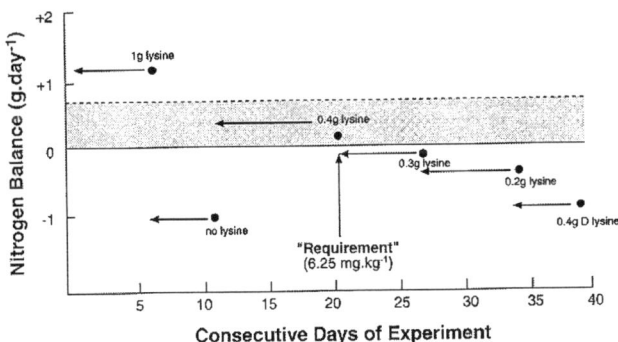

FIGURE 10-1 Rose's Estimate of the Lysine Requirement of Subject R.L.B. Graphic depiction of the design and results of an N balance study conducted by Rose et al (1955b) to estimate the minimum requirement for lysine in Subject RLB. The dashed line refers to the mean N balance for that period of days and the shaded area represents the difference between the measured and estimated N balance if miscellaneous losses of N had been taken into account. In this latter case the only period of zero or positive N balance was during the first period. The requirement for this subject was judged by Rose et al (1995b) to be 0.4g daily (6.25 mg.kg^{-1}).

(Chapter 9) considers sufficient to meet a minimum obligatory metabolic demand, we estimate lysine oxidation to be approximately 21 mg/kg/d or about 10 percent of flux (unpublished data, A.E. El-Khoury and V.R. Young, 1998). On this basis, the estimate of Rose et al. (1955a) would appear to be far too low.

Millward and colleagues' (Millward et al., 1996) estimation (discussed earlier) of the magnitude of the prandial retention of protein necessary to balance subsequent post-absorptive losses appears to be at odds with estimates of the minimum requirement level for lysine; this observation is based on the following line of reasoning. First, for purposes of this argument, the minimum needs for total protein in adults can be estimated from the response of body nitrogen balances measured over relatively short experimental diet periods (e.g., 10 to 14 days) as discussed earlier. Second, at the minimum requirement intake level for total protein, the net prandial retention of protein would be just sufficient to balance the loss of protein during the postabsorptive period. Third, the magnitude of the prandial retention of N (protein = N × 6.25) for subjects consuming 0.77 g/kg/day for 10 days or more was reported to be equivalent to 251 mg/kg/12h (21 mg/kg/h) of protein (Price et al., 1994). These subjects, however, were in a daily negative balance of -9 mg N per kg, so it is uncertain whether a prandial retention of the magnitude observed is actually sufficient to maintain protein homeostasis. Nevertheless, from earlier ^{13}C-leucine tracer studies (Motil et al., 1981), we found an approximate, net whole-body protein synthesis (protein synthesis minus protein breakdown) of about 23 mg/kg/h of protein during the prandial period in healthy adults with an intake of 0.8 g/kg/day of egg protein (Figure 10-2). This

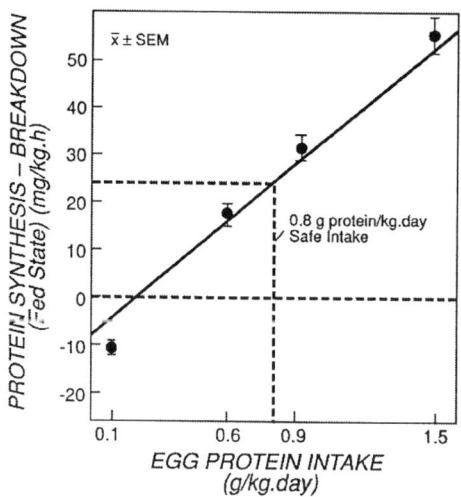

FIGURE 10-2 Net Protein Synthesis at a "Safe" Intake of Protein. The relationship between protein intake and whole body net protein synthesis (protein synthesis-protein breakdown) in healthy young adult men during the absorptive phase of amino acid metabolism. Drawn from Motil et al (1981).

tracer-derived estimate is remarkably close to the N-balance data of Price et al. (1994), and thus a prediction might be made of the amount of dietary lysine minimally required to achieve this level of postprandial protein retention. Assuming that the lysine content of the protein retained is 78 mg/g of crude protein (Reeds, 1990) and that at a requirement level of lysine intake, the efficiency of dietary lysine retention is 80 percent, then the minimum necessary lysine intake to support this gain and subsequent loss of body protein would appear to be 27 mg/kg/d. This value is also comparable to most of the alternative lysine requirement estimates we have summarized (Young and El-Khoury, 1996) except for the far lower estimates based on N balance and the 1985 FAO/WHO/UNU recommendations.

For these reasons, we have not used N balance data to enhance interpretation of the results of our own tracer experiment that was concerned with testing the 1985 FAO/WHO/UNU amino acid requirement pattern (Marchini et al., 1993, 1994). We anticipated that the balance data would be difficult to assess for their nutritional significance and chose, instead, to evaluate the amino acid kinetic and plasma amino acid data. This does not mean, as Millward has incorrectly concluded, that we do not accept N balance data in adults under any circumstances. Rather it implies that we find the N balance technique in relatively short-term studies involving constant N intakes with a variable IAA composition to be of limited value. In contrast, when N intakes are varied over the submaintenance range of total protein needs during different periods, while possibly varying levels and/or sources of IAA, useful N balance data can be obtained. It is also in this context that our N balance data are consistent with our revised estimates of the lysine requirement.

The revised estimate of the lysine requirement in adults, 30 mg/kg/d (Table 10-1), has implications for the nutritional quality of wheat proteins (Young and Pellett, 1985). Thus, the lysine content of wheat products is summarized in Table 10-5, together with the lysine content of a number of FAO/WHO amino acid

TABLE 10-5 Lysine Content of Wheat Flour Compared with other Foods or Amino Acid Requirement Patterns

Food Requirement Pattern	Lysine Content (mg/g protein)
Whole wheat flour	24
Wheat flour (70 percent-80 percent extraction rate)	20
Wheat bran	16
Animal proteins	85 ± 9
Legumes	65 ± 7
1985 FAO/WHO/UNU pattern	
Adults	16
School children	44
Preschool children	58
1991 FAO/WHO pattern	58
MIT-AARP	50

TABLE 10-6. Lysine Content of Whole Wheat Flour in Relation to an Estimate of Protein Quality

Amino Acid Pattern	Amino Acid Score
1985 FAO/WHO/UNU for adults	>100 (L)
1991 FAO/WHO	41 (L)
MIT-AARP	48 (L)
1985 FAO/WHO/UNU preschool child	41 (L)

NOTE: L, lysine, first limiting amino acid, not corrected for digestibility.

scoring patterns. In addition, a usual concentration of lysine in most animal proteins and legumes and that for the MIT requirement pattern are also given for comparison in this table. Hence, if an amino acid score [(amino acid content in the food protein/amino acid content in the reference amino acid requirement pattern) × 100] were calculated for wheat flour, it would be greater than 100 when the 1985 FAO/WHO/UNU amino acid requirement pattern for the adult is used as the reference pattern (Table 10-6). This means that the nutritional value of wheat would be equal to that of high-quality, animal protein foods such as milk, egg, or beef. However, for scoring purposes, the FAO/WHO/UNU (1985) preschool amino acid pattern (or the 1991 FAO/WHO pattern) predicts a relative nutritional quality of 41 percent and with the MIT pattern, the score predicts a slightly higher value of 48 percent. In each case, lysine is predicted to be the most limiting amino acid. These preceding and lower estimates of the nutritional quality of wheat proteins in adults are consistent with the results of nitrogen balance experiments in healthy adults carried out at MIT approximately 20 years ago (Young and Pellett, 1985; Young et al., 1975).

The N balance response to graded intakes of test dietary protein in healthy adults, expressed as relative protein value (RPV = N balance slope using wheat/N balance reference protein × 100), was 54 for whole wheat protein, using beef protein as a reference (Table 10-7). Expressed as relative N requirement (RNR = 1/[amount of wheat protein to achieve N balance in 97.5 percent of population ÷ equivalent amount of beef protein] × 100), the response was about 56 (Table

TABLE 10-7 Biological Assessment of the Nutritional Quality of Whole Wheat Proteins in Young Adults[1]

Measure of Quality	Experimental Value	Predicted from Amino Acid Values:	
		1985 FAO/WHO/UNU	MIT-AARP
Relative protein value	54	>100	48
Relative nitrogen requirement	56	>100	48

[1]Expressed in comparison with beef protein as reference protein (Young et al. 1985).

10-7). The MIT amino acid requirement pattern predicted a value of 48 (Tables 10-6 and 10-7). Hence, there is clearly very good agreement between these experimentally derived (nitrogen balance) values and the predicted nutritional quality of whole wheat proteins. In contrast, use of the 1985 FAO/WHO/UNU adult amino acid pattern gives an invalid estimate of the nutritional value of wheat protein, in that this pattern makes wheat proteins nutritionally equivalent to beef proteins. Notwithstanding the problems faced when attempting to aggregate N balance data across studies carried out in different laboratories or within the same laboratory on different occasions (e.g., Millward et al., 1989), our observations support the view that the 1985 FAO/WHO/UNU lysine requirement value of 12 mg/kg/d for the adult should be discarded. Further, they provide additional justification for the tentative requirement value of 30 mg/kg/d, proposed above (or 50 mg lysine per g of protein), and they strengthen our recommendation that this figure should be used until additional data become available that may make a further change in the recommendation both necessary and desirable.

Millward (Chapter 9) refers to "several long-term trials of both body weight and N balance maintenance on wheat" in support of his view that there is a relatively low requirement for lysine in adult maintenance. He refers to the Minnesota bread study as providing strong support for his proposition. Conducted by Bolourchi et al. (1968), the study reported that N balance was achieved in adults who were given daily, for 50 days, a 12-g N diet (approximately 1g/kg/d of protein) in which wheat proteins supplied about 90 to 95 percent of total N intake. I estimate, from the data presented in their paper, a mean lysine intake of about 18 mg/kg/d. However, to prevent weight loss, these investigators found it necessary to provide their subjects a daily energy intake of 54 kcal/kg. This high energy intake confounds the interpretation of their N balance data, as previously discussed (Young 1987; Young and Marchini, 1990). Hence, this is a serious limitation of this study just as it was in the studies by Rose (1957).

In the North Carolina wheat study, Edwards et al. (1971) measured N balances in adults who were given a diet, for 15 to 29 days, that was based largely on wheat protein but also supplemented with other plant foods so that the level of lysine in the diet approximated 41 mg/g of protein or an intake of about 26 mg/kg/d (not 17 mg/kg as stated by Millward [1997]). Subjects in the study maintained body N equilibrium; these findings support our conclusions if we recognize that the daily lysine intake in this experiment exceeded the FAO/WHO/UNU requirement value (12 mg/kg) by about twofold. It also exceeded the mean requirement estimate (8.8 mg/kg/d) suggested by Rose et al. (1955a) by as much as threefold. For these reasons, I do not find that these N balance studies adequately support Millward's concept of "a relatively low metabolic demand for IAA in the human maintenance requirement."

Stable Isotope Estimates of Requirement Values

The ^{13}C-tracer studies that we conducted to reassess and arrive at new tentative estimates of IAA requirements have been exhaustively examined by Waterlow (1996), who concluded, "Since I have failed to find any source of error large enough to account for the 2-3 fold difference from Rose's estimates, it is logical to look for sources of error in the old rather than the new figures." Millward (Chapter 9) lists the main problems related to the use of ^{13}C-amino acid tracer studies for measurements of amino acid oxidation rates and amino acid balance. Two of the problems are less tractable and possibly insoluble, namely (1) the fact that the amount of tracer given is not massless and may influence balance and (2) the true precursor amino acid enrichment as compared with that measured to determine the oxidation rate.

With respect to the first problem, Fuller and Garlick (1994) and Millward (Chapter 9) appropriately point out that the distribution of the tracer and test diet amino acid over the 24-hour period might well affect the estimate of amino acid requirement. Furthermore, Fuller and Garlick (1994) conclude that this problem might give rise to an appreciable overestimate; this is potentially a real problem.

For example, in our 24-hour tracer studies with labeled tyrosine and phenylalanine (Basile-Filho et al., in press), we gave various tracers in equal amounts during both of the 12-hour fast and fed periods. At the low phenylalanine test intake, this meant that a significant fraction of the total daily intake was given during the postabsorptive phase. This raises the question about the relative utilization of phenylalanine when given in this way versus when it is supplied together with other amino acids. Again, as discussed earlier, the tissue free amino acid pools, particularly muscle, can serve as a reservoir for IAAs when intakes exceed immediate needs for protein synthesis (Bergstrom et al., 1990). Thus, these "stored" amino acids can later be used when there is an intake of the other IAAs. Millward accepts this possibility in his assessment of the recycling of free lysine liberated during the postabsorptive state. However, the extent to which the input of the ^{13}C-labeled tracers during the fast actually determined the whole body fasting-state rate of tyrosine (and aromatic amino acid) oxidation or of leucine and lysine oxidation cannot be easily judged form the data we have gathered to date. Nevertheless, the rate of tyrosine oxidation during the fasting phase of the 24-hour day amounted to 10 μmol/kg/h for a tyrosine-free and an "intermediate" intake level of phenylalanine (38mg/kg/d) and so this observation would seem to be appropriate for further evaluation. Thus, if it can be assumed that the obligatory loss of aromatic amino acids during the fasting period approximates about 7 μmol/kg/h or more (see Table 1 in Millward, Chapter 9), then it does not appear that the two tyrosine and one phenylalanine tracers given simultaneously in this study (total 7.7 μmol/kg/h) had any profound impact on the fasting rate of whole body aromatic amino acid oxidation. Again, the studies by Bergstrom et al. (1990) on the changes in muscle free amino acids in adult subjects following protein-free and protein-rich meals suggest that the tracer doses of tyrosine and phenylalanine given during the fast are retained, at least in part, in tissue free amino acid pools. This effect is also implied in stable isotope tracer studies of amino acid oxidation, protein turnover, and calculated N balance, such as those by Millward and his colleagues (Quevedo et al., 1994), referred to earlier. Our studies with leucine as a tracer further support this contention, in view of the agreement obtained between measured and leucine-derived estimates of N excretion (El-Khoury et al., 1994a). This applies also to the postabsorptive rate of leucine oxidation (El-Khoury et al., 1995). Therefore, Millward's concern about the uncertainty of whether our estimates of leucine requirement are due to excessive postabsorption oxidative loss seems to be lessened by these considerations of the data. However, it remains for future studies to fully resolve this problem of level of isotope administration given during the fast period as well as how best to provide the daily test amino acid intake level during the 24-hour stable isotope tracer protocol.

The second of the less-tractable problems in the ^{13}C-amino acid tracer studies concerns the enrichment of the amino acid precursor pool undergoing oxidation. This is not a problem with respect to leucine and the other branched-chain amino acids since the oxidation is measured in terms of the enrichment of their intracellularly derived keto acids in plasma. This is supported by the agreement between measured N excretion and that predicted from leucine oxidation when the level of trace amino acid intake in the dietary protein (percent w/w) matches that in mixed proteins of the body (El-Khoury et al., 1994a). The lack of agreement between the leucine oxidation and N balance data mentioned by Millward (Chapter 9) and in reference to his study with Price et al. (1994) was very likely due to (1) the choice of the timing of the amino acid oxidation measurements in relation to the daily N excretion determinations, and (2) a mismatch (uncorrected for) between the tracer plus dietary leucine intake per unit of N and the leucine concentration in mixed proteins in the body.

The indicator amino acid oxidation technique (Zello et al., 1990, 1993, 1995) involving ^{13}C-phenylalanine (Zello et al., 1993) and ^{13}C-leucine (Kurpad et al., 1998), as labeled indicators, potentially reduces the severity of the two problems noted above. Using ^{13}C-phenylalanine as an indicator, the Toronto group has estimated the lysine requirement to be on the order of 40 mg/kg/d, using a "break point" analysis of the $^{13}CO_2$ output data (Duncan et al., 1996; Zello et al., 1993). These findings support our tentative new requirement estimates for lysine (Table 10-1). Millward (Chapter 9) criticizes the Toronto studies in human subjects, even though they were designed in relation to the various indicator amino acid studies conducted in experimental animals, which have given amino acid requirement estimates that are consistent with those derived from growth studies (Zello et al., 1995). In a recent paper, Millward (1997) also faulted the lysine requirement values reported by Zello et al. (1993), which were obtained by the indicator oxidation method, because they made no attempt to allow for any adaptation in lysine oxidation due to a reduced intake. The dietary design used by Zello et al. (1993), as well as by Duncan et al. (1996) in a similar followup study, involved giving test intakes of lysine on only one day during a dietary period when subjects were otherwise consuming an experimental diet that supplied a reasonably generous amount of lysine (60 mg/kg/day). Hence, it seems possible that this design would lead to a lower, rather than higher, break point on the lysine intake-indicator amino acid oxidation curve. This is because there could well be a "replete" free lysine pool that would serve as an unaccounted source of utilizable lysine, in addition to the actual intake supplied by the six small hourly meals given beginning 2 hours before and during the 4-hour isotope tracer study. It is also of possible interest that in estimating protein requirements in healthy elderly and young adults from ^{13}C-leucine balance studies, Millward and colleagues (Fereday et al., 1997) followed a somewhat analogous design. They conducted their ^{13}C-leucine balance estimates in subjects who had continued consuming their usual protein intakes until the night before the 9-hour tracer infusion protocol began the following morning.

Finally, in an unpublished study, conducted in collaboration with Anura Kurpad in Bangalore, India, we determined the daily (24-hour) oxidation rate of leucine at graded lysine intakes. Our preliminary results, depicted in Figure 10-3, suggest, once again, an approximate lysine requirement of 30 mg/kg/day.

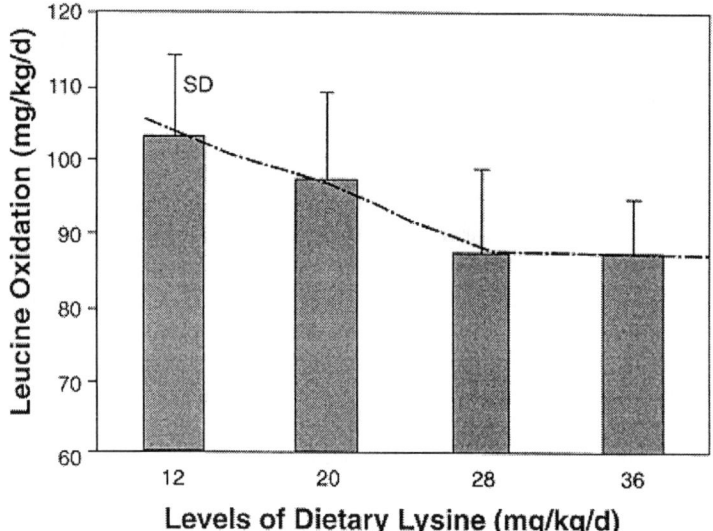

(note: Leucine intake ~ 94mg.kg^{-1}.day^{-1}); Each bar n= 9)

FIGURE 10-3 Relationship between 24h leucine oxidation and graded dietary lysine intakes (at constant N and other IAA's) in healthy Indian adult males. Unpublished, preliminary data of A. Kurpad and coworkers (Bangalore, India). For other experimental details see Kurpad et al. (1998).

The MIT Amino Acid Requirement Pattern and Physical Activity

With respect to physical activity, two final points need to be emphasized regarding the perspective of Millward and coworkers on the MIT amino acid requirement pattern. First, in a critique of one of our papers published in support of our tentative MIT amino acid requirement pattern, Millward (1997) concluded that ". . . given (1) the low minimal obligatory needs for indispensable amino acids,(2) the fact that their metabolic demand reflects the extent to which adaptive changes in oxidation occur, and (3) the growing evidence for the availability of indispensable amino acids, including lysine, deriving from colonic microbial *de novo* amino acid synthesis from salvaged urea . . . , in my view, definition of adult indispensable amino acid requirements for protein quality scoring is not currently possible or likely to be useful in the future." We (Metges et al., 1997) and others (Gibson et al., 1997; Tanaka et al., 1980; Tollardonna et al., 1994; Yeboah et al., 1996) have obtained evidence for the uptake into body tissues of lysine that is derived from intestinal microbial synthesis. However, the extent to which this source of lysine serves as a net contribution to the total lysine intake, supplied from dietary sources, is still to be determined. Assuming there is also a microbial synthesis and uptake of the other IAAs, as well as lysine, our studies with leucine (El-Khoury et al., 1994a, 1996) and those of Millward and colleagues (Price et al., 1994), suggest that this microbial source of IAAs compensates, in large part, for the losses of IAAs via the turnover and secretion of intestinal proteins and the subsequent oxidative catabolism of the liberated amino acids by the activity of the microflora and possibly intestinal tissues. In this case, a cycle of lysine (or IAA) synthesis within the gut lumen and the oxidative catabolism of lumen IAAs could be envisaged as a metabolic cycle that contributes to IAA homeostasis but that would not necessarily complicate estimation of the whole body kinetics and oxidation of the IAAs with the ^{13}C-tracer techniques as currently employed in our laboratories and those of others, including Millward and collaborators.

Second, the above conclusion by Millward reminds me of an editorial by Yates (1983), in which he quotes Murphy (1982), as follows: "The growing point of science is discrepancy . . . the next best thing to uttering the truth is to say something so sufficiently definite that it can be constructively disagreed with." Indeed, I believe that the definitive conclusion drawn by Millward is neither acceptable nor helpful, despite the many real difficulties that he has noted appropriately in relation to defining the amino acid requirements in humans, and I have tried to be constructive in my disagreement.

If a requirement is defined, according to the definition proposed by FAO/IAEA/WHO (1996), as "the lowest continuing level of nutrient intake, that at a specified efficiency of utilization, will maintain the defined level of nutriture in the individual" then it should be possible to define a requirement for any one of the IAAs, just as is the case for many of the trace elements and vitamins. I agree with Millward that the requirement would best be established on the basis of longer term experimental studies and on the functional consequences of variable intakes of protein and/or its constituent IAAs. This is an ideal goal for the future. In the interim, reasonable estimates of intakes minimally necessary to achieve body amino acid balance, after a suitable period of adaptation as discussed above, can be made. These would permit rational and safe decisions and policy concerning the maintenance of a satisfactory state of body protein and amino acid nutriture in both individuals and in defined population groups, including the military, which is the particular focus of concern in this report. Accordingly, it is prudent to accept the MIT amino acid requirement pattern (Table 10-1) in practical considerations of human protein nutrition, until strong data to the contrary are forthcoming. Millward's model (Chapter 9), which views amino acid metabolism and oxidative losses according to (a) a minimum obligatory metabolic demand and (b) an adaptive component, is certainly interesting and possibly might aid in the further scrutiny of the metabolic aspects of amino acid homeostasis at differing intakes of protein or IAA. It does not help solve the problems that are of greatest concern, namely (a) the practical nutritional relevance of the amino acid requirement values proposed initially by Rose and used together with comparable data from other investigators to arrive at the 1985 FAO/WHO/UNU adult indispensable amino acid requirement pattern and (b) determination of amino acid requirement values that are likely to be minimally sufficient to maintain body amino acid homeostasis, and by implication adequate body function, over the long term.

Millward (Chapter 9) refers to the studies by Chittenden (1904), which I have also found to be interesting and relevant (Young, 1997). As I have reviewed (Young, 1997), Millward (Chapter 9) concludes, "Chittenden's views were controversial at the time and have been given less prominence than they deserve in the intervening years of continued controversy. However, the data speak for themselves." I agree that they do. Thus, it might be of interest that the (calculated) lysine intake by Chittenden himself, who consumed a diet supplying about 0.75 g protein per kg body weight daily, approximated 46 mg/kg/day. This intake was possibly somewhat higher than for the soldiers, whose lysine intakes might be estimated to have been around 30 to 40 mg/kg/d, as judged from the limited dietary data given in Chittenden's extensive summary of his now classic experiments (Chittenden, 1904). One might wonder whether the performance of the soldiers, which actually improved over the 5-month experiment, would have deteriorated had the IAA intakes approximated those proposed by the international group (FAO/WHO/UNU, 1985) or those low values that Millward (Chapter 9) proposes as being sufficient to meet the minimum obligatory metabolic demand.

We have not examined this latter question specifically in relation to a further evaluation of the status of amino acid metabolism and balance at IAA intakes that either meet our tentative requirements or those approximating the 1985 FAO/WHO/UNU values. However, I am prepared to speculate from the experiments described in Marchini et al. (1993), that there would be a deterioration in amino acid balance and, consequently, in function at intakes not exceeding the 1985 FAO/WHO/UNU requirement figures or those of Millward that are estimated to meet the minimum obligatory metabolic demand. At least, we can say, based on El-Khoury et al. (1997), that at a more generous IAA and leucine intake (diet and tracer contributing about 126 mg/kg/d) for two 90 min periods of exercise at 45 percent VO$_2$ max, during both the fasting and fed periods of the day, resulted in a small overall increase in the rate of leucine oxidation, amounting to about 5 mg/kg/d. It would be interesting and highly worthwhile to repeat this detailed metabolic/exercise study but at dietary IAA intakes that are far more limiting.

Author's Conclusions and Recommendations

I have addressed a number of the issues raised by Millward (Chapter 9) in order to weigh the merits of our respective views. It is my judgment that the minimum physiological requirements for amino acids to maintain health and body function should be defined and that it would be prudent to apply the tentative requirement values proposed herein for the protein and amino acid nutrition of healthy adults. Clearly, there are difficulties involved in estimating amino acid requirements and limitations to the experiments that have been conducted to date, both by ourselves and by others. I concur with Rose (1957), who stated in his review article: "In view of the difficulties of such studies, it is to be hoped that many more will be made in both men and women and in different laboratories." This is crucial, given the importance of nutrient requirement data for the appropriate planning of world food and dietary protein needs (Young et al., 1998), not to mention the rational design of safe and effective diets for maximizing and maintaining physical performance.

REFERENCES

Atinmo, T., C.M.F. Mbofung, M.A. Hussain, and B.O. Osotimehin. 1985. Human protein requirements: Obligatory urinary and faecal nitrogen losses and the factorial estimation of protein needs of Nigerian male adults. Br. J. Nutr. 54:605–611.

Atinmo, T., C.M.F. Mbofung, G. Eggum, and B.O. Osotimehin. 1988. Nitrogen balance study in young Nigerian adult males using four levels of protein intake. Br. J. Nutr. 60:451–458.

Basile-Filho, A., L. Beaumier, A.E. El-Khoury, Y.-M. Yu, M. Kenneway, R.E. Gleason, and V.R. Young. 1998. Twenty-four hour L-[1-^{13}C]tyrosine and L-[3,3-^{2}H]phenylalanine oral tracer aromatic amino acid requirements in adults. Am. J. Clin. Nutr. 67:640–659

Bergstromm, J., P. Fürst, and E. Vinnars. 1990. Effect of a test meal, without and with protein, on muscle and plasma free amino acids. Clin. Sci. 79:331–337.

Bodwell, C.E., E.M. Schuster, E. Kyle, B. Brooks, M. Womack, P. Steele, and R. Ahrens. 1979. Obligatory urinary and fecal nitrogen losses in young women, older men and young men: Factorial estimation of adult human protein requirements. Am. J. Clin. Nutr. 32:2450–2459.

Bolourchi, S., C.M. Friedeman, and O. Mickelson. 1968. Wheat flour as a source of protein for adult human subjects. Am. J. Clin. Nutr. 21:827–835.

Calloway, D.H., and S. Margen. 1971. Variation in endogenous nitrogen excretion and dietary nitrogen utilization as determinants of human protein requirements. J. Nutr. 101:205–216.

Chittenden, R.H. 1904. Physiological Economy of Nutrition. N.Y.: Frederick A. Stokes Co.

Clugston, G., K.G. Dewey, C. Fjeld, J. Millward, P. Reeds, N.S. Scrimshaw, K. Tontisirin, J.C. Waterlow, and V.R. Young. 1996. Report of the working group on protein and amino acid requirements. Europ. J. Clin. Nutr. 50:S193–S195.

Department of Health, Panel on Dietary Reference Values of the Committee on Medical Aspects of Food Policy. 1991. Report on Health and Social Subjects, Number 41. Dietary Reference Values for Food Energy and Nutrients for the United Kingdom. London: Her Majesty's Stationary Office (HMSO).

Duncan, A.M., R.O. Ball, and P.B. Pencharz. 1996. Lysine requirement of adult males is not affected by decreasing dietary protein intake. Am. J. Clin. Nutr. 64:718–725.

Edwards, C.H., L.K. Booker, C.H. Rmphg, W.G. Wright, and S. Ganapathy. 1971. Utilization of wheat by adult man: Nitrogen metabolism, plasma amino acids and lipids. Am. J. Clin. Nutr. 24:181–193.

El-Khoury, A.E., N.K. Fukagawa, M. Sánchez, R.H. Tsay, R.E. Gleason, T.E. Chapman, and V.R. Young. 1994a. Validation of the tracer balance concept with reference to leucine: 24-h intravenous tracer studies with L-[1-^{13}C]leucine and [^{15}N-^{15}N]urea. Am. J. Clin. Nutr. 59:1000–1011.

El-Khoury, A.E., N.K. Fukagawa, M. Sánchez, R.H. Tsay, R.E. Gleason, T.E. Chapman, and V.R. Young. 1994b.. The 24-h pattern and rate of leucine oxidation, with particular reference to tracer estimates of leucine requirements in healthy adults. Am. J. Clin. Nutr. 59:1012–1020.

El-Khoury, A.E., M. Sánchez, N.K. Fukagawa, and V.R. Young. 1995b. Whole body protein synthesis in healthy adult humans; $^{13}CO_2$ technique vs. plasma precursor approach. Am. J. Physiol. 268:E174–E184.

El-Khoury, A.E., A.M. Ajami, N.K. Fukagawa, T.E. Chapman, and V.R. Young. 1996. Diurnal pattern of the interrelationship among leucine oxidation, urea production, and hydrolysis in humans. Am. J. Physiol. 271:E563–E573.

El-Khoury, A.E., A. Forslund, R. Olsson, S. Branth, A. Sjodin, A. Andersson, A. Atkinson, A. Selvaraj, L. Hambraeus, and V.R. Young. 1997. Moderate exercise at energy balance does not affect 24h leucine oxidation or nitrogen retention in healthy men. Am. J .Physiol. 273:E394–E407.

El-Khoury, A.E., A. Basile, L. Beaumier, S.Y. Wang, H.A. Al-Amiri, A. Selvaraj, S. Wong, A. Atkinson, A.M. Ajami, and V.R. Young. 1998. Twenty-four hour intravenous and oral tracer studies with L-]1-^{13}C]-2-aminoadipic acid and L-[1-^{13}C]lysine as tracers at generous nitrogen and lysine intakes in healthy adults. Am. J. Clin. Nutr. (in press).

FAO/IAEA/WHO (Food and Agriculture Organization of the United Nations/International Atomic Energy Agency/World Health Organization). 1996. Trace Elements in Human Nutrition and Health. Geneva: WHO.

FAO/WHO (Food and Agriculture Organization of the United Nations/World Health Organization). 1991. Protein Quality Evaluation. Report of a Joint FAO/WHO Expert Consultation. FAO Food and Nutrition Paper 51. Rome, Italy: FAO.

FAO/WHO/UNU ((Food and Agriculture Organization of the United Nations/World Health Organization)/United Nations University). 1985. Energy and Protein Requirements. Technical Report Series Number 724. Geneva: WHO.

Fereday A., N.R. Gibson, M. Cox, P.J. Pacy, and D.J. Millward. 1997b. Protein requirements and aging: Metabolic demand and efficiency of utilization. Brit. J. Nutr. 77:685–702.

Fisher H., M.K. Brush, and P. Griminger. 1969. Reassessment of amino acid requirements of young women on low nitrogen diets. 1. Lysine and tryptophan. Am. J. Clin. Nutr. 22:1190–1196.

Fuller, M.F., and P.J. Garlick. 1994. Human amino acid requirements: Can the controversy be resolved. Ann. Rev. Nutr. 14:217–241.

Fuller, M.F., A. Milne, C.I. Harris, T.M.S. Reid, and R. Keenan. 1994. Amino acid losses in ileostomy fluid on a protein-free diet. Am. J. Clin. Nutr. 59:70–73.

Gibson, N., E. Al-Sing, A. Badalloo, T. Forrester, A. Jackson, and D.J. Millward. 1997. Transfer of ^{15}N from urea to the circulating dispensable and indispensable amino acid pool in the human infant. Proc. Nutr. Soc. 56:79A.

Gopalan, C., and B.S. Narasinga Rao. 1966. Effect of protein depletion on urinary nitrogen excretion in undernourished subjects. J. Nutr. 90:213–218.

Goran, M.I., W.H. Beer, R.R. Wolfe, E.T. Poehlman, and V.R. Young. 1993. Variation in total energy expenditure in young healthy free-living men. Metabolism 42:487–496.

Huang, P.C., H.E. Chong, and W.M. Rand. 1972. Obligatory urinary and fecal nitrogen losses in young Chinese men. J. Nutr. 102:1605–1614.

Inoue, G., Y. Fujita, K. Kishi, S. Yamamoto, and Y. Niyama. 1974. Nutritive values of egg protein and wheat gluten in young men. Nutr. Rep. Intl. 10:201–207.

Irwin, M.I., and D.M. Hegsted. 1971. A conspectus of research on amino acid requirements of man. J. Nutr. 101:539–566.

Kurpad, A.V., A.E. El-Khoury, L. Beaumier, A. Srivatsa, R. Kuriyon, T. Raj, S. Borgonha, A.M. Ajami, and V.R. Young. 1998. An initial assessment, using 24-hour ^{13}C-leucine kinetics, of the lysine requirement of healthy adult Indian subjects. Am. J. Clin. Nutr. 67:58–66.

Marchini, J.S., J. Cortiella, T. Hiramatsu, T.E. Chapman, and V.R. Young. 1993. The requirements for indispensable amino acids in adult humans: A longer-term amino acid kinetic study, with support for the adequacy of the Massachusetts Institute of Technology amino acid requirement pattern. Am. J. Clin. Nutr. 58:670–683.

Marchini, J.S., J. Cortiella, T. Hiramatsu, L. Castillo, T.E. Chapman, and V.R. Young. 1994. Phenylalanine and tyrosine kinetics for different patterns and indispensable amino acid in adult humans. Am. J. Clin. Nutr. 60:79–86.

Masud, T., V.R. Young, T. Chapman, and B.J. Maroni. 1994. Adaptive responses to very low protein diets: The first comparison of ketoacids and essential amino acids. Kidney Intl. 45:1182–1192.

McLarney, M.J., P.L. Pellett, and V.R. Young. 1996. Pattern of amino acid requirements in humans: An across-species comparison of published amino acid requirement recommendations. J. Nutr. 126:1871–1882.

Metges, C.C., A.E. El-Khoury, K.J. Petzke, S. Bedri, M.F. Fuller, and V.R. Young. 1997. The quantitative contribution of microbial lysine to lysine flux in healthy male subjects. Abstract. FASEB J. 11(3):A149.

Millward, D.J. 1997. Human amino acid requirements. J. Nutr. 127:1842–1846.

Millward, D.J., and J.P. Rivers. 1989. The nutritional role of indispensable amino acids and the metabolic basis for their requirements. Europ. J. Clin. Nutr. 42:367–393.

Millward, D.J., A.A. Jackson, G. Price, and J.P.W. Rivers. 1989. Human amino acid and protein requirements: Current dilemmas and uncertainties. Nutr. Res. Rev. 2:109–132.

Millward, D.J., A. Fereday, N. Gibson, and P.J. Pacy. 1996. Postprandial protein metabolism. Bailliere's Clin. Endocrin. Metab. 10:533–549.

Motil, K.J., D.E. Matthews, D.M. Bier, J.F. Burke, H.N. Munro, and V.R. Young. 1981. Whole-body leucine and lysine metabolism: Response to dietary protein in young men. Am. J. Physiol. 240:E712–E721.

Murphy, E.A. 1982. Muddling, meddling and modeling. Pp. 333–348 in Genetic Basis of the Epilepsies, V.E. Anderson, W.A. Hauser, J.K. Penry, and C.F. Sing, eds. New York: Raven.

Nicol, B.M., and P.G. Phillips. 1976a. Endogenous nitrogen excretion and utilization of dietary protein. Br. J. Nutr. 35:181–193.

Nicol, B.M., and P.G. Phillips. 1976b. The utilization of dietary protein by Nigerian men. Br. J. Nutr. 36:337–351.

Pineda, O., B. Torun, F.E. Viteri, and G. Arroyove. 1981. Protein quality in relation to estimates of essential amino acid requirements. Pp. 29–42 in Protein Quality in Humans: Assessment and In Vitro Estimation, C.E. Bodwell, J.S. Adkins, and D.T. Hopkins, eds. Westport, Conn.: AVI Publishers.

Price, G.M., D. Halliday, P.J. Pacy, M.R. Quevedo, and D.J. Millward. 1994. Nitrogen homeostasis in man: Influence of protein intake on the amplitude of diurnal cycling of body nitrogen. Clin. Sci. 86:91–102.

Quevedo, M.R., G.M. Price, D. Halliday, P.J. Pacy, and D.J. Millward. 1994. Nitrogen homeostasis in man: Diurnal changes in nitrogen excretion, leucine oxidation and whole body leucine kinetics during a reduction from a high to a moderate protein intake. Clin. Sci. 86:185–193.

Rand, W.M., V.R. Young, and N.S. Scrimshaw. 1976. Change of urinary nitrogen excretion in response to low protein diets in adults. Am. J. Clin. Nutr. 29:639–644.

Rand, W.M., N.S. Scrimshaw, and V.R. Young. 1979. An analysis of temporal patterns in urinary nitrogen excretion of young adults receiving constant diets at two nitrogen intakes for 8 to 11 weeks. Am. J. Clin. Nutr. 32:1408–1414.

Rand, W.M., N.S. Scrimshaw, and V.R. Young. 1981. Conventional ("long-term") nitrogen balance studies for protein quality evaluation in adults: Rationale and limitations. Pp. 61–94 in Protein Quality in Humans: Assessment and In Vitro Estimation, C.E. Bodwell, J.S. Adkin, and D.T. Hopkins, eds. Westport, Conn.: AVI Publishing.

Rand, W.M., N.S. Scrimshaw, and V.R. Young. 1985. A retrospective analysis of long-term metabolic balance studies: Implications for understanding dietary nitrogen and energy utilization. Am. J. Clin. Nutr. 42:1329–1350.

Reeds, P.J. 1990. Amino acid needs and protein scoring patterns. Proc. Nutr. Soc. 49:489–497.

Roberts, S.B., M.B. Heyman, W.J. Evans, P. Fuss, R. Tsay, and V.R. Young. 1990. Dietary requirements of young adult men, determined using doubly labeled water method. Am. J. Clin. Nutr. 54:499–505.

Roberts, S.B., P. Fuss, M.B. Heyman, and V.R. Young. 1995. Influence of age on energy requirements. Am. J. Clin. Nutr. 62:1053S–1058S.

Rose, W.C. 1957. The amino acid requirements of adult man. Nutr. Abstr. Rev. 27:631–647.

Rose, W.C., A. Borman, M.J. Coon, and G.F. Lambert. 1955a. The amino acid requirements of man. X. The lysine requirement. J. Biol. Chem. 214:579–587.

Rose, W.C., B.E. Leach, M.J. Coon, and G.F. Lambert. 1955b. The amino acid requirements of man. IX. The phenylalanine requirement. J. Biol. Chem. 213:913–922.

Scrimshaw, N.S., M.A. Hussein, E. Murray, W.M. Rand, and V.R. Young. 1972. Protein requirements of man: Variations in obligatory urinary and fecal nitrogen losses in young men. J. Nutr. 102:1595–1604.

Tanaka, N., K. Kubo, K. Shiraki, H. Koishi, and H. Yoshimura. 1980. A pilot study on protein metabolism in the Papuan New Guinea Highlanders. J. Nutr. Sci. Vitaminol. 26:247–259.

Tollardona, D., C.I. Harris, E. Milne, and M.F. Fuller. 1994. The contribution of intestinal microflora to amino acid requirements in pigs. Pp. 245–248 in Proceedings of the 6th International Symposium on Digestive Physiology in Pigs, W.B. Souffran, and H. Hagemeister, eds. Dummerstorf, Germany: Forsch. Inst. Biol. Landwirtsch. Nutztiere.

Tom K., V.R. Young, T. Chapman, T. Masud, L. Akpele, and B.J. Maroni. 1995. Long-term adaptive responses to dietary restriction in chronic renal failure. Am. J. Physiol. 268:E668–E677.

Waterlow, J.C. 1996. The requirements of adult man for indispensable amino acids. Europ. J. Clin. Nutr. 50:S151–S179.

Williams, H.H., A.E. Harper, D.M. Hegsted, G. Arroyave, L.E. Hold, Jr. 1974. Nitrogen and amino acid requirements. Pp. 23–63 in Improvement of Protein Nutritive Value. Washington, D.C.: National Academy Press.

Yates, F.E., 1983. Contribution of statistics to ethics of science. Am. J. Physiol 13:R3–R5.

Yeboah, N., E. Al-Sing, A. Bodalloo, T. Forrester, A. Jackson, and D.J. Millward. 1996. Transfer of ^{15}N from urea to the circulating lysine pool in the human infant. Proc. Nutr. Soc. 55:37A.

Yoshida, A. 1986. Nutritional aspects of amino acid oxidation. Pp. 378–382 in Proceedings of the XIII International Congress of Nutrition, T.G. Taylor and N.K. Jenkins, eds. London: John Libbey and Co., Ltd.

Young, V.R. 1987. McCollum Award Lecture: Kinetics of human amino acid metabolism: Nutritional implications and some lessons. Am. J. Clin. Nutr. 46:709–725.

Young, V.R. 1991a. Nutrient interactions with reference to amino acid and protein metabolism in nonruminants: Particular emphasis on protein-energy relations in man. Z. Ernahrungswiss 30:239–267.

Young, V.R. 1992. Protein and amino acid requirements in humans: Metabolic basis and current recommendations. Scand. J. Nutr./Naringsforskning 36:47–56.

Young, V.R. 1994. Adult amino acid requirements: The case for a major revision in current recommendations. J. Nutr. 124:1517S–1523S.

Young, V.R. 1997. Paper 5. Dietary protein standards can be halved (Chittenden, 1904). J. Nutr. 127:1025S–1027S.

Young, V.R., and A.E. El-Khoury. 1995a. Can amino acid requirements for maintenance in adult humans be approximated from the amino acid composition of body mixed proteins? Proc. Natl. Acad. Sci. USA 92:300–304.

Young, V.R., and A.E. El-Khoury. 1996. Human amino acid requirements: A re-evaluation. Food Nutr. Bull. 17:191–203.

Young, V.R., and J.S. Marchini. 1990. Mechanisms and nutritional significance of metabolic responses to altered intakes of protein and amino acids, with reference to nutritional adaptation in humans. Am. J. Clin. Nutr. 51:270–289.

Young, V.R., and P.L. Pellett. 1985. Wheat proteins in relation to protein requirements and availability of amino acids. Am. J. Clin. Nutr. 41:1077–1090.

Young, V.R., and P.L. Pellett. 1990. Current concepts concerning indispensable amino acid needs in adults and their implications for international nutrition planning. Food. Nutr. Bull. 12:289–300.

Young, V.R., and N.S. Scrimshaw. 1968. Endogenous nitrogen metabolism and plasma amino acids in young adults given a "protein-free" diet. Br. J. Nutr. 22:9–20.

Young, V.R., L. Fajardo, E. Murray, W.M. Rand, and N.S. Scrimshaw. 1975. Protein requirements of man: Comparative nitrogen balance response within the submaintenance-to-maintenance range of intakes of wheat and beef proteins. J. Nutr. 105:534–542.

Young, V.R., D.M. Bier, and P.L. Pellett. 1989. A theoretical basis for increasing current amino acid requirements in adult man, with experimental support. Am. J. Clin. Nutr. 50:80–92.

Young, V.R., N.S. Scrimshaw, and P.L. Pellett. 1998. Significance of dietary protein source in human nutrition: Animal or plant proteins. Pp. 205–212 in Feeding a World Population of More than Eight Billion People: A Challenge to Science, J.C. Waterlow, D.G. Armstrong, L. Fowden and R. Riley, eds. Oxford: Oxford University Press.

Zello, G.A., P.B. Pencharz, and R.O. Ball. 1990. Phenylalanine flux, oxidation and conversion to tyrosine in humans studied with [1-^{13}C]phenylalanine. Am. J. Physiol. 259:E835–E843.

Zello, G.A., P.B. Pencharz, and R.O. Ball. 1993. Dietary lysine requirement of young adult males determined by oxidation of L-[1-^{13}C]phenylalanine. Am. J. Physiol. 264:E677–E685.

Zello, G.A., L.J. Wykes, R.O. Ball, and B. Pencharz. 1995. Recent advances in methods of assessing dietary amino acid requirements for adult humans. J. Nutr. 125:2907–2915.

Protein and Amino Acids, 1999
Pp. 243–253. Washington, D.C.
National Academy Press

11

Physical Exertion, Amino Acid and Protein Metabolism, and Protein Requirements

Michael J. Rennie[1]

INTRODUCTION

A whiff of vitalism, the 19th century philosophical doctrine that there is some spiritual essence associated with biological processes, is still discernible in relation to the question of how contractile activity affects protein and amino acid metabolism. The working machine of muscles is, after all, made up of proteins so the idea comes naturally that where machinery works there must be wear and tear, presumably with a greater requirement for maintenance. It is also known that during starvation, i.e. a situation of substantial energy deficit, the lean body mass is the source of gluconeogenic carbon which fuels the central nervous system, erythrocytes and ion pumping in the kidney. Since physical activity requires extra energy it is plausible to imagine that in circumstances of marginal energy intake the lean body mass would also be at risk of diminution.

[1] Michael J. Rennie, Department of Anatomy and Physiology, University of Dundee, Dundee, Scotland DD1 4HN United Kingdom.

Over the past 25 years our understanding of the tidal flows of amino acids between the gut, viscera and the peripheral musculature has improved markedly. So has our understanding of the phenomenology and control mechanisms of protein turnover, the coordination of which with intermediary amino acid metabolism is now rather well understood. This increasing sophistication has led us to understand that the old distinctions between, for example, essential and non-essential amino acids is much less clear cut than it was. Since, by definition, conditionally essential amino acids are those which may become required in greater than normal amounts in special circumstances, it is obvious to ask whether or not muscular activity causes some amino acids to become conditionally essential.

There has been a substantial recent upswing in interest in the investigation of the relationship between contractile activity and amino acid and protein metabolism but much that we require to provide definitive answers is still missing, requiring more research. The gaps will become obvious in the following paper. I propose to discuss the known effects of increased contractile activity in skeletal muscle amino acid oxidation, on protein turnover and interactions with the state of energy balance, to provide a background for discussion of protein requirements.

EXERCISE AND AMINO ACID CATABOLISM

Although this kind of exercise probably only contributes a small fraction to the total daily energy expenditure of soldiers in the field, it may be that repeated bouts of exercise have cumulative effects on protein and amino acid metabolism; we may get some clues as to what these are by investigating exercise under laboratory conditions. There is now a substantial body of work which allows us to make some reasonably firm statements about the relative importance of amino acid metabolism in skeletal muscle during increased contractile activity.

The main fuels for sustained moderate to high intensity exercise are, of course, carbohydrate and fat and the most efficient means of converting these into ATP is via the Krebs cycle and oxidative phosphorylation. Theoretically the rate at which acetyl units may be catabolised in the Krebs cycle (and reducing equivalents fed into oxidative phosphorylation) will be limited by the availability of oxaloacetate since without this citrate cannot be formed nor can the two carbons of acetate be (eventually) transformed to CO_2 as the cycle turns. One of the most pronounced features of amino acid metabolism in muscle (at least in human muscle), i.e. the increase in alanine production as a result of muscular contraction, may be part of a mechanism to ensure the appropriate expansion of the catalytic pool of Krebs cycle intermediates as the drive to increase ATP production switches on.

It is now well established that glutamate concentrations fall in muscle during exercise at 70–80 percent of Vo_{2max} (Katz et al., 1986; Sahlin et al.,

1995). The obvious routes by which this occurs are either transamination with pyruvate (both in the cytosol and in the mitochondria via alanine aminotransferase, to form alanine and α-ketoglutarate) or the glutamate dehydrogenase reaction, which has recently been discovered (Wibom et al., 1990) to have a somewhat greater capacity in human skeletal muscle than hitherto suspected. There is evidence that in the absence of any other mechanism to increase Krebs cycle intermediates (as in patients with McArdle's disease who are unable to generate oxaloacetate from endogenous glycogen stores via the malic enzyme, pyruvic carboxylase, and PEP carboxykinase), muscle glutamate concentration is reduced at rest, and work capacity appears to be limited when glutamate catabolism bottoms out (Sahlin et al., 1995).

The fall in muscle glutamate is puzzling. Branched chain amino acids are transaminated in the cytoplasmic space to form the branched chain ketoacids which are decarboxylated in mitochondria (see below). The carbon from valine and half of that from isoleucine may enter the Krebs cycle as succinyl CoA, processes which are therefore anaplerotic. The puzzling thing is that branched chain amino acids are transaminated at the expense of α-ketoglutarate to produce glutamate, which ought to protect muscle glutamate concentrations and to deplete α-ketoglutarate. The depletion of α-ketoglutarate would be catapleurotic if it extended to the mitochondria. We currently have no information on this.

What other ways can amino acids contribute to the anaplerotic process? Glutamine crosses the inner mitochondrial membrane with much greater ease than glutamic acid and the mitochondrial phosphate-dependent glutaminase would ensure a plentiful supply of glutamate without the necessity of exchanging glutamate for aspartate across the inner mitochondrial membrane. However, one would then expect to see a fall in muscle glutamine concentration, which is, in fact, usually only seen after long term exercise at moderate intensity at least. However, given the very high background of glutamine and the relative imprecision of measurement of glutamine, it is difficult to be sure about the size of the fall which occurs during heavy exercise. The other possible route of glutamine utilisation and production of α-ketoglutarate would be through the action of glutamine transaminase and to produce α-ketoglutaramide which spontaneously deaminates to produce α-ketoglutarate and ammonia. Unfortunately there is no good data to strengthen or weaken this suggestion. More research is needed on this topic.

The branched-chain amino acids are oxidised in muscle during exercise (Figure 11-1) at a rate which appears to be directly proportional to the overall rate of mitochondrial oxidation, and thus the muscle oxygen uptake possibly because oxygen uptake is blood flow dependent; so is catabolism of the

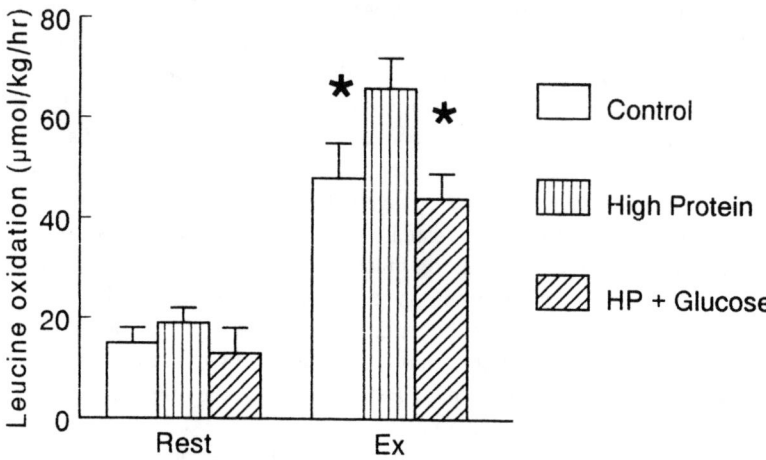

FIGURE 11-1 Effect of exercise in subjects taking a normal diet (control) or a high protein diet (1.5 g protein/kg/day) on leucine oxidation during exercise studied in the absence and presence of exogenous glucose given orally. SOURCE: J. Bowtell and M.J. Rennie (unpublished work).

branched chain amino acids by the transaminase, which has a high K_m (Rennie, 1996). However, the total amount of energy supplied by this process, even at high rates of oxidation, is relatively low, ruling out a major contribution from protein as a metabolic fuel during contractile activity (Millward et al., 1994; Rennie, 1996).

The possible anaplerotic role of the purine nucleotide cycle, which excited a fair amount of interest in the 1970s as a means of generating fumarate, is now thought less likely to be important. The purine nucleotide cycle may not operate fully in contracting muscle and its total capacity is much less than, for example, alanine aminotransferase (Hood et al., 1990; Van Hall et al., 1995b). Ammonia production during exercise is most likely the result of branched chain amino acid catabolism (MacLean et al., 1996; Van Hall et al., 1995b).

The question arises, would repeated muscular exercise lead to a diminution of amino acids from the intramuscular compartment thus possibly limiting any anaplerotic role? If so the corollary is, would exogenous dietary supplementation make sense? These are difficult questions to answer since we do not know what is the lower limit of glutamate concentration before anaplerotic generation would cease to sustain a large enough increase in Krebs cycle intermediates although in long term exercise the pool size of these does fall (Sahlin et al., 1990). It is theoretically possible that repeated bouts of exercise at high intensity might, under circumstances of limited nutrient provision chronically depress muscle glutamate and glutamine concentrations and thereby limit the ability of the Krebs cycle to accelerate rapidly enough. However, this is, at the moment, speculation. More research is needed.

EFFECTS OF CONTRACTILE ACTIVITY ON MUSCLE PROTEIN TURNOVER

It is well accepted that muscular activity has a marked effect on muscle composition and/or muscle bulk, depending upon the type of contractile activity involved. Moderate, repeated, long-term exercise causes increases in mitochondrial mass and alterations in myosin ATPase to isoforms appropriate to long-term aerobic exercise; and high resistance exercise causes increases in expression of actin and myosin and increased mixed muscle protein synthesis (Rennie, 1996) (Figure 11-2). The anabolic phase occurs in the post-exercise period and can be maximised by the provision of exogenous amino acids.

It has been difficult to make measurements of protein synthesis during exercise in human beings because of the difficulty of detecting sufficient change over a short period of time but it seems likely that muscle protein synthesis is depressed (Rennie, 1996; Dohm et al., 1980); thus, the rebound observed post-exercise (Chesley et al., 1992; Biolo et al., 1995) appears to fulfill a homeostatic function as much as anything. The extent of this elevation is limited, with a return to baseline 36 hr after a stimulatory bout of exercise (MacDougal et al., 1995). This ability of contractile activity to stimulate anabolism probably

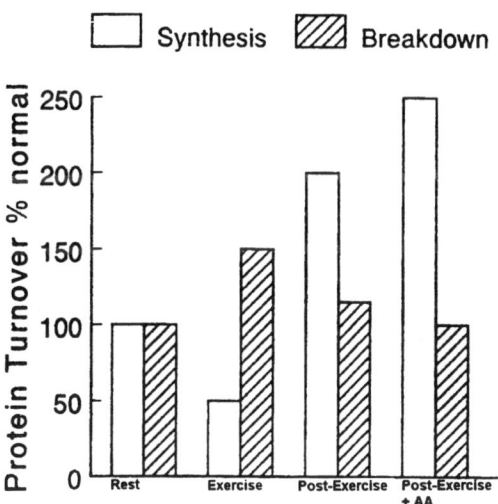

FIGURE 11-2 Notional scheme of changes in muscle protein synthesis and breakdown taking into account most of the current observations in the literature obtained from studies of human muscle.

explains why accustomed exercise increased the efficiency of nitrogen storage in young healthy men (Butterfield and Calloway, 1984).

Protein breakdown during exercise is definitely elevated (Biolo et al., 1995; MacLean et al., 1994) but the increase appears to be confined to the soluble or membrane proteins which are degraded by lysosomal proteases (Kasperek and Snider, 1989; Kasperek et al., 1992). The evidence that myofibrillar protein breakdown occurs during normal exercise is extremely sparse. Nevertheless, there is some evidence that myofibrillar protein breakdown increases as a result of eccentric exercise, i.e. exercise in which muscle is forced to contract as it is stretched, as in walking downhill (Fielding et al., 1991). In the post-exercise period muscle protein breakdown will be elevated to the extent that there is remodeling of muscle (Figure 11-2).

So far as net loss of protein is concerned this can only occur when synthesis occurs at a rate lower than breakdown and the net release of amino acids from muscle in the postabsorptive state and during exercise is a good example of this. However, feeding rapidly reverses the net nitrogen balance (Rennie, 1996) and given the relatively small contribution of amino acids to the fuel economy during exercise, it seems unlikely that the alterations in muscle protein turnover occurring acutely would contribute to increased dietary amino acid requirements. One of the problems for the theory that exercise should increase dietary requirements is of course the fact that eating more protein stimulates the catabolic capacity of the body to oxidise it. This is seen clearly in the results of Figure 11-1.

INTERACTIONS BETWEEN ENERGY SUPPLY AND PROTEIN AND AMINO ACID METABOLISM IN THE CONTEXT OF INCREASED PHYSICAL ACTIVITY

At rest provision of carbohydrate inhibits net protein catabolism, probably mainly by increasing insulin which has inhibitory effects on protein breakdown and stimulatory effects on protein synthesis; in addition the simple provision of carbohydrate inhibits gluconeogenesis from amino acids, diminishing the "pull" from the liver upon the peripheral lean body mass protein. During exercise provision of carbohydrate markedly suppresses leucine oxidation. (Figure 11-1).

There is insufficient evidence available that increased availability of triglycerides and medium chain fatty acids has any effect on the oxidation of amino acids during exercise to make any definitive statements.

What happens when chronic energy expenditure rises to such an extent that a subject is in negative energy balance? Under those circumstances, is muscle mass, for example, at risk? This is a difficult question to answer because the appropriate studies have not actually been done. It would be a reasonable hypothesis, however, that the lean body mass would tend to be preserved as a result of chronic daily exercise, with stores of body fat being used preferentially once the gluconeogenic needs were satisfied by protein breakdown. One study which does throw some light on the situation, although it gives by no means a complete answer, is that carried out on Stroud and Fiennes during their epic, unaided walk across the Antarctic (Stroud et al., 1996). Energy expenditure was measured by the doubly labeled water method and whole body protein turnover was measured using the [^{15}N]glycine/ammonia + urea end-product method. In one subject, whole body protein turnover increased slightly and in the other it decreased slightly but the remarkable thing was that although both subjects were in marked negative energy balance (with total energy expenditures in the range of 50 MJ/day) they did not show the diminution of whole body protein synthesis, which is characteristic of starvation, suggesting that their physical activity actually maintained protein turnover and possibly helped preserve lean body mass. There is no doubt, however, that both men did lose lean body mass and the definitive answer to whether or not this would have been faster or slower without their daily treck in sub-zero temperatures is currently unanswerable.

POSSIBLE BENEFICIAL EFFECTS OF BRANCHED CHAIN AMINO ACID SUPPLEMENTATION

Theoretically, branched chain amino acid supplementation could provide benefits in one of four ways there are to supply anaplerotic intermediates, to decrease the use of other fuels including glycogen, inhibit muscle protein breakdown, and inhibit the transport of tryptophan into the brain, thus limiting the increase in serotonin synthesis which is implicated in the central fatigue hypothesis (Hassmen et al., 1994).

There is, in fact, some evidence that branched chain amino acid supplementation does, to some extent, spare muscle glycogen (MacLean et al., 1996; Blomstrand and Newsholme, 1996) and limit proteolysis (MacLean et al., 1994). There is also said to be a decreased *perceived* exertion during exercise (Blomstrand et al., 1997). The serotonin central fatigue hypothesis is not, however, supported by the results of studies using serotonin receptor antagonists which ought to be the case if the central fatigue hypothesis were true (Pannier et al., 1995). In any case, whatever the mental state of subjects during exercise with or without branched chain amino acid supplementation, there is no evidence that there is any beneficial effect on physical performance (Van Hall et al., 1995a; Pannier et al., 1995; Struder et al., 1996; Madsen et al., 1996) of making branched chain amino acids available or of inhibiting serotonin uptake.

GLUTAMINE, THE OVERTRAINING SYNDROME AND IMMUNE FUNCTION

It has been known for some years that white cells have a specific requirement for glutamine. It has been suggested that the low blood and muscle glutamine observed in athletes in training who suffer the so-called overtraining syndrome is the result of the progressive exhaustion by repeated exercise of the homeostatic system for preserving glutamine (Rowbottom et al., 1995, 1996). This diminishes the supply of glutamine available to white cells and could diminish their capacity for fighting infection. There is some evidence in favour of this proposition (Rohde et al., 1996) but it appears to me that, for the purposes of the present discussion, we need to ask ourselves: are the levels of physical activity which are engaged in by top athletes during training and which lead to the overtraining syndrome similar to those experienced by military personnel on active duty? Personally I doubt this, although it is certainly possible during specific training early after recruitment. There has been a single report of a beneficial effect of glutamine in reducing infection rate of amateur athletes after long distance runs (Castell et al., 1996). We now need to repeat such studies and we need know what the dose-response relationship is between exogenous glutamine and adequate immune function. We also need to understand whether or not small doses of glutamine given at regular intervals would be as efficacious as irregular large doses. The whole area is one which requires further research.

THE CRUCIAL QUESTIONS

Does what we know about physical activity and metabolism suggest to us that protein requirements are increased as a result of exercise? My answer to this would be that in fact there is no evidence that there are increased requirements for protein *per se*. Most foods contain 10–15 percent of protein on an energy basis and the likelihood is that an adequate supply of energy from mixed rations will inevitably supply sufficient protein.

Is there an optimum of protein to carbohydrate + fat ratio and should the composition of rations be altered to this? So far as I can tell, most of the evidence suggests that exercise actually increases the efficiency of protein utilisation and therefore, if anything, the amount of protein in the diet could be reduced without deleterious effects. I do not think we know the lower limit at the present time, and more work is needed to answer the question.

We do know that increasing the protein content of the diet simply increases the activity of amino acid catabolizing enzymes and the capacity for branched chain amino acid oxidation during exercise is substantially increased by increasing total dietary protein (Figure 11-1). In studies we carried out on subjects habituated to a high protein diet, amino acid oxidation was higher than normal during exercise suggesting no net benefit would accrue. In any case it can be calculated that even with an increase of muscle mass of 15 kg over 3 years a 70 kg man would need less than 5 percent of the protein recommended nutritional intake (RNI) to supply the growth. Thus, it is likely that increasing dietary protein over the current US military ration of 100 g/day would simply lead to increased oxidation without any particular benefit. The classic results of

Chittenden, a physiologist at Yale in the early part of this century (Chittenden, 1907), should not be forgotten. Chittenden persuaded a number of young, all-American athletes to eat substantially less protein than they were accustomed to and recorded their performance over the next year. It actually increased rather than decreased!

Is there any evidence for a differential response between the sexes to changes in protein and amino acid metabolism as a result of exercise? Personally I think that there is insufficient data to answer this question definitively but I would be very surprised if there were any major differences.

Is there a likelihood that particular amino acids will be beneficial in terms of performance, or the preservation of the lean body mass, etc? Glutamine is the most promising of the amino acids from this point of view. Nevertheless, there is no clear indication in the literature that glutamine or any other amino acid has a marked effect in stimulating performance *per se* in the short-term nor in maintaining lean body mass in the longer term.

AUTHOR'S CONCLUSIONS AND RECOMMENDATIONS

- Exercise stimulates amino acid catabolism but the extent of the stimulation is too little to have a major effect in contributing to a negative nitrogen balance.
- Most food contains sufficient protein such that so long as energy balance is maintained sufficient protein is delivered to meet the requirements for amino acid oxidation and also probably for preservation and even growth of the lean body mass.
- There is no evidence that supplementation with individual amino acids is of benefit to physical performance or to maintenance or growth of lean body mass, especially muscle.

In summary, therefore, rations for military personnel engaged in a high rate of physical activity should have the following characteristics:

- Will be sufficient in delivery of energy.
- Contain protein in the range of 0.8 g/kg body weight/day.
- Need contain no extra amino acid supplements.

ACKNOWLEDGMENTS

The work described in this article was supported by the Medical Research Council, Ministry of Agriculture, Fisheries and Food, The Wellcome Trust and the University of Dundee.

REFERENCES

Biolo, G., S.P. Maggi, B.D. Williams, K.D. Tipton, and R.R. Wolfe. 1995b. Increased rates of muscle protein turnover and amino acid transport after resistance exercise in humans. Am. J. Physiol. 268:E514–E520.

Blomstrand, E., and E. A. Newsholme. 1996. Influence of ingesting a solution of branched-chain amino-acids on plasma and muscle concentrations of amino-acids during prolonged submaximal exercise. Nutrition 12:485–490.

Blomstrand, E., P. Hassmen, S. Ek, B. Ekblom, and E.A. Newsholme. 1997. Influence of ingesting a solution of branched-chain amino acids on perceived exertion during exercise. Acta Physiol. Scand. 159:41–49.

Butterfield, G.E., and D.H. Calloway. 1984. Physical activity improves protein utilization in young men. Br. J. Nutr. 11:171–184.

Castell, L.M., J.R. Poortmans, and E.A. Newsholme. 1996. Does glutamine have a role in reducing infection in animals. Eur. J. Appl. Physiol. 13:488–490.

Chesley, A., J.D. MacDougall, M.A. Tarnopolsky, S.A. Atkinson, and K. Smith. 1992. Changes in human muscle protein synthesis following resistance exercise. J. Appl. Physiol. 73:1383–1388.

Chittenden, R.H. 1907. The Nutrition of Man. London: Heinemann.

Dohm, G.L., G.J. Kasperek, E.B. Tapscott, and G.R. Beecher. 1980. Effect of exercise on synthesis and degradation of muscle protein. Biochem. J. 188:255–262.

Fielding, R.A., C.N. Meredith, K.P. O'Reilly, W.R. Fontera, J.G. Cannon and N.J. Evans. 1991. Enhanced protein breakdown after eccentric exercise in young and older men. J. Appl. Physiol. 11:674–679.

Hassmen, P., E. Blomstrand, B. Ekblom, and E.A. Newsholme. 1994. Branched-chain amino acid supplementation during 10 km competitive run: Mood and cognitive performance. Nutrition 10:405–410.

Hood, D.A., and R.L. Terjung. 1990. Amino acid metabolism during exercise and following endurance training. Sports Med. 1:23–35.

Kasperek, G.J., and R.D. Snider. 1989. Total and myofibrillar protein degradation in isolated soleus muscles after exercise. Am. J. Physiol. 157:E1–E5.

Kasperek, G.J., G.R. Conway, D.S. Krayeski, and J.J. Lohne. 1992. A reexamination of the effect of exercise on rate of muscle protein degradation. Am. J. Physiol. 163:E1144–E1150.

Katz, A., S. Broberg, K. Sahlin, and J. Wahren. 1986. Muscle ammonia and amino acid metabolism during dynamic exercise in man. Clin. Physiol. 6:365–379.

MacDougall, J.D., M.J. Gibala, M.A. Tarnopolsky, J.R. MacDonald, S.A. Interisano, and K.E. Yarasheski. 1995. The time course for elevated muscle protein synthesis following heavy resistance exercise. Can. J. Appl. Physiol. 10:480–486.

MacLean, D.A., T.E. Graham, and B. Saltin. 1994. Branched-chain amino acids augment ammonia metabolism while attenuating protein breakdown during exercise. Am. J. Physiol. 167:E1010–1022.

MacLean, D.A., T. E. Graham, and B. Saltin. 1996. Stimulation of muscle ammonia production during exercise following branched-chain amino-acid supplementation in humans. J. Physiol. Lond. 193:909–922.

Madsen, K., D.A. MacLean, B. Kiens, and D. Christensen. 1996. Effects of glucose, glucose plus branched-chain amino acids, or placebo on bike performance over 100 km. J. Appl. Physiol. 11:2644–2650.

Millward, D.J., J.L. Bowtell, P. Pacy, and M.J. Rennie. 1994. Physical activity, protein metabolism and protein requirements. Proc. Nutr. Soc. 13:123–240.

Pannier, J.L., J.J. Bouckaert, and R.A. Lefebvre. 1995. The antiserotonin agent pizotifen does not increase endurance performance in humans. Eur. J. Appl. Physiol. 12:175–178.

Rennie, M.J. 1996. Influence of exercise on protein and amino acid metabolism. Pp. 995–1035 in American Physiological Society Handbook of Physiology on Exercise, Chapter 12, Section 12, Control of Energy Metabolism During Exercise, R. L. Terjung, assoc. ed. Bethesda: American Physiological Society.

Rohde, T., D.A. MacLean, A. Hartkopp, and B.K. Pedersen. 1996. The immune-system and serum glutamine during a triathlon. Eur. J. Appl. Physiol. Occupat. Physiol. 14:428–434.

Rowbottom, D.G., D. Keast, C. Goodman, and A.R. Morton. 1995. The hematological, biochemical and immunological profile of athletes. Eur. J. Appl. Physiol. Occupat. Physiol. 10:502–509.

Rowbottom, D.G., D. Keast, and A.R. Morton. 1996. The emerging role of glutamine as an indicator of exercise stress and overtaining. Sports Med. 11:80–97.

Sahlin, K., A. Katz, and S. Broberg. 1990. Tricarboxylic acid cycle intermediates in human muscle during prolonged exercise. Am. J. Physiol. 159:C834–C841.

Sahlin, K., L. Jorfeldt, and K.G. Henriksson. 1995. Tricarboxylic acid cycle intermediates during incremental exercise in healthy subjects and in patients with McArdle's disease. Clin. Sci. 19:687–693.

Stroud, M.A., A.A. Jackson, and J.C. Waterlow. 1996. Protein turnover rates of two human subjects during an unassisted crossing of Antarctica. Br. J. Nutr. 16:165–174.

Struder, H.K., W. Hollmann, P. Platen, J. Duperly, H.G. Fischer, and K. Weber. 1996. Alterations in plasma-free tryptophan and large neutral amino acids. Int. J. Sports Med. 17:73–79.

Van Hall, G., J.S.H. Raaymakers, W.H.M. Saris, and A.J.M. Wagenmakers. 1995a. Ingestion of branched chain amino acids and tryptophan during sustained exercise in man: failure to affect performance. J. Physiol. Lond. 186:789–794.

Van Hall, G., G.J. Van Der Vusse, K. Soderlund, and A.J. Wagenmakers. 1995b. Deamination of amino acids as a source for ammonia production in human skeletal muscle during prolonged exercise. J. Physiol. Lond. 189:151–261.

Wibom, R., and E. Hultman. 1990. ATP production rate in mitochondria isolated from microsamples of human muscle. Am. J. Physiol. 159:E204–E209.

12

Skeletal Muscle Markers

Dympna Gallagher, Steven B. Heymsfield,[1] and Zi-Mian Wang

INTRODUCTION

Skeletal muscle is the largest non-adipose tissue body component and serves both voluntary and involuntary life-sustaining functions. Skeletal muscle composition can be considered in terms of atomic, molecular, cellular, and tissue-system level components (Wang et al., 1992). The traditional concept of skeletal muscle is the non-adipose tissue component shown in Figure 12-1. However, with aging and obesity increased amounts of adipose tissue become interspersed between muscle bundles, and this composite structure is referred to as anatomic skeletal muscle (Wang et al., 1992; Heymsfield et al., 1995). Some methods of measuring skeletal muscle mass rely on quantification of one or more of the components shown in the figure.

Skeletal muscle is almost one-fifth of body weight in the newborn; this proportion doubles in the mature male (Bortz, 1982; Hochachka, 1994; Lexell et al., 1988; Snyder et al., 1975; Tomlinson et al., 1969) (Table 12-1). Both cross-

---

[1] Steven B. Heymsfield, Human Body Composition Laboratory and Weight Control Unit, St. Luke's-Roosevelt Hospital Center, New York, NY 10025.

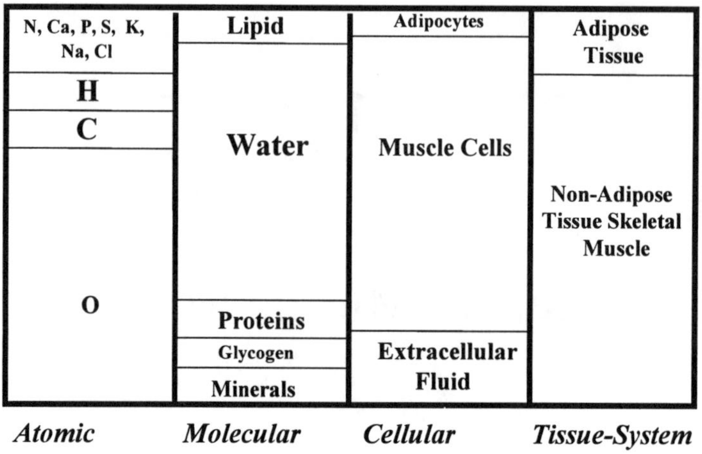

N, Ca, P, S, K, Na, Cl	Lipid	Adipocytes	Adipose Tissue
H			
C	Water	Muscle Cells	
			Non-Adipose Tissue Skeletal Muscle
O	Proteins		
	Glycogen	Extracellular Fluid	
	Minerals		
Atomic	*Molecular*	*Cellular*	*Tissue-System*

FIGURE 12-1 Anatomic skeletal muscle components at the first four body composition levels. The fifth level, whole body, is not shown.

TABLE 12-1 Proportional Contribution of Skeletal Muscle to the Body Weight of a Man at Different Stages of Development

Developmental Stage	Skeletal Muscle (% of Body Weight)
Birth	21
Weaning	18
Adolescent	36
Adult	45
Elderly	27

SOURCE: Adapted from Heymsfield et al., 1995.

TABLE 12-2 Distribution of Total Body Protein in the 70-kg Reference Man

Organ or Tissue	Protein Mass (kg)	% of Total Body Protein
Skeletal muscle	4.8	45.3
Brain	0.11	0.10
Liver	0.32	3.0
Kidney	0.053	0.50
Heart	0.008	0.50
Blood	0.99	9.3
Skin	0.75	7.1
Skeleton	1.9	17.9
Total body	10.5	100

SOURCE: Adapted from Snyder et al., 1975.

sectional and longitudinal studies demonstrate a gradual decline in skeletal muscle mass with advancing age, and "sarcopenia" (as this decline is called) is an important problem in elderly populations (Bortz, 1982; Lexell et al., 1988; Tomlinson et al., 1969). On average, skeletal muscle contains the largest protein pool in the body (~45%), as shown for the Reference Man in Table 12-2 (Snyder et al., 1975). With the exception of adipose tissue, which is found in variable proportions in adults, skeletal muscle is the largest at the tissue-system level body composition component. Skeletal muscle thus plays a central metabolic and functional role, particularly in relation to military activities and physical performance.

Method Organization

At present, most in vivo methods of skeletal muscle assessment are static in that they are designed to quantify skeletal muscle at a single point in time, and dynamic method development is limited (Figure 12-2). Skeletal muscle measurement methods that are applied in vivo are indirect and rely on measurable properties and known components, some of which are outlined in Table 12-3 (Wang et al., 1995). Broadly viewed, these properties are used with two types of models for deriving skeletal muscle mass estimates.

The first model is a descriptive or type I model that shares in common the following characteristics: a reference method is used to estimate skeletal muscle mass in a well-defined subject group in whom the property is also measured; and a statistically derived component prediction equation is developed and then cross-validated in a new subject group (Wang et al., 1995). All methods in this category are formulated conceptually around the following formula:

(1) skeletal muscle mass = Σ [a × (measurable quantities)] + b,

where a and b are the slope and intercept of the prediction formula based on linear regression analysis. The main methods in this category are outlined in Table 12-3. All descriptive methods are population specific and must be cross-validated in new subject groups before they can be applied with confidence.

The second modeling approach is formulated on stable-component relationships, many of which can be understood in terms of underlying mechanisms. These are referred to as mechanistic or type II models (Wang et al., 1995). The methods in this category are formulated conceptually around the following formula:

(2) skeletal muscle = Σ [a × (measurable quantities)],

where a is the stable proportion or other assumed constant biological characteristic. The main methods in this category are outlined in Table 12-3.

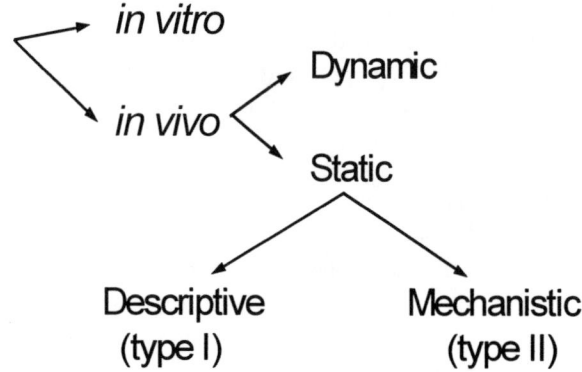

FIGURE 12-2 Organization of body composition methods. SOURCE: Adapted from Wang et al., 1995.

TABLE 12-3 Characteristics of Skeletal Muscle Mass Measurement Methods

Method	Measurable Quantity		Model (or Function) Type	
	Property	Component	Descriptive	Mechanistic
Anthropometry	Skinfolds Circumferences Diameters		+	+
Ultrasound	Reflected sound waves		+	+
BIA	Resistance Reactance		+ +	
Urinary Metabolites	Creatinine 3-Methylhistidine		+ +	
Imaging:				
• CT	X-ray attenuation Cross-sectional area			+
• MRI	Proton relaxation times Cross-sectional area			+
IVNA/Whole-body counting		Total body N Total body K		

NOTE: BIA, bioelectric impedance analysis; CT, computerized axial tomography; FFM, fat-free body mass; IVNA, in vivo neutron activation analysis; MRI, magnetic resonance imaging.

TABLE 12-4 Methods of Measuring Skeletal Muscle Mass In vivo

Method	Level	Measures	Whole Body	Regional
TBK-TBN	Atomic	ATFSM, SMPro	+	–
ASM by DXA	Molecular	FFSM	–	+
Creatinine	Cellular	SMCM	+	–
3MH	Cellular	SMCM	+	–
CT	Tissue system	Anatomic SM, ATFSM	+	+
MRI	Tissue system	Anatomic SM, ATFSM	+	+
Anthropometry	Whole body	Anatomic SM	+	+
Ultrasound	Whole body	Anatomic SM	–	+
BIA	Whole body	FFSM	–	+

NOTE: ASM, appendicular skeletal muscle; ATFSM, adipose tissue-free skeletal muscle; BIA, bioelectric impedance analysis; Creatinine, 24-h urinary creatinine excretion; CT, computerized axial tomography; DXA, dual-energy x-ray absorptiometry; FFSM, fat-free skeletal muscle; 3MH, 24-h urinary 3-methylhistidine excretion; MRI, magnetic resonance imaging; SM, total body skeletal muscle; SMCM, skeletal muscle cell mass; SMPro, skeletal muscle protein; TBK, total body potassium; and TBN, total body nitrogen.

Methods of measuring skeletal muscle can also be organized according to body composition level as shown in Table 12-4. This approach is useful in understanding differences in skeletal muscle components as measured by the various methods. For example, anthropometry quantifies anatomic skeletal muscle whereas urinary creatinine excretion is a marker of muscle cell mass (Heymsfield et al., 1995; Lukasksi et al., 1996).

An overview is now provided that is organized as outlined in Table 12-3. This overview begins with the simpler methods such as anthropometry and progresses to the complex research-based methods of estimating skeletal muscle mass, such as in vivo neutron activation analysis.

AVAILABLE METHODS

Anthropometry

Development of anthropometric skeletal muscle prediction methods are limited. There are also very few model-based anthropometric methods for estimating regional skeletal muscle mass.

An example of a descriptive method is the anthropometric approach suggested by Martin and colleagues (Martin et al., 1990). Anthropometric dimensions were quantified on 12 male Caucasian cadavers who were then dissected and their muscles weighed. The average age (Standard deviation ± SD), weight, and stature of the Caucasian cadavers were 72 ± 8 years, 66.2 ± 12.5 kg, and 169.1 ± 8.2 cm, respectively. A regression equation was then developed in which skeletal muscle (SM) was the dependent variable and anthropometric measurements the independent variables,

$$SM (g) = Stat \times (0.0553 \times CTG^2 + 0.0987 \times FG^2 + 0.0331 \times CCG^2) - 2445,$$

where Stat is stature in cm, CTG is thigh circumference corrected for the front thigh skinfold thickness in cm, FG is uncorrected forearm circumference in cm, and CCG is calf circumference corrected for the medial calf skinfold thickness in cm. The equation has a coefficient of determination R^2 of 0.97 and an Standard Error of the Estimate (SEE) of 1.53 kg. There have not yet been any independent cross-validations of this anthropometric method.

Model-based regional anthropometric methods are widely used for estimating mid-calf, thigh, and upper arm circumferences or areas (Heymsfield et al., 1984). The model used is relatively simple: the limb is assumed to be a concentric set of three cylinders: inner bone, middle muscle, and outer adipose tissue. Limb circumference and a corresponding skinfold thickness are used in the model to calculate muscle area, which includes the enclosed bone. Models are also published for estimating bone-free skeletal muscle areas (Heymsfield et al., 1992)

Although the geometric models are relatively simple, the limb and its associated muscle compartment are complex. For example, calculated arm muscle area overestimates skeletal muscle by 15 to 25 percent in young nonobese subjects (Heymsfield et al., 1982). About one-half of the overestimate is due to the aforementioned inclusion of bone in the estimated area, and the remaining overestimate is due to assumption errors and non muscle tissue (e.g., neurovascular bundle) inclusion in the muscle compartment. Forbes et al. (1988) and Baumgartner et al. (1993) also reported that arm muscle area assumptions are inaccurate in obese and elderly subjects, respectively.

Another problem, particularly in elderly and malnourished subjects, is that atrophic skeletal muscle chemical composition differs from that of normal tissue (4-6). Water, lipid, and collagen are increased per unit tissue mass whereas noncollagen proteins are reduced (Heymsfield et al., 1992). The relative concentration of "functional" proteins is thus lower in atrophied muscle. Abrupt changes in muscle glycogen, as might occur with dieting or glycogen loading, can also change muscle size by approximately 5 to 10 percent as a result of the water-binding properties of glycogen.

The important potential for developing anthropometric muscle estimation equations now exists due to the development of magnetic resonance imaging (MRI) muscle mass measurement methods (discussed in a later section). A simple example demonstrating this potential is shown in Figure 12-3. In this study, multislice whole body MRI was used to estimate total body skeletal muscle volume in 79 healthy adults. Arm, thigh, and calf muscle areas were then calculated using the above mentioned geometric model based on skinfold

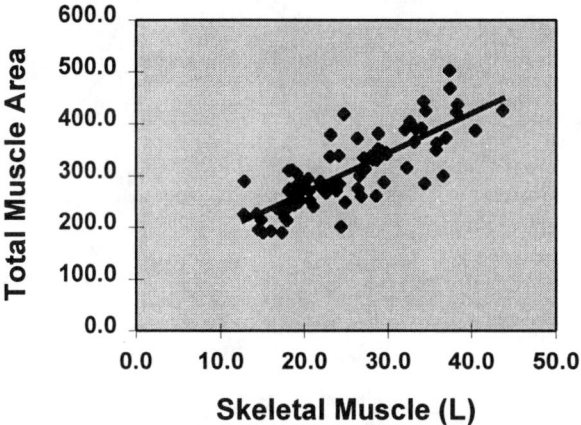

FIGURE 12-3 Sum of anthropometric limb muscle areas [cm^2] versus total body skeletal muscle volume measured by multislice MRI in 79 healthy adults. (Total muscle area [cm^2] = 7.6 × [Total MRI muscle volume, L) +115.4, R^2 = 0.64, p < 0.001).

thickness and limb circumference (Heymsfield et al., 1984). The three anthropometric muscle areas were then summed and plotted against total skeletal muscle volume by MRI. As shown in the figure, there was a good skeletal muscle volume by MRI (Total muscle area [cm^2] = 7.6 × [total MRI muscle volume, in L) + 115.4, r^2 = 0.64, p < 0.001). Anthropometric prediction equations could be developed by systematically extending this technique to specific populations.

Ultrasound

Ultrasound methods can be used to quantify mainly regional skeletal muscle thicknesses or diameters (Sipila et al., 1991), although as with skinfolds, the potential exists to develop whole body prediction formulas. Ultrasound, which was originally developed to detect underwater movement of submarines during World War II, is now used in cardiology, obstetrics, and many other branches of medicine throughout the world.

Two main ultrasound system types are in use, A mode and B mode. The first to be developed, the A mode system, passes an ultrasound signal from the system's probe, which is then reflected off anatomic structures. These reflected sound waves return to the probe as echoes. With A mode systems, tissue interfaces can be identified and the thickness of a muscle established. An important consideration and potential source of error is that with A mode systems, the ultrasound beam must be directed perpendicular to the skin surface, and the muscle site measured must be reproducible.

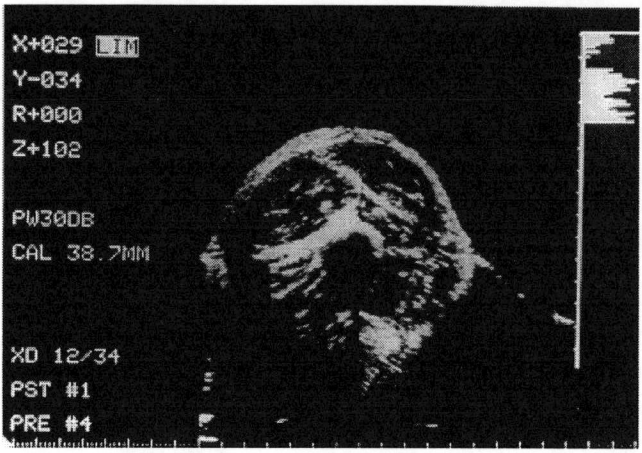

FIGURE 12-4 B-mode ultrasound cross-sectional image of the upper mid-arm in a healthy adult male.

The more-recent B mode systems produce a two-dimensional image of the scanned area, and muscle interfaces are identified as bright lines. An ultrasound cross-sectional image of the biceps muscle is presented in Figure 12-4. Current systems are mainly of the B mode type, operating at frequencies between 3 and 5 MHz.

Bioelectric Impedance Analysis

Bioelectric impedance analysis (BIA) was the subject of a conference and multi-author publication organized by the Nutrition Coordinating Center at the National Institutes of Health (Yanovski et al., 1996). The BIA method involves passing an alternating current at one or more frequencies via electrodes across a tissue bed, and the impedance or voltage drop to electrical flow is measured. Components rich in electrolytes, such as body fluids, impose minimal impedance while lipids and compact minerals impose a high impedance to electrical flow (Chumlea et al., 1994). Tissue composition is therefore the main determinant of impedance and its two components, resistance and reactance. Notably, impedance to electrical flow is determined primarily by the fluid volume present and length of the electrical pathway. The possibility exists to measure impedance of the extremities, skeletal muscle is the main determinant of extremity impedance. Appendicular and total body skeletal muscle can therefore potentially be estimated from measured extremity impedance or resistance by calibration against skeletal muscle mass using a reference method such as MRI or dual x-ray absorptionmetry (DXA).

Important recent advances permit the measurement of extremity impedance and other electrical characteristics such as reactance and phase angle. Baumgartner and colleagues (1989) examined segmental impedance and its relationship to total body impedance (Baumgartner et al., 1989). In this and other studies, investigators applied gel electrodes at predefined anatomical locations on the proximal and distal limb. A typical evaluation required two sets of injector and receiver electrodes, respectively. An important advance was made in 1994 by Organ and colleagues (1994), who developed a lead-switching algorithm for evaluating segmental impedance without the need for electrodes placed on the proximal aspect of the limb. By alternating voltage and current electrodes, the researchers were able to isolate the impedance of each arm and leg using six sets of distal limb gel electrodes. This laboratory expanded on these earlier studies by developing a BIA system for evaluating segmental impedance in the standing subject (Nunez et al., 1997). In this experimental system, conventional gel electrodes are replaced with stainless steel contact electrodes designed for both hands and feet (Figure 12-5). Electrical switching algorithms alternate current and voltage input so as to allow rapid measurement of each extremity impedance. In a recent study, researchers in this lab observed excellent agreement between extremity impedance measured with the new system and extremity impedance quantified using conventionally positioned limb gel electrodes (Nunez et al., 1997). Moreover, there was good agreement between limb or stature-adjusted impedance index (e.g., height squared/ impedance) and limb skeletal muscle mass (Nunez et al., 1997). For example, there was a strong correlation between ht^2/leg-to-leg impedance and leg muscle as quantified by DXA ($R^2 = 0.72$, $p < 0.001$) (Figure 12-6). Although this information is preliminary, the potential for evaluating the muscle compartment of each limb separately and total appendicular skeletal muscle mass is evident. Work in this area is ongoing in this laboratory and at other research centers. The possibility also exists to extend this methodology to multifrequency BIA systems and to limb muscle water compartmentation.

Urinary Metabolites

Two metabolic end products, creatinine and 3-methylhistidine, are excreted in urine and classically have served as indices of skeletal muscle mass. Creatinine is produced by non enzymatic conversion of creatinine, which is distributed mainly in skeletal muscle. Catabolism of actomyosin, the primary skeletal muscle contractile protein, results in 3-methlylhistidine release (Lukasksi et al., 1981). Both terminal products of metabolism are excreted in urine and to a lesser extent into the gastrointestinal tract and other secretions. A summary of the underlying assumptions of these two methods is presented in Table 12-5 (Lukasksi et al., 1981; Heymsfield et al., 1983).

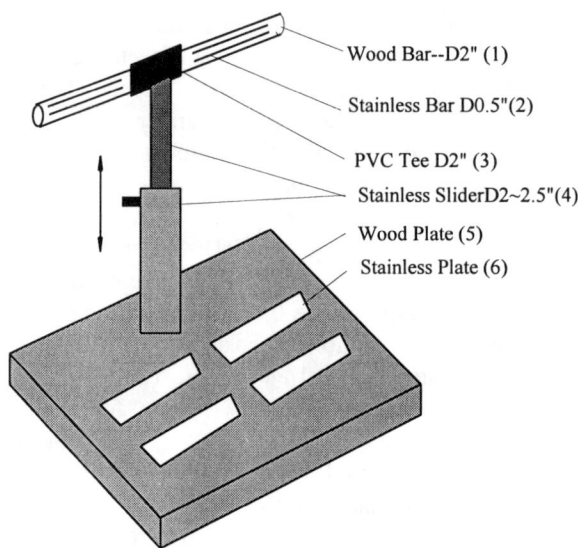

FIGURE 12-5 Electrode stand for measuring appendicular impedance. Four pairs of stainless steel electrodes contact each distal extremity, and electrical switching algorithms are used to measure the impedance of each limb and various limb combinations. This impedance approach can be used to estimate appendicular skeletal muscle mass.

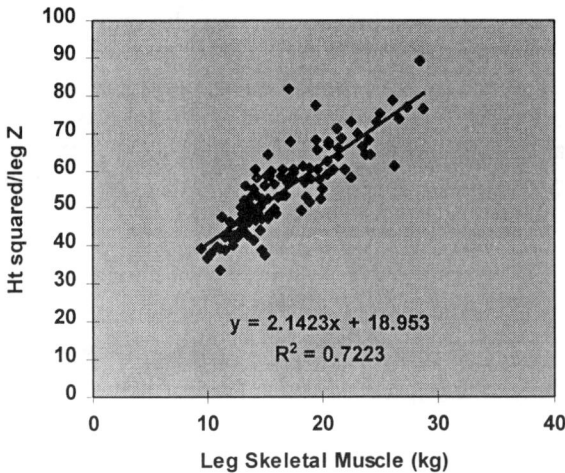

FIGURE 12-6 Stature-adjusted resistance (Ht^2/Z) versus appendicular skeletal muscle estimated by DXA in 103 healthy adults [$R^2 = 0.72$, $p < 0.001$]. Y = H^2/Z; X, leg skeletal muscle; Z, impedance in ohms.

TABLE 12-5 Main Characteristics of Urinary Metabolite Methods

Method	Creatinine	3-Methylhistidine
Model	creatinine = K × [Cr] × SM SM = creatinine/(K × [Cr])	3-MH = K × [protein] × SM SM = 3MH/(K × [protein])
Assumptions • Skeletal Muscle (SM)	Creatine is almost all within skeletal muscle.	Histidine is in two muscle proteins, actin within all muscle fibers, and myosin in white fibers.
• Creatinine [Cr] and protein [protein] concentrations	On a meat-free diet, the average concentration of creatine remains constant, and the total creatine pool remains constant.	On a meat-free diet, the average protein concentration in muscle remains constant, and the total muscle protein pool remains constant.
• Creatine or muscle protein daily fractional breakdown rate (K)	Creatine is converted to creatinine at a constant daily rate.	Muscle protein synthesis and catabolism are in balance, and thus, 3-MH is produced at a constant daily rate.
• Urinary creatinine or 3-MH	Creatinine is neither metabolized nor reused in metabolism and completely excreted in urine.	3-MH is neither metabolized nor reused in metabolism and completely excreted in urine.
• Prediction Equations	SM = 18.9 × creatinine + 4.1; $r = 0.92$, $p < 0.001$.	SM = 0.0887 × 3MH + 11.8; $r = 0.88$. $p < 0.001$.

Ideally, urine is collected under controlled dietary conditions (i.e., meat free) over several days, and the results are averaged. The between-day coefficient of variation for both methods is reasonably high (2–10%), even in careful and well-instructed participants.

The main concept of both methods is that the creatinine and 3-methylhistidine excreted in urine is related to the total endogenous pool and hence total body skeletal muscle mass. Although the data are limited, evidence supporting this hypothesis is reasonably good for healthy, weight-stable adults.

A traditional concept has been that creatinine excretion is directly proportional to total body skeletal muscle mass (Heymsfield et al., 1983). This is shown in Figure 12-7 and can be expressed mathematically as skeletal muscle/creatinine = constant or "K." This led to the concept of a creatinine equivalence, in which 1 g of creatinine excreted in urine per day was purportedly equivalent to a specified weight of skeletal muscle mass. This hypothesis is not supported by experimental observations as the creatinine: skeletal muscle mass ratio is not constant, but varies as a function of muscle mass (Heymsfield et al., 1983). This phenomenon is thought to be due to the presence of non skeletal muscle sources of creatinine (Figure 12-7). For example, creatine is found in brain tissue and contributes to the urinary creatinine pool. Similarly, the 3-methylhistidine:skeletal muscle mass ratio is not constant (Wang et al., 1998), and 3-methylhistidine is also produced by

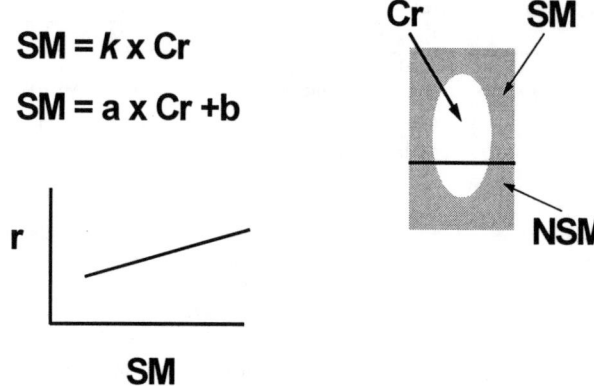

FIGURE 12-7 Muscle Mass: Urinary Creatinine Method. Main features of the 24-h urinary creatinine (Cr) method of estimating total body skeletal muscle mass (SM). The traditional view is that all of urinary creatinine is derived from SM and that the SM:Cr ratio is a constant (k). Experimental observations, however, indicate nonskeletal muscle (NSM) sources of Cr and a nonzero intercept when Cr is plotted against measured SM. Hence, prediction of SM by Cr requires descriptive regression model development with slope a and intercept b.

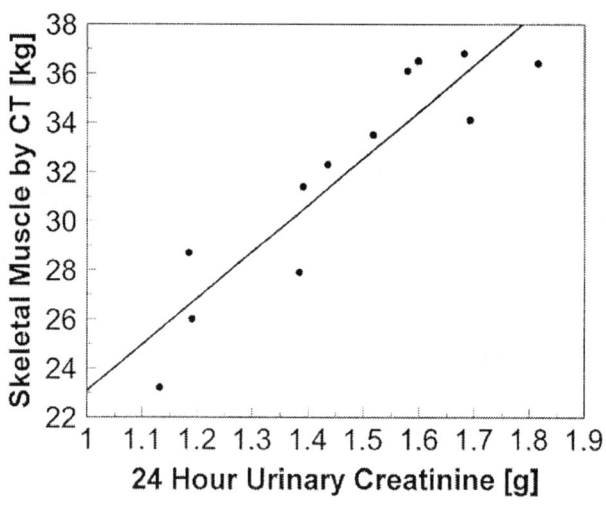

FIGURE 12-8 Total body skeletal muscle mass measured by computerized tomography (CT) (in kg) on the ordinate versus 24-hour urinary creatinine excretion (in g/d) on the abscissa ($n = 12$; SM = $18.9 \times$ Cr + 4.1; $r = 0.92$, $p = 2.55 \times 10^{-5}$; SEE = 1.89 kg). SOURCE: Wang et al., 1996a.

tissues other than skeletal muscle. As a result, both methods can be formulated on descriptive mathematical functions (i.e., skeletal muscle = a × Cr + b) developed using regression analysis with methods such as computerized axial tomography (CT) and MRI as the skeletal muscle reference. For example, this group studied healthy men on a 7-day, meat-free diet (Wang et al., 1998; Wang et al., 1996a). Urine was collected on the last three experimental days, and results creatinine (g) (Wang et al., 1997) and 3-methyl histidine excretion (μmol) (Wang et al., 1998) and skeletal muscle volume by 22 slice CT as shown in Figures 12-8 (SM = 18.9 × Cr + 4.1; r = 0.92, p = 2.55 ×10^{-5}; SEE = 1.89 kg) and 12-9 (SM = 0.0887 × 3MH + 11.8; r = 0.88, p < 0.001; SEE = 2.3 kg), respectively. In both examples, the lack of a "zero" intercept for the urinary metabolite versus skeletal muscle regression line indicates a non constant ratio of urinary metabolite to skeletal muscle mass.

The strong associations observed between urinary metabolites and skeletal muscle mass provide an opportunity to develop skeletal muscle prediction formulas. Additional studies are needed across groups that differ widely in age, gender, and ethnicity. Such studies could provide useful skeletal muscle prediction formulas based on urinary metabolite excretion. The limitation of these methods is the high level of subject participation required for dietary compliance and accurate urine collection. Also, 3-methylhistidine can be measured only in specialized laboratories, whereas creatinine is an almost universally available and inexpensive test.

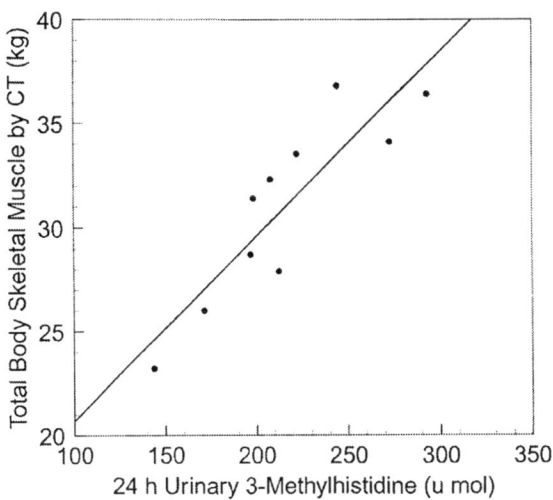

FIGURE 12-9 Total body skeletal muscle mass measured by computerized tomography (CT) (in kg) on the ordinate versus 24-hour urinary 3-methylhistidine excretion (3MH, in μmol) on the abscissa (n = 10; SM = 0.0887 × 3MH + 11.8; r = 0.88, p < 0.001; SEE = 2.3 kg). SOURCE: Wang et al., 1998.

Imaging Methods

Imaging methods represent the most important advance in the evaluation of skeletal muscle mass. Two main types of imaging systems are now in use, CT and MRI. These methods are extensively reviewed in Heymsfield et al., 1997a and Sjöström, 1991. Both CT and MRI are capable of gathering cross-sectional images at predefined anatomic locations. The cross-sectional image *per se* can then be used to quantify selected muscle areas. Multiple cross-sectional images can be used to reconstruct whole muscle groups or total body skeletal muscle mass.

Imaging methods gather data in the form of picture elements or "pixels." Addition of thickness to the pixel creates a volume element or "voxel." When the outer and inner edges of a muscle group are traced on the system scanner, the enclosed pixels include myofibers, nerves, adipocytes, and other components, which collectively are referred to as anatomic skeletal muscle (Figure 12-1).

Historically, CT was the first available imaging method, introduced in 1971 by Hounsfield (1973). Although the underlying physical concepts related to MRI were developed in 1946, it was not until 1984 that in vivo images were first reported (Foster et al., 1984). The CT method relies on x-ray attenuation data collected as the tube and detector rotate in a plane perpendicular to the subject. Images produced by MRI are related to signals measured during alternating magnetic field and radio frequency pulses. The main nuclear effects involve abundant hydrogen and hydrogen proton (i.e) imaging is widely used in the clinical setting. A typical MRI study is shown in Figure 12-10, which demonstrates cross-sectional slices through selected muscle groups and visceral organs (Heymsfield et al., 1997b).

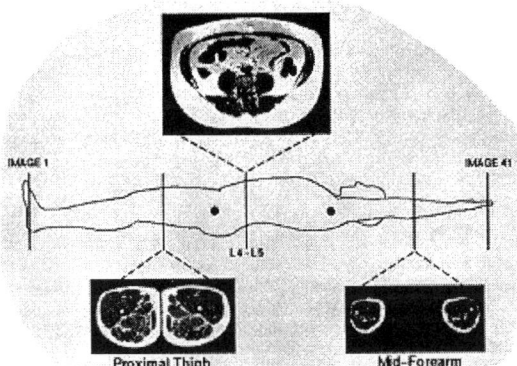

FIGURE 12-10 Three magnetic resonance images in a healthy adult, one from the abdomen and two from the appendicular region. The images were acquired using a T1-weighted, spin-echo pulse sequence. Using this protocol adipose tissue appears white and nonadipose tissue is dark on all images. SOURCE: With permission, from the *Annual Review of Nutrition*, Volume 17, © 1997, by Annual Reviews, http://www.Annual Reviews.org.

Many studies have examined the validity of CT- and MRI- measured cross-sectional image areas and whole body components (Heymsfield et al., 1997a). Volumes of phantoms, isolated cadaver organs, and in situ cadaver organs measured by CT and MRI are in close agreement with actual values (Sjostrom et al., 1991; Heymsfield et al., 1997a). Converting volume estimates to mass requires an assumed tissue density, which for skeletal muscle is 1.04 g/cc. CT and MRI validation studies are reviewed by Heymsfield et al. (1997a).

An important issue not yet totally resolved is the accuracy of CT and MRI in quantifying adipose tissue-free skeletal muscle. The enclosed area within a traced muscle includes entrapped adipose tissue, a component that increases on a relative basis with some diseases and advancing age. Careful manual tracing with the scanner's computerized track-ball device allows exclusion of obvious adipose tissue, and more complex pixel separation algorithms are also possible. Nevertheless, it is difficult to separate out completely all intramuscular adipose tissue using imaging methods, and more advanced techniques may be necessary. With CT, it is possible to identify specific attenuation (i.e., based on Hounsfield units) ranges for adipose tissue and muscle, and this approach appears to provide reasonable separation of the two components. Magnetic resonance spectroscopy methods are becoming increasingly available for in vivo human research, and this may aid in pixel separation (Gadian et al., 1995). Spectroscopic methods based on various nuclei may also eventually be able to quantify skeletal muscle molecular level components, such as water and glycogen in humans (Jue et al., 1989).

At present, most investigators have at their disposal both CT and MRI, although high cost remains a problem at some medical centers. The disadvantage of CT is radiation exposure, which can be substantial for multislice whole body studies. Some subjects become claustrophobic inside the MRI system's magnet bore, although open systems are currently being sold.

Both CT and MRI are based on measured physical properties and well-established reconstruction formulas. Assuming image slices are spaced closely, or even adjacent to one another, there is little question that imaging-derived skeletal muscle can serve as a reference for other methods.

Dual-Energy X-Ray Absorptiometry

DXA systems are the most recent version of methods designed to quantify bone characteristics in the study of osteoporosis (Cameron et al., 1963). Two main photon energy peaks are produced by alternating an x-ray source kVp or by rare-earth filtration. Attenuation of photons occurs as the two energy peaks pass through tissue, and those remaining are counted as they impinge on appropriately positioned detectors. Photon attenuation is characteristic of the applied x-ray energies and elemental characteristics of the tissue under study. These differing characteristics can be expressed mathematically as the ratio, R,

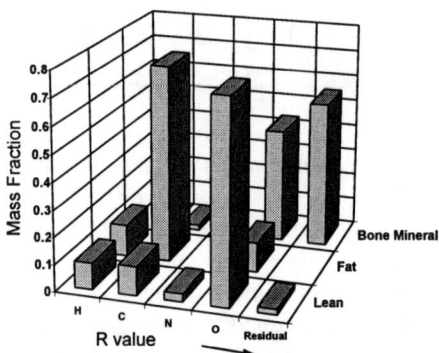

FIGURE 12-11 Mass fraction of main elements in lean soft tissues, fat, and bone mineral based on Reference Man. Residual Mass represents the sum of Na, Mg, P, S, Cl, K, Ca, and trace elements. The respective calculated R values at 40 keV and 70 keV for fat, lean, and bone mineral are 1.21, 1.37, and 2.86, respectively. SOURCE: With permission, from the *Annual Review of Nutrition*, Volume 17, © 1997, by Annual Reviews, http://www.AnnualReviews.org.

attenuation at the low energy to that at the high energy. Fat, bone, mineral, and remaining bone mineral-free lean tissue have characteristic elemental makeup and hence R values (Figure 12-11), which allows for their separation using DXA attenuation values and various mathematical algorithms. Many studies now generally support DXA-measured body composition components in humans and animals (Korht et al., 1995).

Most DXA systems allow isolation of different regions of the body, including the appendages. Lean mass of the upper and lower extremities can be assumed equivalent to appendicular skeletal muscle with a small and relatively constant amount of skin and other non muscle components (Heymsfield et al., 1990). Appendicular skeletal muscle is about 75 percent of total body skeletal muscle mass (Figure 12-12), and this fraction is assumed stable within and

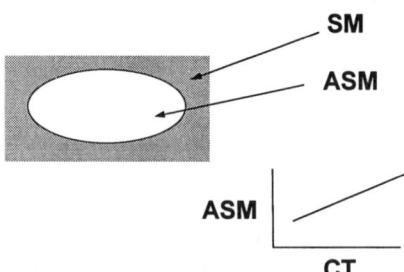

FIGURE 12-12 Appendicular Skeletal Muscle Mass DXA Method. Main features of the Dual X-Ray Absorpiometry method of estimating appendicular skeletal muscle mass (ASM). CT, computerized axial tomography; SM, skeletal muscle mass. See text for additional details.

between individuals. That is, a plot of DXA-estimated appendicular skeletal muscle mass versus total body skeletal muscle mass should show a strong correlation and an intercept not significantly different from zero. In such a study, Wang et al. (1996) measured total body skeletal muscle mass in 25 men using multislice CT. There was a strong correlation between DXA-determined appendicular skeletal muscle and total body skeletal muscle by CT ($r = 0.95$, $p < 0.001$). A plot of DXA appendicular skeletal muscle versus total body skeletal muscle volume by multislice MRI derived from archives in this laboratory supports the study of Wang et al. as shown in Figure 12-13 ($n = 72$, DXA appendicular skeletal muscle = $0.83 \times$ MRI skeletal muscle + 2.7, $R^2 = p < 0.001$). Both studies demonstrate regression line slopes larger than (~.083–.085) the expected (0.75), and the cause of this discrepancy is under investigation. In another study, Baumgartner et al. (in press) compared appendicular skeletal muscle estimated by DXA with muscle volume measured with MRI and found an equally good correlation (r^2 in range of 0.95) between the two.

Considerably more progress can be made in modeling appendicular lean mass as derived by DXA. For example, formulas can be developed for converting lean tissue to muscle and appendicular muscle to total body muscle.

Repeated measurements by DXA of appendicular lean muscle mass indicated technical errors for arm, leg, and total body skeletal muscle of 7.0 ± 2.4 percent, 2.4 ± 0.5 percent, and 3.0 ± 1.5 percent, respectively (Clasey et al., 1997). More research is needed in establishing comparability of body composition estimates across manufacturers and in evaluating if, and to what

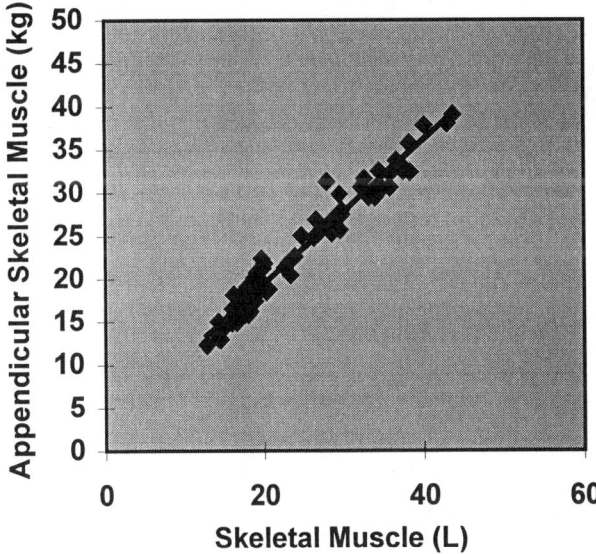

FIGURE 12-13 DXA-measured appendicular skeletal muscle versus total body skeletal muscle volume (L) by multislice magnetic resonance imaging in 72 healthy men and women.

TABLE 12-6 Summary of R Values of Various Human Soft Tissues for Three Age Groups

Soft Tissue	Age Groups			R value
	Newborn	4–7 months	Adults	
Skeletal muscle	1.35237	1.35187	1.35311	n.s.
Heart	1.35617	1.35422	1.35400	n.s.
Liver	1.35568	1.35333	1.35428	n.s
Kidney	1.35778	–	1.35971	n.s

SOURCE: Reprinted from Applied Radiation Isotopes, Volume 49, Pietrobelli et al., Lean value for DXA Two-component Soft-Tissue Model: Influence of Age and Tissue/Organ Type, Pages 743–744, Copyright 1998, with permission from Elsevier Science.

extent, variation in muscle hydration influences lean mass estimates by DXA. There appears to be no influence of age on DXA physical models used for estimating the three main components as shown in Table 12-6. Pietrobelli et al. (1998) observed similar attenuation ratios (i.e., r values) for skeletal muscle based on chemical analysis of biopsies in subject groups ranging in age from newborns to adults. Hence, for a low cost and with minimal radiation exposure, DXA allows quantification of a large proportion of total body muscle mass.

In Vivo Neutron Activation/Whole Body Counting

Skeletal muscle tissue has a relatively small extracellular fluid compartment and large myofibrillar mass (Synder et al., 1975). As a result, skeletal muscle is relatively rich in intracellular potassium when referenced to other components such as total body protein (i.e., nitrogen). Anderson (1963) and later Burkinshaw (1978) and their colleagues exploited this property of muscle tissue to develop multicomponent models. Notably, Burkinshaw's classic model was based on a known and constant ratio of potassium (K) to nitrogen (N) in muscle (3.03 mmol/g) and non-muscle lean tissue (1.33 mmol/g) (Figure 12-14). Burkinshaw, and later Cohn et al. (1980), developed their models based on chemically analyzed tissue samples. A two-component model can be developed:

skeletal muscle (kg) = [TBK (mmol) – 1.33 x TBN (g)] / 51.0
non-skeletal muscle (kg) = [3.03 x TBN (g) – TBK (mmol)] / 61.2

Where TBK is total body potassium and TBN is total body nitrogen, prompt-γ neutron activation analysis is used to measure TBN, and whole-body ^{40}K counting is used to measure TBK (Cohn et al., 1980).

Although the K:N model is of historic interest, there are a number of problems now recognized with the method. First, the K:N ratio is probably not constant in muscle and non muscle lean tissues but may change with age and other factors such as physical activity level. As a result, large between-individual differences most likely exist in the above model constants. Accordingly, Wang et al. (1996) observed a significant correlation between total

FIGURE 12-14 Muscle Mass: TBN-TBK Model. Main features of the neutron activation-whole body counting method of estimating muscle and nonmuscle lean tissue mass. TBK and TBN are total body potassium and nitrogen, respectively. The applied model assumes stable TBK/TBN ratios of muscle and nonmuscle lean mass components.

body skeletal muscle derived by the Burkinshaw-Cohn model and total body skeletal muscle by CT in healthy men and men with acquired immunodeficiency syndrome (AIDS) [$SM_{Burkinshaw} = 0.990 \times SM_{CT} - 6.4$; $r = 0.83$, $p = 0.0001$, SEE = 4.4 kg, n = 25]. Compared with CT, the Burkinshaw method underestimated skeletal muscle by a mean of 6.9 kg (20.1%, $p = 0.0001$) and 6.3 kg (23.2%, $p = 0.01$) in healthy men and men with AIDS, respectively. The hypothesized cause of large between-individual differences in estimated versus measured skeletal muscle (R^2 of 0.65 and SEE of 4.4 kg) and observed bias is model errors (Wang et al., 1996b). Second, the method involves radiation exposure, and alternative methods are available with little or no radiation exposure and with equivalent or superior accuracy. Lastly, very few centers have the necessary facilities for measuring TBN and TBK.

The potential exists to develop improved skeletal muscle mass formulas based on TBN and TBK, and studies such as these should be carried out in the future.

Monitoring Changes in Skeletal Muscle Mass Over Time

An important and inadequately studied topic is selection of a measurement method for monitoring changes in skeletal muscle mass and composition over time. The following is an overview of the potential role of each method in evaluating skeletal muscle changes over time.

Anthropometric and ultrasound methods quantify anatomic skeletal muscle and are thus influenced by all muscle components. Observed temporal changes can reflect, for example, alterations in intramuscular adipose tissue at the tissue system level or in water and glycogen at the molecular level. A change in anthropometrically- or ultrasound-derived muscle mass may therefore not always reflect parallel changes in the main component of interest, muscle proteins.

A second concern related to anthropometry and ultrasound is measurement error, which is relatively large for both methods (e.g., within- and between-observer technical errors in the range of ~5–8%) (Nelson et al., 1996). Both anthropometric and ultrasonic measurements are useful mainly in evaluating large skeletal muscle mass changes over long time periods. The cost of these methods is relatively low and the measurements can be made in field settings.

Urinary metabolites are markers of muscle cell mass, but as already mentioned, there are non muscle sources of both creatinine and 3-methylhistidine. These methods require careful urine collection protocols and subject compliance on a meat-free diet. Even under well-controlled conditions, the between-day coefficient of variation is relatively large (i.e., 4–8%). These methods may therefore be useful in evaluating large muscle mass changes over time or changes in groups of subjects. As creatinine is easy to measure, this method may be applicable in settings without sophisticated analytical equipment.

Bioelectric impedance analysis requires strict attention to measurement protocol (Nelson et al., 1996). Under carefully controlled conditions, the between-measurement technical error of BIA is very small (~1–3%). Measured impedance may, however, reflect changes in electrolyte concentrations, fluid distribution, and adiposity that occur in patients followed over time. Because BIA methods are inexpensive and simple, they have value in field settings. More studies of the ability of BIA methods to quantify small changes in skeletal muscle mass are needed.

Imaging methods have the most potential for quantifying small skeletal muscle mass and composition changes. At present, the CT method is well suited to measure adipose tissue-free muscle (Sjostrom et al., 1991; Heymsfield et al., 1997b), and the same applies to MRI although lingering concerns remain. Specifically, validated procedures for estimating adipose tissue content of muscle by MRI have not yet been published. Both of these imaging methods will record a change in muscle mass related to any one or more components including protein, water, glycogen, and various fluid spaces. With this proviso, the between-measurement technical errors for skeletal muscle areas and volumes by CT and MRI are small (< 2%) (Sjostrom et al., 1991; Heymsfield et al., 1997b; Wang et al., 1996a; Ross et al., 1995), particularly if repeated measurements are made by the same observer. Hence, the likelihood of detecting small muscle volume changes by CT and MRI is good. The disadvantage of these methods is their relatively high cost and limited access.

The DXA method has relatively low technical errors for appendicular skeletal muscle mass measurements (e.g., 2.4 ± 0.5% for leg). As DXA measures total extremity lean mass, changes in mass may be brought about by alterations in skeletal muscle water, fluid spaces, and glycogen. The relative safety and increasing availability of DXA makes it a good choice for longitudinal studies of skeletal muscle mass. However, some concerns are arising on the validity of DXA body composition measurements carried out over time, and more studies are needed to examine the validity of longitudinal DXA skeletal muscle mass estimates.

The in vivo neutron activation method was used in longitudinal studies prior to the development of the newer aforementioned methodologies. The technical error of both TBN (i.e., 3.6%) and TBK (i.e., 3.2%) are high, and method availability is limited. Therefore, this method cannot be recommended for longitudinal studies of skeletal muscle mass.

CONCLUSIONS

Great strides have been made in quantifying skeletal muscle mass over the past decade, and many studies are under way to refine further the many available methods. Moreover, the important potential now exists to link muscle mass with muscle metabolic processes and function using currently available methodology. These advances should prove to be of value in studying issues of importance to military operations.

REFERENCES

Anderson, E.C. 1963. Three component body composition analysis based on potassium and water determinations. Ann NY Acad Sci.110:189–212.

Baumgartner R.N., W.C. Chumlea, and A.F. Roche. 1989. Estimation of body composition from bioelectric impedance of body segments. Am J Clin Nutr. 50:221–226.

Baumgartner R.N., R.L. Rhyne, P.J. Garry, S.B. Heymsfield. 1993. Imaging techniques and anatomical body composition in aging. J Nutr. 123:444–448.

Baumgartner, R.N., R. Ross, S.B. Heymsfield, Z. Wang, D. Gallagher, Y. Martel, J. De Guise, W. Brooks. In Press. Cross-validation of DXA versus MRI methods of quantifying appendicular skeletal muscle. Applied Radiation and Isotopes.

Bortz W.M. II. 1982. Disuse and aging. JAMA 248:1203–1208

Burkinshaw, L., G.L. Hill, D.B. Morgan. 1978. Assessment of the distribution of protein in the human body by in-vivo neutron activation analysis. International Symposium on Nuclear Activation Techniques in the Life Sciences. Vienna: International Atomic Energy Association.

Cameron, J. R., J. Sorenson. 1963. Measurement of bone mineral in vivo. Science 42:230–232.

Chumlea W.C., S.S. Guo. 1994. Bioelectrical impedance and body composition: Present status and future directions. Nutr Rev. 52: 123–131.

Clasey, J.L., M.L. Hartmen, J. Kanaley, L. Wideman, C.D. Teates, C. Bouchard, A. Weltman. 1997. Body composition by DEXA in older adults: Accuracy and influence of scan mode. Med Sci Sports Exerc.29:560–567.

Cohn, S.H., D. Vartsky, S. Yasumura, A. Sawitsky, I. Zanzi, A. Vaswani, K.J. Ellis. 1980. Compartmental body composition based on total-body nitrogen, potassium, and calcium. Am J Physiol. 39: E524–530.

Forbes G.B., M.R. Brown, H.J.L. Griffiths. 1988. Arm muscle plus bone area: Anthropometry and CAT scan compared. Am J Clin Nutr. 47:929–931.

Foster, M.A., J.M.S. Hutchison, J.R. Mallard, and M. Fuller. 1984. Nuclear magnetic resonance pulse sequence and discrimination of high- and low-fat tissues. Mag. Res. Imaging 2:187–192.

Gadian, D.G. 1995. NMR and Its Applications to Living Systems, 2nd edition. New York: Oxford University Press.

Heymsfield S.B., C. McManus, J. Smith, V. Stevens, D.W. Nixon. 1982. Anthropometric measurement of muscle mass: Revised equations for calculating bone-free arm muscle area. Am J Clin Nutr. 36:680–690.

Heymsfield, S.B., C. Arteaga, C. McManus, J. Smith, S. Moffitt. 1983. Measurement of muscle mass in humans: Validity of the 24-hour urinary creatinine method. Am J Clin Nutr. 37: 478–494.

Heymsfield S.B., C. McManus III, S. Seitz, D. Nixon, J. Smith. 1984. Anthropometric assessment of adult protein-energy malnutrition. Pp. 27–82 in RA Wright and SB Heymsfield (eds). Nutritional Assessment of the Hospitalized Patient. Boston: Blackwell Scientific Publications, Inc.

Heymsfield, S.B., R. Smith, M. Aulet, B. Bensen, S. Lichtman, J. Wang, and R.N. Pierson, Jr. 1990. Appendicular skeletal muscle mass: Measurement by dual-photon absorptiometry. Am. J. Clin. Nutr. 52: 214–218.

Heymsfield S.B., A. Tighe, Z. Wang. 1992. Nutritional assessment by anthropometric and biochemical methods. In: Shils ME, Olson JA, Shike M (Eds) Modern Nutrition in Health and Disease. Philadelphia: Lea and Febiger, Eighth Edition.

Heymsfield, S.B., D. Gallagher, M. Visser, C. Nunez, Z. Wang. 1995. Measurement of skeletal muscle: laboratory and epidemiological methods. J. Gerontol. 50A:23–29.

Heymsfield, S.B., R. Ross, Z. Wang, D. Frager. 1997a. Imaging Techniques of Body Composition: Advantages of Measurement and New Uses. Pp. 127–150 in Emerging Technologies for Nutrition Research, S.J. Carlson-Newberry and R.B. Costello, eds. Institute of Medicine. Washington, DC: National Academy Press.

Heymsfield, S.B., Z. Wang, R.N. Baumgartner, R. Ross. 1997b. Human body composition: Advances in models and methods. Ann. Rev. Nutr. 17:527–558.

Hochachka P. 1994. Muscles as Molecular and Metabolic Machines. Boca Raton: CRC Press.

Hounsfield,G.N. 1973. Computerized transverse axial scanning (tomography). Br. J. Radiol. 46:1016.

Jue, T., D.L. Rothman, G.I. Shulman, B.A. Tavitian, R.A. DeFronzo, and R.G. Sulman. 1989. Direct observation of glycogen synthesis in human muscle with ^{13}C NMR. Proc of the National Academy of Sciences. 86:4489–4491.

Kohrt,W. 1995. Body composition by DXA: Tried and true? Med. Sci. Sports Exerc. 27:1349–1353.

Kushner, R.F., R. Gudivaka, D.A. Scholler. 1996. Clinical Characteristics influencing bioelectrical impedance analysis measurements. Am J Clin Nutr. 64(suppl):423S–427S.

Lexell J., C.C. Taylor, M. Sjostrom. 1988. What is the cause of the aging atrophy? J Neurol Sci. 84:275–294.

Lukasksi, H.C. 1996. Estimation of muscle mass. Pps. 109–128 in Human Body Composition, A.F. Roche, S.B. Heymsfield, T.G. Lohman, eds. Champaign, Ill.: Human Kinetics Publishers.

Lukasksi, H.C., J. Mendez, E.R. Buskirk, S.H. Cohn. 1981. Relationship between endogenous 3-methylhistidine excretion and body composition. Am J Physiol. 240:E302–E307.

Martin A.P., L.F. Spenst, D.T. Drinkwater, J.P. Clarys. 1990. Anthropometric estimation of muscle mass in men. Med Sci Sports Exerc. 22:729–733.

Nelson, M.E., M.A. Fiatarone, J.E. Layne, I. Trice, C.D. Economoc, R.A. Fielding, R. Ma, R.N. Pierson, W.J. Evans Nunez C, D. Gallagher, M. Visser, F.X. Pi-Sunyer, Z. Wang, S.B. Heymsfield. 1996. Analysis of body composition techniques and models for detecting change in soft tissue with strength training. Am J Clin Nutr. 63:678–686.

Nuñez, C., D. Gallagher, M. Visser, F.X. Pi-Sunyer, Z. Wang, and S.B. Heymsfield. 1997. Bioimpedance analysis: Evaluation of leg-to-leg system based on pressure contact footpad electrodes. Med. Sci. Sports Exerc. 29:524–531.

Nunez C., J. Grammes, D. Gallagher, R.N. Baumgartner, Z. Wang, S.B. Heymsfield. 1996. Appendicular skeletal muscle: Estimation by total limb impedance measured with new electrode system. Faseb J. 10:3.

Organ L.W., G.B. Bradham, D.T.W. Gore, S.L. Lozier. 1994. Segmental bioelectrical impedance analysis: Theory and Application of a new technique. J. Appl. Physiol. 77:98–112.

Pietrobelli, A., D. Gallagher, R. Baumgartner, R. Ross, S.B. Heymsfield. R. 1998. Lean value for DXA Two-component Soft-Tissue Model: Influence of Age and Tissue/Organ Type. Appl. Radiat. Isot. 49(5–6):743–744.

Ross, R., H. Pedwell, J. Rissanen. 1995. Response of total and regional lean tissue and skeletal muscle to a program of energy restriction and resistance exercise. Int J Obes. 19:781–787.

Sipila S., H. Suominen. 1991. Ultrasound imaging of the quadriceps muscle in elderly athletes and untrained men. Muscle and Nerve 14:527–533.

Sjöström, L.A. 1991. A computer-tomography based multicompartment body composition technique and anthropometric predictions of lean body mass, total and subcutaneous adipose tissue. Int J Obesity. 15:19–30.

Snyder W.S., M.J. Cook, E.S. Nasset, L.R. Karhausen, G.P. Howells, I.H. Tipton. 1975. Report of the Task Group on Reference Man. Oxford: Pergamon Press.

Tomlinson B.E., J.N. Walton, J.J. Rebeiz. 1969. The effects of aging and of cachexia upon skeletal muscle. A histopathological study. J Neurol Sci. 9:321–346.

Wang Z., R.N. Pierson Jr, and S.B. Heymsfield. 1992 . The five-level model: A new approach to organizing body composition research. Am. J. Clin. Nutr. 56:19–28.

Wang Z., S. Heshka, R. Pierson, S.B. Heymsfield. 1995. Systematic organization of body-composition methodology: An overview with emphasis on component-based methods. Am J Clin Nutr. 61: 457–465.

Wang, Z., D. Gallagher, M. Nelson, D. Matthews, S.B. Heymsfield. 1996a. Total body skeletal muscle mass: Evaluation of 24 hour urinary creatinine excretion method by computerized axial tomography. Am. J. Clin. Nutr. 63:863–869.

Wang, Z., M. Visser, R. Ma, R. Baumgartner, D. Kotler, D. Gallagher, and S.B. Heymsfield. 1996b. Skeletal muscle mass: Evaluation of neutron activation and dual-energy x-ray absorptiometry methods. J. Appl. Physiol. 80(3):824–831.

Wang, Z., P. Deurenberg, D. Matthews, S.B. Heymsfield. 1998. Urinary 3-methylhistidine excretion: Association with total body skeletal muscle mass by computerized axial tomography. J. Parenter. Enteral. Nutri. 22:82–86.

Yanovski, S.Z., V.S. Hubbard, S.B. Heymsfield, H.C. Lukasksi. 1996. Bioelectrical Impedance Analysis in Body Composition Measurement. Proceedings of a National Institutes of Health Technology Assessment Conference. Bethesda, Maryland, December 12–14, 1994. Am. J. Clin Nutr. 64(3S):387–532S.

13

Alterations in Protein Metabolism Due to the Stress of Injury and Infection

Robert R. Wolfe[1]

During severe injury or infection an overall metabolic response occurs that results in a loss in lean body mass. However, each tissue has a specific response that may be unique, and net protein synthesis may even be increased in some tissues. Thus, protein synthesis is accelerated in the liver (for the production of acute phase proteins), the immune system, and wound repair requires rapid protein synthesis. The catabolic response largely occurs in the skeletal muscle. Over a short period of time, the muscle has an adequate reserve of protein to maintain normal function despite accelerated catabolism. However, when the catabolic response is extended over several days or weeks, severe debilitation can occur. This is reflected by the fact that currently only 50 percent of patients

---

[1] Robert R. Wolfe, Metabolism Unit, Shriners Burns Institute and University of Texas Medical Branch, Galveston, TX 77550.
This research was supported by Grant 8490 from the Shriners Hospital and National Institutes of Health (NIH) Grant DK33952.

discharged from an intensive care unit return to normal function, including work, within 2 years (Bams and Miranda, 1985). The persistent disability following severe burn injury is also well documented (Chang and Herzog, 1976). Although many factors may contribute to these statistics, loss of muscle strength and function is central to the problem of rehabilitation. This chapter interprets the goal of nutritional and metabolic support during acute hospitalization following severe injury and the period immediately following discharge to be a rapid return to normal physiological function. Therefore, the focus here will be on the response of muscle.

The net synthesis or catabolism of muscle protein depends on the balance between the rate of protein synthesis and breakdown. The precursors for protein synthesis are derived from either protein breakdown or from transmembrane transport from the plasma. The amino acids resulting from protein breakdown can either be re-incorporated into protein or released into plasma. Exogenous amino acids given in nutrition can only be incorporated into protein after being transported into the muscle cells from the blood. Thus, the processes of protein synthesis, breakdown, and transmembrane amino acid transport are linked, and it is necessary to evaluate the response to stress by quantifying these three related processes. Consequently, results will be presented from a technique involving the infusion of tracer amounts of amino acids labeled with heavy stable isotopes of carbon (^{13}C) or hydrogen (^{2}H) and sampling from the femoral artery and vein and from the intramuscular pool of the vastus lateralis (obtained by biopsy) (Biolo et al., 1995a). This approach allows quantification of transmembrane transport rate of various amino acids, as well as the rates of muscle protein synthesis and breakdown.

The negative protein balance caused by severe injury results from a large increase in the rate of protein breakdown. Although synthesis is also increased, the increase is insufficient to offset the increased rate of breakdown. The increase in muscle protein breakdown is coupled with an increase in the outward transport of amino acids, which is consistent with the role of the muscle to provide amino acid precursors for synthesis elsewhere in the body. The negative amino acid balance persists across the muscle even for a person in the fed state. Furthermore, increasing the amount of protein intake has no effect on the rate of muscle protein synthesis.

The alteration in transport kinetics across the muscle cell membrane may be central in the altered muscle protein kinetics. The inward transport of phenylalanine and leucine in severely burned patients is less than half the normal rate. This explains the inefficiency of amino acid or protein intake in stimulating synthesis, because the exogenous amino acids must enter the cells before they can be incorporated into protein. This point can be seen clearly in the case of glutamine. The intramuscular glutamine concentration is decreased in severely burned patients to about one- third its normal value. This is in large part due to an accelerated rate of outward transport, as the intracellular appearance from protein breakdown is double the normal rate and *de novo* synthesis is not markedly suppressed. Nonetheless, the infusion of glutamine directly into the femoral artery of adult burn patients did not increase the intramuscular concentration of glutamine, despite a three-fold increase in the femoral venous concentration of glutamine.

Because of the altered amino acid transmembrane kinetics, it is clear that greater-than-normal protein intake is not effective in curtailing the catabolic response of muscle (Patterson et al, 1997). Therefore, there is no reason to believe that more than 100 g/d of protein will provide any additional benefit. However, with regard to the question of whether protein requirements increase with military operational stressors and what is the optimum protein intake, it is nevertheless clear that protein requirements have increased. This is because normal protein intake is insufficient to alleviate catabolism. Further, it is possible that an optimal formulation of amino acids would be beneficial (Sakurai et al, 1995). However, given the deficiencies in inward transport, it is unlikely any amount or particular mix of amino acids would make any difference without concurrent manipulation of the metabolic state of the tissue. In this author's experience, these observations apply equally to men and women. However, it is possible that metabolic manipulation with hormone therapy could make the muscle more responsive to the stimulatory effect of amino acids on protein synthesis, thereby providing a rationale for increasing protein intake and optimizing the formulation of an amino acid mixture.

ANABOLIC HORMONES

The anabolic action of growth hormone on muscle in children is well established. The acute, intravenous infusion of growth hormone in burned children has been shown to increase muscle protein synthesis to the same extent as a pharmacological dose of insulin, but the effect of the two hormones was not additive (Gore et al., 1991). The effect of growth hormone on adult muscle protein metabolism is less clear. It has been recently shown that growth hormone replacement in growth hormone-deficient adults increased lean body mass after six months, but not after 1 month, of growth hormone supplementation (Solomon et al., 1989). However, it is not clear that this response reflected increased muscle mass, and in fact it could have been due to water retention. In contrast, a period of 5 days of growth hormone treatment in normal volunteers receiving a diet containing only 50 percent of caloric requirement induced a switch from a negative, whole-body N balance to a positive, whole body N balance (Manson and Wilmore, 1986). Furthermore, N balance improved in some (e.g., Wilmore et al., 1974), but not all (e.g., Belcher et al., 1989) studies in which growth hormone was administered to burn patients. However, none of these studies assessed the effect of growth hormone on muscle *per se*. The only studies in which muscle protein synthesis was directly measured failed to show an effect of growth hormone either in normal (Yarasheski et al., 1993b) or elderly volunteers (Yarasheski et al., 1995).

Insulin is the most important anabolic hormone on an hour-to-hour basis in normal human physiology. Local hyperinsulinemia to an extent comparable to that achieved during a normal meal caused a significant increase in muscle protein synthesis and inward transmembrane transport of phenylalanine (Sakurai et al., 1995). To extend this observation to the clinical setting, severely burned adults were infused for 7 days at a rate high enough to maintain plasma insulin concentrations of approximately 500 µU/ml. Additional glucose was given as needed to maintain euglycemia. Patients were studied according to a crossover design, with half of the patients receiving the insulin therapy in the first week, and half receiving the insulin therapy in the second week. Patients were used as their own controls and were studied in the fed state. Insulin therapy reversed the negative amino acid balance across the muscle, primarily by stimulating muscle protein synthesis. Paradoxically, protein breakdown also increased, thereby blunting the net anabolic effect of insulin. It is possible that the stimulation of protein synthesis exceeded the availability of intracellular amino acids, leading to an acceleration of protein breakdown in order to maintain the intracellular pools. If this was the case, it may be that in severely burned patients, an increased intake of amino acids might be more effective during insulin therapy than when insulin is not given, particularly since insulin greatly stimulated the rate of inward amino acid transport. However, the interaction of insulin therapy and the amount of amino acid or protein intake has not been investigated.

In patients given the control diet the first week and insulin therapy the second week, growth hormone was given throughout the third week, without interruption of the insulin. No additional effect of growth hormone beyond that elicited by insulin was observed (Sakurai et al., 1995).

Testosterone is well known for its ability to stimulate muscle anabolism in normal subjects when taken in large doses. However, testosterone is also effective in stimulating muscle protein synthesis when given to normal volunteers in an amount sufficient to increase the plasma concentration to the high-normal range. The effect of testosterone is likely to be exerted on the transcription of mRNA, as suggested by the observations of a significant increase in muscle protein synthesis observed 5 days after the intramuscular injection of 200 mg, while no effect was observed during the intravenous infusion of the same amount of testosterone over 5 hours (Ferrando, et al, 1998). Transcriptional effects are likely to take days to be effective. The effectiveness of testosterone in critically ill or rehabilitating patients is unknown, but it is pertinent that in adult male burn patients the serum concentration of testosterone is reduced more than 80 percent below the normal control value. Further, it is possible that an interactive effect between testosterone and insulin might be anticipated, since they appear to operate via different mechanisms. However, this possibility remains unexplored.

AUTHOR'S CONCLUSIONS AND RECOMMENDATIONS

The catabolic response of muscle is characterized by an outward efflux of amino acids from muscle that minimizes the effectiveness of any nutritional protocol. It is therefore reasonable to explore the interaction of hormonal and nutritional therapy. In adult patients, only insulin therapy has been shown to stimulate muscle protein synthesis, but results from normal volunteers given testosterone are quite promising. Future areas of investigation should involve quantification of hormonal effects on muscle, and interactive effects between hormones and diet. Thus, it is possible that whereas a higher than normal protein intake in injured patients normally doesn't provide an added benefit beyond that achieved with a normal intake, a higher protein intake becomes beneficial when the system is "primed" by testosterone and/or insulin therapy.

Based on currently available data, a diet is recommended for severely injured patients of 1.5 g protein/kg day, with carbohydrate given at a rate that supplies approximately the caloric equivalent of the resting energy expenditure. Fat should be given only as needed to avoid fatty acid deficiency (approximately 2 percent of daily caloric intake). Exogenous insulin should be given to maintain euglycemia. Men and women should be given the same treatment, and it is unlikely the particular source of protein is important provided that it has a reasonable balance of essential amino acids.

REFERENCES

Bams, J.L., and D.R. Miranda. 1985. Outcome and costs of intensive care. Intensive Care Med. 11:234–241.

Belcher, H.J., D. Mercer, K.C. Judkins, S. Shalaby, S. Wise, V. Marks, and N.S. Tanner. 1989. Biosynthetic human growth hormone in burned patients: A pilot study. Burns 15:99–107.

Biolo, G., R.Y.D. Fleming, S.P. Maggi, and R.R. Wolfe. 1995a. Transmembrane transport and intracellular kinetics of amino acids in human skeletal muscle. Am. J. Physiol. 268(31):E75–E84.

Chang, F.C., and B. Herzog. 1976. Burn morbidity: A follow-up study of physical and psychological disability. Ann. Surg. 183:34–37.

Ferrando, H.A., K.D. Tipton, D. Doyle, S.M. Phillips, J. Gortiella, R. R. Wolfe. In Press. Testosterone injection stimulates net protein synthesis but not amino acid transport. Amer. J. of Physiol.

Gore, D.C., D. Honeycutt, F. Jahoor, R.R. Wolfe, and D.N. Herndon. 1991. Effect of exogenous growth hormone on whole-body and isolated-limb protein kinetics in burned patients. Arch. Surg. 126:38–43.

Manson, J.M., and D.W. Wilmore. 1986. Positive nitrogen balance with human growth hormone and hypocaloric intravenous feeding. Surgery 100:188–197.

Patterson, B.W., T. Nguyen, E. Pierre, D.N. Herndon, R.R. Wolfe. 1997. Urea and protein metabolism in burned children: effect of dietary protein intake. Metabolism 46(5):573–578.

Sakurai, Y., A. Aarsland, D.N. Herndon, D.L. Chinkes, E. Pierre, T.T. Nguyen, B.W. Patterson, and R.R. Wolfe. 1995. Stimulation of muscle protein synthesis by long-term insulin infusion in severely burned patients. Ann. Surg. 222(3):283–297.

Solomon, F., R.C. Cuneo, R. Hesp, and P.H. Sonksen. 1989. The effects of treatment with recombinant human growth hormone on body composition and metabolism in adults with growth hormone deficiency. New Engl. J. Med. 321:1797.

Wilmore, D.W., J.A. Moylan, B.F. Bristow, H.D. Mason, Jr., and B.A. Pruitt, Jr. 1974. Anabolic effects of growth hormone and high caloric feedings following thermal injury. Surg. Gynecol. Obstet. 138:875–884.

Yarasheski, K.E., J.J. Zachwieja, T.J. Angelopoulos, and D.M. Bier DM. 1993b. Short term growth hormone treatment does not increase muscle protein synthesis in experienced weight lifters. J. Appl. Physiol. 74:3073–3076.

Yarasheski, K.E., J.J. Zachwieja, J.A. Campbell, and D.M. Bier. 1995. Effect of growth hormone and resistance exercise on muscle growth and strength in older men. Am. J. Physiol. 268:E268–E276.

I

Discussion

ROBERT NESHEIM: I'd like to open this up now for discussion. Dr. Rennie, how would you respond to the questions posed by the Army?

MICHAEL RENNIE: Regarding whether active people need more protein, I think that looking at people who live under circumstances of low protein intake gives us a lot of the answers. If you look at the protein intakes of Kenyan runners, they eat something like 0.6 to 0.7 grams of protein per kilogram per day, and they outperform all of the Western middle- and long-distance runners, right?

So there are natural experiments that suggest that you don't need to eat very much protein to do well in certain kinds of physical performance. Of course, I don't know how well they would do as weight lifters.

I don't want to steal the thunder of Anton Wagenmakers, but it strikes me that all of the studies that have been done by giving individual mixtures of amino acids, either branched-chain amino acids, glutamine, or glutamate, have, without exception, shown no effect on physical performance.

I don't want to steal Harris Lieberman's thunder, either, but I strongly suspect that unless there are kinds of cognitive function that we cannot measure easily, you will not see much in the way of any effect on cognitive performance.

Finally, I think that there are just insufficient data to talk about gender differences so far.

HARRIS LIEBERMAN: I just want to make a comment. With regard to formulating a recommendation, remember that the soldiers who come into the Army are eating an American diet and have eaten an American diet their whole lives, so that a suggestion that I make is not to recommend a change in the American diet or even a change in the soldier's diet in garrison, because we really cannot change diets that drastically; we simply want recommendations on what we are going to feed in those times when we go out into the field, and particularly when we go into combat. What we really need is a recommendation for a combat ration.

ROBERT NESHEIM: Dr. Butterfield?

GAIL BUTTERFIELD: Let me ask my usual question of the speakers. One of the problems with military troops in the field is that their energy intake is dramatically reduced. How do you think that reduction will affect these various factors?

MICHAEL RENNIE: Well, there is some dramatic evidence from Stroud and Fiennes' trans-Antarctic walk, in which they expended huge amounts of energy, greater amounts of energy than Tour de France cyclists expend, and yet their energy intakes were certainly less than that, and they lost very large amounts of weight.

What is quite interesting is that in the study that Stroud and Alan Jackson and John Waterlow published in the *British Journal of Nutrition*, although it cannot really be described as a scientific study, given the way the controls were performed after the event, nevertheless, it is quite interesting that whole body protein turnover did not appear to be markedly depressed, despite the fact that these people were in marked energy deficits, (much more than I imagine any scientist would ever get permission to induce under these circumstances). Presumably, protein balance was maintained because the subjects were taking in a fair amount of nitrogen; I cannot remember exactly how much. But they were also exercising at these very high rates.

So the idea that exercise was somehow making protein utilization more efficient, I think, is borne out by that kind of study.

PART I DISCUSSION

ROBERT NESHEIM: Yes, Nancy?

NANCY BUTTE: Another question we have been asked to address is the potentially detrimental effect, of high protein intake or supplementation of a single amino acid on a woman who is pregnant.

In the military now at any one time, 8 to 9 percent of active duty women are pregnant. Many of these pregnancies are unplanned, so there is a period where a woman may not be aware she is pregnant. Could some of our experts in protein comment on a possible detrimental effect of either a high protein intake, high protein-to-energy ratio, or a single amino acid supplementation?

D. JOE MILLWARD: I don't know of any evidence to suggest that there is a problem. I mean, I cannot ever remember reading any detailed consideration of that question at all, and it is difficult to really identify one off the top of my head.

ROBERT WOLFE: I wanted to just follow up and ask Mike Rennie his opinion. I think that one of the problems with this whole issue of exercise and amino acid requirements, is that because of the limitations of methodologies, where we are limited to certain types of exercise (such as moderate aerobic exercise) that have been studied, these may not always have a parallel in troops sent into combat.

Mike referred very briefly to the situation of overtraining, and it is certainly clear, with overtraining, for example, that you get disruption of normal amino acid concentrations in the muscle and that maybe in this circumstance, the protein requirements are different than for someone who is just doing moderate exercise at a very easy level. But someone who is thrust into repetitive and high-intensity exercise clearly ends up with a different response in their muscles, and this difference is something that we should think a little bit about. Even in more general terms. We need to think about the difference between resistance exercise and aerobic exercise. We tend to move back and forth between the two, but I don't know that that is really justified, and if the goal is to build up strength, then simple maintenance of nitrogen balance may not really be the end goal.

MICHAEL RENNIE: I agree with all of those points.

D. JOE MILLWARD: Can I comment on that? I think part of the difficulty that I see here is that if the ultimate objective is the overall performance of the individual, then one has to avoid thinking about one system in particular.

For example, we know that individuals involved in large amounts of exercise, like marathon runners and certainly overtrained individuals, tend to have a reduced immune function. The research that has shown glutamine supplementation to be effective has been under circumstances where the glutamine is protective against the consequences of reduced immune function, like preventing bacterial translocation across the gut and those sorts of things.

Now, the link [between the immune system and physical activity] would be the problems promoted by leakage of muscle enzymes as a consequence of large amounts of physical activity, which promote an acute phase response, coupled with loss of muscle glutamine and the consequences of that loss.

And so, if you need to think about whether glutamine supplementation would be important in those circumstances, although we think of muscle as the main store of glutamine, you actually have to think of the function of glutamine, which is really on the gut and on the immune function. Thus, unless you actually think about the whole picture, I think you are going to miss some of the outcomes. That is why, ultimately, the studies that must be done must define as many different parameters as possible in terms of performance and include those parameters. So far, we have attacked little bits of the problem with individual studies, and that is why I think it is so hard to put the whole lot together and come up with a sensible answer.

ROBERT NESHEIM: I think that brings us to the next set of presentations, which have to do with cognitive performance, stress, and brain function.

Protein and Amino Acids, 1999
Pp. 289–307. Washington, D.C.
National Academy Press

14

Amino Acid and Protein Requirements: Cognitive Performance, Stress, and Brain Function

Harris R. Lieberman[1]

INTRODUCTION

This chapter addresses amino acid and protein requirements and brain function. A particular focus will be the possibility that central demands for amino acids may modify nutritional requirements when individuals are exposed to extreme environments and other stressors associated with combat and high-intensity military or civilian occupations.

To function adequately, the central nervous system (CNS) requires a number of amino acids found in protein foods. Amino acids such as tryptophan, tyrosine, histidine, and arginine are used by the brain for the synthesis of various neurotransmitters and neuromodulators (Betz et al., 1994). To date, CNS requirements for specific amino acids have not been systematically investigated, perhaps because it has been assumed that brain requirements for precursor amino acids were not critical. Furthermore, appropriate methods of determining

[1] Harris R. Lieberman, U.S. Army Research Institute of Environmental Medicine, Natick, MA 01760.

whether adequate levels of particular precursors are provided to the CNS by the diet do not exist. Although little information on CNS requirements of specific amino acids is available, results from several lines of related research suggest that the peripheral concentration of particular amino acids can be a factor in the regulation of central neurotransmission, cognitive performance, and mood state. For example, if the amino acid tryptophan is either artificially elevated or lowered, changes in brain function and behavior can occur (Young, 1996). Even in normal humans, acute tryptophan depletion produces transient alterations in mood state (presumably by reducing the CNS concentration of serotonin), in particular increased subjective depression and increased aggression (Young, 1996). In contrast, administration of single doses of pure tryptophan to humans increases sleepiness and may reduce pain sensitivity (Hartmann, 1986; Lieberman et al., 1985). These changes are consistent with the various functions attributed to serotonin in the CNS.

CNS requirements for specific amino acids during periods of undernutrition or when individuals are exposed to highly stressful conditions may be particularly critical. For example, among moderately undernourished, but not highly stressed soldiers participating in a field test of an energy deficient ration, decrements in tryptophan were associated with impaired cognitive performance (Lieberman et al., 1997). Furthermore, a series of studies suggests that supplemental administration of tyrosine increases brain catecholaminergic neurotransmission and has beneficial effects on various behavioral parameters associated with resistance to stress (for a recent review, see Lieberman, 1994). Tyrosine is one of the dietary precursors for the synthesis of the catecholamines, dopamine and norepinephrine. The beneficial neurochemical and behavioral consequences of supplemental tyrosine administration are most readily observed when humans and other animals are exposed to various environmental and psychological stressors (Wurtman et al., 1981).

The Blood-Brain Barrier: A Key Determinant of Brain Nutritional Status

Unlike most other organs, the brain is isolated from the general circulation by the blood-brain barrier (BBB). The nature of the barrier is determined by the special properties of the cerebral vasculature, specifically the epithelial cells of the brain capillaries, which selectively prevent the transport of various substances into the brain (Betz et al., 1994 Pardridge 1977). In general, lipophilic compounds typically can passively cross the BBB, but water soluble compounds, such as amino acids, cannot. The BBB, therefore, must contain special mechanisms for selectively transporting key water soluble compounds, such as essential amino acids, into the brain (Pardridge, 1977; Betz et al., 1994). The special status of the brain with regard to its accessibility to nutritional and other systemic metabolic factors has profound implications for the determination of the nutritional requirements of the CNS. For example, in the periphery, a variety of substances can be used as energy substrates, but the brain must rely almost exclusively on glucose (Pardridge, 1977). In examining the brain's requirements for protein and amino acids, it is essential to keep in mind that many substrates available to other organ systems are not available to the brain as a consequence of the BBB's "protection" of the brain. The protected status of the brain could have important implications under conditions of metabolic stress induced by undernutrition, exposure to adverse environmental conditions, or severe physical stress. It is possible that under such conditions, brain nutrient requirements relative to other organ systems may be proportionally much greater.

Although the brain is protected by the BBB, it still requires a number of substrates for adequate function. The importance of various amino acids as precursors for key brain neurotransmitters is well established, and transport mechanisms exist to provide these to the brain. At least three active transport mechanisms convey amino acids into the brain (Betz et al., 1994). Separate mechanisms exist for transport of the large neutral amino acids (LNAA), basic, and acidic amino acids across the BBB. However, small neutral amino acids like glycine and alanine appear to be actively pumped out of the brain (Betz et al., 1994). These three mechanisms and some of the amino acids that they transport are presented in Table 14-1.

Because whole classes of amino acids are actively transported by amino acid-specific carrier mechanisms, compounds from the same class of amino acids actually compete for transport into the brain (Pardridge, 1977). The functional implication of this unique characteristic of the BBB is that the amino acid composition of food is of greater consequence to the brain than perhaps any other organ system. Unlike other organ systems, the brain cannot simply absorb the nutrients it requires from the general circulation but rather only receives those nutrients that are transported across the BBB.

Several of the amino acids that are transported across the BBB and, in some instances, compete for access to the brain, are precursors of important brain neurotransmitters as shown in Table 14-2. There is evidence that most of the

TABLE 14-1 CNS Amino Acid Transport Mechanisms

Large Neutral	Basic	Acidic
Leucine	Lysine	Glutamate
Phenylalanine	Arginine	Aspartate
Trytophan	Ornithine	
Valine		
Isoleucine		

neurotransmitters listed in Table 14-2 are, under certain conditions, precursor sensitive; therefore, as the peripheral concentration of any of the precursors varies, there could be consequences with regard to brain metabolism, function, and behavior (for a recent review see Young, 1996).

As shown in Table 14-3, these neurotransmitters have a variety of important functions in the brain, regulating or modulating a variety of key CNS functions. Of course, the precise relationship between a neurotransmitter and the behavior of an organism is difficult to summarize and is usually the subject of considerable controversy. Clearly, the functions outlined in Table 14-3 are not only critical in a general way for maintaining the normal behavioral status of an organism, but are also closely related to the ability of that organism to function in stressful conditions. In single nonphysiologic doses, or when administered in special diets, all of the amino acid precursors of the transmitter systems shown in Table 14-3 have been found to alter brain activity.

This chapter will focus on the large neutral amino acids (LNAAs), tryptophan and tyrosine, because the data on the effects of these neurotransmitter precursors on brain chemistry and behavior is most convincing. Perhaps one of the most critical issues (and most difficult to answer currently) is: Under what, if any, physiological situations will nutrient availability from food affect brain function?

TABLE 14-2 Amino Acids and Their Neurotransmitter/Neuromodulator Products

Amino Acid	Product
Tryptophan	Serotonin
	Melatonin*
Tyrosine/phenylalanine	Dopamine
	Norepinephrine
	Epinephrine
Histidine	Histamine
Arginine	Nitric Oxide
Threonine (etc.)	Glycine

*Melatonin is not believed to be a neurotransmitter but is a behaviorally active metabolite of tryptophan synthesized by the pineal gland. There is evidence that administration of tryptophan increases the release of melatonin in humans (Hajak et al., 1991).

TABLE 14-3 Putative Functions of Various Neurotransmitter Systems (with Amino Acid Precursors)

Serotonin	Mood, pain, food intake, arousal
Dopamine	Motor function, mood, arousal
Norepinephrine	Arousal, attention, anxiety
Histamine	Food intake, arousal, Thermoregulation
Nitric oxide	Arousal, anxiety, memory
Glycine	Motor function

TRYPTOPHAN

Tryptophan is the rarest of the essential amino acids found in food and, as noted above, is the precursor of serotonin. Evidence from a number of animal studies suggests strongly that changes in the protein: carbohydrate ratio of food consumed can alter brain serotonin. As the proportion of carbohydrate relative to protein increases, so does the level of brain serotonin. Consistent with the functions of serotonin listed in Table 14-3, the availability of tryptophan to the brain can alter behavioral factors such as alertness, level of depression, aggression, and pain sensitivity (for a recent review, see Young, 1996).

The effects of the protein: carbohydrate ratio of meals on brain serotonin is well documented, although the physiologic relevance of the relationship is controversial. The association between protein and carbohydrate intake and brain serotonin concentration was first observed by Fernstrom and Wurtman (1971). Subsequently, numerous animal studies have appeared that confirm and extend the initial observation. A recent paper by Fernstrom and Fernstrom (1995a) carefully documented the effects of meals that varied in protein:carbohydrate ratio on serotonin concentration in several brain regions in the rat. It was clearly demonstrated that as the concentration of protein in the test meals increased, brain concentration of serotonin fell parametrically. In both the hypothalamus and cortex, the highest levels of serotonin were observed following a pure carbohydrate meal. The authors noted that variation in hypothalamic serotonin induced by meals differing in protein: carbohydrate ratio could play a role in regulating food intake since the hypothalamus performs a critical role in appetite regulation. However, the functional significance of cortical changes is difficult to explain.

The phenomenon of peripheral modulation of brain serotonin occurs as a consequence of both the previously noted competition of the LNAA for transport across the BBB and the selective uptake by muscle of the branched-chain amino acids following a carbohydrate meal. Following consumption of carbohydrate, which elicits the secretion of insulin, the concentration of the branched-chain amino acids in the plasma falls as they move into muscle, while tryptophan levels remain relatively unchanged. Therefore, since more tryptophan is available for transport by the LNAA carrier mechanism, tryptophan transport to the brain increases and more tryptophan is available for the synthesis of serotonin. In humans, it has been established that changes in the protein:carbohydrate ratio of foods can alter plasma tryptophan levels in the same way as in experimental animals (Lieberman et al., 1986a).

To date, an association of changes in protein and carbohydrate intake with changes in behavior has not been clearly established, but several behavioral functions have been shown to be sensitive to substantial changes in plasma tryptophan levels. When tryptophan is administered in pure form or in higher than physiologic concentrations in meals, or is artificially lowered by administering tryptophan-free meals, substantial changes in a variety of behavioral parameters can be observed. One of the most frequently documented effects of tryptophan administration is an increase in self-reported mental fatigue. Tryptophan's effects on arousal and alertness appear to be similar to those produced by mild over-the-counter sedatives such as antihistamines. These effects have been documented with self-reported mood questionnaires (Figure 14-1) and by polysomnography (Hartmann and Greenwald, 1984; Lieberman et al., 1985). The hypnotic-like effects observed in humans after tryptophan administration are consistent with reports implicating brain serotonin neurons in the regulation of alertness. Although tryptophan is clearly not as potent as prescription hypnotic drugs, it was sold as an over-the-counter, natural sleep aid until several years ago. It was withdrawn from the market when a large number of cases of a rare disease called eosinophilia-myalgia syndrome (EMS) were seen in individuals who were taking it. The exact cause of the disease has never been conclusively demonstrated, although it has been suggested that a contaminant was present in the tryptophan produced by one manufacturer.

Another well-established effect of tryptophan relates to the role of serotonin in the regulation of mood, in particular, level of depression. Most antidepressant drugs increase serotonergic neurotransmission, although many have other central effects as well. Consistent with this role of serotonin in the brain, it appears that single meals that are deficient in tryptophan have substantial acute effects on self-reported level of depression. A series of studies have been conducted in several laboratories to examine the effects of specially formulated tryptophan-free meals on mood state. These studies have consistently observed substantial increases in depression when these low- tryptophan meals are ingested (Young, 1996) when measured with standard self-reported mood questionnaires in normal individuals, as shown in Figure 14-2 (Smith et al., 1987), and in individuals suffering from various clinical syndromes associated with high levels of depression.

Panel A

Panel B

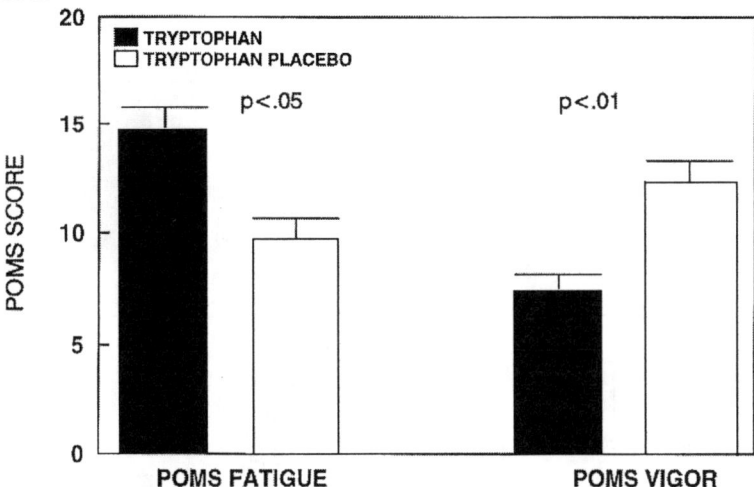
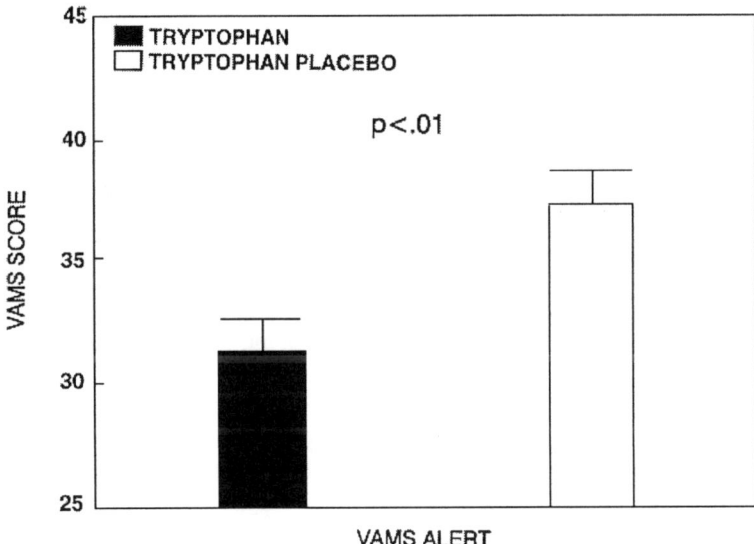

FIGURE 14-1 Effects of a single dose of tryptophan (50 mg/kg) on the self-reported fatigue, vigor, and alertness of healthy volunteers as assessed with the Profile of Mood States (POMS) Questionnaire (Panel A) and Visual Analog Mood Scale (VAMS) (Panel B). SOURCE: Adapted from Lieberman et al. (1986).

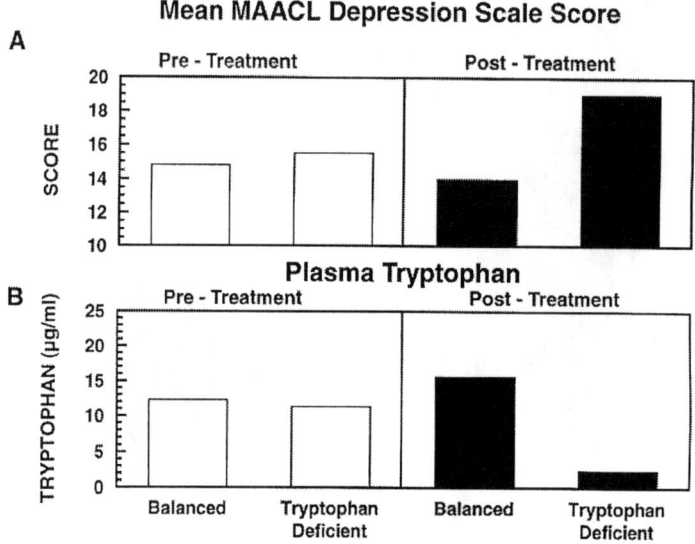

FIGURE 14-2 Effects of a single tryptophan deficient meal on self-reported depression and plasma tryptophan levels of healthy volunteers. SOURCE: Adapted from Smith et al. (1987).

Tryptophan-deficient meals also affect the level of aggression displayed by normal volunteers. When normal individuals are classified as belonging to a high-aggressive personality type based on standardized personality questionnaires, administration of tryptophan-deficient meals increases not only aggressive mood state, but also the level of aggressive behavior overtly displayed (Figure 14-3) (Cleare and Bond, 1995). Administration of a high- tryptophan meal to such individuals decreases aggressive mood and behavior (Cleare and Bond, 1994).

Tryptophan has also been reported to decrease pain sensitivity in animal models, normal humans, and patients suffering from certain clinical conditions where pain is present (Lieberman et al., 1983; Seltzer et al., 1983). The changes in aggression and pain induced by artificially altering plasma levels of tryptophan are consistent with data implicating serotonin in the regulation of aggression and pain sensitivity.

Overall, there is little doubt that substantial variations in plasma tryptophan levels can have a major impact on the behavior of humans and other animals. However, it should be noted that the doses of tryptophan that have been shown to be unequivocally psychoactive may produce changes in brain tryptophan that are larger than those produced by any food that increases or decreases brain serotonin. Currently, the smallest change in levels of plasma or brain tryptophan that will have an impact on brain function or behavior is unknown. Therefore, it is not currently possible to determine whether nutritional requirements for tryptophan are in any way related to brain demands for this amino acid.

TYROSINE

Another amino acid that has been extensively examined for behavioral effects is tyrosine, the precursor of three neurotransmitters: norepinephrine, dopamine, and epinephrine (see Table 14-2). Tyrosine is not typically considered to be an essential amino acid since it can be synthesized by humans from phenylalanine; however, it has been suggested by some investigators that the brain may not be able to synthesize sufficient tyrosine from phenylalanine to meet its needs (Pardridge, 1977). Tyrosine is generally found in larger quantities than tryptophan in most protein foods. Since tyrosine is a LNAA, it competes with tryptophan and the other LNAAs for transport across the BBB.

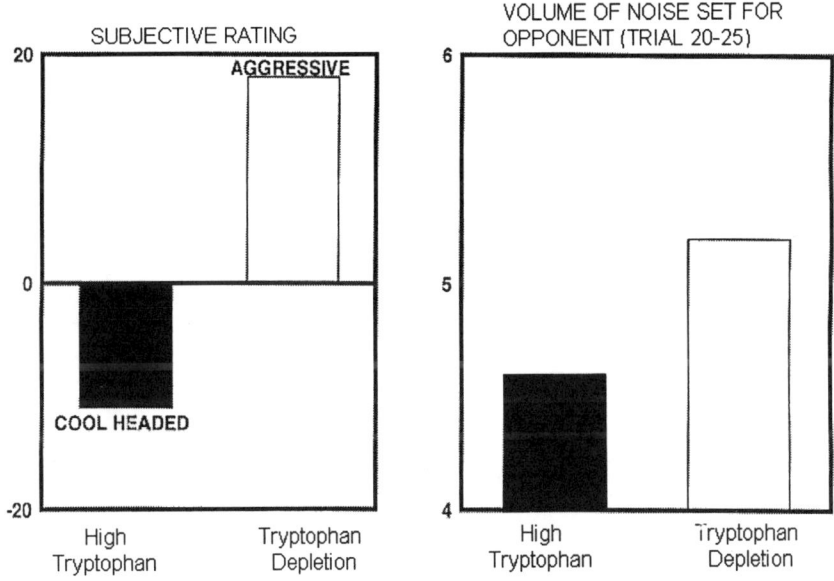

FIGURE 14-3 Effects of tryptophan-depleted and -supplemented meals on subjective and objective measures of aggression among normal, high-trait aggressive volunteers. On the left, the effects of such meals on a bipolar scale measuring aggression/cool headedness are plotted. On the right, an objective measure of aggression, (willingness of the tested individual to deliver aversive [noise] stimuli to another individual) is plotted. SOURCE: Adapted from Cleare and Bond (1995).

Under certain conditions, it appears that administration of tyrosine can affect brain neurotransmission. Specifically, it has been hypothesized that when central catecholaminergic neurons are very active (as occurs during exposure to acute stress), they will become precursor sensitive (Wurtman et al., 1981). Although these neurons are not normally believed to be affected by the availability of tyrosine, they may require additional tyrosine to function optimally when they are firing frequently (Lieberman, 1994; Wurtman et al., 1981). Norepinephrine is believed to play a critical role in the response of the brain to acute stress. Exposure to heat, cold, cardiovascular stressors, and electric shock all produce significant increases in brain catecholaminergic activity (Stone, 1975). Central noradrenergic neurons seem to be critical for regulating key behavioral parameters such as attention, arousal level, and mood state (Lieberman, 1994). Although norepinephrine appears to be particularly critical for the brain's response to stress, another brain catecholamine, dopamine, also appears to be involved in certain aspects of the acute response to various stressors.

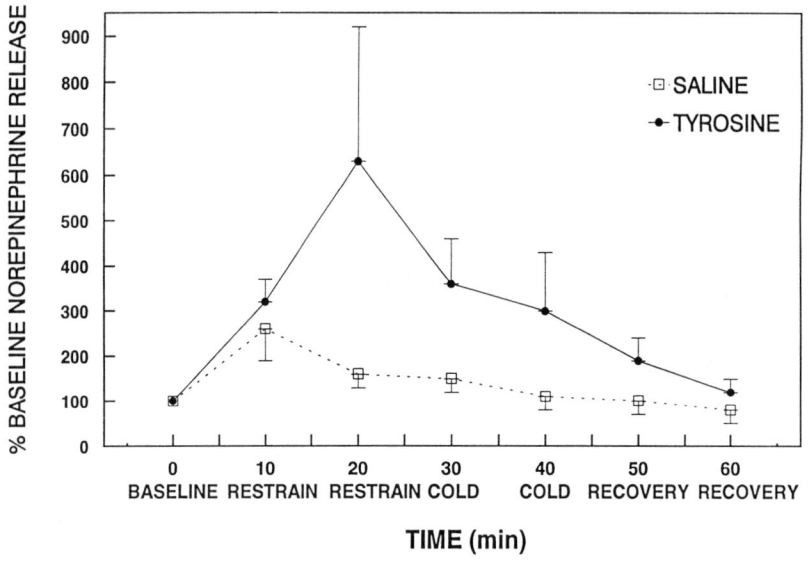

FIGURE 14-4 Effect of immobilization stress alone and in combination with cold stress on hippocampal norepinephrine release in tyrosine- (400 mg/kg) and placebo-treated rats. Testing conditions, in 10-min intervals, are specified on the x-axis. SOURCE: Adapted from Lieberman and Shukitt-Hale (1996).

The neurochemical consequences of exposure to stress and the effects of supplemental tyrosine under such conditions have been examined in animal models. Two recent studies examined the effects of a combination of cold and restraint stress on release of norepinephrine in the rat hippocampus (Luo et al., 1993; and Shukitt-Hale, 1996). The technique of microdialysis was used to assess norepinephrine release since it permits continuous assessment of neurotransmitter release in vivo from a specific brain region (Ungerstedt et al., 1982). Immediately prior to exposure to the stressors, some animals received an intraperitoneal injection of 400 mg/kg of tyrosine while others received a placebo injection. Figure 14-4 illustrates the effects of stress and supplemental tyrosine under these conditions. The combination of cold and restraint stress substantially increased the release of norepinephrine in the hippocampus over baseline levels; when animals were pretreated with tyrosine, the magnitude of the increase was substantially amplified (Lieberman and Shukitt-Hale, 1996). To evaluate the hypothesis that supplemental tyrosine can prevent some adverse behavioral and physiological effects of exposure to various acute stressors, a number of animal and human studies have been conducted (for reviews, see Lieberman, 1994; Owasoyo et al., 1992; Salter, 1989). In general, the results of these studies suggest that tyrosine administration, particularly when the stress is severe, will have beneficial effects on the ability of the organism to function adequately.

In some of the initial studies, rats were exposed to foot shock, and their spontaneous behavior was assessed (Lehnert et al., 1984a; Lehnert et al., 1984b). In these studies, rats that were pretreated with tyrosine were more active and appeared to be less debilitated following exposure to the stressor. In other animal studies in which high doses of tyrosine were administered, learning and memory, as well as other aspects of performance in the cold and under high-altitude conditions, were improved (Ahlers, et al., 1994; Rauch and Lieberman, 1990; Shukitt-Hale et al., 1996; Shurtleff et al., 1993). In addition, tyrosine has been shown to have beneficial effects in animals exposed to heat stress by reducing immobility, the dependent measure in the Porsolt swim test (Yeghiayan, in press; Figure 14-5). In human studies, tyrosine has been found to have positive effects on cognitive performance during exposure to a combination of cold and high-altitude stress as well as cold stress alone (Banderet and Lieberman, 1989; Shurtleff et al., 1994). Tyrosine also appears to enhance performance of individuals exposed to psychological stress (Figure 14-6) (Deijen and Orlebeke, 1994). These human studies are consistent with neurochemical and behavioral studies of animals that also suggest that tyrosine has beneficial effects on the ability of animals to cope with acute stress and can improve performance on tasks requiring attention and learning. Although not directly addressing the issue of dietary requirements for tyrosine, these studies indicate that there may be an increased CNS requirement for this amino acid during periods of intense stress.

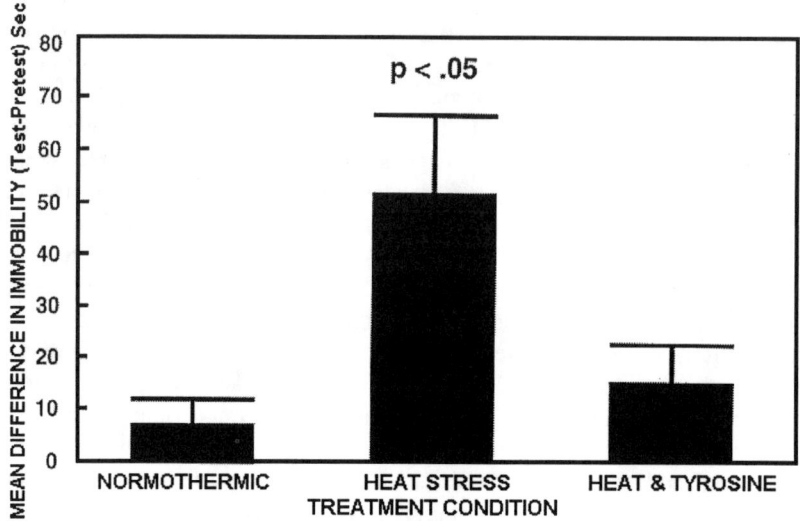

FIGURE 14-5 Effect of heat stress and tyrosine on performance of rats in the Porsolt swim test. Increased immobility (mean difference in immobility) indicates inability of the animal to respond appropriately to the heat stressor. SOURCE: Adapted from Yeghiayan, in press.

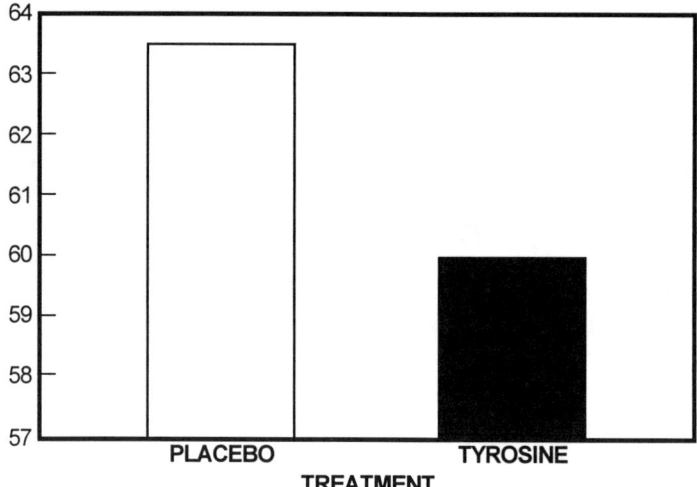

FIGURE 14-6 Performance on a stressful test of attention and memory (Stroop Task) in humans treated with placebo or 100 mg/kg of tyrosine. SOURCE: Adapted from Deijen and Orlebeke (1994).

CHANGES IN AMINO ACIDS DURING FIELD STUDIES: UNDERNUTRITION AND MENTAL PERFORMANCE

Several years ago, as part of a U.S. Army Research Institute of Environmental Medicine (USARIEM) field study of an experimental lightweight ration, the relationship between plasma amino acid levels and mental performance was assessed. The light weight ration tested, termed the Ration, LightWeight-30 (RLW-30), was intended to be the sole source of nutrition for soldiers operating without logistical support for up to 30 days (Askew et al., 1987). The ration was nutritionally balanced but calorie energy deficient since it provided only 2,000 kcal of energy per day. In this study, which was conducted under temperate climatic conditions, the RLW-30 was compared with the standard Army field ration—the Meal, Ready-to-Eat (MRE-Version VI). The macronutrient intake of soldiers consuming the two types of rations, as well as the actual mean daily energy expenditure of the soldiers in each group can be found in Table 14-4. The individuals receiving the RLW-30 ration had a substantial daily energy deficit of over 1,300 kcal, while the control group's energy intake was only several hundred kcal below their daily energy expenditure level. At the start and conclusion of the study, two standard tests of cognitive performance previously shown to be sensitive to the effects of nutritional parameters (simple visual reaction time and four-choice visual reaction time) were administered to the soldiers. In addition, on the same day that performance was assessed, blood samples were drawn and plasma amino acids determined. Plasma levels of both tryptophan and tyrosine were reduced substantially (Figures 14-7 and 14-8) over the course of the study among the soldiers consuming the RLW-30 ration. To ascertain whether changes in plasma levels of either tryptophan or tyrosine were related to behavioral function during this field study, changes in the ratio of tryptophan and tyrosine to the other

TABLE 14-4 Mean Daily Nutrient Intakes of the Standard Field Ration and Lightweight Ration Groups for 30 Days of a Field Study

Nutrient	Standard Field Ration Group (+SEM)	Lightweight Ration Group (+SEM)
Energy, kcal	2,782 ± 42	1,946 ± 15
Protein, g	112 ± 2	64 ± 1
Carbohydrate, g	318 ± 6	197 ± 2
Fat, g	119 ± 2	100 ± 1
Energy Expenditure (kcal)	3,250	3,275

SOURCE: Adapted from Askew et al., 1987.

LNAAs were computed and correlated with changes in performance. Plasma ratios are believed to be a better indicator of transport of amino acids across the BBB than absolute levels of an amino acid because of the previously noted competition of similar amino acids for a common carrier (Pardridge, 1977). There were significant correlations between both types of performance and the tryptophan to other LNAAs ratio but not the tyrosine ratio (Figure 14-9) (Lieberman et al., 1997). This indicates that under conditions of undernutrition, tryptophan may be the best indicator of changes in mental performance. Therefore, maintaining adequate tryptophan levels may be particularly important when the optimal amino acid content of field rations is under consideration. This is consistent with the data discussed above, indicating that decrements in plasma tryptophan induced by administration of a single tryptophan-deficient meal can substantially increase depression and aggression and alter arousal in normal volunteers (Cleare and Bond, 1995; Smith et al., 1987). Although a significant correlation between tyrosine levels and performance was not observed during this field study, the research was conducted in a relatively nonstressful environment, not under conditions where the influence of tyrosine on central catecholamines is likely to be important.

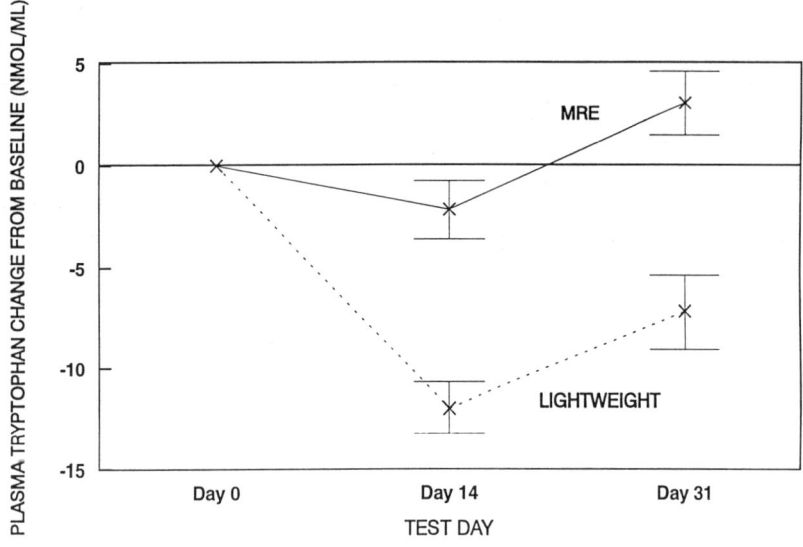

FIGURE 14-7 Plasma tryptophan levels in soldiers consuming either a lightweight ration or standard field rations (the MRE) over the course of a 31-day field study conducted in a temperate climate. SOURCE: Adapted from Lieberman et al., 1997.

COGNITIVE PERFORMANCE, STRESS, AND BRAIN FUNCTION 303

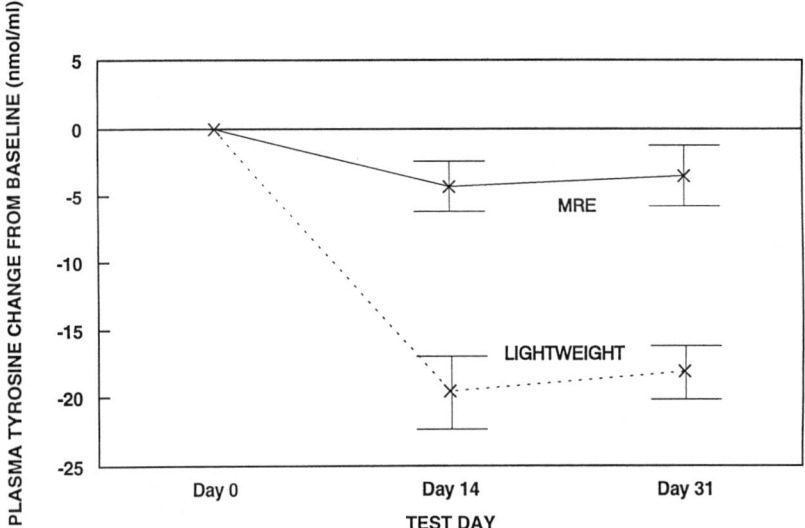

FIGURE 14-8 Plasma tyrosine levels in soldiers consuming either a lightweight ration or standard field rations (the MRE) over the course of a 31-day field study conducted in a temperate climate. SOURCE: Adapted from Lieberman et al., 1997.

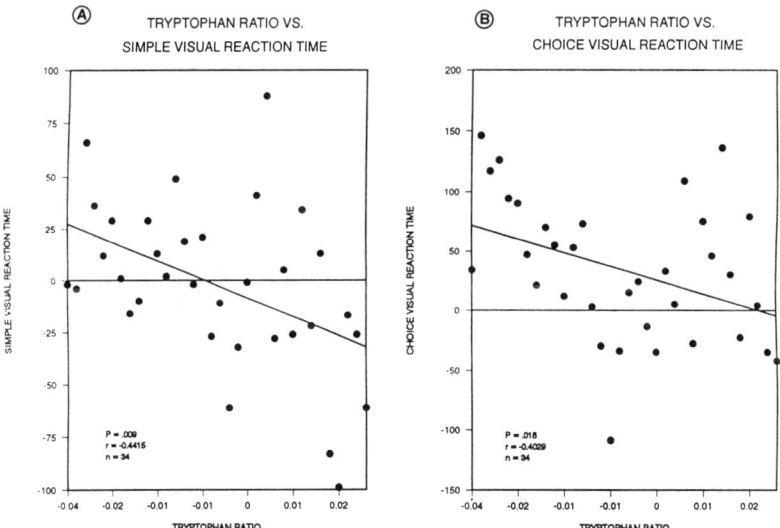

FIGURE 14-9 Relationship between changes in plasma: tryptophan ratio and two tests of cognitive performance in soldiers consuming either a lightweight ration or standard field rations (the MRE) over the course of a 31-day field study conducted in a temperate climate. Reductions in tryptophan ratio were significantly correlated with degraded performance on both simple (A) and choice (B) visual reaction time tasks. SOURCE: Adapted from Lieberman et al., 1997.

AUTHOR'S CONCLUSION AND RECOMMENDATIONS

Maintenance of appropriate plasma concentration of at least one amino acid, tryptophan, the precursor of serotonin, is essential for optimal brain function and cognitive performance. Substantial decreases or increases in the typical levels of tryptophan present in the plasma will substantially disrupt normal behavior and brain function. Reduced plasma tryptophan increases depression and aggression, while increases in this amino acid induce drowsiness and decrease pain sensitivity. The optimal range for plasma and brain tryptophan levels has not been established in humans or any other species, nor has the daily requirement for this amino acid been determined with respect to its effects on brain function.

Administration of tyrosine, precursor of the catecholamines including norepinephrine, has been shown to prevent some of the adverse neurochemical and behavioral effects of exposure to acute stress. Optimal plasma and brain levels of this amino acid may be less critical than that of tryptophan, except under stressful conditions. Of course, such conditions are of great relevance to the development of optimal military rations. The possible importance of other amino acids such as histidine, arginine, or threonine to the regulation of behavior is currently not known.

Given the importance of optimal cognitive function to soldiers and the documented relationship between several amino acids and brain function, studies to quantify CNS requirements for specific amino acids under conditions of metabolic, environmental, and psychological stress are required. Such studies could provide the basis for optimizing the amino acid content of field rations intended for use in extremely stressful combat conditions. Development of methods to evaluate CNS requirements for specific amino acids under normal and adverse circumstances is also necessary. Consideration should be given to conducting further animal research using techniques such as microdialysis to assess release of brain transmitters under various environmentally and nutritionally stressful conditions, including undernutrition, thermal stress, hypoxia, and psychological stress.

A recent consensus report by an international working group on protein and amino acid requirements concluded that "Amino acid requirements at all ages require further investigation. Such studies should include consideration of amino acid use for processes other than protein deposition" (Working Group, 1996). Given the importance of the neurotransmitter precursors for the CNS, it is recommended that some of these functional measures be behavioral. Specifically, functional outcome measures based on behavioral and other CNS end points should be considered as potentially critical measures of amino acid and protein requirements, particularly when the amino acid in question is known to affect brain function. When humans are exposed to stressors such as extreme environmental conditions, intense exercise, or psychological stress, the importance of brain requirements for amino acids may be relatively greater than under optimal physiological conditions.

REFERENCES

Ahlers, S.T., J.R. Thomas, J. Schrot, and D. Shurtleff. 1994. Tyrosine and glucose modulation of cognitive deficits. Pp. 301–320 in Food Components to Enhance Performance, B. Marriott, ed. Institute of Medicine. Washington, D.C.: National Academy Press.

Askew, E.W., I. Munro, M.A. Sharp, S. Siegel, R. Popper, M.S. Rose, R.W. Hoyt, K. Reynolds, H.R. Lieberman, D. Engell, and C.P. Shaw. 1987. Nutritional status and physical and mental performance of soldiers consuming the Ration, Lightweight or the Meal, Ready-to-Eat military field ration during a 30 day field training exercise (RLW-30). Technical Report No. T7-87. Natick, Mass.: U.S. Army Research Institute of Environmental Medicine.

Banderet, L.E., and H.R. Lieberman. 1989. Treatment with tyrosine, a neurotransmitter precursor, reduces environmental stress in humans. Brain Res. Bull. 22:759–762.

Betz, A.L., G.W. Goldstein, and R. Katzman. 1994. Blood-brain-cerebrospinal fluid barriers. Pp. 681–698 in Basic Neurochemistry: Molecular, Cellular, and Medical Aspects, 5th ed., G.J. Siegal, ed. New York: Raven Press

Cleare, A.J., and A.J. Bond. 1994. Effects of alterations in plasma tryptophan levels on aggressive feelings. Arch. Gen. Psychiatry 51(12):1004-1005.

Cleare, A.J., and A.J. Bond. 1995. The effect of tryptophan depletion and enhancement on subjective and behavioral aggression in normal male subjects. Psychopharmacology. 118(1):72–81.

Deijen, J.B., and J.F. Orlebeke. 1994. Effect of tyrosine on cognitive function and blood pressure under stress. Brain Res. Bull. 33:319–323.

Fernstrom, H.M., and J.D. Fernstrom. 1995a. Brain tryptophan concentrations and serotonin synthesis remain responsive to food consumption after the ingestion of sequential meals. Am. J. Clin. Nutr. 61:312–319.

Fernstrom, J.D., and R.J. Wurtman. 1971. Brain serotonin content: Physiological dependence on plasma tryptophan levels. Science 173:149–152.

Hajak, G., G. Huether, J. Blanke, M. Blömer, C. Freyer, B. Poeggler, A. Reimer, A. Rodenbeck, M. Schulz-Varszegli, and E. Rüther. 1991. The influence of intravenous l-tryptophan on plasma melatonin and sleep in men. Pharmaco- psychiatry. 24:17–20.

Hartmann, E.L. 1986. Effect of l-tryptophan and other amino acids on sleep. Nutr. Rev. May Suppl. 44:70-73.

Hartmann, E., and D. Greenwald. 1984. Tryptophan and human sleep: An analysis of 43 studies. Pp. 297–304 in Progress in Tryptophan and Serotonin Research, H.G. Schlossberger, W. Kochen, B. Linzen, and H. Steinhart, eds. Berlin: Walter de Gruyter.

Lehnert, H.R., D.K. Reinstein, B.W. Strowbridge, and R.J. Wurtman. 1984a. Neurochemical and behavioral consequences of acute, uncontrollable stress: effects of dietary tyrosine. Brain Res. 303:215–223.

Lehnert, H.R., D.K. Reinstein, and R.J. Wurtman. 1984b. Tyrosine reverses the depletion of brain norepinephrine and the behavioral deficits caused by tail-shock stress in rats. Pp. 81–91 in Stress: The Role of Catecholamines and Other Neurotransmitters. E. Usdin and R. Kvetnansky, eds. New York: Gordon and Beach.

Lieberman, H.R., 1994. Tyrosine and stress: Human and animal studies. Pp. 277–299 in Food Components to Enhance Performance, An Evaluation of Potential Performance—Enhancing Food Components for Operational Rations, B.M. Marriott, ed. Institute of Medicine. Washington, D.C.: National Academy Press.

Lieberman, H.R. and B. Shukitt-Hale. 1996. Food components and other treatments that may enhance performance at high altitude and in the cold. Pp. 453–465 in Nutritional Needs in Cold and in High Altitude Environments, B. Marriott and S. Newberry, eds. Institute of Medicine. Washington, D.C.: National Academy Press.

Lieberman, H.R., S. Corkin, B.J. Spring, J.H. Growdin, and R.J. Wurtman. 1983. Mood, performance, and pain sensitivity: Changes induced by food constituents. J. Psychiatr. Res. 17(2):135–145.

Lieberman, H.R., S. Corkin, B.J. Spring, R.J. Wurtman, and J.H. Growdon. 1985. The effects of dietary neurotransmitter precursors on human behavior. Am. J. Clin. Nutr. 42:366–370.

Lieberman, H.R., B. Caballero, and N. Finer. 1986a. The composition of lunch determines afternoon tryptophan ratios in humans. J. Neural Transmission. 65:211–217.

Lieberman, H.R., E.W. Askew, R.W. Hoyt, B. Shukitt-Hale, and M.A. Sharp. 1997. Effects of thirty days of undernutrition on plasma neurotransmitters, other amino acids and behavior. J Nutr. Biochem. 8:119-126.

Luo, S., E.T.S. Li, and H.R. Lieberman. 1993. Tyrosine increases hypothermia-induced norepinephrine (NE) release in rat hippocampus assessed by in vivo microdialysis. Soc. Neurosci. Abstr. 19:1452.

Owasoyo, J.O., D.F. Neri, and J.G. Lamberth. 1992. Tyrosine and its potential use as a countermeasure to performance decrement in military sustained operations. Aviat. Space Environ. Med. 63:364–369.

Pardridge, W.M. 1977. Regulation of amino acid availability to the brain. Pp. 141–190 in Nutrition and the Brain, vol. 1, R.J. Wurtman and J.J. Wurtman, eds., New York: Raven Press.

Rauch, T.M., and H.R. Lieberman. 1990. Pre-treatment with tyrosine reverses hypothermia induced behavioral depression. Brain Res. Bull. 24:147–150.

Salter, C. 1989. Dietary tyrosine as an aid to stress resistance among troops. Milit. Med. 154:144–146.

Seltzer, S., D. Dewart, R. L. Pollack, and E. Jackson. 1983. The effects of dietary tryptophan on chronic maxillo facial pain and experimental pain tolerance. J. Psychiat. Res. 17(2):181-186.

Shukitt-Hale, B., M.J. Stillman, and H.R. Lieberman. 1996. Tyrosine administration prevents hypoxia-induced decrements in learning and memory. Physiol. Behav. 59:867–871.

Shurtleff, D., J.R. Thomas, S.T. Ahlers, and J. Schrot. 1993. Tyrosine ameliorates a cold-induced delayed matching-to-sample performance decrement in rats. Psychopharmacology. 112:228–232.

Shurtleff, D., J.R. Thomas, J. Schrot, K. Kowalski, and R. Harford. 1994. Tyrosine reverses a cold-induced working memory deficit in humans. Pharmacol. Biochem. Behav. 47(4):935–941.

Smith, S.E., R.O. Pihl, S.N. Young, and F.R. Ervin. 1987. A test of possible cognitive and environmental influences on the mood lowering effect of tryptophan depletion in normal males. Psychopharmacology. 91:451–457.

Stone, E. A. 1975. Stress and catecholamines. Pp. 31–71 in Catecholamines and Behavior, A.J. Freidhoff, ed. New York: Plenum Press.

Ungerstedt, U., C. Forster, M. Herrera-Marschitz, I. Hoffman, U. Jungnelius, U. Tossman, and T. Zetterstrom. 1982. Brain dialysis: A new in vivo technique for studying neurotransmitter release and metabolism. Neuroscience Lett, 10(suppl):S493.

Working Group (G. Clugston, K.G. Dewey, C. Fjeld, J. Millward, P. Reeds, N.S. Scrimshaw, K. Tontisirin, J.C. Waterlow, and V.R. Young). 1996. Report of the working group on protein and amino acid requirements. Eur. J. Clin. Nutr. 50(suppl. 1):S193–S195.

Wurtman, R.J., F. Hefti, and E. Melamed. 1981. Precursor control of neurotransmitter synthesis. Pharmacol. Rev. 32:315–335.

Yeghiayan, S.K., C. Amendola, T.H. Wu, T.J. Maher, and H.R. Lieberman. In Press. Hyperthermia and tyrosine administration affect behavioral performance but not hippocampal catecholamines. Society for Neuroscience abstracts.

Young, S.N. 1996. Behavioral effects of dietary neurotransmitter precursors: Basic and clinical aspects. Neurosci. Behav. Rev. 20(2):313–323.

DISCUSSION

ROBERT NESHEIM: I will take one question if anybody has a quick one, and then I think we need to take a break. Yes?

GERALD COMBS: Since tryptophan is the least abundant amino acid in most proteins and would be the most constant, its relationship to total protein would be more nearly the same than any other amino acid. Do you think that might be part of the reason why it was the only one that was correlated with function?

HARRIS LIEBERMAN: It is hard to say, because tryptophan has other unique characteristics. I think the fact that it is precursor dependent with regard to variations in protein to carbohydrate ratio, could also be an important factor. But, yes, it is quite possible. You are right there.

Protein and Amino Acids, 1999
Pp. 309–329. Washington, D.C.
National Academy Press

15

Supplementation with Branched-Chain Amino Acids, Glutamine, and Protein Hydrolysates: Rationale for Effects on Metabolism and Performance

Anton J. M. Wagenmakers[1]

INTRODUCTION

Endurance athletes in periods of intense training with a high daily workload tend to believe that they need more protein and amino acids in the diet than sedentary subjects. According to Munro (1964b), this belief originates from the 1840s when the German physiologist Von Liebig hypothesized that muscle protein was the main fuel used to achieve muscular contraction. After Von Liebig's hypothesis had been invalidated around 1870 by the first experimental data on urinary nitrogen excretion during and following prolonged demanding exercise, many exercise physiologists took the opposite stand and for the next 100 years disregarded the amino acid pool in muscle as playing any role of significance in exercise and energy metabolism. Despite this 100 percent change in scientific opinion, the general public continued to believe that a high

[1] Anton J. M. Wagenmakers, Department of Human Biology, Nutrition Research Institute and Stable Isotope Research Centre, Maastricht University, 6200 MD Maastricht, The Netherlands.

workload leads to an increase in dietary protein needs. Industrial workers were and are served bigger steaks the more demanding their daily workload is. Milk is promoted as the "white motor" in advertisements. For these reasons, tons of protein are sold each year to strength athletes, as they believe they can only make their muscles grow when they ingest protein supplemental to their usual diet, and to endurance athletes, because they believe they break down muscle protein during training and competition.

The high-purity crystalline amino acids now available have created an easy new health food and supplementation market among the athletes who already believed in the beneficial effect of protein. Advertisements claim that free amino acids and small peptides are a better source of amino acids to support high net rates of protein deposition because they are more readily absorbed than whole dietary protein. The rapid absorption of amino acids leads to high plasma and tissue amino acid concentrations, In fact, it seems to drive the amino acids primarily into the oxidation pool, since the Michaelis constant (K_m) of the oxidative enzymes is much higher than the K_m of the enzymes involved in protein synthesis. By definition, increased oxidation means a reduced net rate of protein deposition, at least in the first hours after ingestion. Apart from serving as building blocks for protein synthesis, several amino acids have unique functions in human physiology, and they are marketed as supplements to support these functions. However, the scientific evidence to support the claim that they are needed in addition to the habitual ingestion of dietary proteins is usually lacking.

This chapter focuses on several aspects of protein and amino acid metabolism in endurance exercise and on the effect of amino acid and protein supplementation on performance-related aspects of metabolism. First presented will be the author's opinion on whether endurance exercise leads to net protein catabolism, increased amino acid oxidation, and therefore an increased protein requirement. A detailed overview is given elsewhere in this volume (see also Rennie, 1996). Subsequently, the rationale for supplementation with branched-chain amino acids (BCAAs) and glutamine and the effects on metabolism and performance will be evaluated. Finally, suggestions will be presented that the glycogen resynthesis rates in muscle following exercise can be accelerated by the combined ingestion of carbohydrate and protein (Zawadzki et al., 1992) and by glutamine infusion (Varnier et al., 1995).

PROTEIN AND AMINO ACID METABOLISM IN ENDURANCE EXERCISE

Recent stable isotope tracer studies have reopened the discussion on whether net protein breakdown (protein synthesis < protein degradation) and increased amino acid oxidation occur during prolonged endurance exercise at the whole body level. If the answer is yes, the next question is whether net protein breakdown occurs in muscle or in the gut and liver area. Stable isotope tracer methods, although valuable in other areas of nutrition and medicine, do not clarify the issue. Some amino acid tracers and some studies report a net protein breakdown during exercise at 40 to 70 percent maximal oxygen consumption (VO_{2max}) for 1 or 2 or maximally 4 hours, whereas others do not (for a review see Rennie, 1996).

When considering whether net protein break-down occurs during prolonged high workload exercise in the field, it is important to know whether or not carbohydrates are ingested and whether energy balance can be maintained around the clock. Many laboratory studies using tracers were conducted in subjects after overnight fasting without carbohydrate ingestion, a condition that favors net protein breakdown during exercise but bears little relevance for athletes. A triathlete or a Tour de France cyclist with an energy expenditure of over 30 MJ/(7143 kcalld) in a stage with five to six mountain passes at altitudes of over 2,000 m in the mountains (Saris, 1997) would never begin exercise without a proper breakfast and without carbohydrate ingestion before, during, and after competition (see glycogen resynthesis section of this chapter). Carbohydrate ingestion during exercise prevents activation of the branched-chain α-keto acid dehydrogenase in muscle (the rate-determining step in the oxidation of the branched-chain amino acids, Wagenmakers et al, 1991) and leucine oxidation in several tracer studies (see Rennie in this volume for references). Recently this laboratory conducted the most demanding exercise study ever performed in a laboratory setting. It was observed that 6 hours of exercise (2.5 h of cycling at 50% W_{max} followed by 1 h of running at 11 km/hr and then again by 2.5 h of cycling at 50% W_{max}; W_{max} = maximum power output) by elite cyclists while ingesting carbohydrate led to an imbalance between protein synthesis and degradation (synthesis < degradation) and an increase in amino acid oxidation when a L-[1-^{13}C]leucine tracer was used (Pannemans et al., 1997). In contrast to expectations, net protein degradation and leucine oxidation did not increase in duration, possibly because carbohydrate was ingested. Without carbohydrate ingestion, subjects could not have exercised for 6 hours. Other tracers, [$^{2}H_{5}$]phenylalanine and [$^{15}N_{2}$]urea, did not show an increase of amino acid oxidation and net protein breakdown during exercise. Therefore, it is unclear whether proteins are broken down in highly demanding endurance and ultra-endurance exercise under laboratory conditions and in the field. It is difficult to believe that during a stage in the Tour de France with a cyclist's energy expenditure of > 20 MJ within 8 h of cycling, that there will be no increase in protein breakdown and amino acid oxidation. During such events, the rate of carbohydrate oxidation far exceeds the maximal amount of carbohydrate ingestion, and these athletes, at least during the last hours of the race, are in a net negative energy balance. The combination of a negative energy balance and highly demanding muscle contractions leads to net protein breakdown in the muscle. Evidence of this has been obtained during one-leg knee-extensor exercise with normal and low muscle glycogen content (Van Hall et al., 1995a; Wagenmakers et al., 1996; Van Hall, 1996). In one-leg knee-extensor exercise, approximately 3 kg of skeletal muscle contracts. Maximal oxygen consumption and the metabolic rate of the active muscle was two- to three-fold higher under these conditions than for whole body dynamic exercise (Andersen and Saltin, 1985). The net muscle production rate of threonine, lysine, and tyrosine (amino acids that are not transaminated and metabolized in muscle) and of the sum of the non-metabolized amino acids was 9- to 20-fold higher during one-leg exercise than at rest, implying that the one-leg exercise led to a massive increase in the net muscle protein degradation rate (protein degradation minus protein synthesis). During exercise with the low-glycogen-content leg, the release of these amino acids was again 1.5- to 2.5-fold higher than during exercise with the normal-content-leg. This indicates that net muscle protein degradation increased further when the active muscle was glycogen depleted (Van Hall, 1996).

Another observation of a high release of non-metabolized amino acids from the leg muscles, reflecting net protein catabolism, occurred during two-legged cycling in patients with McArdle's disease (Wagenmakers et al., 1990). These patients have no or no active glycogen phosphorylase in muscle and, therefore, cannot use muscle glycogen as an energy source during exercise. During incremental exercise they reached a maximal workload of only 40 to 100 W. Due to the energy deficiency in their muscles these subjects rapidly broke down their muscle proteins, even at relatively low-intensity exercises. These data suggest that during demanding endurance or ultra-endurance exercise in the field, where muscle glycogen depletion cannot be prevented despite oral carbohydrate ingestion, net protein breakdown will occur in skeletal muscle once the glycogen stores have been emptied. Demanding endurance exercise may also lead to net protein breakdown in gut and liver, but direct evidence in humans has not been reported.

BRANCHED-CHAIN AMINO ACIDS

Central Fatigue Hypothesis

Prolonged exercise that leads to glycogen depletion inevitably causes fatigue. Fatigue is defined physiologically as the inability to maintain power output. To the endurance athlete, this means a gradually increasing sense of effort, sometimes even discomfort and pain, and the need to reduce the speed. Both peripheral (i.e., muscle [Fitts, 1994]) and central (brain) mechanisms have been suggested to play a role in the development of fatigue in endurance exercise. In 1987, Newsholme and colleagues launched the "central fatigue hypothesis" as an important mechanism contributing to the development of fatigue during prolonged moderate-intensity exercise. During exercise, free fatty acids (FFA) are mobilized from adipose tissue and via the blood transported to muscle to serve as fuel. As a consequence, the blood FFA concentration will increase. FFA and the amino acid tryptophan both bind to albumin and compete for the same binding sites. Tryptophan will be displaced from binding to albumin by the increasing FFA concentration, and therefore, the free tryptophan concentration in the blood will rise. Simultaneously, the increased oxidation of BCAA in muscle (Wagenmakers et al., 1991) leads to a decrease of the sum concentration of the BCAA in the blood, and the free tryptophan:BCAA ratio, therefore, will increase substantially. The increase in this ratio leads to increased tryptophan transport across the blood-brain barrier, since BCAA and tryptophan compete for carrier-mediated entry into the central nervous system by the large neutral amino acid (LNAA) transporter (Knott and Curzon, 1972; Chaouloff et al., 1985). Once taken up, conversion of tryptophan to serotonin (5-hydroxytryptamine) occurs and leads to a local increase of this neurotransmitter (Knott and Curzon, 1972; Chaouloff et al., 1985). This increase has been found in certain brain areas in the rat, but it has not been established in humans. According to the central fatigue hypothesis, the increase in serotoninergic activity subsequently leads to central fatigue, forcing athletes to stop exercise or reduce running or cycling speed. Neurobiologists have established that serotonin plays a role in the onset of sleep and is a determinant of mood and aggression. It is uncertain, however, whether it also plays a role in fatigue experienced during prolonged exercise as suggested in the central fatigue hypothesis. One of the implications of the central fatigue hypothesis is that ingestion of BCAA could reduce the exercise-induced increase in brain tryptophan uptake and thus delay fatigue and give athletes the ability to push on for a more prolonged period even when the peripheral fatigue mechanism has come into operation. BCAA, in other words, would be ergogenic. Another implication is that ingestion of tryptophan prior to exercise would reduce time to exhaustion. Tryptophan, in other words, would be ergolytic.

Effects of Ingestion of Branched-Chain Amino Acids and Tryptophan on Endurance Performance

The effect of BCAA ingestion on physical performance was investigated for the first time in a field test by Blomstrand et al., 1991. Male subjects (n = 193) were studied during a marathon in Stockholm. Subjects were randomly (without matching) divided into an experimental group that received 16 g of BCAA in plain water during the race and a placebo group that received flavored water. Subjects also had ad libitum access to carbohydrate (CHO)-containing drinks. No difference was observed in the marathon time of the two groups. However, when the original subject group was divided into fast and slower runners, a small significant reduction in marathon time was observed in the slower runners only. In retrospect, this first study has been the only one to claim a positive effect of BCAA ingestion during exercise. However, three criticisms can be raised against its design, and each of these points may have biased the data obtained: 1. In a performance test investigating a potentially ergogenic effect, subjects in the two groups should have been matched for previous performance; 2. CHO intake and nutritional status should have been controlled and matched in the two groups; and 3. division of subjects in a group of fast and slower runners, taking an arbitrary marathon time as selection criterion is not in accordance with accepted statistical methods.

Varnier et al. (1994) investigated six moderately trained subjects after glycogen-depleting exercise, followed by overnight fasting. Subjects were investigated the morning after the fast during graded incremental exercise to exhaustion and received an intravenous infusion of BCAA (260 mg/kg/ h for 70 min) or saline only. No significant differences were observed between the tests in total work performed.

Blomstrand and colleagues (1995) also investigated performance in the laboratory in five male endurance-trained subjects during exhaustive exercise on a cycle ergometer at a work rate corresponding to 75 percent of VO_{2max} after reduction of their muscle glycogen stores. Subjects were given, during exercise and in random order, a 6 percent CHO solution containing 7 g/liter of BCAA, a 6 percent CHO solution, and flavored water. The positive effect of the field test was not confirmed in this controlled laboratory study, because no difference in performance was seen when the subjects were given CHO + BCAA or only CHO.

Madsen and colleagues (1996) recently investigated performance in nine trained cyclists in a 100-km timed trial in the laboratory. Subjects used their own bikes at a freely chosen power output, simulating field conditions, and were studied while ingesting flavored water only (placebo), a 5 percent CHO solution (66 gram/h), and CHO (66 gram/h) plus BCAA (6.8 gram/h). There was no difference between treatments in the time needed to finish the 100-km trial.

Based on the two implications of the central fatigue hypothesis, (BCAA ingestion improves performance and tryptophan ingestion reduces time to exhaustion), Van Hall and colleagues (1995b) designed an experiment in which both aspects of the central fatigue hypothesis were investigated. Ten endurance-trained male athletes were studied during cycle exercise at 70 to 75 percent of W_{max}, while ingesting, in random order and double blinded, drinks that contained 6 percent sucrose (control) or 6 percent sucrose supplemented with (1) tryptophan (3 g/liter), (2) a low dose of BCAA (6g/liter, comparable to doses used by Blomstrand et al., [1991]), and (3) a high dose of BCAA (18 g/liter). These treatments greatly increased the plasma concentration of the respective amino acids to values well outside the normal physiological range. By measuring the concentration of all amino acids competing for transport by the LNAA carrier, Van Hall and colleagues (1995b) were able to calculate the rate of unidirectional influx of circulating plasma tryptophan into the brain using kinetic parameters of transport of human brain capillaries reported by Hargreaves and Pardridge (1988). These calculations showed that the administered BCAAs only reduced tryptophan transport at exhaustion by 8 to 12 percent, while tryptophan ingestion caused an increase of 600 to 1,900 percent (depending on the use of free or total tryptophan concentration in the calculations). Despite these massive differences in tryptophan transport, time to exhaustion was not different between the four treatments. Van Hall and colleagues, therefore, concluded that manipulation of tryptophan supply to the brain by ingestion of BCAA and tryptophan in CHO containing drinks either did not change the serotonin concentration in relevant local areas in the brain or a change in serotoninergic activity during prolonged exercise contributed little to mechanisms of fatigue. Two earlier studies (Segura and Ventura, 1988; Stensrud et al., 1992) investigated the effect of lower doses of tryptophan. Segura and Ventura (1988) reported that 1.2 g of L-tryptophan supplementation taken in 300-mg doses over a 24-hour period before exercise increased total exercise time by 49 percent in 12 subjects who were running at 80 percent of VO_{2max} (which fully contradicts to the central fatigue hypothesis). The results of this study were questioned by Stensrud et al., (1992), who studied 49 well-trained males in a randomized, double-blind, placebo experiment. Subjects in the tryptophan group (n = 24) and placebo group (n = 25) were matched for performance (maximal oxygen uptake, anaerobic threshold, and speed during an all-out run). Tryptophan ingestion (again 1.2 g over a 24-h period prior to the run) had no effect on running performance, when subjects ran until exhaustion at a speed corresponding to 100 percent of their VO_{2max}.

Summary of Performance Studies with BCAA and Tryptophan

Neither BCAA nor tryptophan ingestion has an effect on endurance performance in healthy subjects. Thus, performance studies have not provided experimental support for the central fatigue hypothesis as playing a dominant role in fatigue mechanisms during prolonged exercise.

Interaction of the BCAA-Aminotransferase Reaction with the Tricarboxylic Acid (TCA) Cycle in Muscle

The carbon flux in the tricarboxylic acid (TCA) cycle is one of the processes which determines the rate of ATP turnover in skeletal muscle. One possible way to achieve an increase in TCA cycle activity when going from rest to exercise (to meet the increased energy demand of exercise) is to increase the concentration of the TCA cycle intermediates in muscle such that more substrate is available for the individual enzymatic reactions. This increase in concentration has been observed for the most abundant TCA cycle intermediates during early exercise (Essen and Kaijser, 1978; Sahlin et al., 1990). It is achieved by rapid conversion of the muscle glutamate pool into α-ketoglutarate (Van Hall et al., 1995a). The reaction used to achieve that increase is the alanine aminotransferase reaction: glutamate + pyruvate $\leftrightarrow$ α-ketoglutarate + alanine (Van Hall et al., 1995a). The alanine aminotransferase reaction is a near-equilibrium reaction, which implies that the increase in muscle pyruvate concentration that occurs at the start of exercise due to an acceleration of glycolysis will automatically lead to production of alanine and α-ketoglutarate and consumption of glutamate. After observing the early increase in concentration of TCA cycle intermediates, Sahlin et al., (1990) observed a subsequent gradual decrease in the intermediate in human subjects exercising until exhaustion at 75 percent VO_{2max}. This author and colleagues (Wagenmakers et al., 1990, 1991) have hypothesized that increased oxidation of the BCAAs plays an important role in that subsequent decrease of the concentration of the TCA cycle intermediates. Branched-chain α-keto acid dehydrogenase (BCKAD) is increasingly activated during prolonged exercise, leading to glycogen depletion (Wagenmakers et al., 1990, 1991; Van Hall et al., 1996), and an increase in oxidation by definition will increase the flux through the BCAA aminotransferase step. In the case of leucine, this reaction will put a net carbon drain on the TCA cycle as the carbon skeleton of leucine is oxidized to 3 acetyl-coenyzme A (CoA) molecules, and the aminotransferase step removes α-ketoglutarate: leucine + α-ketoglutarate ↔ 3 acetyl-CoA + glutamate. Increased oxidation of valine and isoleucine will not lead to net removal of TCA cycle intermediates, as the carbon skeleton of valine is oxidized to succinyl-CoA and that of isoleucine to both succinyl-CoA and acetyl-CoA. Net removal of α-ketoglutarate via leucine transamination can be compensated for by the alanine aminotransferase reaction (see above) as long as muscle glycogen is available and the muscle pyruvate concentration is kept high. However, because activation of the BCKAD complex is highest in glycogen-depleted muscle, this mechanism eventually will lead to a decrease in the concentration of TCA cycle intermediates; a reduction in TCA cycle activity; and a reduction of the ATP turnover rates, which via cellular mediators, will offset the known muscle fatigue mechanisms (Fitts, 1994).

BCAAs after oral ingestion are rapidly taken up in the blood and extracted by the leg muscles, and this is accompanied by activation of the BCKAD complex at rest and increased activation during exercise (Van Hall et al., 1996). This could imply that the indicated carbon drain on the TCA cycle is greater after BCAA ingestion and that by this mechanism, BCAA ingestion leads to premature fatigue during prolonged exercise, leading to glycogen depletion.

Wagenmakers et al., (1990) investigated the effect of BCAA ingestion (20 g) 30 minutes prior to graded incremental exercise in two patients with McArdle's disease (muscle glycogen phosphorylase deficiency). Due to their disease, these patients can be regarded as an experiment of nature. From the metabolism in their muscle, one learns what happens in a muscle without glycogen. BCAA ingestion reduced the W_{max} that was reached during the incremental exercise test by about 20 percent, and heart rate and perceived exertion were higher at the same workload. Ingestion of the keto acids of the BCAA (without the amino group) improved performance in the patients. These observations might suggest that, by the proposed mechanism, BCAA supplementation has a negative effect on performance in healthy subjects in conditions where the glycogen stores have been completely emptied by highly demanding endurance exercise. However, with co-ingestion of carbohydrate, BCAA ingestion did not change the time to exhaustion in healthy subjects (Van Hall et al., 1995b; Blomstrand et al., 1995; Madsen et al., 1996).

Effect of BCAA Ingestion on Plasma Ammonia Concentrations and Muscle Ammonia Production During Exercise

BCAA ingestion increased the production of ammonia by muscle and plasma ammonia concentration during exercise both in patients with McArdle's disease (Wagenmakers et al., 1990) and in healthy controls (Van Hall et al., 1995a; MacLean et al., 1996; Madsen et al., 1996). Because it has been suggested that ammonia leads to central fatigue and loss of motor coordination (Banister and Cameron, 1990), great care is indicated with the use of BCAA supplements in sport (e.g., tennis and soccer) and military activities where performance depends on motor coordination.

GLUTAMINE

Glutamine is the main end product of muscle amino acid metabolism both in the overnight fasted state and during feeding (Marliss et al., 1971; Chang and Goldberg 1978a; Wagenmakers et al., 1985; Elia et al., 1989; Wagenmakers and Soeters, 1995; Nurjhan et al., 1995). Alanine most likely only serves to export part of the amino groups (Chang and Goldberg 1978b; Wagenmakers et al., 1985). Glutamine is the most abundant amino acid in human plasma (600–700 µmol) and in the muscle free amino acid pool (20 mmol; 60% of the intramuscular pool excluding the nonprotein amino acid taurine). The synthesis rate of glutamine in muscle is higher than that of any other amino acid. Extrapolations of limb production rates in the fed and fasted state support that between 10 and 25 g of glutamine are synthesized in the combined human skeletal muscles per day (Marliss et al., 1971; Elia et al., 1989). Tracer dilution studies even indicate that 80 g of glutamine are produced per day (Darmaun et al., 1986), but this may be a methodological overestimation due to slow mixing of the glutamine tracer with the large endogenous glutamine pool in muscle (Van Acker et al., 1996). Furthermore, although muscle is the main glutamine-producing tissue, other tissues (e.g., adipose tissue, liver, and brain) may also contribute to the rate of appearance of glutamine in the plasma pool that is measured by tracer dilution techniques.

The reason for this high rate of glutamine production in muscle is probably that glutamine plays an important role in human metabolism in other organs. Sir Hans Krebs (1975) wrote: "Maybe the significance of glutamine synthesis is to be sought in the role of glutamine in other organs, as a precursor of urinary ammonia and as a participant in the biosynthesis of purines, NAD^+, amino sugars and proteins. Glutamine is an important blood constituent, present in higher concentrations than any other amino acid, presumably to serve these various functions. Muscle may play a role in maintaining the high plasma concentration of glutamine." Glutamine has been shown to be an important fuel for cells of the immune system (Ardawi and Newsholme, 1983) and for mucosal cells of the intestine (Windmueller and Spaeth, 1976; Souba, 1991). Low muscle and plasma glutamine concentrations are observed in patients with sepsis and trauma (Vinnars et al., 1975; Rennie et al., 1986; Lacey and Wilmore, 1990), conditions that also are attended by mucosal atrophy, loss of the gut barrier function (bacterial translocation), and a weakened immune response. Although the link between the reduced glutamine concentrations and these functional losses has not been fully underpinned by experimental evidence, the possibility should seriously be considered that it is a causal relationship. Due to its numerous metabolic key functions and a potential shortage in patients with sepsis and trauma, glutamine has recently been proposed to be a conditionally essential amino acid (Lacey and Wilmore, 1990), that should be added to the nutrition of long-term hospitalized, critically-ill and depleted patients. These patients have a reduced muscle mass due to continuous muscle wasting and as muscle is the dominant site of glutamine production, they probably also have a reduced capacity for glutamine production.

Plasma Glutamine Concentrations Following Prolonged Exercise

During exercise, plasma glutamine concentrations may increase, decrease, or remain unchanged depending on the type, duration, and intensity of exercise. However, lowered plasma glutamine concentrations were observed for a period of several hours following demanding endurance exercise leading to glycogen depletion (e.g., Van Hall, 1996). The lowest concentration is seen some 2 hours after exercise. It takes more than 7 hours before the concentration has returned to the pre-exercise resting level.

Plasma Glutamine Concentrations in Overtrained Athletes and in Tour de France Cyclists

Parry-Billings et al., (1992) observed lower plasma glutamine concentrations (503 ± 12 µmol) in 40 international standard athletes diagnosed as overtrained than in 36 matched control athletes (550 ± 14 µmol, mean ± SD). During continued intense competition without symptoms of overtraining, decreases in plasma glutamine have also been observed. Saris (1997) measured plasma glutamine in Tour de France cyclists. During the tour, a cyclist pedals about 4,000 km (2,500 miles) over a period of 22 days with only one rest day during the race. The mean energy expenditure per day is about 25 MJ and peak expenditures of 38 MJ/day have been observed. Plasma glutamine concentration fell gradually from 568 ± 56 µM (m ± SEM, n = 8) at the start to 479 ± 65 µM on day 22 (Figure 15-1).

Link to Immune System

It has been suggested in the literature that there is a high incidence of upper respiratory tract infections in athletes after strenuous prolonged exercise and that overtrained athletes have an even more severe impairment of the immune system, which makes them more prone to infections (Brenner et al., 1994; Hoffman-Goetz and Pedersen, 1994). Parry-Billings et al., (1992) suggested that the decrease in plasma glutamine following endurance exercise created an "open time window" in which athletes are more susceptible to catching a common cold or infection. It was also concluded that the lower plasma glutamine conentration in overtrained athletes contributed to the "weakened immune system" in this condition, given the proposed importance of glutamine for cells of the immune system (Ardawi and Newsholme, 1983; Parry-Billings et al., 1990). However, the link between glutamine metabolism and the immune system is not as strong as in critically ill or burn patients who tend to have much lower plasma glutamine concentrations and also low muscle glutamine stores. These patients have a reduced muscle mass and, therefore, may have a lower endogenous capacity to synthesize glutamine. They often completely lack the delayed-type hypersensitivity response, whereas this aspect of the immune function has not yet been reported to be abnormal in overtrained or heavily training athletes.

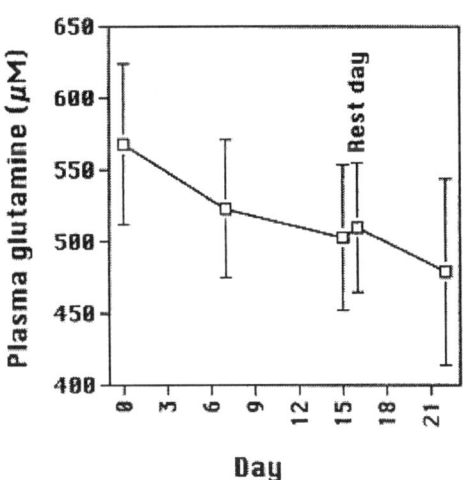

FIGURE 15-1 Plasma glutamine concentrations were measured (by HPLC) in 8 Tour de France cyclists on day 0, 7, 15 (resting day), 16 and 22. Values are given as means ± SEM.
SOURCE: Data from Saris (1997).

Castell et al., (1996) gave glutamine supplements or a placebo immediately after exercise and 2 hours later to athletes participating in a marathon or ultramarathon and reported lower infection rates in the glutamine group in the week following the race. However, infection levels were monitored by questionnaires. All symptoms reported by subjects under the heading of cold, cough, sore throat, or influenza occurring within 1 week after the race/training session were taken to reflect an infection. More valid, verifiable, and objective evidence with good functional measurements of the immune system will be needed before it can be concluded that: (1) the reported decreases in plasma glutamine concentration are linked to immune system impairments in overtrained athletes and in the period following demanding exercise, and (2) glutamine supplementation reduces infections. Interestingly, an increase of the protein content of the diet to 2.2 g/kg/day in healthy subjects also decreased the plasma glutamine concentration to 411 ± 28 µmol (mean $\pm$ SEM; n = 5; Matthews and Campbell, 1992). However, there is no reason to believe that this decrease, which is larger than the decreases mentioned above, impairs the immune system in these individuals. If anything, a high protein intake strengthens the immune system.

GLYCOGEN RESYNTHESIS FOLLOWING EXERCISE

During prolonged exercise at intensities between 50 and 85 percent VO_{2max}, muscle glycogen from a quantitative point of view is by far the most important energy source (Romijn et al., 1993). In elite male endurance athletes, between 600 and 700 g of glycogen is present in the combined skeletal muscles. Most of this glycogen is lost during prolonged moderate-to-high intensity exercise. Resynthesis of muscle glycogen after exercise is an important factor determining the time needed by such athletes to recover. This is especially important for individuals (male and female) performing demanding exercise on a daily basis (e.g., professional cyclists, marathon skaters, mountaineers). If a Tour de France cyclist fails to replenish the muscle glycogen stores fully before the next stage begins on the following day, it is likely he can not finish in the group (peleton).

To obtain maximal glycogen synthesis rates in the first 4 hours following exercise, adequate amounts of CHO should be ingested (Bergström and Hultman, 1967; Ivy et al., 1988, Reed et al., 1989). Maximal synthesis rates were obtained in healthy subjects who cycled regularly when 1.5 g/kg body weight of glucose was ingested as a glucose/glucose polymer drink immediately after glycogen-lowering exercise and 2 hours later (Ivy et al., 1988; Table 15-1). Doubling of the rate of glucose ingestion did not influence the glycogen resynthesis rate. Similar rates of glycogen resynthesis also were obtained when 3 g/kg body weight of glucose was taken orally as a drink immediately after and 2 hours after exercise, in solid form at the same points in time, and by continuous intravenous infusion (Reed et al., 1989).

TABLE 15-1 Muscle Glycogen Resynthesis Rates Following Exercise

Mean CHO Ingestion Rate g/kg/h	Glycogen Synthesis Rate µmol/g Wet Weight/h
0.00*	0.5
0.75*	5.2 ± 0.9
1.50*	5.8 ± 0.7
1.50†	6.1
1.50 (solid)†	6.3 ± 1.6
1.50 (continuous infusion)†	7.0 ± 0.9
0.8‡	5.1
0.8 + protein 0.3 g/kg/h‡	7.0 (+ 37%; $P < 0.05$)
Only protein 0.3 g/kg/h‡	1.5
0.00 + glutamine infusion (50 mg/kg/h)§	1.4 ± 0.3 ($P < 0.05$)
0.00 + alanine + glycine infusion§	0.4 ± 0.2
0.00 + saline infusion§	0.5 ± 0.2
0.8#	6.5 ± 1.2
0.8 + glutamine 0.3 g/kg/h#	5.4 ± 1.6
0.8 + wheat hydrolysate 0.3 g/kg/h#	8.9 ± 1.6 (+ 36%; $P = 0.061$)
0.8 + whey hydrolysate 0.3 g/kg/h#	7.7 ± 0.9 (+ 18%; $P = 0.068$)

NOTE: CHO, amino acids and protein are taken as drinks unless indicated differently.
* Ivy et al., J. Appl. Physiol. 65:2018–2023, 1988.
† Reed et al., J. Appl. Physiol. 66:720–726, 1989.
‡ Zawadzki et al., J. Appl. Physiol. 72:1854–1859, 1992.
§ Varnier et al., Am. J. Physiol. 269:E309–E315, 1995.
# Van Hall, thesis, Maastricht University, 1996.

Two interventions were recently reported that appeared to help increase glycogen resynthesis rates following exercise. Zawadzki et al. (1992) reported that addition of protein to the ingested CHO increased glycogen resynthesis rates (Table 15-1). Because higher plasma insulin concentrations were observed, it was suggested that the insulin effect on glycogen synthase was responsible for the observed effect. Varnier et al., (1995) reported that intravenous infusion of glutamine increased glycogen resynthesis rates (Table 15-2). However, glycogen resynthesis was measured at submaximal rates in the absence of ingested glucose. It was suggested that glutamine may directly stimulate glycogen synthase by activation of glycogen synthase phosphatase. The latter observation raised the possibility that the effect observed by Zawadzki was mediated in part by glutamine, since plasma glutamine concentration increases following protein ingestion.

In another study, Van Hall (1996) and Wagenmakers et al., (1997) investigated whether the stimulatory effect of protein on glycogen resynthesis rates observed after CHO ingestion was mediated by insulin, glutamine, or both and whether co-ingestion of glutamine and glucose also stimulated glycogen resynthesis rates. Eight trained subjects were studied during 3 hours of recovery in four tests each consuming one of four drinks in random order. Drinks were divided into and ingested in three 500-ml portions, the first immediately after exercise and the others after 1 and 2 hours of recovery. Composition of the drinks is specified in Table 15-1. The wheat hydrolysate had a glutamine content of 26 percent and the whey hydrolysate had 6.6 percent. The plasma glutamine concentration fell from about 650 µmol at the end of exercise to 500 µmol after 2 hours in the control treatment. Both protein hydrolysates prevented that decrease (plasma concentration maintained between 600 and 700 µmol). The glutamine increased plasma glutamine to values between 1,000 and 1,200 µmol. Plasma insulin concentrations were twice as high with the protein hydrolysates as with glucose only and glucose plus glutamine. Glutamine alone did not increase glycogen resynthesis rates, while both protein hydrolysates did, though not significantly.

AUTHOR'S CONCLUSIONS AND RECOMMENDATIONS

- No conclusive evidence supports net protein breakdown and increased amino acid oxidation during demanding endurance exercise in the laboratory and under field conditions. When energy balance is maintained, 100 g of protein should be more than enough to cover the protein requirements of subjects regularly involved in high workload exercise.
- No evidence suggests that BCAA ingestion would optimize athletic performance during demanding endurance exercise. A potential risk exists that BCAA ingestion could lead to premature fatigue or loss of motor coordination under conditions where glycogen stores have been emptied (e.g., high workload with insufficient food intake).
- Glutamine may be useful to support the immune system during high workload conditions, but solid evidence of the link between glutamine and the immune system and of the usefulness of glutamine supplementation has not yet been published.
- Glycogen resynthesis immediately following demanding exercise can be accelerated by the addition of protein to the ingested CHO solutions. This is particularly relevant when subjects maintain a high daily energy expenditure (> 20–25MJ) for several days or weeks (e.g., Tour de France cyclists and those engaged in military operations). Failure to replenish the glycogen stores overnight will impede performance on the following day.

REFERENCES

Andersen, P., and B. Saltin. 1985. Maximal perfusion of skeletal muscle in man. J. Physiol. 366:233–249.
Ardawi, M.S.M., and E.A. Newsholme. 1983. Glutamine metabolism in lymphocytes of the rat. Biochem. J. 212:835–842.
Banister, E.W., and B.J.C. Cameron. 1990. Exercise-induced hyperammonemia: Peripheral and central effects. Int. J. Sports Med. 11:S129–S142.
Bergström, J., and E. Hültman. 1967. Muscle glycogen synthesis after exercise: An enhancing factor localized to the muscle cells in man. Nature 201:309–310.
Blomstrand, E., S. Andersson, P. Hassmén, B. Ekblom, and E.A. Newsholme. 1995. Effect of branched-chain amino acid and carbohydrate supplementation on the exercise induced change in plasma and muscle concentration of amino acids in human subjects. Acta Physiol. Scand. 153:87–96.
Blomstrand, E., P. Hassmén, B. Ekblom, and E.A. Newsholme. 1991. Administration of branched-chain amino acids during sustained exercise-effects on performance and on plasma concentration of some amino acids. Eur. J. Appl. Physiol. 63:83–88.
Brenner, I.K.M., P.N. Shek, and R.J. Shepherd. 1994. Infection in athletes. Sports Med 17:86–107.
Castell, L.M., J.R. Poortmans, and E.A. Newsholme. 1996. Does glutamine have a role in reducing infections in athletes? Eur. J. Appl. Physiol. 73:488–490.
Chaouloff, F., J.L. Elghozi, Y. Guezennec, and D. Laude. 1985. Effect of conditioned running on plasma, liver and brain tryptophan and on brain 5-hydroxytryptamine metabolism of the rat. Brit. J. Pharmacol. 86:33–41.
Chang, T.W., and A.L. Goldberg. 1978a. The origin of alanine produced in skeletal muscle. J. Biol. Chem. 253:3677–3684.
Chang, T.W., and A.L. Goldberg. 1978b. The metabolic fates of amino acids and the formation of glutamine in skeletal muscle. J. Biol. Chem. 253:3685–3695.
Darmaun, D., D. Matthews, and D. Bier. 1986. Glutamine and glutamate kinetics in humans. Am. J. Physiol. 251:E117–E126.
Elia, M., A. Schlatmann, A. Goren, and S. Austin. 1989. Amino acid metabolism in muscle and in the whole body of man before and after ingestion of a single mixed meal. Am. J. Clin. Nut. 49:1203–1210.
Essen, B., and L. Kaijser. 1978. Regulation of glycolysis in intermittent exercise in man. J. Physiol. 281:499–511.
Fitts, R.H. 1994. Cellular mechanisms of muscle fatigue. Physiol. Rev. 74:49–94.
Hargreaves, K.M., and W.M. Pardridge. 1988. Neutral amino acid transport at the human blood-brain barrier. J Biol. Chem. 263:19392–19397.
Hoffman-Goetz, L., and B.K. Pedersen. 1994. Exercise and the immune system. Immunol. Today 15:382–385.
Ivy, J.L., M.C. Lee, J.T. Brozinick, and M.J. Reed. 1988. Muscle glycogen storage after different amounts of carbohydrate ingestion. J. Appl. Physiol. 65:2018–2023.
Knott, P.J., and G. Curzon. 1972. Free tryptophan in plasma and brain tryptophan metabolism. Nature 239:452–453.
Krebs, H.A. 1975. The role of chemical equilibria in organ function. Adv. Enzyme Regulation 15:449–472.
Lacey, J.M., and D.W. Wilmore. 1990. Is glutamine a conditionally essential amino acid? Nutr. Rev. 48:297–309.
MacLean, D.A., T.E. Graham, and B. Saltin. 1996. Stimulation of muscle ammonia production during exercise following branched-chain amino acid supplementation in humans. J. Physiol. 493:909–922.

Madsen, K., D.A. MacLean, B. Kiens, and D. Christensen. 1996. Effects of glucose, glucose plus branched-chain amino acids or placebo on bike performance over 100 km. J. Appl. Physiol. 81:2644–2650.
Marliss, E.B., T.T. Aoki, T. Pozefsky, A.S. Most, and G.F. Cahill. 1971. Muscle and splanchnic glutamine and glutamate metabolism in postabsorptive and starved man. J. Clin. Invest. 50:814–817.
Matthews, D.E., and R.G. Campbell. 1992. The effect of dietary protein intake on glutamine and glutamate nitrogen metabolism in humans. Am. J. Clin. Nutr. 55:963–970.
Munro, H.N. 1964b. Historical introduction: The origin and growth of our present concept of protein metabolism. Pp. 1-29 in Mammalian Protein Metabolism, H.N. Munro and J.B. Allison, ed. New York: Academic Press.
Newsholme, E.A., I.N. Acworth, and E. Blomstrand. 1987. Amino acids, brain neurotransmitters and a functional link between muscle and brain that is important in sustained exercise. Pp. 127–138 in Advances in Myochemistry, G. Benzi, ed. London: John Libby Eurotext.
Nurjhan, N., A. Bucci, G. Perriello, N. Stumvoll, G. Dailey, D.M. Bier, I. Toft, T.G. Jenssen, and J.E. Gerich. 1995. Glutamine: A major gluconeogenic precursor and vehicle for interorgan carbon transport in man. J. Clin. Invest. 95:272–277.
Pannemans, D.L.E., A.J.M. Wagenmakers, A.E. Jeukendrup, A.P. Gijsen, J.M.G. Senden, and W.H.M. Saris. 1997. The effect of prolonged moderate intensity exercise on protein metabolism of trained men. Med. Sci. Sport Ex. 29: S224.
Parry-Billings M., R. Budgett, Y. Koutedakis, E. Blomstrand, S. Brooks, C. Williams, P.C. Calder, S. Pilling, R. Baigrie, and E.A. Newsholme. 1992. Plasma amino acid concentrations in the overtraining syndrome: possible effects on the immune system. Med. Sci. Sports. Exerc. 24:1353–1358.
Parry-Billings M., J. Evans, P.C. Calder, and E.A. Newsholme. 1990. Does glutamine contribute to immunosuppression after major burns? Lancet 336:523–525.
Reed, M.J., J.T. Brozinick, M.C. Lee, and J.L. Ivy. 1989. Muscle glycogen storage postexercise: effect of mode of carbohydrate administration. J. Appl. Physiol. 66:720–726.
Rennie, M.J. 1996. Influence of exercise on protein and amino acid metabolism. Pp. 995–1035 in Handbook of Physiology, Section 12, Exercise: Regulation and Integration of Multiple Systems, L.B. Rowell and J.T. Shepherd, eds. New York: Oxford University Press.
Rennie, M.J., P. Babij, P.M. Taylor, H.S. Hundal, P. MacLennan, P.W. Watt, M.M. Jepson, and D.J. Millward. 1986. Characteristics of a glutamine carrier in skeletal muscle have important consequences for nitrogen loss in injury, infection and chronic disease. Lancet 2:1008–1012.
Romijn, J.A., E.F. Coyle, L. Sidossis, A. Castaldelli, J.F. Horowitz, E. Endert, and R.R. Wolfe. 1993. Regulation of endogenous fat and carbohydrate metabolism in relation to exercise intensity. Am. J. Physiol. 265:E380–E391.
Sahlin, K., A. Katz, and S. Broberg. 1990. Tricarboxylic acid cycle intermediates in human muscle during prolonged exercise. Am. J. Physiol. 269:C834–C841.
Saris, W.H.M. 1997. Limits of human endurance: Lessons from the Tour de France. Pp. 451–462 in Physiology, Stress, and Malnutrition: Functional Correlates, Nutritional Intervention, J.M. Kinney and H.N. Tucker, eds. New York: Lippincott-Raven Publishers.
Segura, R., and J.L. Ventura. 1988. Effect of L-tryptophan supplementation on exercise performance. Int. J. Sports Med. 9:301-305.
Souba, W.W. 1991. Glutamine: a key substrate for the splanchnic bed. Ann. Rev. Nut. 11:285–308.

Stensrud, T., F. Ingjer, H. Holm, S.B. Strømme. 1992. L-tryptophan supplementation does not improve running performance. Int. J. Sports Med. 13:481-485.

Van Acker, B.A.C, K.W.E. Hulsewé, A.J.M. Wagenmakers, N.E.P. Deutz, B.K. Van Kreel, P.B. Soeters, and M.F. Von Meijenfeldt. 1996. Measurement of glutamine metabolism in gastrointestinal cancer patients. Clin. Nutr. 15(suppl. 1):1.

Van Hall, G. 1996. Amino acids, ammonia and exercise in man. Thesis, Maastricht University, The Netherlands.

Van Hall, G., D.A. MacLean, B. Saltin, and A.J.M. Wagenmakers. 1996. Mechanisms of activation of muscle branched-chain α-keto acid dehydrogenase during exercise in man. J. Physiol. 494:899–905.

Van Hall, G., J.S.H. Raaymakers, W.H.M. Saris, and A.J.M. Wagenmakers. 1995a. Ingestion of branched-chain amino acids and tryptophan during sustained exercise: Failure to affect performance. J. Physiol. 486:789–794.

Van Hall, G., B. Saltin, G.J. Van der Vusse, K. Söderlund, and A.J.M. Wagenmakers. 1995b. Deamination of amino acids as a source for ammonia production in human skeletal muscle during prolonged exercise. J. Physiol. 489:251–261.

Varnier, M., G.P. Leese, J. Thompson, and M.J. Rennie. 1995. Stimulatory effect of glutamine on glycogen accumulation in human skeletal muscle. Am. J. Physiol. 269:E309–E315.

Varnier, M., P. Sarto, D. Martines, L. Lora, F. Carmignoto, G.P. Leese, and R. Naccarato. 1994. Effect of infusing branched-chain amino acids during incremental exercise with reduced . muscle glycogen content. Eur. J. Appl. Physiol. 69:26–31.

Vinnars, E., J. Bergström, and P. Fürst. 1975. Influence of the postoperative state on the intracellular free amino acids in human muscle tissue. Annals Surg. 182:665–671.

Wagenmakers, A.J.M. 1996. Muscle glycogen depletion and fatigue. The Physiologist 39:A68.

Wagenmakers, A.J.M., and P.B. Soeters. 1995. Metabolism of branched-chain amino acids. Pp. 67–83 in Amino Acid Metabolism and Therapy in Health and Nutritional Disease, L.A. Cynober, ed. New York, CRC Press, Inc.

Wagenmakers, A.J.M., E.J. Beckers, F. Brouns, H. Kuipers, P.B. Soeters, G.J. van der Vusse, and W.H.M. Saris. 1991. Carbohydrate supplementation, glycogen depletion, and amino acid metabolism during exercise. Am. J. Physiol. 260:E883–E890.

Wagenmakers, A.J.M., J.H. Coakley, and R.H.T. Edwards. 1990. Metabolism of branched-chain amino acids and ammonia during exercise: Clues from McArdle's disease. Int. J. Sports Med. 11:S101–S113.

Wagenmakers, A.J.M., H.J.M. Salden, and J.H. Veerkamp. 1985. The metabolic fate of branched-chain amino acids and 2-oxo acids in rat muscle homogenates and diaphragms. Int. J. Biochem. 17:957–965.

Wagenmakers, A.J.M., G. Van Hall, P.A.I. Van de Schoor, and W.H.M. Saris. 1997. Glutamine does not stimulate glycogen resynthesis in human skeletal muscle. Med. Sci. Sport Ex. 29:S280

Wagenmakers, A.J.M., G. Van Hall, and B. Saltin. 1996. Excessive muscle proteolysis during one leg exercise is exclusively attended by increased *de novo* synthesis of glutamine, not of alanine. Clin. Nutr. 15(suppl. 1):1.

Windmueller, H.G., and A.E. Spaeth. 1974. Uptake and metabolism of plasma glutamine by the small intestine. J. Biol. Chem. 249:5070–5079.

Zawadzki, K.M., B.B. Yaspelkis, and J.L. Ivy. 1992. Carbohydrate-protein complex increase the rate of muscle glycogen storage after exercise. J. Appl. Physiol. 72:1854–1859.

DISCUSSION

JOAN CONWAY: I have a simple question. Were these studies done only in men, or were they done in both men and women?

ANTON WAGENMAKERS: I am afraid that they were all done in top male athletes.

JOAN CONWAY: And would you think that gender would have any effect?

ANTON WAGENMAKERS: I don't have a reason to assume that there will be a difference. But the studies should be done in women, too, I think.

JOAN CONWAY: Thank you.

ROBERT NESHEIM: Doug?

DOUGLAS WILMORE: A very nice presentation. Thank you.
Joe Cannon and some of his associates did biopsies following eccentric exercise and found increased levels of interleukin-1 in muscle and later found rises in interleukin-1 in the bloodstream.
I think, in subsequent studies, they gave vitamin E to try to modify those responses. Do you have any thoughts about the use of antioxidants in athletes who have these heavy performance levels where muscle damage and inflammation may play a role?

ANTON WAGENMAKERS: When were these interleukin levels increased, immediately after or three or four days after?

DOUGLAS WILMORE: No, several hours after. And the bloodstream levels were elevated two to four hours after. Bob, is that right?

ROBERT WOLFE: I think it was at least that much later. It definitely was some time later.

ANTON WAGENMAKERS: So my impression is that there is certainly free radical damage in the muscle, but it probably is a couple of days later, when there are inflammatory cells to clear damaged fibres; certainly after eccentric exercise, there will be inflammatory cells after three days in the muscle, and they can produce free radicals. That is for sure.

But I am not sure whether exercise per se leads to free radical damage. If that would be the case, then you would expect that vitamin E supplementation or antioxidant supplementation could improve performance. There is no definitive study yet, I don't think, that has shown an effect. Many studies have been done, but everybody has failed to find an effect on performance.

I think an area that is much more interesting is the central area, the gut-liver area. Demanding exercise, certainly in the heat, leads to a massive reduction of blood flow to the gut, and that can really imply the development of a severe underperfusion (gut ischemia). I really wonder whether, when athletes stop after a strenous event like the Hawaii Iron Man, reperfusion damage may occur. Quite a few of these athletes anecdotally have been reported to become quite ill within 15 minutes after they finish. I think the gut is a more interesting tissue to direct your antioxidants at than the muscle.

ROBERT NESHEIM: Dennis?

DENNIS BIER [off microphone]: Is it possible that the mechanism for increased glycogen storage after the addition of amino acid to the glucose may be repartitioning, in the sense that you now supply the nonessential amino acids that might have had to be synthesized from carbohydrate if you didn't supply them for muscle repair, so that in fact there is more glucose available for glycogen?

ANTON WAGENMAKERS: I think that it is a major assumption that muscle repair is needed. Over a 24-hour day, the Tour de France cyclists are in energy balance. After exercise, they cannot be in energy balance, because their total carbohydrate oxidation rate always exceeds their capacity for oral ingestion of carbohydrates.

But still, my impression is that as long as they are in reasonable energy balance, they don't break a lot of protein down during the exercise. They would not be able to perform at this high intensity for 22 days when they are in negative protein balance.

So I am not sure whether the mechanism you propose is the answer, because it really means, then, that repair and high protein synthesis rates are necessary following exercise. Certainly it is something we should look at, because these proteins, of course, may also have an effect on protein synthesis and degradation in the muscle in the first few hours after such demanding exercise.

ROBERT NESHEIM: Sree?

K. SREEKUMARAN NAIR [off microphone]: It is also possible that protein may be providing substrates for gluconeogenesis in the liver, and that the glucose thus formed is providing substrates for the glycogen synthase enzyme.

Do you have any data that protein alone also can increase glycogen synthesis after exercise?

ANTON WAGENMAKERS: Yes, there is. In the study done by Zawadzki, protein was given alone, and—you are right in that the glycogen resynthesis rate was a little bit higher than in basal conditions but much lower than with glucose plus protein. I don't think gluconeogenesis is extremely important in these studies, because the insulin concentration is so high that it probably will reduce gluconeogenesis. So I think it is more the higher insulin and probably the upgraded glycogen synthase activity that is causing that effect.

ROBERT NESHEIM: Michael Rennie and then Harris Lieberman.

MICHAEL RENNIE: We just got exactly the same result. Although we find this effect in the basal state, we don't get any difference when we add glucose. So I suspect that the whole effect is overridden by insulin. So do you think that the effects of the protein are insulin-dependent? Was the area under the insulin curve higher when you give glucose and protein together?

ANTON WAGENMAKERS: Yes, it is about double the effect. It probably is better to give the supplements every half hour because we got these big spikes because of the hourly doses. But at the high points, the insulin was twice as high.

MICHAEL RENNIE: So you could just give insulinogenic amino acids and see whether there is a beneficial effect.

ANTON WAGENMAKERS: Yes. We are doing that at the moment. We don't have the answers yet.

ROBERT NESHEIM: Harris?

HARRIS LIEBERMAN: Several years ago, there was a report by a Spanish group—Segura, I believe, was the principal investigator—which suggested that if you gave pharmacologic doses of tryptophan, you actually would improve endurance performance. Would you care to comment on that?

ANTON WAGENMAKERS: Yes. There has been a repeat of that study by a group from Norway (Stensrud et al., 1988) and they repeated that study very carefully, with exactly the same doses, in a very big group of subjects, about 30, I believe, and they did not see any effect of tryptophan at all on performance.

When you look at the Spanish data, some of the subjects were exhausted with a heart rate of 142, and as an exercise physiologist, I question such results.

ROBERT NESHEIM: Thank you very much. I think we need to move on.

16

Dietary Supplements Aimed at Enhancing Performance: Efficacy and Safety Considerations

Timothy J. Maher[1]

INTRODUCTION

Consumers have used amino acids as dietary supplements for many years in the hope of producing a wide variety of effects, including enhanced physical performance, improved quality of sleep, analgesic relief, and accelerated muscle mass development. Despite continued evidence of the potential for adverse effects associated with imbalances in amino acid ingestion (Benevenga and Steele, 1984), this practice of consuming supranutritional doses (that is, amounts in excess of those believed required to support normal growth and sustain health) of amino acids continues to a significant extent due to the highly effective advertising in the lay literature and press. Although 7 years have passed since the initial outbreak of the L-tryptophan-associated eosinophilia-myalgia syndrome (EMS), little progress has been made in determining the exact cause of this disease (Hertzman et al., 1990; Kamb et al., 1992). The

---

[1] Timothy J. Maher, Massachusetts College of Pharmacy and Allied Health Sciences, Boston, MA 02115.

major research emphasis continues to be placed on one contaminant, the so-called Peak E (1,1'-ethylene-*bis*[tryptophan]), which has been routinely identified in implicated lots of manufactured trytophan. Interestingly, recent studies have identified a plethora of potentially active compounds including other indole derivatives (e.g., dioxindoylalanine), phenyl derivatives (e.g., anthranilic acid), and 1,2,3,4-tetrahydro-β-carboline-3-carboxylic acid derivatives, some of the latter of which are known to act as benzodiazepine inverse agonists (Simat et al., 1996). Attempts to reproduce in experimental animals the symptomatology associated with EMS in humans have generally failed, and thus, understanding of the mechanisms responsible for EMS associated with contaminated tryptophan products remains elusive.

SAFETY OF AMINO ACIDS AS DIETARY SUPPLEMENTS: THE FASEB/LSRO STUDY

Following the tragic EMS epidemic that resulted from the use of L-tryptophan, the U.S. Food and Drug Administration (FDA) contracted with the Life Sciences Research Office (LSRO) of the Federation of the American Societies for Experimental Biology (FASEB) to perform an extensive review of the extant scientific literature to determine the safety of amino acids used by consumers as dietary supplements (Food and Drug Administration, 1990; FDA Contract No. 223-88-2124, Task Order No. 8). Not only was tryptophan use assessed for safety, but all amino acids identified as being available to consumers were evaluated since prolonged daily ingestion of supranutritional quantities of these compounds as dietary supplements was known to be commonplace.

LSRO initiated their study in the fall of 1990 by first searching the extant scientific literature for reports that related to safety of amino acids. This was followed by an open meeting, in which interested parties presented information and views related to this issue. Additionally, an invitation was extended to the public to submit written materials for consideration (FDA Docket No. 90N-0379). An ad hoc expert panel consisting of nine scientists met subsequently to advise LSRO and prepare a final report. The report, made available in 1992 (FASEB/LSRO, 1992), also contained suggested guidelines for future safety testing (Anderson and Raiten, 1992).

Because consumers primarily used supplemental amino acids presumably to enhance physiological functions or produce pharmacological responses, rather than to affect any nutritional function, a significant dilemma was faced by the expert panel. No credible evidence was available in the scientific literature indicating that a normal, healthy individual would benefit nutritionally in any way from supplementation of the diet with any single amino acid. Furthermore, even in those individuals with a less-than-ideal diet, the practice of supplementing with single amino acids was considered potentially dangerous, since the literature was replete with studies demonstrating "antinutritional" effects (i.e., depressed growth and other adverse effects) associated with the intake of imbalanced amino acid diets (Benevenga and Steele, 1984).

A survey of products in the marketplace revealed a wide diversity of label information that generally failed to provide adequate information regarding chemical composition, isomeric identification, purity, shelf life and contraindications to use. While some labels of products containing L-phenylalanine warned patients with phenylketonuria, others failed to do so despite the documented potential for adverse effects associated with the ingestion of this amino acid in subjects with this inherited metabolic abnormality (Matalon et al., 1991). Labeling and advertising carefully avoided legal "drug claim" language, while effectively suggesting that these products provided pharmacological rather than nutritional benefit. The use of D-amino acids as dietary supplements was viewed as an especially dangerous practice since these enantiomers have typically been demonstrated not to provide nutritional support for humans. In fact, in many cases, they are potentially toxic (Friedman, 1991). Concern was also expressed regarding the potential for interaction with numerous over-the-counter and prescription drugs, as there are many documented interactions in the literature (e.g., antidepressants, monoamine oxidase inhibitors, opioid analgesics, and sympathomimetics) (Glassman and Platman, 1969; Hull and Maher, 1990; Hull et al., 1994).

The expert panel concluded, based on the scientific evidence available, that the only safe form of amino acid ingestion was via protein in the diet and that there was no evidence to support a safe upper level of any individual amino acid intake in the form of a dietary supplement beyond that found in typical protein foods. Additionally, as a result of the lack of available information bearing on the safety of any of the amino acids used as dietary supplements, it was concluded that a systematic approach to safety testing was required prior to the rational use of such compounds by the general public. Safety testing, which would involve both animal and human studies, would be comprehensive and utilize sophisticated techniques. The effects of both acute and chronic ingestion of amino acids on weight changes, food intake, neurologic and behavioral changes, liver function, routine blood chemistry and hematologic parameters, hormonal changes, and pharmacokinetic profiles following oral administration of the amino acid with and without food in both genders would be determined. Following extensive studies in animals, acute and chronic testing in humans would be required to satisfy additional safety concerns. Growth, neurological and behavioral function, hematologic parameters, pharmacokinetic profiles, and hormonal changes would be monitored following subject's exposure to various doses of the selected amino acids. Careful selection of experimental subjects for these studies was emphasized with specific guidelines set forth.

Despite the findings in the FASEB/LSRO report (1992) the U.S. Congress passed the Dietary Supplement Health and Education Act (DSHEA), which was subsequently signed into law on October 25, 1994. The DSHEA significantly reduced the barriers to the sale of dietary supplements to the general public. On a promising note, the FDA recently has begun to consider proposed guidelines for the manufacture of dietary supplements. The proposed rule, "Current Good Manufacturing Practice (CGMP) in Manufacturing, Packing, or Holding Dietary Supplements", is currently open for public comment (Food and Drug Administration, 1997). The development of CGMP is critical to ensure some governmental control over the quality of products now more easily available to the general public.

STUDIES ON PERFORMANCE WITH SELECTED AMINO ACIDS AND PROTEIN

Individual amino acids have been administered in supranutritional doses in a variety of studies where effects on performance have been determined. Some of these studies have documented adverse effects with such pharmacological doses, while others failed to document any adverse effects. The difficulty with making decisions regarding the safety of such amino acids from these negative studies may result from the study design, as many of the studies were not carefully designed to detect and monitor adverse effects (Anderson and Raiten, 1992; Bucci, 1989). Similarly, some studies have investigated the effects of enhanced protein intake on physical performance. As reviewed by Hickson and Wolinsky (1989), there is no current consensus on the efficacy of such dietary protein interventions. Carefully designed studies with individual amino acids and protein supplementation should be pursued by the military in their quest to optimize performance. However, careful attention must be paid to the monitoring of adverse effects, realizing that some of these may be subtle or delayed in nature. Although there may be benefits to gain from such dietary approaches, the risks involved must be minimized for the military subjects involved.

ALTERNATIVE PROTEIN SOURCES

While most studies in the past involving manipulations of protein in the diet have utilized very high-quality protein from animal sources, increasing evidence supports the potential benefits from incorporating proteins of vegetable origin. The protein obtained from soybeans (*Glycine max*) can provide a high- to intermediate-quality protein as determined by growth pattern and nitrogen balance studies (Young, 1991). The use of highly processed soybean protein isolates and concentrates as part of the diet can provide all, or a portion of, the daily protein requirements in an increasingly palatable form. Additionally, differences between countries in the incidence of many diseases including heart disease, hormone-dependent cancers, and osteoporosis have been epidemiologically associated with the degree of dietary soybean protein consumption (Knight and Eden, 1996). For example, Japanese women have approximately one-sixth the incidence of breast cancer compared with their Western counterparts. Similarly, the incidence of premenstrual symptoms and menopausal complaints is significantly lower in populations where soybeans comprise a significant portion of the daily protein intake.

Contained within raw soybeans, and carried through intact in the processing of protein concentrates/isolates and flours, are a group of compounds chemically classified as isoflavones (Wang and Murphy, 1994). Among the isoflavones are the phytoestrogenic glucosides, genestin and daidzin, along with their corresponding aglycones, genestein and daidzein, respectively. These weakly estrogenic compounds are approximately .001 to .0001 as potent as estradiol. However, in a typical Eastern soy-based diet, the levels of circulating genestein and daidzein are approximately 1,000 to 10,000 times higher than usual estradiol levels. Thus, as supported by many recent studies, the dietary consumption of such phytoestrogens could favorably influence the incidence of estrogenic-mediated processes. More recently, reports have demonstrated the association between estrogen replacement therapy and the decreased incidence of certain dementias. Much more research is needed to explore fully the utility of soybean and alternate protein sources on the performance and overall health of subjects in the general public, as well as in the military.

DIETARY SUPPLEMENTS AND PREMENSTRUAL SYNDROME

Numerous studies have demonstrated the ability of serotoninergic agonists to affect somatic, appetitive, behavioral, and cognitive changes that recur monthly in many women during the late luteal phase of the menstrual cycle (Brzezinski et al., 1990; Wood et al., 1992). Although administration of L-tryptophan, the precursor of serotonin, has been shown to increase serotonin in the brain of experimental animals and affect numerous serotonergic-mediated behaviors, this amino acid is no longer available for unrestricted use due to the above described association with the EMS epidemic. However, another approach to increasing central serotonergic function without the exogenous administration of the individual amino acid precursor involves the consumption of carbohydrate which, by virtue of its ability to increase insulin release, enhances the uptake from the circulation of large neutral amino acids (LNAA) that compete with L-tryptophan for transport into the brain (Fernstrom and Wurtman, 1972). Studies have demonstrated that carbohydrate ingestion enhances the synthesis of serotonin in the brain of experimental animals and alters serotonin-mediated behaviors. Similarly, studies in humans have also demonstrated that various carbohydrate meals enhance the serum tryptophan: LNAA ratio (a predictor of tryptophan entry into the brain) and affect behavior (Lieberman et al., 1986b; Maher et al., 1984).

This approach was recently used in a study to determine the influence of a dietary supplement to alter premenstrual symptoms in women (Sayegh et al., 1995). Three isocaloric beverages were formulated and administered in a double-blind crossover fashion to 24 women confirmed to suffer from premenstrual symptoms. One beverage containing 47.5 g of a mixture of dextrose and maltodextrin led to a 29 percent increase in the serum tryptophan:LNAA ratio and significantly decreased (as determined by an abbreviated Profile of Mood States test) self-reports of tension, anger, depression, confusion, and increased cognitive performance in recognition memory measures (as determined by the Auditory Consonants Trigrams Recognition test). The other beverages tested (15 g casein with 32.5 g dextrose; and 47.5 g galactose plus dextrose) did not significantly alter the serum tryptophan:LNAA ratio or the behavioral and cognitive performance parameters monitored. The ability of this dietary supplement to alter the mood and performance decrements associated with the premenstrual period in some women may offer an alternative strategy to addressing this syndrome with a relatively benign intervention.

CHOLINE AND ENDURANCE PERFORMANCE

Although it is not an amino acid or protein, choline has recently been demonstrated to enhance certain types of endurance performance. Initial studies in subjects completing the Boston Marathon demonstrated a decrease in plasma choline levels of approximately 40 percent from prerace values (Conlay et al., 1986). Subsequent studies (Sandage et al., 1992) demonstrated the ability of supplemental choline administration in the form of a beverage to prevent the dramatic decrease in plasma choline following a similar physical endurance event (20-mi run). Additionally, the performance of individuals was significantly increased as indicated by reduced time required to complete the run when subjects consumed the choline-containing beverage versus the placebo, choline-free beverage. This double-blind, crossover study has led to numerous other trials that may help to further characterize the potential beneficial effects of this dietary supplement in a variety of other physical endurance activities. The use of choline in endurance events should be investigated in greater detail by the military.

AUTHOR'S SUMMARY AND RECOMMENDATIONS

The FASEB/LSRO report on the safety of amino acids as dietary supplements concluded the following:

- There is no nutritional rationale for the use of amino acids as dietary supplements, and such a practice can be dangerous.
- Supplemental amino acids should be used for pharmacological rather than nutritional purposes.

- The extant scientific literature fails to support a safe upper limit for supplementation of any amino acid beyond that found in dietary protein.
- Appropriate testing in animals and humans is required before the safety of supplemental amino acids can be adequately assessed.

Several additional recommendations may be made:

- Alternative proteins of vegetable origin should be considered when designing long-term studies.
- Manipulation of serotonergic function with carbohydrate may constitute a relatively benign intervention for lessening the symptoms of the premenstrual period in some women.
- Choline supplementation for endurance activities should be considered and requires careful investigation under a variety of applicable military conditions.

It is therefore recommended that approaches to fortify military rations with supplemental amino acids, protein, or additional compounds be initiated. However, appropriate safeguards for the welfare of subjects must be treated as such interventions may result in adverse effects, especially if individual amino acids are used at supranutritional doses over a long period of time.

REFERENCES

Anderson, S.A., and D.J. Raiten (eds.). 1992. Safety of amino acids as dietary supplements. Bethesda, MD.: Federation of the American Societies for Experimental Biology.

Benevenga, N.J., and R.D. Steele. 1984. Adverse effects of excessive consumption of amino acids. Ann. Rev. Nutr. 4:157–181.

Brzezinski, A., J. Wurtman, R.J. Wurtman, R. Gleason, T. Nader, and B. Laferrere. 1990. d-Fenfluramine suppresses the increased calorie and carbohydrate intakes and improves the mood of women with premenstrual syndrome. Obstet. Gynecol. 76:206–211.

Bucci, L.R. 1989. Nutritional ergogenic aids. Pp. 107–184 in Nutrition in Exercise and Sport, J.F. Hickson, Jr., and I. Wolinsky, eds. Boca Raton, Fla.: CRC Press.

Conlay, L.A., R.J. Wurtman, J.K. Blusztajan, I.L.G. Coviella, T.J. Maher, and G.E. Evoniuk. 1986. Marathon running decreases plasma choline concentrations. N. Engl. J. Med. 315:892.

FASEB/LSRO. 1992. Safety of amino acids used as dietary supplements. Center for Food Safety and Applied Nutrition. FDA Contract No. 223-88-2124, Task No. 8.

Fernstrom, J.D., and R.J. Wurtman. 1972. Brain serotonin content: Physiological regulation by plasma neutral amino acids. Science 178:414–416.

FDA (Food and Drug Administration). 1990. Evaluation of the safety of amino acids and related products; announcement of study; request for scientific data and information; announcement of open meeting. Fed. Regist. 55:49141–49142.

FDA (Food and Drug Administration). 1997. Current good manufacturing practice in Manufacture, Packing, or holding dietary supplements; proposed rule. Fed. Regist. 62:5699–5709.

Friedman, M. 1991. Formation, nutritional value, and safety of D-amino acids. Pp. 447–482 in Nutritional and Toxicological Consequences of Food Processing. New York: Plenum Press.

Glassman, A.H., and S.R. Platman. 1969. Potentiation of a monoamine oxidase inhibitor by tryptophan. J. Psychiatr. Res. 7:83–88.

Hertzman, P.A., W.L. Blevins, J. Mayer, B. Greenfield, M. Ting, and G.J. Gleich. 1990. Association of the eosinophilia-myalgia syndrome with the ingestion of tryptophan. N. Engl. J. Med. 322:869–873.

Hickson, J.F., Jr., and I Wolinsky. 1989. Human protein intake and metabolism in exercise and sport. Pp. 5–36 in Nutrition in Sport and Exercise, J.F. Hickson, Jr., and I. Wolinsky, eds. Boca Raton, Fla.: CRC Press.

Hull, K.M., and T.J. Maher. 1990. L-Tyrosine potentiates the anorexia induced by mixed-acting sympathomimetic drugs in hyperphagic rats. J. Pharmacol. Exp. Ther. 255:403–409.

Hull, K.M., D.E. Tolland, and T.J. Maher. 1994. L-Tyrosine potentiation of opioid-induced analgesia in mice utilizing the hot plate test. J. Pharmacol. Exp. Ther. 269:1190–1195.

Kamb, M.L., J.L. Murphy, J.L. Jones, J.C. Caston, K. Nederlof, L.F. Horney, L.A. Swygert, H. Falk, and E.M. Kilbourne. 1992. Eosinophilia-myalgia syndrome in L-trytophan-exposed patients. J. Am. Med. Assoc. 267:77-82.

Knight, D.C., and J.A. Eden. 1996. A review of the clinical effects of phytoestrogens. Obstet. Gynecol. 87:897–904.

Lieberman, H., J. Wurtman, and B. Chew. 1986b. Changes in mood after carbohydrate consumption among obese individuals. Am J. Clin. Nutr. 44:772–778.

Maher, T.J., B.S. Glaeser, and R.J. Wurtman. 1984. Diurnal variations in plasma concentrations of basic and neutral amino acids and in red cell concentrations of aspartate and glutamate: Effects of dietary protein. J. Clin. Nutr. 39:722–729.

Matalon, R., K. Michals, C. Azen, E.G. Friedman, R. Koch, E. Wenz, H. Levy, F. Rohr, B. Rouse, L. Castiglioni, W. Hanley, V. Austin, and F. de la Cruz. 1991. Maternal PKU collaborative study: The effect of nutrient intake on pregnancy outcome. J. Inherit. Dis. 14:371-374.

Sandage, B.W., Jr., L. Sabounjian, R. White, and R.J. Wurtman. 1992. Choline citrate may enhance athletic performance. Paper presented at the American Physiological Society Conference on Integrative Biology of Exercise, Colorado Springs, Colo.

Sayegh, R., I. Schiff, J. Wurtman, P. Spiers, J. McDermott, and R.J. Wurtman. 1995. The effect of a carbohydrate-rich beverage on mood, appetite, and cognitive function in women with premenstrual syndrome. Obstet. Gynecol. 86:520–528.

Simat, T., B. van Wickern, H. Eulitz, and H. Steinhart. 1996. Contaminants in biotechnologically manufactured L-tryptophan. J. Chromatogr.B. Biomed .685:41–51.

Wang, H., and P.A. Murphy. 1994. Isoflavone composition of American and Japanese soybeans in Iowa: Effects of variety, crop year and location. J. Agric. Food Chem. 42:1674–1677.

Wood, S., J. Mortola, Y.F. Chan, F. Moossazadeh, and S. Yen. 1992. Treatment of premenstrual syndrome with fluoxetine: A double-blind, placebo-controlled, crossover study. Obstet. Gynecol. 80:339–344.

Young, V.R. 1991. Soy protein in relation to human and amino acid nutrition. J. Am. Diet. Assoc. 91:828–835.

DISCUSSION

ROBERT NESHEIM: Any questions for Dr. Maher? Yes?

ROBERT WOLFE: It seems like you are always better safe than sorry, but I guess I have a little problem with the conclusion that that entire list of tests should be run for testing safety of any amino acid given orally.

In light of the practicality of any studies actually being completed, and in light of most of the data you showed—and I did review that Document 12 for the NIH symposium on protein supplements that was held, and I was not impressed that there was really much evidence of detrimental effects, and with the layers of proposed tests that would be run before any studies would be done, I am concerned that any further investigation would be precluded without a sufficient basis in many cases for implementing these additional screening tests.

TIMOTHY MAHER: I think that we have the two extremes. On the one hand, we have the FASEB-LSRO study that suggests treating these compounds similar to pharmacological agents, and performing extensive safety testing, as you would with any other drug.

On the other hand, we can go down the street to any of the health food stores, and we can buy many of these things, and there has been no testing done whatsoever.

I think it is important to find a happy medium somewhere in between, maybe not as extensive as the FASEB-LSRO study but definitely more than what is currently known about these compounds.

I would hope that whoever decides to investigate amino acids as performance-enhancing strategies would at least do some of those studies. I don't know that the financial impetus is there on the part of the health food industry, but I hope that the military or whatever other organization wants to look at this in a proper way will do some of those studies.

ROBERT NESHEIM: Any other questions for Tim, or questions on anything that was covered today?

MARITZA RUBIO-STIPEC: I just had a minor question. I just wanted to know, all of the studies you mentioned except one were done only on men?

TIMOTHY MAHER: Are you referring to studies of individual amino acids or protein? The PMS studies were obviously with women.

MARITZA RUBIO-STIPEC: Yes, I know. What about the others.

TIMOTHY MAHER: With studies of choline, we have used both men and women.

MARITZA RUBIO-STIPEC: Do you have the distribution by gender?

TIMOTHY MAHER: No. But, most of the studies of the individual amino acids have been in men, especially the ones that have dealt with athletics and performance.

ROBERT NESHEIM: A question?

MACKENZIE WALSER: I wanted to comment. I was on that same committee that Dr. Maher was on, and I think one point that you did not bring out was that we were concerned about quality control. You can put sugar on the shelf in a grocery store, and it is going to have to meet certain standards. I have no idea what they are, but it does have to meet certain standards, or the FDA will impound it.

On the other hand, you can put lysine on the shelf, and it can actually be sodium cyanide. I am not kidding. There is absolutely no control of quality.

I will never forget that while working in this committee, we got some tablets—not sodium cyanide—but we did get some tablets of lysine and put them in a beaker of water to see what the dissolution time was. Three days later, they were still there, intact.

So there are no criteria, either, for dissolution, which is a minimum requirement for any capsule that is sold as a drug.

So I don't understand why the FDA, cannot design standards for quality control, at least for amino acids, before we have these tremendous studies of efficacy.

II

Discussion

ROBERT NESHEIM: I would like to open the session for a general discussion. Gail?

GAIL BUTTERFIELD: Alana (Cline), did you have an actual measure of protein intake in the women who responded to your survey?

ALANA CLINE: No, we don't have an actual number of grams of protein, but we have the information on food category intake, so we will be able to take a look at that. Again, we are still compiling the data.

GAIL BUTTERFIELD: Do you have a caloric intake yet?

ALANA CLINE: No.

GAIL BUTTERFIELD: Okay. The reason I asked is, it is very clear from all the presentations, that we have very little information on protein requirements in women, and I suspect from work that I have done with female athletes that in fact women as a group may be somewhat at risk for inadequate intakes. So it would be interesting to see what the protein and energy intakes were for the female soldiers.

ALANA CLINE: We do have data available from the study we did at Fort Sam Houston on iron deficiency in women, so we do have some data from women's studies.

PARTICIPANT: And we are planning a field study in which we will have a significantly larger women's population than we have had in the past, too. That will be done in April and May.

ROBERT NESHEIM: Any other questions? Yes?

DOUGLAS WILMORE: A question for Harris Lieberman and for Karl Friedl. We have heard some really fairly simple and practical strategies about repletion of muscle glycogen. In practice, in Ranger training, are any of those strategies utilized at all, for example, carbohydrate loading, with or without protein, at the end of a march or at the end of an exercise period?

KARL FRIEDL [off microphone]: It is not used in Ranger training. The point there is to create this energy deficit as one of the deliberate stressors. That is their training strategy.

We have a new carbohydrate drink that has just been fielded, and maybe Pat Dunne can talk about that, a maltodextran drink, which was really the result some of the field studies done by ARIEM, and they finally recommended that it get type-classified. Actually, the USARIEM folks and the Natick folks have been ahead of the game there. The product is already available.

So there is at least a carbohydrate supplement, an energy supplement, which, in some of the Pennington studies of Special Forces soldiers on the treadmill, was shown to enhance and prolong the endurance time by 15 percent when soldiers received the supplement during exercise.

DOUGLAS WILMORE [off microphone]: We have examined the effects of protein supplements, but didn't consider using potassium, sodium, and some of the other factors.

ANTON WAGENMAKERS [off microphone]: Why do you deliberately conduct training in a state of energy deficiency [inaudible]?

KARL FRIEDL: Oh, that's a long story that we will have to talk about over dinner.

ROBERT NESHEIM: Jim?

JAMES HODGDON: I just wanted to respond to the question. We currently have a program, and I suspect more people in the Navy are interested in studying this. We have been working with a group at Yale so that we can use the magnet to measure the amount of glycogen in the muscles rather than doing biopsies, so we have the opportunity to look at depletion rates.

DOUGLAS WILMORE: Have you looked at possible additions?

JAMES HODGDON: No. We will now, I think, since that issue has been raised.

PARTICIPANT: We are doing that.

ROBERT NESHEIM: John?

JOHN VANDERVEEN: We didn't really address the issue of long-term health and protein quality today, but one of the questions we had asked in planning this was, is there any benefit from changing protein sources? There is some literature out there, and maybe someone would like to address the issue of animal versus plant proteins relative to lipids and cardiovascular disease and perhaps other conditions. There was a little discussion, perhaps, of the benefit of phytoestrogens and so forth. But is there anyone who wants to address or talk about that area?

DOUGLAS WILMORE: I will just start the discussion. You can change animal protein for soy or a vegetable protein mix on a gram for gram basis and keep the fat contents the same in the diet and observe lipid-lowering effects. There is a fair amount of epidemiology supporting the belief that these effects and the phytoestrogen effects may be the reason you can get protection for prostate cancer in men and for breast cancer in women as you move over to plant protein sources.

It is also why you don't see perimenopausal symptoms in Asian cultures where soy protein is the dominant protein source, because prior to menopause, the exogenous estrogens from food sources block some of the endogenous estrogen effects, and during the postmenopausal period, exogenous estrogen is still stimulating those receptors, so estrogen withdrawal is not a major factor. So there is a growing amount of evidence that would suggest that long-term health effects are improved with vegetable, particularly soy, protein sources.

D. JOE MILLWARD: Can I just add one comment to that? As was just shown to us, the principal phytoestrogens from soy are genestein and daidzein, and certainly those compounds have been quite intensively investigated now.

But the lignins, which come from other plant sources, also have phytoestrogen effects, and actually, the intake of lignins is higher than that of some phytoestrogens. Of course, there are lots of societies where soya is not a normal part of the diet but the vegetarian background of the diet involves high intakes of lignins.

So I think that the phytoestrogen story is something that is spreading from soya right across the spectrum, and in many ways the key issues of plant versus animal protein may well reflect this issue of other travelers (as yet unidentified compounds). It is a complicated issue, because the micronutrient bioavailability is high from animal protein and low from plant protein sources. But these other phytoprotectants in plants are not present in animal protein, so it is quite a complicated issue. Thus it is probably that phyto protectant element that is important rather than the amino acid issue.

DOUGLAS WILMORE: And you cannot substitute these things with vitamin pills. That is the other thing that society must learn. You know, you've got to eat your spinach if you want to preserve your eyes and your late vision. You cannot take a vitamin pill and preserve that, and that has actually been shown.

The most common operations done in the United States are cataract operations, and if you take the people over 65 and divide them into quartiles, the people who eat five or more vegetables a day have an incidence of cataract surgery of 5 percent or less, and the people who eat zero to one helping a day have an incidence of cataract surgery of something like 25 or 30 percent. This is presumably preventable through long-term nutrition.

ROBERT NESHEIM: There are some interesting studies. There was the three-country study that was conducted in Mexico, Kenya, and Egypt, looking at the effect of diet on infants, toddlers, and so on.

One of the results of that study that was rather interesting was that they found that among those young toddlers and school children, the ones who got some animal protein in their diet actually had better cognitive performance and better growth than did those who were not getting that. It was probably related to B12 for one thing—at least that was one of the hypotheses, which could be more important in certain stages than others. I only mention that because I don't think we are ever going to get to a totally vegetarian diet devoid of B_{12} sources in this country, but there is a balance of all these factors that needs to be looked at in the total population. Yes?

RICK LEVINE: In line with the idea that Dr. Wilmore discussed, supplements versus foods, he previously had talked about antioxidants and supplementation. Should we also be looking at high sources of antioxidant foods to be incorporating those into operational rations as opposed to looking at supplemental sources?

DOUGLAS WILMORE: That might be ideal. Clearly, what would be ideal would be to genetically manipulate plant protein so that we would have a high-glutamine plant protein source, for example, or a profile of amino acids that would be more desirable for certain kinds of conditions. The iron issue that Wanda raised is a very timely issue, particularly because iron supplementation is now being carried out, I think, in a number of women because of anemia.

So I think it is a desirable goal to consider food sources of nutrients. My concern has always been those periods of time when people don't eat enough, and I have always had concern, and have concern now, about the recommendations that we have in our notebooks about people going out on 10-day missions with inadequate diets, and those diets are inadequate across the board. What concerns me is that there may be very small, lightweight supplements that might give soldiers some additional protection.

KARL FRIEDL: That is a reasonable point, too, because we have a new generation survival ration.

PARTICIPANT: Actually, the LRPR (Long-Range Patrol Ration) would be the one you mean.

KARL FRIEDL: Yes, the LRPR.

ROBERT NESHEIM: There is a concern with survival rations containing too much protein. Extra protein increases the requirement for water, which may be limited in a situation where survival rations are needed.

We issued a report not too long ago on the subject of not eating enough, and in that report the committee made a strong recommendation that, just as the military has a water doctrine, there ought to be a food doctrine; in other words, that soldiers and military personnel should be educated to understand that food is the fuel that runs them, just as diesel is the fuel that runs their motorized equipment, and that they need to be aware of the potential adverse effects of energy deficits in the field.

Now, we know there are a lot of things that contribute to why soldiers don't eat enough. I am sure that any of us, in a similarly stressful situation, would find ourselves not wanting to eat or not being in a position to eat even though we know that we are supposed to eat. But I think the emphasis on a food doctrine is important, just as it is on a water doctrine, and I think we probably need to continue to emphasize that.

ROBERT NESHEIM: I think we have had a full day, a very full day. I just wanted to thank all of the speakers for their fine contributions. I think we have had an excellent day. The speakers made interesting presentations. They were pretty well on target all the way through. We had good discussion, and you have presented our committee with a lot of challenges in terms of writing a report.

I do want to urge the speakers to get your papers to us, because we are under a real tight constraint in trying to get a report completed within the time frame of what we are committed to doing. So please help us out by getting papers to us as quick as you can if you haven't got it done already.

So I just wanted to close this by thanking all of you for your participation in this particular meeting. I appreciate all our visitors, military and civilian, that are with us, and I hope that you have had an enjoyable day of listening to all of this particular presentation. You have put some real challenges to us as a committee, to try to come up with some recommendations and some meaningful input to the military.

HARRIS LIEBERMAN: Okay. Bob, I cannot even begin to convey it, but I wanted to thank you for the tremendous contribution that you have made to the Committee on Military Nutrition, to the health of soldiers, to the advancement of the science of military nutrition in your many years as the leader of the Committee on Military Nutrition Research. So thank you very much.

ROBERT NESHEIM: Thank you, Harris. It could not have happened without the input of all of the people on the committee and all of the speakers who have presented data to us so we were able to get it together. So it has been great.

APPENDIXES

A Workshop Agenda
B Biographical Sketches
C Acronyms and Abbreviations
D Proteins and Amino Acids—A Selected Bibliography
E Protein and Energy Content of Selected Operational Rations

A

Workshop Agenda

THE ROLE OF PROTEIN AND AMINO ACIDS IN SUSTAINING AND ENHANCING PERFORMANCE

A Workshop Sponsored by
Committee on Military Nutrition Research and Subcommittee on Body Composition, Nutrition, and Health of Military Women
Food and Nutrition Board, Institute of Medicine

March 13–14, 1997
The Foundry Building
1055 Thomas Jefferson Street, N.W.
Washington, DC 20007

THURSDAY, MARCH 13, 1997   **ROOM 2004**

I.   WELCOME AND INTRODUCTION: IMPACT OF MILITARY OPERATIONS ON PROTEIN NEEDS

8:00 a.m.–8:05 a.m.   Welcome and Introductions
Robert O. Nesheim, CMNR Chair and BCNH Vice Chair

8:05 a.m.–8:10 a.m.   Military Overview
LTC Karl E. Friedl, USARMRMC

II. OVERVIEW OF PROTEIN REQUIREMENTS

8:10 a.m.–8:30 a.m.	Overview of Garrison, Field, and Supplemental Protein Intake by U.S. Military Personnel *LTC Alana D. Cline, USARIEM*
8:30 a.m.–9:00 a.m.	Overview of Protein Metabolism: Lean and Mean on Uncle Sam's Team *Dennis M. Bier, Children's Nutrition Research Center, Baylor College of Medicine*
9:00 a.m.–9:30 a.m.	Regulation of Muscle Mass and Functions *K. Sreekumaran Nair, Mayo Clinic*
9:30 a.m.–10:00 a.m.	Effects of Protein Intake on Renal Function and on the Development of Renal Disease *Mackenzie Walser, The Johns Hopkins School of Medicine*
10:00 a.m.–10:15 a.m.	*Break*
10:15 a.m.–10:45 a.m.	Infection and Injury—Effect on Whole-Body Protein Metabolism *Douglas W. Wilmore, Brigham and Women's Hospital*
10:45 a.m.–11:15 a.m.	Amino Acid Flux and Requirements: Point Inherent Difficulties in Defining Amino Acids Requirements *D. Joe Millward, University of Surrey*
11:15 a.m.–11:45 a.m.	Amino Acid Flux and Requirements: Counterpoint An Argument for New Requirement Values *Vernon R. Young, Massachusetts Institute of Technology* **CANCELLED**
11:45 a.m.–12:15 p.m.	Discussion
12:15 p.m.–1:00 p.m.	*Lunch*

APPENDIX A

III. PROTEIN REQUIREMENTS AND OPERATIONAL STRESSORS

1:00 p.m.–1:30 p.m.	Physical Exertion and Amino Acid and Protein Metabolism Requirements *Michael J. Rennie, University of Dundee*
1:30 p.m.–2:00 p.m.	Muscle Markers *Steven B. Heymsfield, St. Luke's-Roosevelt Hospital Center*
2:00 p.m.–2:30 p.m.	Alterations in Protein Metabolism Due to the Stress of Injury and Infection *Robert R. Wolfe, Shriner's Burn Institute and University of Texas Medical Branch at Galveston*
2:30 p.m.–3:00 p.m.	Cognitive Performance, Stress, and Brain Function *Harris R. Lieberman, USARIEM*
3:00 p.m.–3:20 p.m.	Discussion
3:20 p.m.–3:35 p.m.	*Break*

IV OPTIMIZATION OF PROTEIN AND AMINO ACID INTAKES FOR PERFORMANCE

3:35 p.m.–4:05 p.m.	Branched-Chain Amino Acids, Glutamine, and Protein Hydrolysates: Rationale for Supplementation, Effects on Metabolism and Performance *Anton J. M. Wagenmakers, Maastricht University*
4:05 p.m.–4:35 p.m.	Dietary Supplements Aimed at Enhancing Performance: Efficacy and Safety Considerations *Timothy J. Maher, Massachusetts College of Pharmacology and Allied Health Sciences*
4:35 p.m.–5:15 p.m.	Discussion
5:15 p.m.	Concluding Remarks and Adjournment

B

Biographical Sketches

COMMITTEE ON MILITARY NUTRITION RESEARCH

ROBERT O. NESHEIM (*Former Chair, through June 30, 1998*) was vice president of research and development and later science and technology for the Quaker Oats Company; he retired in 1983. Before his retirement in 1992, he was vice president of science and technology and president of the Advanced Health Care Division of Avadyne, Inc. During World War II, he served as captain in the U.S. Army. Dr. Nesheim has served on the Institute of Medicine's Food and Nutrition Board, currently chairing the Committee on Military Nutrition Research and formerly chairing the Committee on Food Consumption Patterns and serving as a member of several other committees. He also was active in the Biosciences Information Service (as board chairman), American Medical Association, American Institute of Nutrition, Institute of Food Technologists, and *Food Reviews International* editorial board. Dr. Nesheim's academic services included professor and head of the Department of Animal Science at the University of Illinois, Urbana.

He is a fellow of the American Institute of Nutrition and American Association for the Advancement of Science and a member of several professional organizations. Dr. Nesheim holds a B.S. in agriculture, an M.S. in animal science, and a Ph.D. in nutrition and animal science from the University of Illinois.

JOHN E. VANDERVEEN (*Chair*) is the former director of the Food and Drug Administration's (FDA) Office of Plant and Dairy Foods and Beverages in Washington, D.C. His previous position at the FDA was director of the Division of Nutrition at the Center for Food Safety and Applied Nutrition. He also served in various capacities at the U.S. Air Force (USAF) School of Aerospace Medicine at Brooks Air Force Base, Texas. He has received accolades for service from the FDA and the USAF. Dr. Vanderveen is a member of the American Society for Clinical Nutrition, American Institute of Nutrition, Aerospace Medical Association, American Dairy Science Association, Institute of Food Technologists, and American Chemical Society. In the past, he was the treasurer of the American Society of Clinical Nutrition and a member of the Institute of Food Technology, National Academy of Sciences Advisory Committee. Dr. Vanderveen holds a B.S. in agriculture from Rutgers University in New Jersey and a Ph.D. in chemistry from the University of New Hampshire.

LAWRENCE E. ARMSTRONG is an associate professor of exercise science at the University of Connecticut. He has joint appointments in the Department of Physiology and Neurobiology and the Department of Nutritional Sciences. Dr. Armstrong received his Ph.D. in human bioenergetics–exercise physiology from Ball State University. His research interests include thermoregulation, fluid-electrolyte balance, energy metabolism, exercise physiology, and the human heat illnesses. He previously served as a research physiologist at the U.S. Army Research Institute of Environmental Medicine. He is a fellow of the American College of Sports Medicine and a member of the Federation of American Societies for Experimental Biology and the Aerospace Medical Association.

WILLIAM R. BEISEL is adjunct professor in the Department of Immunology and Infectious Diseases at the Johns Hopkins University School of Hygiene and Public Health. He held several positions at the U.S. Army Medical Research Institute for Infectious Diseases at Fort Detrick, Maryland, including in turn, chief of the Physical Sciences Division, scientific adviser, and deputy for science. He then became special assistant for biotechnology to the Surgeon General. After serving in the U.S. military during the Korean War, Dr. Beisel was the chief of medicine at the U.S. Army Hospital in Ft. Leonard Wood, Missouri, before becoming the chief of the Department of Metabolism at the Walter Reed Army Hospital. He was awarded a Commendation Ribbon, Bronze Star for the Korean War, Hoff Gold Medal at the Walter Reed Army Institute of Research, B.L. Cohen Award of the American Society for Microbiology, the

Robert Herman Award from the American Association for Clinical Nutrition, and Department of Army Decoration for Exceptional Civilian Service. He was named a diplomate of the American Board of Internal Medicine and a fellow of the American College of Physicians. In addition to his many professional memberships, Dr. Beisel is a contributing editor to Clinical Nutrition and an associate editor of the Journal of Nutritional Immunology. He received his A.B. from Muhlenberg College in Allentown, Pennsylvania, and M.D. from the Indiana University School of Medicine.

GAIL E. BUTTERFIELD is director of nutrition research for Palo Alto Veterans Affairs Health Care System in California. Concurrently, she is lecturer in the Department of Medicine, Stanford University Medical School; visiting assistant professor in the Program of Human Biology, Stanford University; and director of nutrition in the Program in Sports Medicine, Stanford University Medical School. Her previous academic appointments were at the University of California, Berkeley. Dr. Butterfield belongs to the American Institute of Nutrition, American Society for Clinical Nutrition, American Dietetic Association, and American Physiological Society. As a fellow of the American College of Sports Medicine (ACSM), she serves as chair of the Pronouncements Committee and is on the Board of Trustees; she also was president and executive director of the Southwest Chapter of that organization. She is a member of the Respiratory and Applied Physiology Study Section of the National Institutes of Health (NIH) and is on the editorial boards of the following journals: *Medicine and Science in Sports and Exercise, American Journal of Clinical Nutrition, Health and Fitness Journal of ACSM, Canadian Journal of Clinical Sports Medicine,* and *International Journal of Sports Nutrition.* Dr. Butterfield received her A.B. in biological sciences, M.A. in anatomy, and M.S. and Ph.D. in nutrition from the University of California, Berkeley. Her current research interests include nutrition in exercise, effect of growth factors on protein metabolism in the elderly, and metabolic fuel use in women exposed to high altitude.

WANDA L. CHENOWETH is professor in the Department of Food Science and Human Nutrition at Michigan State University. Previously she held positions as teaching associate at the University of Iowa and University of California, Berkeley. Other work experience includes positions as research dietitian and head clinical dietitian at University of Iowa Hospitals and as research dietitian at Mayo Clinic. She is a member of the American Society for Nutritional Sciences, American Dietetic Association, and Institute of Food Technology. She serves as a reviewer for several journals, including the *Journal of the American Dietetic Association, American Journal of Clinical Nutrition,* and *Journal of Nutrition,* and is a member of the Associate Editorial Board of *Plant Foods for Human Nutrition.* She has served on a technical review committee for the Diet, Nutrition, and Cancer Program of the National Cancer

Institute and as a site evaluator, Commission on Evaluation of Dietetic Education of the American Dietetic Association. Her research interests are in the areas of mineral bioavailability and clinical nutrition. Dr. Chenoweth completed a B.S. in dietetics from the University of Iowa, dietetic internship and M.S. in nutrition at the University of Iowa, and Ph.D. in nutrition at the University of California, Berkeley.

JOHN D. FERNSTROM is professor of psychiatry, pharmacology, and behavioral neuroscience at the University of Pittsburgh School of Medicine, and director of the Basic Neuroendocrinology Program at the Western Psychiatric Institute and Clinic. He received his B.S. in biology and his Ph.D. in nutritional biochemistry from the Massachusetts Institute of Technology (MIT). He was a postdoctoral fellow in neuroendocrinology at the Roche Institute for Molecular Biology in Nutley, New Jersey. Before coming to the University of Pittsburgh, Dr. Fernstrom was an assistant and then associate professor in the Department of Nutrition and Food Science at MIT. He has served on numerous governmental advisory committees. He presently is a member of the National Advisory Council of the Monell Chemical Senses Center, chairman of the Neurosciences Section of the American Society for Nutritional Sciences (ASNS), and a member of the ASNS Council. He is a member of numerous professional societies, including the American Institute of Nutrition, the American Society for Clinical Nutrition, the American Physiological Society, the American Society for Pharmacology and Experimental Therapeutics, the American Society for Neurochemistry, the Society for Neuroscience, and the Endocrine Society. Among other awards, Dr. Fernstrom received the Mead-Johnson Award of the American Institute of Nutrition, a Research Scientist Award from the National Institute of Mental Health, a Wellcome Visiting Professorship in the Basic Medical Sciences, and an Alfred P. Sloan Fellowship in Neurochemistry. His current major research interest concerns the influence of the diet and drugs on the synthesis of neurotransmitters in the central and peripheral nervous systems.

G. RICHARD JANSEN (*through August 31, 1997*) is professor emeritus in the Department of Food Science and Human Nutrition at Colorado State University, where he was head of the department from 1969 to 1990. He was a research fellow at the Merck Institute for Therapeutic Research and senior research biochemist in the Electrochemical Department at E.I. DuPont de Nemours. Prior to his stint in private industry, he served in the U.S. Air Force. Dr. Jansen is a past member of the U.S. Department of Agriculture (USDA) Human Nutrition Board of Scientific Counselors and of the editorial boards of the *Journal of Nutrition*, *Nutrition Reports International*, and *Plant Foods for Human Nutrition*. His research interests deal with protein-energy relationships during lactation and new foods for developing countries based on low-cost extrusion cooking. He received the Babcock-Hart Award of the Institute of Food

Technologists (IFT) and a Certificate of Merit from the USDA's Office of International Cooperation and Development for his work on low-cost extrusion cooking; he is also an IFT fellow. Dr. Jansen is a member of the American Institute of Nutrition, Institute of Food Technologists, and American Society for Biochemistry and Molecular Biology among others. Dr. Jansen holds a B.A. in chemistry and Ph.D. in biochemistry from Cornell University in Ithaca, New York.

ROBIN B. KANAREK is professor of psychology and adjunct professor of nutrition at Tufts University in Medford, Massachusetts, where she also is the chair of the Department of Psychology. Her prior experience includes research fellow, Division of Endocrinology, University of California, Los Angeles (UCLA) School of Medicine, and research fellow in nutrition at Harvard University. In addition to reviewing for several journals, including *Science, Brain Research Bulletin, Journal of Nutrition, American Journal of Clinical Nutrition,* and *Annals of Internal Medicine,* she is a member of the editorial boards of *Physiology and Behavior* and the *Tufts Diet and Nutrition Newsletter* and is a past editor-in-chief of *Nutrition and Behavior.* Dr. Kanarek has served on ad hoc review committees for the National Science Foundation, National Institutes of Health, and USDA nutrition research, as well as the Member Program Committee of the Eastern Psychological Association. She is a fellow of the American College of Nutrition, and her other professional memberships include the American Institute of Nutrition, New York Academy of Sciences, Society for the Study of Ingestive Behavior, and Society for Neurosciences. Dr. Kanarek received a B.A. in biology from Antioch College in Yellow Springs, Ohio, and an M.S. and a Ph.D. in psychology from Rutgers University in New Brunswick, New Jersey.

ORVILLE A. LEVANDER is research leader for the U.S. Department of Agriculture Nutrient Requirements and Functions Laboratory in Beltsville, Maryland. He was research chemist at the USDA's Human Nutrition Research Center, resident fellow in biochemistry at Columbia University's College of Physicians and Surgeons, and research associate at Harvard University's School of Public Health. Dr. Levander served on the Food and Nutrition Board's Committee on Dietary Allowances. He also served on panels of the National Research Council's Committee on Animal Nutrition and Committee on the Biological Effects of Environmental Pollutants. He was a member of the U.S. National Committee for the International Union of Nutrition Scientists and temporary adviser to the World Health Organization's (WHO's) Environmental Health Criteria Document on Selenium. Dr. Levander was awarded the Osborne and Mendel Award for the American Institute of Nutrition. His society memberships include the American Institute of Nutrition, American Chemical Society, and American Society for Clinical Nutrition. Dr. Levander received his

B.A. from Cornell University and his M.S. and Ph.D. in biochemistry from the University of Wisconsin, Madison.

ESTHER STERNBERG is chief of the Section on Neuroendocrine Immunology and Behavior and associate branch chief of the Clinical Neuroendocrinology Branch of the National Institutes of Mental Health Intramural Research Program at the National Institute of Health (NIH). Dr. Sternberg received her M.D. degree and trained in rheumatology at McGill University, Montreal, Canada. She did postdoctoral training at Washington University, Barnes Hospital, St. Louis, Missouri, in the Division of Allergy and Immunology. She was subsequently a Howard Hughes associate and instructor in medicine at Washington University and Barnes Hospital before joining NIH. Dr. Sternberg is internationally recognized for her ground-breaking discoveries in the area of central nervous system–immune system interactions. She has received the Arthritis Foudation William R. Felts Award for Excellence in Rheumatology Research Publications, has been awarded the Public Health Service Superior Service Award, and has been elected to the American Society for Clinical Investigation in recognition of this work. Dr. Sternberg is also internationally recognized as a foremost authority on the l-tryptophan eosinophilia–myalgia syndrome (L-TRP-EMS). She was the first to describe this syndrome in relation to a similar drug l-5-hydroxytryptophan, and published this landmark article in the *New England Journal of Medicine* in 1980.

DOUGLAS W. WILMORE, the Frank Sawyer Professor of Surgery at Harvard Medical School, is a senior staff scientist and surgeon at Brigham and Women's Hospital, Boston, Massachusetts. Concurrently, he is also a consultant for the Dana-Farber Cancer Center, Children's Hospital Medical Center, the Beth Israel–Deaconess Hospital, Wrentham State School, and Youville Hospital and Rehabilitation Center. Dr. Wilmore's main interests are related to metabolic and nutritional means to support critically ill patients and enhance recovery. His basic research has been applied to patients with thermal and accidental injury, patients with infectious complications, and those with multiple organ failure. He worked with the team that developed the current method of intravenous nutrition used for patients throughout the world. This technique has been improved in Dr. Wilmore's laboratory: new amino acid solutions have been developed utilizing the amino acid glutamine, and anabolic factors such as growth hormone have been incorporated in this new feeding program with dramatic therapeutic results. Dr. Wilmore serves on the Advisory Board of the Tufts Pediatric Trauma Center; the International Editorial Committee of the *Chinese Nutritional Sciences Journal*; Chinese Academy of Medical Sciences; and the editorial boards of *Annals of Surgery* and *Journal of the American College of Surgeons*. He is senior editor of *Scientific American Surgery*, the surgical text published by the American College of Surgeons that serves as the basis for care of general surgical patients. He also has published more than 300 scientific papers and 4 books. Among his professional memberships, Dr. Wilmore includes the American College of Surgeons, American Surgical Association, American Medical Association, Society of University Surgeons, and American Society for Enteral and Parenteral Nutrition. He holds a B.A. and an honorary Ph.D. from Washburn University of Topeka, an M.D. from the University of Kansas School of Medicine in Kansas City, and an honorary M.S. from Harvard University.

JOHANNA T. DWYER (*FNB Liaison*) is the director of the Frances Stern Nutrition Center at New England Medical Center, professor of medicine and community health at the Tufts University School of Medicine, and professor of nutrition at Tufts University School of Nutrition in Boston. She is also senior scientist at the Jean Mayer/USDA Human Nutrition Research Center on Aging at Tufts. Dr. Dwyer is the author or coauthor of more than 100 research articles and 185 review articles published in scientific journals. Her work centers on life-cycle-related concerns such as the prevention of diet-related disease in children and adolescents and maximization of quality of life and health in the elderly. She also has a longstanding interest in vegetarian and other alternative life-styles.

Dr. Dwyer is a past president of the American Institute of Nutrition, past secretary of the American Society for Clinical Nutrition, and past president and current fellow of the Society for Nutrition Education. She served on the Program Development Board of the American Public Health Association from 1989 to 1992 and is a member of the Food and Nutrition Board of the National Academy of Sciences, the Technical Advisory Committee of the Nutrition Screening Initiative, and the Board of Directors of the American Institute of Wine and Food. As the Robert Wood Johnson Health Policy Fellow (1980–1981), she served on the personal staffs of Senator Richard Lugar (R-Indiana) and Senator Barbara Mikulski (D-Maryland).

Dr. Dwyer has received numerous honors and awards for her work in the field of nutrition, including the 1996 W.O. Atwater Award of the U.S. Department of Agriculture and the J. Harvey Wiley Award from the Society for Nutrition Education. She gave the Lenna Frances Cooper Lecture at the annual meeting of the American Dietetic Association in 1990. Dr. Dwyer is currently on the editorial boards of *Family Economics* and *Nutrition Review* and the advisory board of *Clinics in Applied Nutrition*, and she is a contributing editor to *Nutrition Reviews*, as well as a reviewer for the *Journal of the American Dietetic Association, American Journal of Clinical Nutrition*, and *American Journal of Public Health*. She received her D.Sc. and M.Sc. from the Harvard School of Public Health, an M.S. from the University of Wisconsin, and her undergraduate degree with distinction from Cornell University.

SUBCOMMITTEE ON BODY COMPOSITION, NUTRITION, AND HEALTH OF MILITARY WOMEN

BARBARA O. SCHNEEMAN (*Chair*) serves as dean, College of Agricultural and Environmental Sciences, and professor of nutrition in the Departments of Nutrition and of Food Science and Technology and in the Division of Clinical Nutrition and Metabolism (School of Medicine), University of California, Davis. Her professional activities include membership on the Dietary Guidelines Advisory Committee; the Board of Trustees of the International Life Sciences Institute America; and the editorial boards of the *Proceedings of the Society of Experimental Biology* and the Medicine, Food and Nutrition Series of Academic Press, *Nutrition Reviews*, *Journal of Nutrition*, and *California Agriculture*. Dr. Schneeman's professional honors include the Samuel Cate Prescott award for research, the Future Leader Award, and several honorary lectureships. She received her B.S. in food science and technology from the University of California, Davis; Ph.D. in nutrition from the University of California, Berkeley; and postdoctoral training in gastrointestinal physiology at Children's Hospital in Oakland. Dr. Schneeman's research areas include fat absorption, complex carbohydrates, and gastrointestinal function, and she has a strong interest in and appreciation of nutritional issues that affect women throughout the life cycle.

NANCY F. BUTTE is associate professor of pediatrics, Children's Nutrition Research Center, Department of Pediatrics, Baylor College of Medicine, Houston, Texas. She is a current member of the International Dietary Energy Consultancy Group Steering Committee, Executive Committee for the International Society for Research on Human Milk and Lactation, and Society for International Nutrition Research and a former member of the Institute of Medicine Subcommittee on Nutritional Status and Weight Gain during Pregnancy and of the Expert WHO Committee on Physical Status: The Use and Interpretation of Anthropometry. Dr. Butte received her B.S. in food and nutritional sciences, M.P.H. in public health nutrition, and Ph.D. in nutritional sciences from the University of California, Berkeley; she is a registered dietitian. Her research experience includes nutritional needs during pregnancy and lactation, including her current focus on military women.

JOAN M. CONWAY is a research chemist with the Beltsville Human Nutrition Research Center, Diet and Human Performance Laboratory, Beltsville, Maryland and a member of the graduate adjunct faculty, Department of Human Nutrition, University of Maryland, College Park. Dr. Conway has a B.A. in chemistry from St. Joseph's College in Brooklyn, New York, master's degree in science education from City College of New York, master's in human nutrition from Columbia University College of Physicians and Surgeons, and Ph.D. in nutritional biochemistry and metabolism from the Massachusetts Institute of Technology; she is a registered dietitian. Her research activities for USDA include ethnic or racial influences on body composition, physical activity, and free-living energy metabolism in women; body composition methodology; and stable isotope studies among Navy divers and other military Special Forces. Currently, she is a consultant at the Food and Nutrition Division of the Food and Agriculture Organization (FAO) in Rome, where she is facilitating the planning and execution of a proposed Joint FAO/WHO Expert Consultation on Vitamin and Minerals.

STEVEN B. HEYMSFIELD is professor of medicine at Columbia University, College of Physicians and Surgeons in New York. He also currently serves as deputy director of the New York Obesity Research Center and is director of the Human Body Composition Laboratory. Dr. Heymsfield is immediate past president of the American Society of Parenteral and Enteral Nutrition and is an active member of the American Society of Clinical Nutrition and the North American Society for the Study of Obesity. He was recently made an honorary member of the American Dietetic Association. He received his B.A. in chemistry from Hunter College of the City University of New York and his M.D. from Mt. Sinai School of Medicine. Dr. Heymsfield has done extensive research and has clinical experience in the areas of body composition, weight cycling, nutrition, and obesity, especially as they relate to women.

ANNE LOOKER is senior research epidemiologist, National Center for Health Statistics, Division of Health Examination Statistics, where she serves as the center's expert consultant on calcium and iron status data for the National Health and Nutrition Examination Surveys. She is currently serving as director of research projects for the National Osteoporosis Foundation and is a member of the National Institute of Arthritis, Musculoskeletal, and Skin Diseases National Osteoporosis Data Group. Dr. Looker received a B.A. in zoology from Miami University, and M.S. and Ph.D. degrees in nutrition from the Pennsylvania State University; she is a registered dietitian. She has done work in areas that are of special concern to women, such as iron nutrition and osteoporosis.

MARY Z. MAYS is currently the director of Eagle Creek Research Services, San Antonio, Texas. From 1993 to 1995, Dr. Mays served as the planner for science and technology Programs at the U.S. Army Medical Research and Materiel Command. Prior to this, she was the director of the Military Performance and Neuroscience Division at the U.S. Army Research Institute of Environmental Medicine (1990–1993). Dr. Mays earned a B.A. in psychology from Trinity College in San Antonio, Texas, and an M.S. and a Ph.D. in experimental psychology from the University of Oklahoma. Her research interests are focused on the influence of nutrition on cognitive performance.

MARITZA RUBIO-STIPEC is professor, Department of Economics, University of Puerto Rico. She is co-investigator/statistical consultant for a study of the psychiatric epidemiology of mental disorders in Puerto Rico; co-principal investigator/statistical consultant for a child psychiatry epidemiologic study in Puerto Rico; and principal investigator, of the WHO/NIH Joint Project on Diagnosis and Classification of Mental Disorders and Drug Related Problems. Ms. Rubio-Stipec earned her B.A. in economics from the University of Puerto Rico and M.A. in economics from New York University. She has an extensive background in study design, data analysis, epidemiology, and statistics.

JANET C. KING (*FNB Liaison*) is director, U.S. Department of Agriculture Human Nutrition Research Center, Presidio of San Francisco, and professor in the Graduate School, University of California, Berkeley. Prior to her university experience, she worked for the U.S. Department of Defense. She is a member of the Institute of Medicine (IOM) and served as chair of the IOM's Food and Nutrition Board (FNB), and the Subcommittee on Nutrition Status and Weight Gain During Pregnancy. Dr. King received a B.S. in dietetics from Iowa State University and Ph.D. in nutrition from the University of California, Berkeley; she is a registered dietitian.

REBECCA B. COSTELLO (*FNB Staff, through May 1998*) is the former project director for the Committee on Military Nutrition Research and Committee on Body Composition, Nutrition, and Health of Military Women. Prior to joining the FNB staff, she served as research associate and program director for the Risk Factor Reduction Center, a referral center for the detection, modification, and prevention of cardiovascular disease through dietary and/or drug interventions, at the Washington Adventist Hospital in Takoma Park, Maryland. She received her B.S. and M.S. in biology from the American University, Washington, D.C., and a Ph.D. in clinical nutrition from the University of Maryland at College Park. She has active membership in the American Institute of Nutrition, American College of Nutrition, American Dietetic Association, and American Heart Association Council on Epidemiology. Dr. Costello's areas of research interest include mineral nutrition, dietary intake methodology, and chronic disease epidemiology.

MARY I. POOS (*FNB Staff, Study Director*) is project director for the Committee on Military Nutrition Research (CMNR). She joined the Food and Nutrition Board of the Institute of Medicine in November 1997. She has been a project director for the National Academy of Sciences since 1990. Prior to officially joining the FNB staff, she served as a project director for the National Research Council's Board on Agriculture for more than seven years, two of which were spent on loan to the FNB. Her work with the FNB includes senior staff officer for the IOM report *The Program of Research for Military Nursing* and study director for the reports *A Review of the Department of Defense's Program for Breast Cancer Research* and *Vitamin C Fortification of Food Aid Commodities*. Currently, she also serves as study director to the Subcommittee on Interpretation and Uses of Dietary Reference Intakes, and directs the planning activities in global food and nutrition. While working with the Board on Agriculture, Dr. Poos was responsible for the Committee on Animal Nutrition and directed the production of seven reports in the *Nutrient Requirements of Domestic Animals* series, including a letter report to the commissioner of FDA concerning the importance of selenium in animal nutrition. Prior to joining the National Academy of Sciences she was consultant/owner of Nutrition Consulting Services of Greenfield, , Massachusetts; assistant professor in the Department of Veterinary and Animal Sciences at the University of Massachusetts, Amherst; and adjunct assistant professor in the Department of Animal Sciences, University of Vermont. She received her B.S. in biology from Virginia Polytechnic Institute and State University, and a Ph.D. in animal sciences (nutrition/biochemistry) from the University of Kentucky; she completed a postdoctoral fellowship in the Department of Animal Sciences Area of Excellence Program at the University of Nebraska. Dr. Poos's areas of research interest include protein and nitrogen metabolism and nutrition–reproduction interactions.

SYDNE J. NEWBERRY (*FNB Staff*) was the former program officer for the Committee on Military Nutrition Research and Subcommittee on Body Composition, Nutrition, and Health of Military Women. She is currently employed at the University of California, Los Angeles School of Public Health/Center for Human Nutrition. Prior to joining the FNB staff, she served as project director for the Women's Health Project and adjunct assistant professor in the Department of Family Medicine, Wright State University School of Medicine; as a behavioral health educator for a hospital-based weight management program in Dayton, Ohio; and as a research associate at the Ohio State University Biotechnology Center. She received her B.A. from Brandeis University and her Ph.D. in nutritional biochemistry and metabolism from the Massachusetts Institute of Technology; she completed an NIH postdoctoral fellowship in the Departments of Biochemistry and Molecular Genetics at Ohio State. Dr. Carlson-Newberry's areas of research interest include eating disorders and diabetes management.

AUTHORS

DENNIS M. BIER is professor of pediatrics, director of the USDA Children's Nutrition Research Center, and program director of the Pediatric Clinical Research Center at Baylor College of Medicine. He is also president of the International Pediatric Research Foundation, associate editor of the *Annual Review of Nutrition*, and chair of the USDA Human Studies Review Committee.

Previously, Dr. Bier was professor of Pediatrics and Medicine at Washington University in St. Louis, where he was director of the Mass Spectrometry Facility and codirector of the Pediatric Endocrinology and Metabolism Division. He has been editor-in-chief of *Pediatric Research*, chair of the NIH Nutrition Study Section, chair of the NIH General Clinical Research Center Committee, chair of the NIH/NICHD Expert Panel Five-Year Plan for Nutrition Research and Training, and president of the American Society for Clinical Nutrition. He also has served as a member of the various additional scientific advisory panels, including the HHS/USDA Dietary Guidelines Advisory Committee, the FNB, the FDA Food Advisory Committee, the Medical Science Advisory Board of the Juvenile Diabetes Foundation, the Steering Committee of the Pediatric Scientist Development Program, and the Advisory Board of the National Stable Isotopes Resource at Los Alamos National Laboratory.

ALANA D. CLINE is a research dietitian in the Military Nutrition and Biochemistry Division, U.S. Army Research Institute of Environmental Medicine (USARIEM), Natick, Massachusetts. She has been an Army dietitian for 25 years, assigned to a variety of military installations. Her graduate education includes an M.Ed. in nutrition education from Incarnate Word College, San Antonio, Texas, in 1982 and a Ph.D. in applied nutrition from Colorado State University, Fort Collins, in 1993. She currently is a member of the Department of Defense (DoD) Nutrition Committee, contributing toward the decision-making process that determines the DoD Strategic Plan for Nutrition Programs for all the military services. Her research interests include women's nutritional health issues, cardiovascular disease, and nutrition and performance.

KARL E. FRIEDL is the research area manager for the Army Operational Medicine Research Program at the U.S. Army Medical Research and Materiel Command, Fort Detrick, Maryland. Prior to this assignment, he was an Army research physiologist in the Occupational Physiology Division at USARIEM, where he specialized in physical and biochemical limits of prolonged, intensive military training. Previously, LTC Friedl worked in the Department of Clinical Investigation at Madigan Army Medical Center in Tacoma, Washington, performing studies in endocrine physiology. He received his Ph.D. in physiology in 1984 from the Institute of Environmental Stress at the University of California, Santa Barbara.

HARRIS R. LIEBERMAN is deputy chief of the Military Nutrition and Biochemistry Division of the U.S. Army Research Institute of Environmental Medicine in Natick, Massachusetts. Dr. Lieberman is an internationally recognized expert in the area of nutrition and behavior and has published more than 90 original, full-length papers in scientific journals and edited books. He has been an invited lecturer at numerous national and international conferences, government research laboratories, and universities. Dr. Lieberman received his

Ph.D. in physiological psychology in 1977 from the University of Florida. Upon completing his graduate training he was awarded an NIH fellowship to conduct postdoctoral research at the Department of Psychology and Brain Science at the Massachusetts Institute of Technology. In 1980 he was appointed to the research staff at MIT and established an interdisciplinary research program in the Department of Brain and Cognitive Sciences to examine the effects of various food constituents and drugs on human behavior and brain function. Key accomplishments of the laboratory included the development of appropriate methods for assessing the effects of food constituents and other subtle environmental factors on human brain function and the determination that specific foods and hormones reliably alter human performance.

In 1990, Dr. Lieberman joined the civilian research staff of USARIEM where he has continued his work in nutrition and behavior. He has addressed the effects of various nutritional factors, diets, and environmental stress on animal and human performance, brain function, and behavior. His research program has focused on development and application of emerging technologies to sustaining and enhancing human performance.

TIMOTHY J. MAHER is Sawyer Professor of Pharmaceutical Sciences and director of the Division of Pharmaceutical Sciences at the Massachusetts College of Pharmacy and Allied Health Sciences, in addition to being on the faculty at the Massachusetts Institute of Technology in the Department of Brain and Cognitive Sciences. His area of research involves the investigation of the ability of amino acids and other dietary components to influence the neurochemical composition of the central nervous system and affect function in a pharmacological fashion. He served on the Federation of American Societies for Experimental Biology Life Science Research Office's ad hoc expert panels that investigated the safety of amino acids as dietary supplements and has analyzed the adverse reactions to monosodium glutamate for the U.S. Food and Drug Administration.

D. JOE MILLWARD is professor of nutrition and director of the Center for Nutrition and Food Safety at the University of Surrey in the United Kingdom. He trained as a biochemist, obtaining his initial degree and D.Sc. from the University of Wales and his Ph.D. from the University of the West Indies. He worked for many years with John Waterlow, initially at the Medical Research Council's (MRC's) Tropical Metabolism Research Unit in Jamaica and subsequently in the Department of Human Nutrition at the London School of Hygiene and Tropical Medicine, as reader in nutritional biochemistry. He has taught nutrition and metabolism for more than 25 years. His research has focused on the regulation of protein metabolism and turnover, and the metabolic basis of protein and amino acid requirements. He is an editor of the *American Journal of Physiology*, *British Journal of Nutrition*, *Nutrition Research Reviews*, and *Clinical Science*.

K. SREEKUMARAN NAIR has been a professor of medicine and a consultant in endocrinology, metabolism and diabetes at Mayo Clinic and Foundation since 1994. He received his MBBS degree from the University of Kerala, India in 1993, an MD degree from New York State University in 1998, and a Ph.D. from the Council of National Academic Awards, London in 1984. He is a fellow of the Royal College of Physicians, London and the American College of Physicians. He was elected in 1992 to the American Society of Clinical Investigation and in 1999 to the Association of Physicians. He served on NIH Metabolism and Aging Study sections and served on NASA Biological Review Panels.

MICHAEL J. RENNIE is Symers Professor of Physiology, Department of Anatomy and Physiology at the University of Dundee. Dr. Rennie received his B.S. in biological chemistry/zoology from Hull, his M.S. from Manchester, and his Ph.D. from Glasgow. Dr. Rennie's major research interests are protein, amino acids, and metabolism. He has published more than 200 papers and chapters. He was head of the Department of Anatomy and Physiology at the University of Dundee; Wellcome senior lecturer, Department of Medicine, and lecturer in human metabolism at the University College London Medical School. Dr. Rennie is a member of MRC Physiological Medicine and Infections Board Grants Committee and chairman of the Committee of Heads of Department of Physiology for the United Kingdom. Dr. Rennie is also Honorary Secretary of the Clinical Nutrition and Metabolism Group of the Nutrition Society. His honors include the Arvid Wretlind Lecturer of the European Society of Parenteral and Enteral Nutrition, recipient of the Rank Prize Funds Award, and fellow of the Royal Society of Edinburgh. Dr. Rennie is also the editor of the *British Journal of Intensive Care* and executive editor of the *International Journal of Intensive Care*.

ANTON J. M. WAGENMAKERS studied chemistry (organic chemistry and biochemistry) in Nijmegen from 1972 to 1979. In 1984, he obtained the Ph.D. at the same university, where his Ph.D. thesis was "Branched-Chain Amino Acid Degradation in Muscle." From 1984 until 1988, Dr. Wagenmakers was a senior research fellow of the Muscular Dystrophy Group of Great Britain in Liverpool. In 1988, he moved to the Department of Human Biology, Faculty of Health Sciences, Maastricht University, where he was appointed an associate professor in 1996. Dr. Wagenmakers' main research interests are amino acid, carbohydrate, and fat metabolism in health (rest and exercise) and disease. He has long-standing (5–10 years) collaborations with international experts in stable isotope methodology (Professors D. Halliday and M. Rennie) and exercise physiology (Professor B. Saltin). Dr. Wagenmakers is scientific coordinator of the Stable Isotope Research Centre in Maastricht and, as such, occupies a central position in the metabolism- and nutrition-related research within several divisions of Nutrition and Toxicology Research Institute

Maastricht. He is a member of the working groups on nutrition and diabetes and metabolism of the Netherlands Organisation for Scientific Research. He is an international expert in amino acid biochemistry and physiology and acts as a referee for many peer-reviewed journals (*Journal of Physiology* (UK), *American Journal of Physiology*, *Journal of Applied Physiology*, *Biochimica Biophysica Acta*, and *Clinical Science*). He acts as a peer reviewer for the Netherlands Organisation for Scientific Research and the Wellcome Trust in the United Kingdom and is a member of the external review committee of the Copenhagen Muscle Research Centre, Denmark, and of the Macronutrient Metabolism Group of the Nutrition Society in the United Kingdom.

MACKENZIE WALSER received a B.A. from Yale in 1944 and an M.D. from Columbia College of Physicians and Surgeons in 1948. After residency training in medicine at Massachusetts General Hospital, he joined the Department of Internal Medicine at Southwestern Medical School. He then served two years at the Naval Medical Research Institute in Bethesda, Maryland. From 1954 to 1957, he was an investigator in the Renal and Electrolyte Section of the National Heart Institute. In 1957, he joined the Department of Pharmacology and Experimental Therapeutics at the Johns Hopkins School of Medicine with a secondary appointment in the Department of Medicine. He has chaired the local chapter of the American Association of University Professors, the Medical School Council, the Joint Committee on Clinical Investigation, the University-wide Invention Committee, and the Nutrition Advisory Committee.

Dr. Walser received the Experimental Therapeutics Award of the American Society for Pharmacology and Experimental Therapeutics, the Herman Award of the American Society of Clinical Nutrition, and the first Addis Award of the International Society for Renal Nutrition and Metabolism. Dr. Walser has served on editorial boards of the *American Journal of Kidney Diseases*, *American Journal of Physiology*, *Calcified Tissue International*, *Clinical Nutrition*, *Clinical Science*, *Journal of Clinical Investigation*, *Kidney International*, *Mineral and Electrolyte Metabolism*, *International Yearbook of Nephrology*, and *Renal Physiology and Biochemistry*.

JOHN P. WARBER is a research dietitian assigned to the Military Nutrition and Biochemistry Division of the U.S. Army Research Institute of Environmental Medicine, Natick, Massachusetts. He received his doctorate in public health nutrition in 1994 from Loma Linda University School of Public Health, Loma Linda, California. Upon completing his doctoral training, he was assigned to his current position for a postdoctoral utilization tour. LTC Warber has dual master of science degrees in nutrition and exercise physiology from the University of North Carolina. He is a member of the American Dietetic Association, maintains credentials as a certified health fitness instructor from the American College of Sports Medicine, and is a certified strength and conditioning specialist of the National Strength and Conditioning Association.

Previously, he was assigned as part of the initial faculty and staff at the U.S. Army Soldier Physical Fitness School at Fort Benjamin Harrison, Indiana (now relocated to Fort Benning, Georgia) where he held a dual position on the teaching faculty and training and doctrine staff. He later was assigned as fitness coordinator for the 4th Infantry Division, Fort Carson, Colorado. LTC Warber's research interests include nutrition monitoring, dietary assessment, and performance-enhancing nutritional aids.

ROBERT R. WOLFE has been professor of surgery and nutritional biochemistry at the University of Texas Medical Branch, Galveston, and chief of metabolism at the Shriners Burns Institute since 1983. From 1976 to 1983 he was assistant and associate professor of surgery at Harvard Medical School and the Shriners Burns Institute, Boston. Dr. Wolfe has served as associate editor of the *American Journal of Physiology, Endocrinology, and Metabolism* and as an editorial board member of several other journals. He has been a regular member of the Surgery, Anesthesiology, and Trauma Study Section and many other ad hoc study sections of the NIH, as well as a member of the Committee on Research Review of the American Diabetes Association. He has been an active member of the American Physiological Society and American Institute of Nutrition. Dr. Wolfe received a B.A. in biology from the University of California, Berkeley and a Ph.D. in biology from the University of California, Santa Barbara.

VERNON R. YOUNG is a professor of nutritional biochemistry at the Massachusetts Institute of Technology and director of the Mass Spectrometry Facility, Shriners Burns Hospital, Boston. Dr. Young received a B.Sc. in agriculture from the University of Reading, United Kingdom, and a Ph.D. in nutrition from the University of California, Davis. He later received a D.Sc. from the University of Reading for his research on various aspects of muscle and whole-body protein metabolism. Dr. Young has served as president of the American Institute of Nutrition (now American Society for Nutritional Sciences). He is a member of the Food and Nutrition Board, the American Society of Clinical Nutrition, and the Nutrition Society (UK). He has served on numerous editorial boards, including the *Journal of Nutrition* and the *American Journal of Clinical Nutrition*. In 1990, Dr. Young was elected to the National Academy of Sciences and in 1993 to the Institute of Medicine. His research has focused mainly on human protein and amino acid metabolism and nutritional requirements. He is the recipient of numerous awards including the Mead-Johnson and Borden Awards from the ASNS, the McCollum Award from the American Society of Clinical Nutrition, the Rank PRIZE in Nutrition (UK), the Bristol-Myers Squibb/Mead Johnson Award for Excellence in Nutrition Research (U.S.) and the Danone International Prize for Nutrition (France). He also received an M.D. (honoris causa) from Uppsala University, Sweden.

C

Acronyms and Abbreviations

3MH	3-methylhistidine
AA	amino acid
AARP	Amino Acid Requirement Pattern
ACSM	American College of Sports Medicine
AIDS	acquired immunodeficiency syndrome
AMS	acute mountain sickness
AR	Army regulation
ASM	appendicular skeletal muscle
ASNS	American Society for Nutritional Sciences
ATFSM	adipose tissue-free skeletal muscle
ATP	adenosine 5′-triphosphate
BBB	blood-brain barrier
BC	branched chain
BCKAD	branched-chain α-keto acid dehydrogenase
BCT	basic combat training
BCAA	branched chain amino acid

BCNH	Subcommittee on Body Composition, Nutrition and Health of Military Women
BHA	butylated hydroxyanisole
BHT	butylated hydroxytoluene
BIA	bioelectric impedance analysis
BV	biological value
BW	body weight
CGMP	Current Good Manufacturing Practice
CHO	carbohydrate
CMNR	Committee on Military Nutrition Research
CNS	central nervous system
CoA	coenzyme A
COX	cytochrome c oxidase
Cr	creatine
CT	computerized tomography
CV	coefficient of variation
DoD	Department of Defense
DRI	dietary reference intake
DRV	dietary reference values
DSHEA	Dietary Supplement Health and Education Act
DXA	dual-energy x-ray absorptiometry
EMS	eosinophilia-myalgia syndrome
FAO	Food and Agriculture Organization
FASEB	Federation of the American Societies for Experimental Biology
FDA	Food and Drug Administration
FFA	free fatty acid
FFM	fat-free body mass
FFSM	fat-free skeletal muscle
FNB	Food and Nutrition Board
FSR	fractional synthesis rate
GFR	glomerular filtration rate
GH	growth hormone
GDP	guanosine 5′-diphosphate
GTP	guanosine 5′-triphosphate
GTW	Go to War Ration
HHS	Department of Health and Human Services
HPLC	high-pressure liquid chromatography
IAA	indispensable amino acid
IFT	Institute of Food Technologists
IGF-I	insulin-like growth factor-I
IL-1	interleukin-1
IL-4	interleuking-4
IL-8	interleukin-8

IL-10	interleukin-10
IL-13	interleukin-13
IOM	Institute of Medicine
IVNA	in vivo neutron activation analysis
LNAA	large neutral amino acid
LRPR	Long-Range Patrol Ration
LSRO	Life Sciences Research Office
L-TRP-EMS	L-Tryptophan Eosinophilia Myalgia Syndrome
MCI	Meal, Combat Individual
MD	metabolic demand
MIT	Massachusetts Institute of Technology
MIT-AARP	Massachusetts Institute of Technology Amino Acid Requirement Pattern
MRC	Medical Research Council
MRDA	Military Recommended Dietary Allowance
MRE	Meal, Ready-to-Eat
MRI	magnetic resonance imaging
mRNA	messenger RNA (ribonucleic acid)
NAD^+	nicotinamide adenine dinucleotide
NAIDS	nutritionally acquired immune dysfunction syndromes
NAS	National Academy of Sciences
NASA	National Aeronautics and Space Administration
NCHS	National Center for Health Statistics
NEFA	nonesterified fatty acid
NCO	Noncommissioned officer
NHANES	National Health and Nutrition Examination Survey
NIAMS	National Institute of Arthritis, Musculoskeletal, and Skin Diseases
NICHD	National Institute of Child Health and Human Development
NIH	National Institutes of Health
NO	nitric oxide
NPU	net protein utilization
NRC	National Research Council
NRDEC	Natick Research, Development and Engineering Center
NSM	nonskeletal muscle
OAAL	obligatory amino acid losses
OCS	Officer Candidate School
OMD	obligatory metabolic demand
ONL	obligatory nitrogen loss
OOL	oxidative amino acid loss
PDCAAS	protein-digestibility corrected amino acid score
PEM	protein energy malnutrition
POMS	Profile of Mood States
PP	protein powders

PX	post exchange
RCW	Ration, Cold Weather
RDA	Recommended Dietary Allowance
RLW-30	Ration, LightWeight-30
RNA	ribonucleic acid
RNI	recommended nutritional intake
RNR	relative nitrogen requirement
RPV	relative protein value
SAA	sulfur amino acids
SD	standard deviation
SEE	Standard Error of the Estimate
SEM	standard error of the mean
SFAS	Special Forces Assessment and Selection
SM	skeletal muscle
SMCM	skeletal muscle cell mass
SMPro	skeletal muscle protein
T_3	triiodothyronine
TAA	total aromatic amino acid
TBK	total body potassium
TBN	total body nitrogen
TCA	tricarboxylic acid
TNF	tumor necrosis factor
TRNA	transfer RNA (ribonucleic acid)
TSA	total sulfur amino acid
UCLA	University of California, Los Angeles
UGR	unitized group ration
UN	United Nations
UNU	United Nations University
USAF	U.S. Air Force
USAMRMC	U.S. Army Medical Research and Materiel Command
USARIEM	U.S. Army Research Institute of Environmental Medicine
USASSC	U.S. Army Soldier Systems Center
USDA	U.S. Department of Agriculture
USMA	U.S. Military Academy
VAMS	Visual Analog Mood Scale
VO_{2max}	maximal oxygen uptake
WHO	World Health Organization

D

Protein and Amino Acids— A Selected Bibliography

On the following pages is a selection of references dealing with the role of protein and amino acids in sustaining and enhancing performance. This bibliography was compiled from the joint reference lists of the chapters in this report, selected references from a limited literature search, and references recommended by the invited speakers as background reading for workshop participants. As a result, references that are historical in nature are included in this listing along with the most current studies.

Ahlers, S.T., J.R. Thomas, J. Schrot, and D. Shurtleff. 1994. Tyrosine and glucose modulation of cognitive deficits. Pp. 301–320 in Food Components to Enhance Performance, B. Marriott, ed. Washington, D.C.: National Academy Press.

Albino, J.E., and R.B. Mateo. 1995. Nitric oxide. Pp. 99–123 in Amino Acid Metabolism and Therapy in Health and Nutritional Disease, L.A. Cynober, ed. Boca Raton, Fla.: CRC Press.

Allen, L.H., E.A. Oddoye, and S. Margen. 1979. Protein-induced hypercalciuria: A longer term study. Am. J. Clin. Nutr. 32:741–749.

Allison, J.B., J.A. Anderson, and R.D. Seeley. 1947. Some effects of methionine on the utilization of nitrogen in the adult dog. J. Nutr. 33:361–370.

Andersen, P., and B. Saltin. 1985. Maximal perfusion of skeletal muscle in man. J. Physiol. 366:233–249.

Anderson, C.F., J.A. Velosa, P.P. Frohnert, V.E. Torres, K.P. Offord, J.P. Vogel, J.V. Donadio, Jr., and D.M. Wilson. 1985. The risks of unilateral nephrectomy: Status of kidney donors 10 to 20 years postoperatively. Mayo Clin. Proc. 60:367–374.

Anderson, E.C. 1963. Three component body composition analysis based on potassium and water determinations. Ann NY Acad Sci.110:189–212.

Anderson, J.W., B.M. Johnstone, and M.E. Cooke-Newell. 1995. Meta-analysis of the effects of soy protein intake on serum lipids. N. Engl. J. Med. 333(5): 276–282.

Anderson, S., and B.M. Brenner. 1987. The aging kidney: structure, function, mechanisms, and therapeutic implications. J. Am. Geriatr. Soc. 35:590–593.

Anderson, S.A., and D.J. Raiten (eds.). 1992. Safety of amino acids as dietary supplements. Bethesda, Md.: Federation of the American Societies for Experimental Biology.

Andrews, P.M., and S.B. Bates. 1986. Dietary protein prior to renal ischemia dramatically affects postischemic kidney damage. Kidney Int. 30:299–303.

Anggard, E. 1994. Nitric oxide: Mediator, murderer and medicine. Lancet 343:1199–1206.

AR (Army Regulation) 40-250. 1947. See U.S. Department of the Army, 1947.

AR (Army Regulation) 40-25. 1985. See U.S. Departments of the Army, the Navy, and the Air Force.

AR (Army Regulation) 40-25. 1989. See U.S. Departments of the Army, the Navy, and the Air Force, 1985.

Ardawi, M.S.M., and E.A. Newsholme. 1983. Glutamine metabolism in lymphocytes of the rat. Biochem. J. 212:835–842.

Artuson, G. 1961. Pathophysiological aspects of the burn syndrome. Acta Chirg. Scand. 274:7–64.

Askew, E.W. 1989. Nutrition for a cold environment. Phys. Sportsmed. 17:77–89.

Askew, E.W., J.R. Claybaugh, S.A. Cucinell, A.J. Young, and E.G. Szeto. 1986. Nutrient intakes and work performance of soldiers during seven days of exercise at 7,200 ft. altitude consuming the Meal, Ready-to-Eat ration. Technical Report No. T3-87. Natick, Mass.: U.S. Army Research Institute of Environmental Medicine.

Askew, E.W., I. Munro, M.A. Sharp, S. Siegel, R. Popper, M.S. Rose, R.W. Hoyt, K. Reynolds, H.R. Lieberman, D. Engell, and C.P. Shaw. 1987. Nutritional status and physical and mental performance of soldiers consuming the Ration, Lightweight or the Meal, Ready-to-Eat military field ration during a 30 day field training exercise (RLW-30). Technical Report No. T7-87. Natick, Mass.: U.S. Army Research Institute of Environmental Medicine.

Atinmo, T., C.M.F. Mbofung, M.A. Hussain, and B.O. Osotimehin. 1985. Human protein requirements: Obligatory urinary and faecal nitrogen losses and the factorial estimation of protein needs of Nigerian male adults. Br. J. Nutr. 54:605–611.

Atinmo, T., C.M.F. Mbofung, G. Eggum, and B.O. Osotimehin. 1988. Nitrogen balance study in young Nigerian adult males using four levels of protein intake. Br. J. Nutr. 60:451–458.

Aulick, L.H., and D.W. Wilmore. 1979. Increased peripheral amino acid release following burn injury. Surgery 30:196–197.

Baker-Fulco, C.J., J.C. Buchbinder, S.A. Torri, and E.W. Askew. 1992. Dietary Status of Marine Corps officer candidates. Fed. Am. Soc. Exp. Biol. J. [FASEB J] 6(4):A1682.

Baker-Fulco, C.J., S.A. Torri, J.E. Arsenault, and J.C. Buchbinder. 1994. Impact of menu changes designed to promote a training diet [abstract]. J. Am. Diet. Assn. 94:A9.

Baker-Fulco, C.J. 1995. Overview of dietary intakes during military exercises. Pp. 121–149 in Not Eating Enough, Overcoming Underconsumption of Military Operational Rations, B.M. Marriott, ed. Institute of Medicine. Washington, D.C.: National Academy Press.

Balagopal, P., O.E. Rooyackers, D.B. Adey, P.A. Ades, and K.S. Nair. 1997. Effects of aging on in vivo synthesis of skeletal muscle myosin heavy-chain and sarcoplasmic protein in humans. Am. J. Physiol. 273:E790–E800

Ballevre, O., A. Cadenhead, A.G. Calder, W.D. Rees, G.E. Lobley, M.F. Fuller, and P.J. Garlick. 1990. Quantitative partition of threonine oxidation in pigs: Effect of dietary threonine. Am. J. Physiol. 259:E483–E491.

Bams, J.L., and D.R. Miranda. 1985. Outcome and costs of intensive care. Intensive Care Med. 11:234–241.

Banderet, L.E., and H.R. Lieberman. 1989. Treatment with tyrosine, a neurotransmitter precursor, reduces environmental stress in humans. Brain Res. Bull. 22:759–762.

Banister, E.W., and B.J.C. Cameron. 1990. Exercise-induced hyperammonemia: Peripheral and central effects. Int. J. Sports Med. 11:S129–S142.

Bararc-Nieto, M., G.B. Spurr, H. Lotero, and M.G. Maksud. 1978. Body composition in chronic under nutrition. Am. J. Clin. Nutr. 31:23–41.

Barbul, A., S.A. Lazarou, D.T. Efron, H.L. Wasserkrug, and G. Efron. 1990. Arginine enhances wound healing and lymphocyte immune responses in humans. Surgery 108(2):331–336.

Basile-Filho, A., L. Beaumier, A.E. El-Khoury, Y.M. Yu, M. Kenneway, R.E. Gleason, and V.R. Young. 1998. Twenty-four hour L-[L-^{13}C]tyrosine and L-[3,3-^{2}H]phenylalanine oral tracer aromatic amino acid requirements in adults. Am. J. Clin. Nutr. 67:640–659

Baumgartner, R.N., W.C. Chumlea, and A.F. Roche. 1989. Estimation of body composition from bioelectric impedance of body segments. Am. J. Clin. Nutr. 50(2):221–226.

Baumgartner R.N., R.L. Rhyne, P.J. Garry, S.B. Heymsfield. 1993. Imaging techniques and anatomical body composition in aging. J Nutr. 123:444–448.

Baumgartner, R.N., R. Ross, S.B. Heymsfield, Z. Wang, D. Gallagher, Y. Martel, J. De Guise, W. Brooks. In Press. Cross-validation of DXA versus MRI methods of quantifying appendicular skeletal muscle. Applied Radiation and Isotopes.

Bay, W.H., and L.A. Hebert. 1987. Kidney donors and protein intake [letter to the editor]. Ann. Intern. Med. 107:427.

Bean, WB., J.B. Youmans, W.F. Ashe, N. Nelson, D.M. Bell, L.M. Richardson, C.E. French, C.R. Henderson, R.E. Johnson, G.M. Ashmore, K.N. Halverson, and J. Wright. 1944. Project No. 30: Test of Acceptability and Adequacy of U.S. Army C, K, and 10-in-1, and Canadian Army Mess Tin Rations. Fort Knox, Ky.: Armored Medical Research Laboratory.

Bean, W.B., C.R. Henderson, R.E. Johnson, and L.M. Richardson. 1946. Nutrition Survey in Pacific Theater of Operations. Report to the Surgeon General. Bull. U.S. Army Med. Dept. 5(6):697.

Begum, A., A.N. Radhakrishnan, and S.M. Pereira. 1970. Effect of amino acid composition of cereal-based diets on growth of preschool children. Am. J. Clin. Nutr. 23:1175–1183.

Beisel, W.R. 1992. Metabolic responses of the host to infections. Pp. 1–13 in Textbook of Pediatric Infectious Diseases, Vol. I, 3rd ed., R.D. Feigin and J.D. Cherry, eds. Philadelphia: W.B. Saunders Co.

Beisel, W.R., W.D. Sawyer, E.D. Ryll, and D. Crozier. 1967. Metabolic effects of intracellular infection in man. Ann. Int. Med. 67:744–779.

Belcher, H.J., D. Mercer, K.C. Judkins, S. Shalaby, S. Wise, V. Marks, and N.S. Tanner. 1989. Biosynthetic human growth hormone in burned patients: A pilot study. Burns 15:99–107.

Bender, A.E. 1961. Determination of the nutritive value of proteins by chemical analysis. Pp. 407–415 in Progress in Meeting Protein Needs of Infants and Preschool Children. Pub. 843. Washington, D.C.: National Academy of Sciences and National Research Council.

Benevenga, N.J., and R.D. Steele. 1984. Adverse effects of excessive consumption of amino acids. Ann. Rev. Nutr. 4:157–181.

Benevenga, N.J., M.J. Gahl, T.D. Crenshaw, and M.D. Fink. 1994. Protein and amino acid requirements for maintenance and growth of laboratory rats. J. Nutr. 124:451–453.

Bergström, J., and E. Hültman. 1967. Muscle glycogen synthesis after exercise: An enhancing factor localized to the muscle cells in man. Nature 201:309–310.

Bergström, J., P. Fürst, and E. Vinnars. 1990. Effect of a test meal, without and with protein, on muscle and plasma free amino acids. Clin. Sci. 79:331–337.

Berneis, K., R. Ninnis, J. Girard, B.M. Frey, and U. Keller. 1997. Effects of insulin-like growth factor I combined with growth hormone on glucocorticoid-induced whole-body protein catabolism in man. Clin. Endocrinol. Metab. 82:2528–2534.

Bessey, P.Q., and K.A. Lowe. 1993. Early hormonal changes affect the catabolic response to trauma. Ann. Surg. 218:476–489.

Bessey, P.Q., J.M. Watters, T.T. Aoki, and D.W. Wilmore. 1984. Combined hormonal infusions simulate the metabolic response to injury. Ann. Surg. 200:264–281.

Betz, A.L., G.W. Goldstein, and R. Katzman. 1994. Blood–brain–cerebrospinal fluid barriers. Pp. 681–698 in Basic Neurochemistry: Molecular, Cellular, and Medical Aspects, 5th ed., G.J. Siegal, ed. New York: Raven Press

Bhasin, S., T. Storer, N. Berman, C. Callegari, B. Clevenger, J. Phillips, T.J. Bunell, R. Tricker, A. Shirari, and R. Casaburi. 1996. The effects of supraphysiologic doses of testosterone on muscle size and strength in normal men. N. Engl. J. Med. 335:1–7.

Bier, D.M., D.E. Matthews, and V.R. Young. 1985. Interpretation of amino acid kinetic studies in the context of whole body protein metabolism. Pp. 27–36 in Substrate and Energy Metabolism in Man, J.S. Garrow and D. Halliday, eds. London and Paris: John Libbey.

Bingham, S., B. Pignatelli, J. Pollock, A. Ellul, C. Mallaveille, G. Gross, S. Runswick, J.H. Cummings, and I.K. O'Neill. 1996. Does increased formation of endogenous N nitroso compounds in the human colon explain the association between red meat and colon cancer? Carcinogenesis 17:515–523.

Biolo, G., R.Y.D. Fleming, S.P. Maggi, and R.R. Wolfe. 1995a. Transmembrane transport and intracellular kinetics of amino acids in human skeletal muscle. Am. J. Physiol. 268(31):E75–E84.

Biolo, G., R.Y.D. Fleming, and R.R. Wolfe. 1995b. Physiologic hyperinsulinemia stimulates protein synthesis and enhances transport of selected amino acids in human skeletal muscle. J. Clin. Invest. 95:811–819.

Biolo, G., S.P. Maggi, B.D. Williams, K.D. Tipton, and R.R. Wolfe. 1995c. Increased rates of muscle protein turnover and amino acid transport after resistance exercise in humans. Am. J. Physiol. 268:E514–E520.

Black, P.R., D.C. Brooks, P.Q. Bessey, R.R. Wolfe, and D.W. Wilmore. 1982. Mechanisms of insulin resistance following surgery. Ann. Surg. 196:420–435.

Block, G.D., R.J. Wood, and L.H. Allen. 1980. A comparison of the effects of feeding sulfur amino acids and protein on urine calcium in man. Am. J. Clin. Nutr. 33:2128–2136.

Blomstrand, E., and E. A. Newsholme. 1996. Influence of ingesting a solution of branched-chain amino acids on plasma and muscle concentrations of amino acids during prolonged submaximal exercise. Nutrition 12:485–490.

Blomstrand, E., P. Hassmén, B. Ekblom, and E.A. Newsholme. 1991. Administration of branched-chain amino acids during sustained exercise—Effects on performance and on plasma concentration of some amino acids. Eur. J. Appl. Physiol. 63:83–88.

Blomstrand, E., S. Andersson, P. Hassmén, B. Ekblom, and E.A. Newsholme. 1995. Effect of branched-chain amino acid and carbohydrate supplementation on the exercise induced change in plasma and muscle concentration of amino acids in human subjects. Acta Physiol. Scand. 153:87–96.

Blomstrand, E., P. Hassmen, S. Ek, B. Ekblom, and E.A. Newsholme. 1997. Influence of ingesting a solution of branched-chain amino acids on perceived exertion during exercise. Acta Physiol. Scand. 159:41–49.

Bodwell, C.E., E.M. Schuster, E. Kyle, B. Brooks, M. Womack, P. Steele, and R. Ahrens. 1979. Obligatory urinary and fecal nitrogen losses in young women, older men and young men: Factorial estimation of adult human protein requirements. Am. J. Clin. Nutr. 32:2450–2459.

Bolourchi, S., C.M. Friedeman, and O. Mickelson. 1968. Wheat flour as a source of protein for adult human subjects. Am. J. Clin. Nutr 21:827–835.

Borchers, J., and G.E. Butterfield. 1992. The effect of meal composition on protein utilization following an exercise bout. Med. Sci. Sports. Exer. 24:S51.

Bortz W.M. II. 1982. Disuse and aging. J. Amer. Med. Assoc. 248:1203–1208

Bourgoignie, J.J., G. Gavellas, S.G. Sabnis, and T.T. Antonovych. 1994. Effect of protein diets on the renal function of baboons (*Papio hamadryas*) with remnant kidneys: A 5-year follow-up. Am. J. Kidney Dis. 23:199–204.

Bovée, K.C. 1991. Influence of dietary protein on renal function in dogs. J. Nutr. 121:S128–S139.

Brenner, B.M., T.W. Meyer, and T.W. Hostetter. 1982. Dietary protein intake and the progressive nature of kidney disease: The role of hemodynamically moderated glomerular injury in the pathogenesis of progressive glomerular sclerosis in aging, renal ablation and intrinsic renal disease. New Engl. J. Med. 307:652–659.

Brenner, I.K.M., P.N. Shek, and R.J. Shepherd. 1994. Infection in athletes. Sports Med. 17:86–107.

Breslau, N.A., L. Brinkley, K.D. Hill, and C.Y.C. Pak. 1988. Relationship of animal protein-rich diet to kidney stone formation and calcium metabolism. J. Clin. Endocrinol. Metab. 66:140–146.

Brodsky, I.G., P. Balagopal, and K.S. Nair. 1996. Effects of testosterone replacement on muscle mass and muscle protein synthesis in hypogonadal men. J. Clin. Endocrinol. 81:3469–3475.

Brotherhood, J.R. 1984. Nutrition and sports performance. Sports Med. 1:350–389.

Brzezinski, A., J. Wurtman, R.J. Wurtman, R. Gleason, T. Nader, and B. Laferrere. 1990. d-Fenfluramine suppresses the increased calorie and carbohydrate intakes and improves the mood of women with premenstrual syndrome. Obstet. Gynecol. 76:206–211.

Bucci, L.R. 1989. Nutritional ergogenic aids. Pp. 107–184 in Nutrition in Exercise and Sport, J.F. Hickson, Jr., and I. Wolinsky, eds. Boca Raton, Fla.: CRC Press.

Burkinshaw, L., G.L. Hill, D.B. Morgan. 1978. Assessment of the distribution of protein in the human body by in-vivo neutron activation analysis. International Symposium on Nuclear Activation Techniques in the Life Sciences. Vienna: International Atomic Energy Association.

Burns, J.E., and H. Kacser. 1987. Genetic effects on susceptibility to histidine induced teratogenesis in the mouse. Genet. Res. 50(2):147–153.

Buskirk, E.R. 1993. Energetics and climate with emphasis on heat: A historical perspective. Pp. 97–116 in Nutritional Needs in Hot Environments: Applications for Military Personnel in Field Operations. B.M. Marriott, ed. Institute of Medicine. Washington, D.C.: National Academy Press.

Buskirk, E.R. 1996. Exercise. Pp. 420–429, Chapter 41 in Present Knowledge in Nutrition, E.E. Ziegler and L.J. Filer, Jr., eds. Washington, D.C.: ILSI Press.

Butterfield, G.E. 1987. Whole-body protein utilization in humans. Med. Sci. Sports. Exerc. 19:S157–165.

Butterfield, G.E. 1996. Maintenance of body weight at high altitudes: In search of 500 kcal/day. Pp. 357–378 in Nutritional Needs in Cold and in High-Altitude Environments, B.M. Marriott and S.J. Carlson, eds. Institute of Medicine. Washington, D.C.: National Academy Press.

Butterfield, G.E., and D.H. Calloway. 1984. Physical activity improves protein utilization in young men. Br. J. Nutr. 11:171–184.

Butterfield, G.E., J. Gates, S. Fleming, G.A. Brooks, J.R. Sutton, and J.T. Reeves. 1992. Increased energy intake minimizes weight loss in men at high altitude. J. Appl. Physiol. 72:1741–1748.

Butterfield, G.E., J. Thompson, M.J. Rennie, R. Marcus, R.L. Hintz and A.R. Hoffman. 1997. Effect of rhGH and IGF-1 treatment on protein utilization in elderly women. Am. J. Physiol. 272:E94–E99.

Calloway, D.H. 1975. Nitrogen balance of men with marginal intakes of protein and energy. J. Nutr. 105:914–923.

Calloway, D.H., and S. Margen. 1971. Variation in endogenous nitrogen excretion and dietary nitrogen utilization as determinants of human protein requirements. J. Nutr. 101:205–216.

Calloway, D.H., and H. Spector. 1954. Nitrogen balance as related to caloric and protein intake in active young men. Am. J. Clin. Nutr. 2:405–412.

Calloway, D.H., A.C.F. Odell, and S.J. Margen. 1971. Sweat and miscellaneous nitrogen losses in human balance studies. J. Nutr. 101:775–786.

Cameron, J. R., J. Sorenson. 1963. Measurement of bone mineral in vivo. Science 42:230–32.

Campbell, W.W., M.C. Crim, V.R. Young, L.J. Joseph, and W. Evans. 1995. Effects of resistance training and dietary protein intake on protein metabolism in older adults. Am. J. Physiol. 268:E1143–1153.

Cannon, J.G., R.G. Tompkins, J.A. Gelfand, H.R. Michie, G.G. Stanford, J.W.M. van der Meer, S. Endres, G. Lonnemann, J. Corsetti, B. Chernow, D.W. Wilmore, S.M. Wolff, J.F. Burke, and C.A. Dinarello. 1990. Circulating interleukin-1 and tumor necrosis factor in septic shock and experimental endotoxin fever. J. Infect. Dis. 161:79–84.

Carlson, D.E., T.B. Dugan, J.C. Buchbinder, J.D. Allegretto, and D.D. Schnakenberg. 1987. Nutritional assessment of the Ft. Riley Non-commissioned Officer Academy dining facility. Technical Report T14-87. Natick, Mass.: U.S. Army Research Institute of Environmental Medicine.

Carraro, F., W.H. Hartl, C.A. Stuart, D.K. Layman, F. Jahoor, and R.R. Wolfe. 1990. Whole body and plasma protein synthesis in exercise and recovery in human subjects. J. Appl. Physiol. 258:E821–E831.

Castell, L.M., J.R. Poortmans, and E.A. Newsholme. 1996. Does glutamine have a role in reducing infection in animals. Eur. J. Appl. Physiol. 13:488–490.

Celejowa, I., and M. Homa. 1970. Food intake, nitrogen, and energy balance in Polish weightlifters during training camp. Nutr. Metab. 12:259–274.

Champagne, C.M., J.P. Warber, and H.R. Allen. 1997. Dietary intake of U.S. Army Rangers: Estimation of nutrient intake before and after field training. Presented at the XVI International Congress of Nutrition, Montreal.

Chang, F.C., and B. Herzog. 1976. Burn morbidity: A follow-up study of physical and psychological disability. Ann. Surg. 183:34–37.

Chang, T.W., and A.L. Goldberg. 1978a. The origin of alanine produced in skeletal muscle. J. Biol. Chem. 253:3677–3684.

Chang, T.W., and A.L. Goldberg. 1978b. The metabolic fates of amino acids and the formation of glutamine in skeletal muscle. J. Biol. Chem. 253:3685–3695.

Chaouloff, F. 1989. Physical exercise and brain monoamines: a review. Acta. Physiol. Scand. 137:1–113.

Chaouloff, F., J.L. Elghozi, Y. Guezennec, and D. Laude. 1985. Effect of conditioned running on plasma, liver and brain tryptophan and on brain 5-hydroxytryptamine metabolism of the rat. Brit. J. Pharmacol. 86:33–41.

Charlton, M.R., D.B. Adey, and K.S. Nair. 1996. Evidence for a catabolic role of glucagon during an amino acid load. J. Clin. Invest. 98:90–99.

Chesley, A., J.D. MacDougall, M.A. Tarnopolsky, S.A. Atkinson, and K. Smith. 1992. Changes in human muscle protein synthesis following resistance exercise. J. Appl. Physiol. 73:1383–1388.

Chittenden, R.H. 1904. Physiological Economy of Nutrition. N.Y.: Frederick A. Stokes Co.

Chittenden, R.H. 1907. The Nutrition of Man. London: Heinemann.

Chumlea W.C. and S.S. Guo. 1994. Bioelectrical impedance and body composition: Present status and future directions. Nutr Rev. 52: 123–31.

Churchill, D.N. 1987. Medical treatment to prevent recurrent calcium urolithiasis. Miner. Electrolyte Metab. 13:294–304.

Clark, H.E., L.L. Reitz, T.S. Vacharotayan, and E.T. Mertz. 1962. Effect of certain factors on nitrogen retention and lysine requirements of adult human subjects. J. Nutr. 78:173–178.

Clarke, J.T.R., and D.M. Bier. 1982. The conversion of phenylalanine to tyrosine in man. Direct measurement by continuous intravenous tracer infusions of L-[$ring$-$^2H_5$] phenylalanine and L-[l-^{13}C] tyrosine in the postabsorptive state. Metabol. 31:999–1005.

Clasey, J.L., M.L. Hartmen, J. Kanaley, L. Wideman, C.D. Teates, C. Bouchard, A. Weltman. 1997. Body composition by DEXA in older adults: Accuracy and influence of scan mode. Med Sci Sports Exerc.29:560–567.

Cleare, A.J., and A.J. Bond. 1994. Effects of alterations in plasma tryptophan levels on aggressive feelings. Arch. Gen. Psychiatry 51(12):1004–1005.

Cleare, A.J., and A.J. Bond. 1995. The effect of tryptophan depletion and enhancement on subjective and behavioral aggression in normal male subjects. Psychopharmacology 118(1):72–81.

Cline, A.D., and A.E. Pusateri. 1996. Comparisons of iron status, physical activity, and nutritional intake of women entering Army officer and enlisted basic training. Proceedings of 21st National Nutrient Databank Conference, Baton Rouge, La.

Cline, A.D., J.F. Patton, W.J. Tharion, S.R. Strowman, C.M. Champagne, J. Arsenault, K.L. Reynolds, J.P. Warber, C. Baker-Fulco, J. Rood, R.T. Tulley, and H.R. Lieberman. 1998. Assessment of the relationship between iron status, dietary intake, performance, and mood state of female Army officers in a basic training population. Technical Report No. T98-24. Natick, Mass.: U.S. Army Research Institute of Environmental Medicine.

Cline, A.D., J.P. Warber, and C.M. Champagne. 1997. Assessment of meal consumption behaviors by physically active young men dining in a cafeteria setting. Presented at Experimental Biology Annual Meeting, New Orleans, La. FASEB J; 11:A184.

Clugston, G., K.G. Dewey, C. Fjeld, J. Millward, P. Reeds, N.S. Scrimshaw, K. Tontisirin, J.C. Waterlow, and V.R. Young. 1996. Report of the working group on protein and amino acid requirements. Europ. J. Clin. Nutr. 50:S193–S195.

Coe, F.L. 1978. Calcium–uric acid nephrolithiasis. Arch. Intern. Med. 138:1090–1093.

Cohn, S.H., D. Vartsky, S. Yasumura, A. Sawitsky, I. Zanzi, A. Vaswani, K.J. Ellis. 1980. Compartmental body composition based on total-body nitrogen, potassium, and calcium. Am J Physiol. 39: E524–530.

Coleman, G.L., W. Barthold, G.W. Osbaldiston, S.J. Foster, and A.M. Jonas. 1977. Pathological changes during aging in barrier-reared Fischer 344 male rats. J. Gerontol. 32:258–278.

Coleman, W., and E.F. DuBois. 1915. Clinical calorimetry. VII. Calorimetric observations on the metabolism of typhoid patients with and without food. Ann. Int. Med. 15:887–938.

Collins, D.M., C.T. Rezzo, J.B. Kopp, P. Ruiz, T.M. Coffman, and P.E. Klotman. 1990. Chronic high protein feeding does not produce glomerulonephrosis or renal insufficiency in the normal rat. J. Am. Soc. Nephrol. 1:624.

Collipp, P.J., J. Thomas, V. Curti, R.K. Sharma, V.T. Maddaiah, S.E. Cohn. 1973. Body composition changes in children receiving human growth hormone. Metabolism 22:589–595.

Conlay, L.A., R.J. Wurtman, J.K. Blusztajan, I.L.G. Coviella, T.J. Maher, and G.E. Evoniuk. 1986. Marathon running decreases plasma choline concentrations. N. Engl. J. Med. 315:892.

Consolazio, C.F., and R. Shapiro. 1964. Energy requirements of men in extreme heat. Pp. 121–124 in Environmental Physiology and Psychology in Arid Conditions: Proceedings of the Lucknow Symposium. Liege, Belgium: United Nations, UNESCO.

Consolazio, C.F., H.L. Johnson, R.A. Nelson, R. Dowdy, and H.J. Krzywicki. 1979. The relationship of diet to the performance of the combat soldier's minimal calorie intake during combat patrols in a hot humid environment (Panama). Technical Report 76. San Francisco, Calif.: Letterman Army Institute of Research.

Copeland, K.C., and K.S. Nair. 1994. Acute growth hormone effects on amino acid and lipid. J. Clin. Endocrinol. Metab. 78:1040–1047.

Costa, G., L. Ullrich, F. Kantor, and J.F. Holland. 1968. Production of elemental nitrogen by certain mammals including man. Nature 281:546–551.

Cummings, J.H., S.A. Bingham, K.W. Heaton, and M.A. Eastwood. 1992. Quantitative estimates of bowel habit, bowel disease risk and the role of diet. Gastroenterol. 103:1783–1789.

Curhan, G.C., W.C. Willett, E.B. Rimm, and M.J. Stampfer. 1993. A prospective study of dietary calcium and other nutrients and the risk of symptomatic kidney stones. New Engl. J. Med. 328:833–838.

Cuneo, R.C., F. Salomon, C.M. Wiles, R. Hesp, P.H. Sonksen. 1991. Growth hormone treatment in growth hormone-deficient adults. II. Effects on exercise performance. J. Appl. Physiol. 70:695–700.

Cuthbertson, D. and H.N. Munro. 1937. A study of the effect of over-feeding on the protein metabolism of man. III. The protein-saving effect of carbohydrate and fat when superimposed on a diet adequate for maintenance. Biochem. J. 31:694–705.

Cuthbertson, D.P. 1932. Observations on disturbance of metabolism produced by injury to the limbs. Quart. J. Med. 25:233–246.

Darmaun, D., D. Matthews, and D. Bier. 1986. Glutamine and glutamate kinetics in humans. Am. J. Physiol. 251:E117–E126.

Davis, J.M., S.P. Bailey, J.A. Woods, F.J. Galiano, M.T. Hamilton, and W.P. Bartoli. 1992. Effects of carbohydrate feedings on plasma free tryptophan and branched chain amino acids during prolonged cycling. Eur. J. Appl. Physiol. 65:513–519.

Deijen, J.B., and J.F. Orlebeke. 1994. Effect of tyrosine on cognitive function and blood pressure under stress. Brain Res. Bull. 33:319–323.

Delgado, P.L., D.S. Charney, L.H. Price, G.K. Aghajanian, H. Landis, and G.R. Heninger. 1990. Serotonin function and the mechanism of antidepressant action. Reversal of antidepressant-induced remission by rapid depletion of plasma tryptophan.. Arch. Gen. Psychiatry 47:411–418.

Department of Health, Panel on Dietary Reference Values of the Committee on Medical Aspects of Food Policy. 1991. Report on Health and Social Subjects, Number 41. Dietary Reference Values for Food Energy and Nutrients for the United Kingdom. London: Her Majesty's Stationery Office.

Dewey, K.G., G. Beaton, C. Fjeld, B. Lonnerdal, and P. Reeds. 1996. Protein requirements of infants and children. Eur. J. Clin. Nutr. 50(Suppl. 1):S119–S147.

Dohm, G.L., G.J. Kasperek, E.B. Tapscott, and G.R. Beecher. 1980. Effect of exercise on synthesis and degradation of muscle protein. Biochem. J. 188:255–262.

Duncan, A.M., R.O. Ball, and P.B. Pencharz. 1996. Lysine requirement of adult males is not affected by decreasing dietary protein intake. Am. J. Clin. Nutr. 64:718–725.

Edwards, C.H., L.K. Booker, C.H. Rmphg, W.G. Wright, and S. Ganapathy. 1971. Utilization of wheat by adult man: Nitrogen metabolism, plasma amino acids and lipids. Am. J. Clin. Nutr. 24:181–193.

Edwards, J.S.A., D.E. Roberts, T.E. Morgan, and L.S. Lester. 1989. An evaluation of the nutritional intake and acceptability of the Meal, Ready-To-Eat consumed with and without a supplemental pack in a cold environment. Technical Report No. T18-89. Natick, Mass.: U.S. Army Research Institute of Environmental Medicine.

Edwards, J.S.A., E.W. Askew, N. King, C.S. Fulco, R.W. Hoyt, and J.P. DeLany. 1991. An assessment of the nutritional intake and energy expenditure of unacclimatized U.S. Army soldiers living and working at high altitude. Technical Report No. T10-91. Natick, Mass.: U.S. Army Research Institute of Environmental Medicine.

Elahi, D., M. McAloon-Dyke, N.K. Fukugawa, A.L. Sclater, G.A. Wong, R.P. Shannon, K.L. Minaker, J.M. Miles, A.H. Rubenstein, and C.J. Vandepol. 1993. Effects of recombinant human IGF-I on glucose and leucine kinetics in men. Am. J. Physiol. 265:E831–E838.

Elia, M., A. Schlatmann, A. Goren, and S. Austin. 1989. Amino acid metabolism in muscle and in the whole body of man before and after ingestion of a single mixed meal. Am. J. Clin. Nut. 49:1203–1210.

El-Khoury, A.E., N.K. Fukagawa, M. Sanchez, R.H. Tsay, R.E. Gleason, T.E. Chapman, and V.R. Young. 1994a. Validation of the tracer-balance concept with reference to leucine: 24-h intravenous tracer studies with l-[1-^{13}C]leucine and [^{15}N-^{15}N]urea. Am. J. Clin. Nutr. 59(5):1000–1011.

El-Khoury, A.E., N.K. Fukagawa, M. Sanchez, R.H. Tsay, R.E. Gleason, T.E. Chapman, and V.R. Young. 1994b. The 24-h pattern and rate of leucine oxidation, with particular reference to tracer estimates of leucine requirements in healthy adults. Am. J. Clin. Nutr. 59(5):1012–1020.

El-Khoury, A.E., M. Sanchez, N.K. Fukagawa, R.E. Gleason, R.H. Tsay, and V.R. Young. 1995a. The 24-h kinetics of leucine oxidation in healthy adults receiving a generous leucine intake via three discrete meals. Am. J. Clin. Nutr. 62(3):579–590.

El-Khoury, A.E., M. Sánchez, N.K. Fukagawa, and V.R. Young. 1995b. Whole body protein synthesis in healthy adult humans; $^{13}CO_2$ technique vs. plasma precursor approach. Am. J. Physiol. 268:E174–E184.

El-Khoury, A.E., A.M. Ajami, N.K. Fukagawa, T.E. Chapman, and V.R. Young. 1996. Diurnal pattern of the interrelationship among leucine oxidation, urea production, and hydrolysis in humans. Am. J. Physiol. 271:E563–E573.

El-Khoury, A.E., A. Forslund, R. Olsson, S. Branth, A. Sjodin, A. Andersson, A. Atkinson, A. Selvaraj, L. Hambraeus, and V.R. Young. 1997. Moderate exercise at energy balance does not affect 24 h leucine oxidation or nitrogen retention in healthy men. Am. J. Physiol. 273:E394–E407.

El-Khoury, A.E., A. Basile, L. Beaumier, S.Y. Wang, H.A. Al-Amiri, A. Selvaraj, S. Wong, A. Atkinson, A.M. Ajami, and V.R. Young. 1998. Twenty-four hour intravenous and oral tracer studies with 1-[1-^{13}C]-2-aminoadipic acid and 1-[1-^{13}C]lysine as tracers at generous nitrogen and lysine intakes in healthy adults. Am. J. Clin. Nutr. 68:827–839.

Essen, B., and L. Kaijser. 1978. Regulation of glycolysis in intermittent exercise in man. J. Physiol. 281:499–511.

Ettinger, B., J.T. Citron, A. Tang, and B. Livermore. 1985. Prophylaxis of calcium oxalate stones; Clinical trials of allopurinol, magnesium hydroxide, and chlorthalidone. Pp. 549–552 in Urolithiasis and Related Clinical Research, P.O. Schwille, L.H. Smith, W.G. Robertson, and W. Vahlensieck, eds. New York: Plenum.

Fairbrother, B., R.E. Shippee, T.R. Kramer, E.W. Askew, and M.Z. Mays. 1995. Nutritional and immunological assessment of soldiers during the Special Forces Assessment and Selection Course. Report USARIEM-T95-22, AD-A299 556. Natick, Mass.: U.S. Army Research Institute of Environmental Medicine.

FAO/IAEA/WHO (Food and Agriculture Organization of the United Nations/International Atomic Energy Agency/World Health Organization). 1996. Trace Elements in Human Nutrition and Health. Geneva: WHO.

FAO/WHO (Food and Agriculture Organization of the United Nations/World Health Organization). 1973. Energy and Protein Requirements. Report of a joint FAO/WHO Ad Hoc expert committee. WHO Technical Report Series No. 522. Geneva: World Health Organization.

FAO/WHO (Food and Agriculture Organization of the United Nations/World Health Organization). 1991. Protein Quality Evaluation. Report of a Joint FAO/WHO Expert Consultation. FAO Food and Nutrition Paper 51. Rome: FAO.

FAO/WHO/UNU (Food and Agriculture Organization of the United Nations/World Health Organization/United Nations University). 1985. Energy and protein requirements. Report of a joint expert consultation. World Health Organization Technical Report Series No. 724. Geneva: World Health Organization.

FASEB/LSRO. 1992. Safety of amino acids used as dietary supplements. Center for Food Safety and Applied Nutrition. FDA Contract No. 223-88-2124, Task No. 8.

FDA (Food and Drug Administration). 1990. Evaluation of the safety of amino acids and related products; Announcement of study; Request for scientific data and information; Announcement of open meeting. Fed. Regist. 55:49141–49142.

FDA (Food and Drug Administration). 1997. Current good manufacturing practice in manufacture, packing, or holding dietary supplements; Proposed rule. Fed. Regist. 62:5699–5709.

Fellström, B., B.G. Danielson, B. Karlström, H. Lithell, S. Ljunghall, and B. Vessby. 1983. Dietary animal protein and urinary supersaturation in renal stone formers. Proc. Eur. Dial. Transplant. Assoc. 20:411–416.

Fellström, B., B.G. Danielson, B. Karlström, H. Lithell, S.L. Ljunghall, B. Vessby, and L. Wide. 1984. Effects of high intake of dietary animal protein on mineral metabolism and urinary supersaturation of calcium oxalate in renal stone formers. Brit. J. Urol. 56:263–269.

Fereday, A., N.R. Gibson, M. Cox, D. Halliday, P.J. Pacy, and D.J. Millward. 1994. Postprandial utilization of wheat protein in normal adults. Proc. Nutr. Soc. 53:201a.

Fereday, A., N.R. Gibson, M. Cox, D. Halliday, P.J. Pacy, and D.J. Millward. 1997a. Postprandial protein utilization of wheat protein from a single meal in normal adults. Proc. Nutr. Soc. 56:80a.

Fereday A., N.R. Gibson, M. Cox, P.J. Pacy, and D.J. Millward. 1997b. Protein requirements and aging: Metabolic demand and efficiency of utilization. Brit. J. Nutr. 77:685–702.

Fernstrom, J.D. 1990. Aromatic amino acids and monoamine synthesis in the central nervous system: Influence of the diet. J. Nutr. Biochem. 1:508–517.

Fernstrom, J.D. 1994. Stress and monoamine neurons in the brain. Pp. 161–175 in Food Components to Enhance Performance, B.M. Marriott, ed. Institute of Medicine. Washington, D.C.: National Academy Press.

Fernstrom, J.D., and M.J. Hirsch. 1975. Rapid repletion of brain serotonin in malnourished, corn-fed rats following L-tryptophan injection. Life Sciences 17:455–464.

Fernstrom, J.D., and L.D. Lytle. 1976. Corn malnutrition, brain serotonin, and behavior. Nutr. Rev. 34:257–262.

Fernstrom, J.D., and R.J. Wurtman. 1971. Brain serotonin content: Physiological dependence on plasma tryptophan levels. Science 173:149–152.

Fernstrom, J.D., and R.J. Wurtman. 1972. Brain serotonin content: Physiological regulation by plasma neutral amino acids. Science 178:414–416.

Fernstrom, M. H., and J.D. Fernstrom. 1995a. Brain tryptophan concentrations and serotonin synthesis remain responsive to food consumption after the ingestion of sequential meals. Am. J. Clin. Nutr. 61:312–319.

Fernstrom, M.H., and J.D. Fernstrom. 1995b. Effect of chronic protein ingestion on rat central nervous system tyrosine levels and in vivo tyrosine hydroxylation rate. Brain Res. 672: 97–103.

Ferrando, A.A., K.D. Tipton, D. Doyle, S.M. Phillips, J. Cortiella, R.R. Wolfe. 1998. Testosterone injection stimulates net protein synthesis but not tissue amino acid transport. Amer. J. of Physiol. 275:E864–871.

Feskanich, D., W.C. Willett, M.J. Stampfer, and G.A. Colditz. 1996. Protein consumption and bone fractures in women. Am. J. Epidemiol. 143:472–479.

Fielding, R.A., C.N. Meredith, K.P. O'Reilly, W.R. Fontera, J.G. Cannon and N.J. Evans. 1991. Enhanced protein breakdown after eccentric exercise in young and older men. J. Appl. Physiol. 11:674–679.

Fisher H., M.K. Brush, and P. Griminger. 1969. Reassessment of amino acid requirements of young women on low nitrogen diets. 1. Lysine and tryptophan. Am. J. Clin. Nutr. 22:1190–1196.

Fitts, R.H. 1994. Cellular mechanisms of muscle fatigue. Physiol. Rev. 74:49–94.

Fliser, D., E. Franek, M. Joest, S. Block, E. Mutschler, and E. Ritz. 1997. Renal function in the elderly: Impact of hypertension and cardiac function. Kidney Int. 51:1196–1204.

Forbes, G.B. 1973. Another source of error in the metabolic balance method. Nutr. Rev. 31:297–300.

Forbes G.B., M.R. Brown, H.J.L. Griffiths. 1988. Arm muscle plus bone area: Anthropometry and CAT scan compared. Am J Clin Nutr. 47:929–931.

Foster, M.A., J.M.S. Hutchison, J.R. Mallard, and M. Fuller. 1984. Nuclear magnetic resonance pulse sequence and discrimination of high- and low-fat tissues. Mag. Res. Imaging 2:187–192.

Fouque, D., M. Laville, J.P. Boissel, R. Chifflet, M. Labeeuw, and P.Y. Zech. 1992. Controlled low protein diets and chronic renal insufficiency: Meta-analysis. Br. Med. J. 304:216–220.

Frieder, B., and V.E. Grimm. 1984. Prenatal monosodium glutamate (MSG) treatment given through the mother's diet causes behavioral deficits in rat offspring. Int. J. Neurosci. 23(2):117–126.

Friedl, K.E. 1997. Variability of fat and lean tissue loss during physical exertion with energy deficit. Pp. 431–450 in Physiology, Stress, and Malnutrition: Functional Correlates, Nutritional Intervention, J.M. Kinney and H.N. Tucker, eds. Philadelphia: Lippincott-Raven Publishers.

Friedman, M. 1991. Formation, nutritional value, and safety of d-amino acids. Pp. 447–482 in Nutritional and Toxicological Consequences of Food Processing. New York: Plenum Press.

Friedman, S.A., A.E. Raizner, H. Rosen, N.A. Solomon, and W. Sy. 1972. Functional defects in the aging kidney. Ann. Intern. Med. 76:41–45.

Frontera, W.R., C.N. Meredith, and W.J. Evans. 1988. Dietary effects on muscle strength gain and hypertrophy during heavy resistance training in older men [abstract]. Can. J. Sport Sci. 13:13P.

Fryburg, D.A. 1994. Insulin-like growth factor I exerts growth hormone and insulin-like actions on human muscle protein metabolism. Am. J. Physiol. 267:E331–E336.

Fryburg, D.A. 1996. NG-monomethyl-L-arginine inhibits the blood flow but not the insulin-like response of forearm muscle to IGF- I: possible role of nitric oxide in muscle protein synthesis. J. Clin. Invest. 97:1319–1328.

Fryburg, D.A., and E.J. Barrett. 1993. Growth hormone acutely stimulates skeletal muscle but not whole-body protein synthesis in humans. Metabolism 42:1223–1227.

Fryburg, D.A., R.A. Gelfand, and E.J. Barrett. 1991. Growth hormone acutely stimulates forearm protein synthesis in normal subjects. Am. J. Physiol. 260:E499–E504

Fryburg, D.A., L.A. Jahn, S.A. Hill, D.M. Oliveras, and E.J. Barrett. 1995. Insulin and insulin-like growth factor I enhance human skeletal muscle protein anabolism during hyperaminoacidemia by different mechanisms. J. Clin. Invest. 96:1722–1729.

Fuller, M.F., and P.J. Garlick. 1994. Human amino acid requirements: Can the controversy be resolved? Ann. Rev. Nutr. 14:217–241.

Fuller, M.F., R. McWilliam, T.C. Wang, and L.R. Giles. 1989. The optimum dietary amino acid pattern for growing pigs; Requirements for maintenance and for tissue protein accretion. Brit. J. Nutr. 62:255–267.

Fuller, M.F., A. Milne, C.I. Harris, T.M. Reid, and R. Keenan. 1994. Amino acid losses in ileostomy fluid on a protein-free diet. Am. J. Clin. Nutr. 59(1):70–73.

Funk, D.N., B. Worthington-Roberts, and A. Fantel. 1991. Impact of supplemental lysine or tryptophan on pregnancy course and outcomes in rats. Nutr. Res. 11:501–512.

Gadian, D.G. 1995. NMR and Its Applications to Living Systems, 2nd edition. New York: Oxford University Press.

Gahl, M.J., M.D. Finke, T.D. Crenshaw, and N.J. Benevenga. 1991. Use of a four-parameter logistic equation to evaluate the response of growing rats to ten levels of each indispensable amino acid. J. Nutr. 121(11):1720–1729

Gallagher, D., D. Belmonte, P. Deurenberg, Z. Wang, N. Krasnow, F.X. Pi-Sunyer, S.B. Heymsfield. 1998. Organ-tissue mass measurement allows modeling of REE and metabolically active tissue mass. Am. J. Physiol. 275:E249–E258.

Gelfand, R.A., and E.J. Barrett. 1987. Effect of physiologic hyperinsulinemia on skeletal muscle protein synthesis and breakdown in man. J. Clin. Invest. 80:1–6.

Gibson, N., E. Al-Sing, A. Badalloo, T. Forrester, A. Jackson, and D.J. Millward. 1997. Transfer of ^{15}N from urea to the circulating dispensable and indispensable amino acid pool in the human infant. Proc. Nutr. Soc. 56:79A.

Glassman, A.H., and S.R. Platman. 1969. Potentiation of a monoamine oxidase inhibitor by tryptophan. J. Psychiatr. Res. 7:83–88.

Goldfarb, S. 1988. Dietary factors in the pathogenesis and prophylaxis of calcium nephrolithiasis. Kidney Int. 34:544–555.

Gontzea, I., P. Sutzescu, and S. Dumitrache. 1975. The influence of adaptation to physical effort on nitrogen balance in man. Nutrition Reports International 22:231–236.

Gopalan, C., and B.S. Narasinga Rao. 1966. Effect of protein depletion on urinary nitrogen excretion in undernourished subjects. J. Nutr. 90:213–218.

Goran, M.I., W.H. Beer, R.R. Wolfe, E.T. Poehlman, and V.R. Young. 1993. Variation in total energy expenditure in young healthy free-living men. Metabolism 42:487–496.

Gordon, C.C., T. Churchill, C.E. Clauser, B. Bradtmiller, J.T. McConville, I. Tebbetts, and R.A. Walker. 1989. 1988 Anthropometric survey of U.S. Army personnel: Methods and summary statistics. Technical Report No. TR-89/044. Natick, Mass.: U.S. Army Natick Research, Development and Engineering Center.

Gore, D.C., D. Honeycutt, F. Jahoor, R.R. Wolfe, and D.N. Herndon. 1991. Effect of exogenous growth hormone on whole-body and isolated-limb protein kinetics in burned patients. Arch. Surg. 126:38–43.

Gottesman, S., M.R. Maurizi, and S. Wickner. 1997. Regulatory subunits of energy-dependent proteases. Cell 91:435–438.

Graham, G.G., E. Morales R.P. Placko, and W.C. Maclean. 1979. Nutritive value of brown and black beans for infants and small children. Am. J. Clin. Nutr 32:2362–2366.

Graham, G.G., J. Lembcke, and E. Morales. 1991. Quality-protein maize as the sole source of dietary protein and fat for rapidly growing young children. Pediatrics 85(1):85–91.

Griffiths, R.D., C. Jones, and T.E.A. Palmer. 1997. Six-month outcome of critically ill patients given glutamine-supplemented parenteral nutrition. Nutrition 13(4):295–302.

Grimble, R.F. 1993. The maintenance of antioxidant defenses during inflammation. Pp. 347–366 in Metabolic Support of the Critically Ill Patient, D.W. Wilmore and Y.A. Carpentier, eds. New York: Springer-Verlag.

Hajak, G., G. Huether, J. Blanke, M. Blömer, C. Freyer, B. Poeggler, A. Reimer, A. Rodenbeck, M. Schulz-Varszegli, and E. Rüther. 1991. The influence of intravenous *l*-tryptophan on plasma melatonin and sleep in men. Pharmacopsychiatry 24:17–20.

Hakim, R.M., R.C. Goldszer, and B.M. Brenner. 1984. Hypertension and proteinuria: Long-term sequelae of uninephrectomy in humans. Kidney Int. 25:930–936.

Hargreaves, K.M., and W.M. Pardridge. 1988. Neutral amino acid transport at the human blood–brain barrier. J Biol. Chem. 263:19392–19397.

Hartl, F.U. 1966. Molecular chaperones in cellular protein folding. Nature 381:571–580.

Hartl, W.H., H. Miyoshi, F. Jahoor, S. Klein, D. Elahi, and R.R. Wolfe. 1990. Bradykinin attenuates glucagon-induced leucine oxidation in humans. Am. J. Physiol. 259:E239–E245.

Hartley, H. 1951. Origin of the word "protein." Nature 168:244.

Hartmann, E., and D. Greenwald. 1984. Tryptophan and human sleep: An analysis of 43 studies. Pp. 297–304 in Progress in Tryptophan and Serotonin Research, H.G. Schlossberger, W. Kochen, B. Linzen, and H. Steinhart, eds. Berlin: Walter de Gruyter.

Hartmann, E.L. 1986. Effect of *l*-tryptophan and other amino acids on sleep. Nutr. Rev. 44:70–73.

Hassmen, P., E. Blomstrand, B. Ekblom, and E.A. Newsholme. 1994. Branched-chain amino acid supplementation during 10 km competitive run: Mood and cognitive performance. Nutrition 10:405–410.

Hayashida, M., B.P. Yu, E.J. Masoro, K. Iwasaki, and T. Ikeda. 1986. An electron microscopic examination of age-related changes in the rat kidney: The influence of diet. Exp. Gerontol. 21:535–553.

Hegsted, D.M. 1963. Variation in requirements of nutrients: Amino acids. Fed. Proc. 22:1424–1430.

Hegsted, D.M. 1973. The amino acid requirements of rats and human beings. Pp. 275–292 in Proteins in Human Nutrition, J.W.G. Porter and B.A. Rolls, eds. London: Academic Press.

Hegsted, D.M. 1976. Balance studies. J. Nutr. 106:307–311.

Hegsted, M., and H.M. Linkswiler. 1981. Long-term effects of level of protein intake on calcium metabolism in young adult women. J. Nutr. 111:244–251.

Hegsted, M., S.A. Schuette, M.B. Zemel, and H.M. Linkswiler. 1981. Urinary calcium and calcium balance in young men as affected by level of protein and phosphorus intake. J. Nutr. 111:553–562.

Hertzman, P.A., W.L. Blevins, J. Mayer, B. Greenfield, M. Ting, and G.J. Gleich. 1990. Association of the eosinophilia–myalgia syndrome with the ingestion of tryptophan. N. Engl. J. Med. 322:869–873.

Heymsfield, S.B., C. Arteaga, C. McManus, J. Smith, S. Moffitt. 1983. Measurement of muscle mass in humans: Validity of the 24-hour urinary creatinine method. Am J Clin Nutr. 37: 478–494.

Heymsfield S.B., C. McManus, J. Smith, V. Stevens, D.W. Nixon. 1982. Anthropometric measurement of muscle mass: Revised equations for calculating bone-free arm muscle area. Am J Clin Nutr. 36:680–690.

Heymsfield, S.B., R. Smith, M. Aulet, B. Bensen, S. Lichtman, J. Wang, and R.N. Pierson, Jr. 1990. Appendicular skeletal muscle mass: Measurement by dual-photon absorptiometry. Am. J. Clin. Nutr. 52: 214–218.

Heymsfield S.B., A. Tighe, Z. Wang. 1992. Nutritional assessment by anthropometric and biochemical methods. In: Shils ME, Olson JA, Shike M (Eds) Modern Nutrition in Health and Disease. Philadelphia: Lea and Febiger, Eighth Edition.

Heymsfield, S.B., D. Gallagher, M. Visser, C. Nuñez, and Z.M. Wang. 1995. Measurement of skeletal muscle: Laboratory and epidemiological methods. J. Gerontol. 50A:23–29.

Heymsfield S.B., C. McManus III, S. Seitz, D. Nixon, J. Smith. 1984. Anthropometric assessment of adult protein-energy malnutrition. Pp. 27–82 in Nutritional Assessment of the Hospitalized Patient, R.A. Wright and S.B. Heymsfield, eds. Boston: Blackwell Scientific Publications, Inc.

Heymsfield, S.B., R. Ross, Z. Wang, D. Frager. 1997a. Imaging Techniques of Body Composition: Advantages of Measurement and New Uses. Pp. 127–150 in Emerging Technologies for Nutrition Research, S.J. Carlson-Newberry and R.B. Costello, eds. Washington, DC: National Academy Press.

Heymsfield, S.B., Z. Wang, R.N. Baumgartner, R. Ross. 1997b. Human body composition: Advances in models and methods. Ann. Rev. Nutr. 17:527–558.

Hickson, J.F., Jr., and I. Wolinsky. 1989. Human protein intake and metabolism in exercise and sport. Pp. 5–36 in Nutrition in Sport and Exercise, J.F. Hickson, Jr., and I. Wolinsky, eds. Boca Raton, Fla.: CRC Press.

Hill, A.G., L. Jacobson, J. Gonzalez, J. Rounds, J.A. Majzoub, and D.W. Wilmore. 1996. Chronic central nervous system exposure to interleukin-1β causes catabolism in the rat. Am. J. Physiol. 271:R1142–R1148.

Hirsch, E., W. Johnson, P. Dunne, C. Shaw, N. Hotson, W. Tharion, H. Lieberman, R. Hoyt, and D. Dacumos. In press. The effects of diet composition on food intake, food selection, and water balance in a hot environment. Technical Report. Natick, Mass.: Natick Research, Development and Engineering Center.

Hochachka P. 1994. Muscles as Molecular and Metabolic Machines. Boca Raton: CRC Press.

Hoffer, L.J., and R.A. Forse. 1990. Protein metabolic effects of a prolonged fast and hypocaloric refeeding. Am. J. Physiol. 258:E832–E840.

Hoffman-Goetz, L., and B.K. Pedersen. 1994. Exercise and the immune system. Immunol. Today 15:382–385.

Hood, D.A., and R.L. Terjung. 1990. Amino acid metabolism during exercise and following endurance training. Sports Med. 1:23–35.

Horber, F.F., and M.W. Haymond. 1990. Human growth hormone prevents the protein catabolic side effects of prednisone in humans. J. Clin. Invest. 86:265–272.

Hounsfield, G.N. 1973. Computerized transverse axial scanning (tomography). Br. J. Radiol. 46:1016.

Howe, P.E., and G.H. Berryman. 1945. Average food consumption in the training camps of the United States Army (1941–1943). Am. J. Physiol. 144:588–594.

Hoyt, R.W., and A. Honig. 1996. Body fluid and energy metabolism at high altitude. Pp. 1277–1289 in Handbook of Physiology, Section 4: Environmental Physiology, C.M.

Blatteis and M.J. Fregly, eds. New York: Oxford University Press for the American Physiological Society.

Huang, P.C., H.E. Chong, and W.M. Rand. 1972. Obligatory urinary and fecal nitrogen losses in young Chinese men. J. Nutr. 102:1605–1614.

Huether, G., F. Thomke, and L. Adler. 1992. Administration of tryptophan-enriched diets to pregnant rats retards the development of the serotonergic system in their offspring. Brain Res. Dev. Brain Res. 68(2):175–181.

Hull, K.M., and T.L. Maher. 1990. l-Tyrosine potentiates the anorexia induced by mixed-acting sympathomimetic drugs in hyperphagic rats. J. Pharmacol. Exp. Ther. 255:403–409.

Hull, K.M., D.E. Tolland, and T.J. Maher. 1994. l-Tyrosine potentiation of opioid-induced analgesia in mice utilizing the hot plate test. J. Pharmacol. Exp. Ther. 269:1190–1195.

Hunter, J. 1794. A Treatise on the Blood Inflammation and Gunshot Wounds. London.

Iguchi, M., T. Umekawa, Y. Ishikawa, Y. Katayama, M. Kodama, M. Takada, Y. Katoh, K. Kataoka, K. Kohri, and T. Kurita. 1989. [Dietary habits of Japanese stone formers and clinical effects of prophylactic dietary treatment]. Hinyokika Kiyo 35:2115–2128.

Iguchi, M., T. Umekawa, Y. Ishikawa, Y. Katayama, M. Kodama, M. Takada, Y. Katoh, K. Kataoka, K. Kohri, and T. Kurita. 1990. Dietary intake and habits of Japanese renal stone patients. J. Urol. 143:1093–1095.

Inoue, G., Y. Fujita, K. Kishi, S. Yamamoto, and Y. Niyama. 1974. Nutritive values of egg protein and wheat gluten in young men. Nutr. Rep. Intl. 10:201–207.

Interagency Board (Interagency Board for Nutrition Monitoring and Related Research). 1993. Chartbook I: Selected Findings from the National Nutrition Monitoring and Related Research Program. DHHS Publication No. (PHS) 93-1255-2. Hyattsville, Md.: Public Health Service.

Interagency Board (Interagency Board for Nutrition Monitoring and Related Research). 1995. Third Report on Nutrition Monitoring in the United States, Vol. 2. Life Sciences Research Office, Federation of American Societies for Experimental Biology. Washington, D.C.: U.S. Government Printing Office.

IOM (Institute of Medicine). 1992. A Nutritional Assessment of U.S. Army Ranger Training Class 11/91. March 23. Washington, D.C.

IOM. 1993a. Nutritional Needs in Hot Environments, Applications for Military Personnel in Field Operations, B.M. Marriott, ed. Washington, D.C.: National Academy Press.

IOM. 1993b. Review of the Results of Nutritional Intervention, U.S. Army Ranger Training Class 11/92 (Ranger II), B.M. Marriott, ed. Washington, D.C.: National Academy Press.

IOM. 1994. Food Components to Enhance Performance, An Evaluation of Potential Peformance-Enhancing Food Components for Operational Rations, B.M. Marriott, ed. Washington, D.C.: National Academy Press.

IOM. 1995. Not Eating Enough, Overcoming Underconsumption of Military Operational Rations, B.M. Marriott, ed. Washington, D.C.: National Academy Press.

IOM. 1996. Nutritional Needs in Cold and in High-Altitude Environments, Applications for Military Personnel in Field Operations, B.M. Marriott and S.J. Carlson, eds. Washington, D.C.: National Academy Press.

IOM. 1997. Dietary Reference Intakes: Calcium, Phosphorus, Magnesium, Vitamin D and Fluoride. Washington D.C.: National Academy Press.

IOM. 1998. Assessing Readiness in Military Women: The Relationship of Body Composition, Nutrition, and Health. Washington, D.C.: National Academy Press.

IOM. 1999. Military Strategies for Sustainment of Nutrition and Immune Function in the Field. Washington, D.C.: National Academy Press.

Irwin, M.I., and D.M. Hegsted. 1971. A conspectus of research on amino acid requirements of man. J. Nutr. 101:539–566.

Ivy, J.L., M.C. Lee, J.T. Brozinick, and M.J. Reed. 1988. Muscle glycogen storage after different amounts of carbohydrate ingestion. J. Appl. Physiol. 65:2018–2023.

Jacobs, B.L. and C.A. Fornel. 1993. 5-Hydroxytryptamine and motor control: a hypothesis. Trends in Neurosciences. 16:346–352.

Jacobi, C.A., J. Ordemann, F. Wenger, K. Zuckerman, H.D. Volk, and J.M. Muller. 1997. The influence of glutamine substitution in postoperative parenteral nutrition on immunologic function. First results of a prospective randomized trial (abstract). Shock 7(S):605.

Jaeger, P., L. Portmann, J.M. Ginalski, and P. Burckhardt. 1988. [So-called renal idopathic hypercalciuria most often has a dietary origin]. Schweiz. Med. Wochenschr. 118:15–17.

Jepson, M.M., P.C. Bates, and D.J. Millward. 1988. The role of insulin and thyroid hormones in the regulation of muscle growth and protein turnover in response to dietary protein. Brit. J. Nutr. 59:397–415.

Johnson, C.M., D.M. Wilson, W.M. O'Fallon, R.S. Malek, and L.T. Kurland. 1979. Renal stone epidemiology: A 25-year study in Rochester, Minnesota. Kidney Int. 16:624–631.

Johnson, R.E., and R.M. Kark. 1946. Feeding problems in man as related to environment: An analysis of United States and Canadian Army ration trials and surveys, 1941–1946. Research Report. Chicago, Ill.: Quartermaster Food and Container Institute for the Armed Forces, Research and Development Branch, Office of the Quartermaster General.

Jones E.M., C.A. Bauman and M.S. Reynolds. 1956. Nitrogen balances in women maintained on various levels of lysine. J. Nut. 60:549–559.

Jones, P.J.H., and I.K.K. Lee. 1996. Macronutrient requirements for work in cold environments. Pp. 189–202 in Nutritional Needs in Cold and in High-Altitude Environments: Applications for Military Personnel in Field Operations, B.M. Marriott and S.J. Carlson, eds. Institute of Medicine. Washington, D.C.: National Academy Press.

Jones, T.E., S.H. Mutter, J.M. Aylward, J.P. DeLany, and R.L. Stephens. 1993. Nutrition and hydration status of aircrew members consuming the Food Packet, Survival, General Purpose, improved during a simulated survival scenario. Report USARIEM-T1-93. Natick, Mass.: U.S. Army Research Institute of Environmental Medicine.

Jorgensen, J.O.L., L. Thuesen, T. Ingemann-Hansen, S.A. Pedersen, J. Jorgensen, N.E. Skakkebaek, and J.S. Christiansen. 1989. Beneficial effects of growth hormone treatment in GH deficient adults. Lancet 1:1221–1225.

Jue, T., D.L. Rothman, G.I. Shulman, B.A. Tavitian, R.A. DeFronzo, and R.G. Sulman. 1989. Direct observation of glycogen synthesis in human muscle with ^{13}C NMR. Proc of the National Academy of Sciences. 86:4489–4491.

Jungers, P., M. Daudon, C. Hennequin, and B. Lacour. 1993. [Correlations between protein and sodium intake and calciuria in calcium lithiasis]. Nephrologie 14:287–290.

Kamb, M.L., J.L. Murphy, J.L. Jones, J.C. Caston, K. Nederlof, L.F. Horney, L.A. Swygert, H. Falk, and E.M. Kilbourne. 1992. Eosinophilia–myalgia syndrome in l-tryłophan-exposed patients. J. Am. Med. Assoc. 267:77–82.

Kaneko, K, U. Masaki, M. Aikyo, K. Yabuki, A. Haga, C. Matoba, H. Sasaki, and G. Koike. 1990. Urinary calcium and calcium balance in young women affected by high protein diet of soy protein isolate and adding sulfur-containing amino acids and/or potassium. J. Nutr. Sci. Vitaminol. (Tokyo) 36:105–116.

Kaplan, C., S. Pasternack, H. Shah, and G. Gallo. 1975. Age-related incidence of sclerotic glomeruli in human kidneys. Am. J. Pathol. 80:227–234.

Kasperek, G.J., and R.D. Snider. 1989. Total and myofibrillar protein degradation in isolated soleus muscles after exercise. Am. J. Physiol. 157:E1–E5.

Kasperek, G.J., G.R. Conway, D.S. Krayeski, and J.J. Lohne. 1992. A reexamination of the effect of exercise on rate of muscle protein degradation. Am. J. Physiol. 163:E1144–E1150.

Katz, A., S. Broberg, K. Sahlin, and J. Wahren. 1986. Muscle ammonia and amino acid metabolism during dynamic exercise in man. Clin. Physiol. 6:365–379.

Kenyon, A.T., K. Knowlton, I. Sandiford, F.C. Koch, and G. Lotwin. 1940. A comparative study of the metabolic effects of testosterone proprionate in normal men and women and in eunuchoidism. Endocrinology 26:26–45.

Kerr, G.R., E.S. Lee, M.M. Lan, R.J. Lorimor, E. Randall, R.N.Forthofer, M.A. Davis, and S.M. Magnetti. 1982. Relationships between dietary and biochemical measures of nutritional status in NHANES I data. Am. J. Clin. Nutr. 35:294–308.

Kerstetter, J.E., and L.H. Allen. 1990. Dietary protein increases urinary calcium. J. Nutr. 120:134–136.

Kimball, S.R., and L.S. Jefferson. 1988. Cellular mechanisms involved in the action of insulin on protein synthesis. Diabetes Metab. Rev. 4:773–787.

King, N., K.E. Fridlund, and E.W. Askew. 1993. Nutrition issues of military women. J. Am. Coll. Nutr. 12:344–348.

King, N., J.E. Arsenault, S.H. Mutter, C.M. Champagne, T.C. Murphy, K.A. Westphal, and E.W. Askew. 1994. Nutritional intake of female soldiers during the U.S. Army basic combat training. Technical Report No. T94-17. Natick, Mass.: U.S. Army Research Institute of Environmental Medicine.

King, R.W., R.J. Deshaies, J.-M. Peters, and M.W. Kirschner. 1996. How proteolysis drives the cell cycle. Science: 274:1652–1659.

Kinney, J.M. 1966. Energy deficits in acute illness and injury. P. 174 in Proceedings of a Conference on Energy Metabolism and Body Fuel Utilization. A.P. Morgan, ed. Cambridge: Harvard University Press.

Kinney, J.M., and D.H. Elwyn. 1995. Amino acid metabolism in health and nutritional disease. Pp. 1–12 in Amino Acid Metabolism in Health and Nutritional Disease, L.A. Cynober, ed. Boca Raton, Fla.: CRC Press.

Kishi, K., S. Miyatani, and G. Inoue. 1978. Requirements and utilization of egg protein by Japanese young men with marginal intakes of energy. J. Nutr. 108:658–669.

Klicka, M.V., D.E. Sherman, N. King, K.E. Friedl, and E.W. Askew. 1993. Nutritional assessment of cadets at the U.S. Military Academy: Part 2. Assessment of nutritional intake. Technical Report T94-1. Natick, Mass.: U.S. Army Research Institute of Environmental Medicine.

Knight, D.C., and J.A. Eden. 1996. A review of the clinical effects of phytoestrogens. Obstet. Gynecol. 87:897–904.

Knott, P.J., and G. Curzon. 1972. Free tryptophan in plasma and brain tryptophan metabolism. Nature 239:452–453.

Kohrt, W. 1995. Body composition by DXA: Tried and true? Med. Sci. Sports Exerc. 27:1349–53.

Kramer, T.R., R.J. Moore, R.L. Shippee, K.E. Friedl, L.E. Martinez-Lopez, M.M. Chan, and E.W. Askew. 1997. Effects of food restriction in military training on T-lymphocyte responses. Int. J. Sports Med. 18(1):S84–S90.

Krebs, H.A. 1975. The role of chemical equilibria in organ function. Adv. Enzyme Regulation 15:449–472.

Kreider, R.B., V. Miriel, amd E. Bertum. 1993. Amino acid supplementation and exercise performance—Analysis of the proposed ergogenic value. Sports Med. 16:190–209.

Kretsch, M.J., P.M. Conforti, and H.E. Sauberlich. 1986. Nutrient intake evaluation of male and female cadets at the United States Military Academy, West Point, New York, Report No. 218. Presidio of San Francisco, Calif. Letterman Army Institute of Research.

Kurpad, A.V., A.E. El-Khoury, L. Beaumier, A. Srivatsa, R. Kuriyon, T. Raj, S. Borgonha, A.M. Ajami, and V.R. Young. 1998. An initial assessment, using 24-hour ^{13}C-leucine kinetics, of the lysine requirement of healthy adult Indian subjects. Am. J. Clin. Nutr. 67:58–66.

Kurzer, M.S., and D.H. Calloway. 1981. Nitrate and nitrogen metabolism in men. Am. J. Clin. Nutr. 34:1305–1313.

Kurzer, M.S. and D.H. Calloway. 1986. Effects of energy deprivation on sex hormone patterns in healthy menstruating women. Am. J. Physiol. 251:E483–E488.

Kushner, R.F., R. Gudivaka, D.A. Scholler. 1996. Clinical Characteristics influencing bioelectrical impedance analysis measurements. Am J Clin Nutr. 64(suppl):423S–427S.

Lacey, J.M., and D.W. Wilmore. 1990. Is glutamine a conditionally essential amino acid? Nutr. Rev. 48:297–309.

Lalau, J.D., J.M. Achard, P. Bataille, C. Bergot, I. Jans, B. Boudailliez, J. Petit, G. Henon, P.F. Westeel, N. El Esper, M.A. Laval-Jeantet, R. Bouillon, J.L. Sebert, and A. Fournier. 1992. [Vertebral density in hypercalciuric lithiasis: Relationship with calcium and protein consumption and vitamin D metabolism]. Ann. Med. Interne 143:293–298.

Laritcheva, K.A., N.I. Yalavaya, V.I. Shubin, and P.V. Smornov. 1978. Study of energy expenditure and protein needs of top weight lifters. In Nutrition, Physical Fitness, and Health, J. Parizkova and V.A. Rogozkin, eds. Baltimore, Md.: University Park.

LeBlanc, J.A. 1996. Cold exposure, appetite, and energy balance. Pp. 203–214 in Nutritional Needs in Cold and in High-Altitude Environments: Applications for Military Personnel in Field Operations, B.M. Marriott and S.J. Carlson, eds. Institute of Medicine. Washington, D.C.: National Academy Press.

Lehnert, H.R., D.K. Reinstein, B.W. Strowbridge, and R.J. Wurtman. 1984a. Neurochemical and behavioral consequences of acute, uncontrollable stress: Effects of dietary tyrosine. Brain Res. 303:215–223.

Lehnert, H.R., D.K. Reinstein, and R.J. Wurtman. 1984b. Tyrosine reverses the depletion of brain norepinephrine and the behavioral deficits caused by tail-shock stress in rats. Pp. 81–91 in Stress: The Role of Catecholamines and Other Neurotransmitters, E. Usdin and R. Kvetnansky, eds. New York: Gordon and Beach.

Lemann, J., Jr., J.R. Litzow, and E.J. Lennon. 1967. Studies of the mechanism by which chronic metabolic acidosis augments urinary calcium excretion in man. J. Clin. Invest. 46:1318–1328.

Lemann, J., Jr., J.A. Pleuss, E.M. Worcester, L. Hornick, D. Schrab, and R.G. Hoffmann. 1996. Urinary oxalate excretion increases with body size and decreases with increasing dietary calcium intake among healthy adults. Kidney Int. 49:200–208.

Lemann, J., Jr., J.A. Pleuss, R.W. Gray, and R.G. Hoffmann. 1991. Potassium administration reduces and potassium deprivation increases urinary calcium excretion in healthy adults. Kidney Int. 39:973–983.

Lemon, P.R., M.A. Tarnopolsky, J.D. MacDougall, and S.A. Atkinson. 1992. Protein requirements and muscle mass/strength changes during intensive training in novice bodybuilders. J. Appl. Physiol. 73:767–775.

Lemon, P.W. 1996. Is increased dietary protein necessary or beneficial for individuals with a physically active lifestyle? Nutr. Rev. 54:S169–S175.

LeRoith, D. 1997. Insulin-like growth factors. N. Engl. J. Med. 336:633–640.

Levey, A.S., S. Adler, A.W. Caggiula, B.K. England, T. Greene, L.G. Hunsicker, J.W. Kusek, N.L. Rogers, and P.E. Teschan, for the Modification of Diet in Renal Disease Study Group. 1996. Effects of dietary protein restriction on the progression of advanced renal disease in the Modification of Diet in Renal Disease Study. Am. J. Kidney Dis. 27:652–663.

Lew, S.Q., and J.P. Bosch. 1991. Effect of diet on creatinine clearance and excretion in young and elderly healthy subjects and in patients with renal disease. J. Am. Soc. Nephrol. 2:856–865.

Lexell J., C.C. Taylor, M. Sjostrom. 1988. What is the cause of the aging atrophy? J Neurol Sci. 84:275–294.

Lieberman, H.R., 1994. Tyrosine and stress: Human and animal studies. Pp. 277–299 in Food Components to Enhance Performance, An Evaluation of Potential Performance—Enhancing Food Components for Operational Rations, B.M. Marriott, ed. Washington, D.C.: National Academy Press.

Lieberman, H.R. and B. Shukitt-Hale. 1996. Food components and other treatments that may enhance performance at high altitude and in the cold. Pp. 453–465 in Nutritional Needs in Cold and in High Altitude Environments, B. Marriott and S. Newberry, eds. Washington, D.C.: National Academy Press.

Lieberman, H.R., B. Caballero, and N. Finer. 1986a. The composition of lunch determines afternoon tryptophan ratios in humans. J. Neural Transmission 65:211–217.

Lieberman, H.R., J. Wurtman, and B. Chew. 1986b. Changes in mood after carbohydrate consumption among obese individuals. Am J. Clin. Nutr. 44:772–778.

Lieberman, H.R., E.W. Askew, R.W. Hoyt, B. Shukitt-Hale, and M.A. Sharp. 1997. Effects of thirty days of undernutrition on plasma neurotransmitters, other amino acids and behavior. J Nutr. Biochem. 8:119–126.

Lieberman, H.R., S. Corkin, B.J. Spring, J.H. Growdin, and R.J. Wurtman. 1983. Mood, performance, and pain sensitivity: Changes induced by food constituents. J. Psychiatr. Res. 17(2):135–145.

Lieberman, H.R., S. Corkin, B.J. Spring, R.J. Wurtman, and J.H. Growdon. 1985. The effects of dietary neurotransmitter precursors on human behavior. Am. J. Clin. Nutr. 42:366–370.

Lieberman, S.A., G.E. Butterfield, D. Harrison, and A.R. Hoffman. 1994. Anabolic effects of recombinant insulin-like growth factor-I in cachectic patients with the acquired immunodeficiency syndrome. J. Clin. Endocrinol. Metab. 78:404–410.

Lindeman, R.D. 1990. Overview: Renal physiology and pathophysiology of aging. Am. J. Kidney Dis. 16:276–282.

Lindeman, R.D., J. Tobin, and N.W. Shock. 1984. Association between blood pressure and the decline in renal function with age. Kidney Int. 26:861–268.

Lindeman, R.D., J. Tobin, and N.W. Shock. 1985. Longitudinal studies on the rate of decline in renal function with age. J. Am. Geriatr. Soc. 33:278–285.

Ling, P.R., J.H. Schwartz, and B.R. Bistrian. 1997. Mechanisms of host wasting induced by administration of cytokines in rats. Am. J. Physiol. 272:E333–E339.

Linkswiler, H.M., C.L. Joyce, and R. Anand. 1974. Calcium retention of young adult males as affected by level of protein and of calcium intake. Proc. N.Y. Acad. Sci. 36:333–340.

Linkswiler, H.M., M.B. Zemel, M. Hegsted, and S. Schuette. 1981. Protein-induced hypercalciuria. Fed. Proc. 40:2429–2433.

Longenecker, J.B. 1961. Relationship between plasma amino acids and clinical chemistry of dogs. Pp. 469–485 in Progress in Meeting Protein Needs of Infants and Pre-school Children. Publ. 843. Washington, D.C.: National Academy of Sciences.

Longenecker, J.B. 1963. Utilization of dietary protein. Pp. 113–144, Chapter 2, in Newer Methods of Nutritional Biochemistry, A.A. Albanese, ed. New York: Academic Press.

Longenecker, J.B., and N.L. Hause. 1959. Relationship between plasma amino acids and composition of the ingested protein. Arch. Biochem. Biophys. 84:46–60.

Longenecker, J.B., and N.L. Hause. 1961. Relationship between plasma amino acids and composition of the ingested protein. II. A shortened procedure to determine plasma amino acid (PAA) ratios. Am. J. Clin. Nutr. 9:356–363.

Loucks A.B., and E.M. Heath. 1994. Induction of low-T3 syndrome in exercising women occurs at a threshold of energy availability. Am. J. Physiol. 266:R817–R823.

Lukaski, H.C., J. Mendez, E.R. Buskirk, and S.H. Cohn. 1981. Relationship between endogenous 3-methylhistidine excretion and body composition. Am. J. Physiol. 240: E302–E307.

Lukasksi, H.C. 1996. Estimation of muscle mass. Pps. 109–128 in Human Body Composition, A.F. Roche, S.B. Heymsfield, and T.G.Lohman, eds. Champaign, Ill.: Human Kinetics Publishers.

Luo, S., E.T.S. Li, and H.R. Lieberman. 1993. Tyrosine increases hypothermia-induced norepinephrine (NE) release in rat hippocampus assessed by *in vivo* microdialysis. Soc. Neurosci. Abstr. 19:1452.

Lutz, J. 1984. Calcium balance and acid-base status of women as affected by increased protein intake and by sodium bicarbonate ingestion. Am. J. Clin. Nutr. 39:281–288.

MacDougall, J.D., M.J. Gibala, M.A. Tarnopolsky, J.R. MacDonald, S.A. Interisano, and K.E. Yarasheski. 1995. The time course for elevated muscle protein synthesis following heavy resistance exercise. Can. J. Appl. Physiol. 10:480–486.

MacFarlane, G., and J.H. Cummings. 1991. The colonic flora, fermentation, and large bowel digestive function. Pp. 51–92 in The Large Intestine: Physiology, Pathophysiology and Disease, S. Phillips, J.H. Peniberton, and R.G. Shorter, eds. New York: Raven.

MacLean, D.A., T.E. Graham, and B. Saltin. 1994. Branched-chain amino acids augment ammonia metabolism while attenuating protein breakdown during exercise. Am. J. Physiol. 167:E1010–1022.

MacLean, D.A., T. E. Graham, and B. Saltin. 1996. Stimulation of muscle ammonia production during exercise following branched-chain amino-acid supplementation in humans. J. Physiol. Lond. 193:909–922.

Maclean, W.C., G.L. De Romana, R.P., Placko, and G.G. Graham. 1981. Protein quality and digestibility of sorghum in pre school children: Balance studies and plasma free amino acids. J. Nutr. 111:1928–1936.

Madsen, K., D.A. MacLean, B. Kiens, and D. Christensen. 1996. Effects of glucose, glucose plus branched-chain amino acids, or placebo on bike performance over 100 km. J. Appl. Physiol. 11:2644–2650.

Maher, T. J. 1994. Safety concerns regarding supplemental amino acids: Results of a study. Pp. 455–460 in Food Components to Enhance Performance, An Evaluation of Potential Performance—Enhancing Food Components for Operational Rations, B.M. Marriott, ed. Washington, D.C.: National Academy Press.

Maher, T.J., B.S. Glaeser, and R.J. Wurtman. 1984. Diurnal variations in plasma concentrations of basic and neutral amino acids and in red cell concentrations of aspartate and glutamate: Effects of dietary protein. J. Clin. Nutr. 39:722–729.

Manson, J.M., and D.W. Wilmore. 1986. Positive nitrogen balance with human growth hormone and hypocaloric intravenous feeding. Surgery 100:188–197.

Marangella, M., O. Bianco, C.V. Martini, M. Petrarulo, C. Vitale, and F. Linari. 1989. Effect of animal and vegetable protein intake on oxalate excretion in idiopathic calcium stone formers. Br. J. Urol. 63:348–351.

Marchini, J.S., J. Cortiella, T. Hiramatsu, T.E. Chapman, and V.R. Young. 1993. Requirements for indispensable amino acids in adult humans: Longer term amino acid kinetic study with support for the adequacy of the Massachusetts Institute of Technology amino acid requirement pattern. Am. J. Clin. Nutr. 58:670–683.

Marchini, J.S., J. Cortiella, T. Hiramatsu, L. Castillo, T.E. Chapman, and V.R. Young. 1994. Phenylalanine and tyrosine kinetics for different patterns and indispensable amino acid intakes in adult humans. Am. J. Clin. Nutr. 60:79–86.

Marliss, E.B., T.T. Aoki, T. Pozefsky, A.S. Most, and G.F. Cahill. 1971. Muscle and splanchnic glutamine and glutamate metabolism in postabsorptive and starved man. J. Clin. Invest. 50:814–817.

Martin A.P., L.F. Spenst, D.T. Drinkwater, J.P. Clarys. 1990. Anthropometric estimation of muscle mass in men. Med Sci Sports Exerc. 22:729–733.

Martin-Castellanos, C., and S. Moreno. 1997. Recent advances on cyclins, CDKs and CDK inhibitors. Trends in Cell Biol. 7:95–98.

Martinez-Lopez, L.E., K.E. Friedl, R.J. Moore, and T.R. Kramer. 1993. A prospective epidemiological study of infection rates and injuries of Ranger students. Mil. Med. 158:433–437.

Martini, L.A., I.P. Heilberg, L. Cuppari, F.A. Medeiros, S.A. Draibe, H. Ajzen, and N. Schor. 1993. Dietary habits of calcium stone formers. Braz. J. Med. Biol. Res. 26:805–812.

Masoro, E.J., and B.P. Yu. 1989. Diet and nephropathy [editorial]. Lab. Invest. 60:165–167.

Masud, T., V.R. Young, T. Chapman, and B.J. Maroni. 1994. Adaptive responses to very low protein diets: The first comparison of ketoacids and essential amino acids. Kidney Intl. 45:1182–1192.

Matalon, R., K. Michals, C. Azen, E.G. Friedman, R. Koch, E. Wenz, H. Levy, F. Rohr, B. Rouse, L. Castiglioni, W. Hanley, V. Austin, and F. de la Cruz. 1991. Maternal PKU collaborative study: The effect of nutrient intake on pregnancy outcome. J. Inherit. Dis. 14:371–374.

Matsueda, S., and Y. Niiyama. 1982. The effects of excess amino acids on maintenance of pregnancy and fetal growth in rats. J. Nutr. Sci. Vitaminol. (Tokyo). 28:557–573.

Matthews, D.E., and R.G. Campbell. 1992. The effect of dietary protein intake on glutamine and glutamate nitrogen metabolism in humans. Am. J. Clin. Nutr. 55:963–970.

McLarney, M.J., P.L. Pellett, and V.R. Young. 1996. Pattern of amino acid requirements in humans: An interspecies comparison using published amino acid requirements recommendations. J. Nutr. 126:1871–1882.

Meguid, M.M., D.E. Matthews, D.M. Bier, C.N. Meredith, J.S. Soeldner, and V.R. Young. 1986a. Leucine kinetics at graded leucine intakes in young men. Am. J. Clin. Nutr. 43:770–780.

Meguid, M.M., D.E. Matthews, D.M. Bier, C.N. Meredith, and V.R. Young. 1986b. Valine kinetics at graded valine intakes in young men. Am. J. Clin. Nutr. 43:781–786.

Meredith, C.N., D.M. Bier, M.M. Meguid, D.E. Matthews, Z. Wen, and V.R. Young. 1982. Whole body amino acid turnover with ^{13}C tracers: A new approach for estimation of human amino acid requirements. Pp. 42–59 in Clinical Nutrition '81, R.I.C. Wesdorp and P.B. Soeters, eds. Edinburgh and London: Churchill Livingstone.

Meredith, C.N., Z.M. Wen, D.M. Bier, D.E. Matthews, and V.R. Young. 1986. Lysine kinetics at graded lysine levels in young men. Am. J. Clin. Nutr. 43:787–794.

Meredith, C.N., M.J. Zackin, W.R. Frontera, and W.J. Evans. 1989. Dietary protein requirements and body protein metabolism in endurance-trained men. J. Appl. Physiol. 66:2850–2856.

Messing, R.B., and L.D. Lytle. 1977. Serotonin-containing neurons: their possible role in pain and analgesia. Pain 4:1–21.

Metges, C.C., A.E. El-Khoury, K.J. Petzke, S. Bedri, M.F. Fuller, and V.R. Young. 1997. The quantitative contribution of microbial lysine to lysine flux in healthy male subjects [abstract]. FASEB J. 11(3):A149.

Michie, H.R., K.R. Manogue, D.R. Spriggs, A. Revhaug, S. O'Dwyer, C.A. Dinarello, A. Cerami, S.M. Wolff, and D.W. Wilmore. 1988. Detection of circulating tumor necrosis factor after endotoxin administration. N. Engl. J. Med. 318:1481–1486.

Millward, D.J. 1990. Amino acid requirements in adult man. Am. J. Clin. Nutr. 51:492–493.

Millward, D.J. 1992. The metabolic basis of the amino acid requirement. Pp. 31–57 in Protein-Energy Interactions I/D/E/C/G, N.W. Scrimshaw and B. Schurch, eds. Lausanne, Switzerland: Nestle Foundation.

Millward, D.J. 1993. Stable-isotope-tracer studies of amino acid balance and human indispensable amino acid requirements Am. J. Clin. Nutr. 57(1):81–86.

Millward, D.J. 1994. Can we define indispensable amino acid requirements and assess protein quality in adults? J. Nutr. 124:1509S–1516S.

Millward, D.J. 1997. Human amino acid requirements. J. Nutr. 127:1842–1846.

Millward, D.J., and J.C. Waterlow. 1996. Letter to the editor. Eur. J. Clin. Nutr. 50:832–833.

Millward, D.J., and J.P. Pacy. 1995. Postprandial protein utilization and protein quality assessment in man. Clin. Sci. 88:597–606.

Millward, D.J., and J.P. Rivers. 1988. The nutritional role of indispensable amino acids and the metabolic basis for their requirements. Eur. J. Clin. Nutr. 42:367–393.

Millward, D.J., and J.P. Rivers. 1989. The need for indispensable amino acids: The concept of the anabolic drive. Diab. Metab. Rev. 5(2):191–211.

Millward, D.J., and S.R. Roberts. 1996. Protein requirement of older individuals. Nutr. Res. Rev. 9:67–88.

Millward, D.J., A.A. Jackson, G. Price, and J.P.W. Rivers. 1989. Human amino acid and protein requirements: Current dilemmas and uncertainties. Nutr. Res. Rev. 2:109–132.

Millward, D.J., G.M. Price, P.J.H. Pacy, and D. Halliday. 1990. Maintenance protein requirements: The need for conceptual revaluation. Proc. Nutr. Soc. 49:473–487.

Millward, D.J., G. Price, P.J.H. Pacy, and D. Halliday. 1991. Whole body protein and amino acid turnover in man: What can we measure with confidence? Proc. Nutr. Soc. 50:197–216.

Millward, D.J., J.L. Bowtell, P. Pacy, and M.J. Rennie. 1994. Physical activity, protein metabolism and protein requirements. Proc. Nutr. Soc. 53(1):223–240.

Millward, D.J., A. Fereday, N. Gibson, and P.J. Pacy. 1996. Postprandial protein metabolism. Bailliere's Clin. Endocrin. Metab. 10:533–549.

Mitch, W.E. 1989. Dietary restrictions for a single normal kidney. Pediatr. Nephrol. 3:129.

Mitch, W.E., and A.L. Goldberg. 1996. Mechanisms of muscle wasting. N. Engl. J. Med. 335: 1897–1905.

Mitchell, H.H., and M. Edman. 1949. Nutrition and Resistance to Climatic Stress, with Reference to Man. Chicago, Ill.: Quartermaster Food and Container Institute for the Armed Forces.

Mitchell, H.H., and M. Edman. 1951. Nutrition and Resistance to Climatic Stress, with Particular Reference to Man. Springfield, Ill.: Charles C. Thomas.

Moller, N., J.O.L. Jorgensen, J. Moller, P. Ovesen, O. Schmitz, J.S. Christiansen, and H. Orskov. 1995. Metabolic effects of growth hormone in humans. Metabolism 44(Suppl.):33–36.

Moore, R.J., K.E. Friedl, T.R. Kramer, L.E. Martinez-Lopez, R.W. Hoyt, R.E. Tulley, J.P. DeLany, E.W. Askew, and J.A. Vogel. 1992. Changes in soldier nutritional status and immune function during the Ranger training course. Report USARIEM-T13-92, AD-A257 437. Natick, Mass.: U.S. Army Research Institute of Environmental Medicine.

Morgan, D.B., H.M. Newton, C.J. Schorah, M.A. Jewitt, M.R. Hancock, and R.P. Hullin. 1986. Abnormal indices of nutrition in the elderly: A study of different clinical groups. Age Ageing 15:65–76.

Morrison, W.L., J.N.A. Gibson, R.T. Jung, and M.J. Rennie. 1988. Skeletal muscle and whole body protein turnover in thyroid disease. Eur. J. Clin. Invest. 18:62–68.

Motil, K.J., C.M. Montandon, M. Thotathuchery, and C. Garza. 1990. Dietary protein and nitrogen balance in lactating and nonlactating women. Am. J. Clin. Nutr. 51:378–384.

Motil, K.J., D.E. Matthews, D.M. Bier, J.F. Burke, H.N. Munro, and V.R. Young. 1981. Whole-body leucine and lysine metabolism: Response to dietary protein in young men. Am. J. Physiol. 240:E712–E721.

Motil, K.J., T.A. Davis, C.M. Montandon, W.W. Wong, and P.D. Klein. 1996. Whole-body protein turnover in the fed state is reduced in response to dietary protein restriction in lactating women. Am. J. Clin. Nutr. 64:32–39.

Muhlbacher, F., C.R. Kapadia, M.F. Colpys, R.J. Smith, and D.W. Wilmore. 1984. Effects of glucocorticoids on glutamine metabolism in skeletal muscle. Am. J. Physiol. 247:E75–E83.

Muldowney, F.P., R. Freaney, and M.F. Moloney. 1982. Importance of dietary sodium in the hypercalciuria syndrome. Kidney Int. 22:292–296.

Mulligan, K., and G.E. Butterfield. 1990. Discrepancies between energy intake and expenditure in physically active women. Br. J. Nutr. 64(1):23–36.

Munro, H.N. 1964a. General aspects of the regulation of protein metabolism by diet and hormones. Pp. 381–481 in Mammalian Protein Metabolism, Vol. 1., Munro H.N. and J.B. Allison, eds. New York: Academic press.

Munro, H.N. 1964b. Historical introduction: The origin and growth of our present concept of protein metabolism. Pp. 1–29 in Mammalian Protein Metabolism, H.N. Munro and J.B. Allison, ed. New York: Academic Press.

Munro, H.N., and M.C. Crim. 1994. Protein and amino acids. In Modern Nutrition in Health and Disease, M.E. Shils, J.A. Olson, and M. Shike, eds. Philadelphia: Lea and Febiger.

Murphy, E.A. 1982. Muddling, meddling and modeling. Pp. 333–348 in Genetic Basis of the Epilepsies, V.E. Anderson, W.A. Hauser, J.K. Penry, and C.F. Sing, eds. New York: Raven.

Murphy, T.C., R.W. Hoyt, T.E. Jones, C.L. Gabaree, and E.W. Askew. 1994. Performance enhancing ration components program: Supplemental carbohydrate test. Report USARIEM-T95-2, AD-A288 560. Natick Mass.: Army Research Institute of Environmental Medicine.

Nair, K.S., D. Halliday, D.E. Matthews, and S.L. Welle. 1987. Hyperglucagonemia during insulin deficiency accelerates protein catabolism. Am. J. Physiol. 253:E208–E213.

Nair, K.S., S. Welle, D. Halliday, and R.G. Campell. 1988. Effect of 3-hydroxybutyrate on whole-body leucine kinetics and fractional mixed skeletal muscle protein synthesis in humans. J. Clin. Invest. 82:198–205.

Nair, K.S., R.G. Schwartz, and S. Welle. 1992. Leucine as a regulator of whole body and skeletal muscle protein metabolism in humans. Am. J. Physiol. 263:E928–E934.

Nair, K.S., G.C. Ford, K. Ekberg, E. Fernqvist-Forbes, and J. Wahren. 1995. Protein dynamics in whole body and in splanchnic and leg tissues in Type I diabetic patients. J. Clin. Invest. 95:2926–2937.

Nelson, M.E., M.A. Fiatarone, J.E. Layne, I. Trice, C.D. Economoc, R.A. Fielding, R. Ma, R.N. Pierson, W.J. Evans Nunez C, D. Gallagher, M. Visser, F.X. Pi-Sunyer, Z. Wang, S.B. Heymsfield. 1996. Analysis of body composition techniques and models for detecting change in soft tissue with strength training. Am J Clin Nutr. 63:678–686.

Nelson. R.A., J.D. Jones, H.W. Wahner, D.B. McGill, and C.F. Code. 1975. Nitrogen metabolism in bears: Urea metabolism in summer starvation and in winter sleep and role of urinary bladder in water and nitrogen conservation. Mayo Clin. Proc. 50:141–146.

Neri, D.F., D. Wiegmann, R.R. Stanny, S.A. Shappell, A. McCardie, and D.L. McKay. 1995. The effects of tyrosine on cognitive performance during extended wakefulness. Aviat. Space Environ. Med. 66:313–319.

Newsholme, E.A., I.N. Acworth, and E. Blomstrand. 1987. Amino acids, brain neurotransmitters and a functional link between muscle and brain that is important in sustained exercise. Pp. 127–138 in Advances in Myochemistry, G. Benzi, ed. London: John Libby Eurotext.

Nicol, B.M., and P.G. Phillips. 1976a. Endogenous nitrogen excretion and utilization of dietary protein. Br. J. Nutr. 35:181–193.

Nicol, B.M., and P.G. Phillips. 1976b. The utilization of dietary protein by Nigerian men. Br. J. Nutr. 36:337–351.

Nindl, B.C., K.E. Friedl, P.N. Frykman, L.J. Marchitelli, R.L. Shippee, and J.F. Patton. 1997. Physical performance and metabolic recovery among lean, healthy men following a prolonged energy deficit. Int. J. Sports Med. 18:1–8.

NRC (National Research Council). 1941. Recommended Dietary Allowances. Washington, D.C.: National Academy Press.

NRC. 1945. Recommended Dietary Allowances. Washington, D.C.: National Academy Press.

NRC. 1980. Recommended Dietary Allowances, 9th ed. Washington, D.C.: National Academy Press.

NRC. 1986. Cognitive Testing Methodology. Washington, D.C: National Academy Press.

NRC. 1989. Recommended Dietary Allowances, 10th ed. Washington, D.C.: National Academy Press.

Nuñez, C., D. Gallagher, and S.B. Heymsfield. 1995. Appendicular skeletal muscle mass: Measurement with single frequency bioimpedance analysis. FASEB J. 9(4):A1012.

Nuñez C., J. Grammes, D. Gallagher, R.N. Baumgartner, Z. Wang, S.B. Heymsfield. 1996. Appendicular skeletal muscle: Estimation by total limb impedance measured with new electrode system. FASEB J. 10:3.

Nuñez, C., D. Gallagher, M. Visser, F.X. Pi-Sunyer, Z. Wang, and S.B. Heymsfield. 1997. Bioimpedance analysis: Evaluation of leg-to-leg system based on pressure contact foot-pad electrodes. Med. Sci. Sports Exerc. 29:524–531.

Nurjhan, N., A. Bucci, G. Perriello, N. Stumvoll, G. Dailey, D.M. Bier, I. Toft, T.G. Jenssen, and J.E. Gerich. 1995. Glutamine: A major gluconeogenic precursor and vehicle for interorgan carbon transport in man. J. Clin. Invest. 95:272–277.

Organ L.W., G.B. Bradham, D.T.W. Gore, S.L. Lozier. 1994. Segmental bioelectrical impedance analysis: Theory and Application of a new technique. J. Appl. Physiol. 77:98–112.

O'Riordain, M., K.C. Fearon, J.A. Ross, P. Rogers, J.S. Falconer, D.C. Bartolo, O.J. Garden, and D.C. Carter. 1994. Glutamine supplemental parenteral nutrition enhances T-lymphocyte response in surgical patients undergoing colorectal resection. Ann Surg. 220:212–221.

Osborne, T.B., and L.B. Mendel. 1916. The amino-acid minimum for maintenance and growth, as exemplified by further experiments with lysine and tryptophan. J. Biol. Chem. 25:1–8.

Owasoyo, J.O., D.F. Neri, and J.G. Lamberth. 1992. Tyrosine and its potential use as a countermeasure to performance decrement in military sustained operations. Aviat. Space Environ. Med. 63:364–369.

Owen, O.E., K.J. Smalley, D.A. D'Alessio, M.A. Mozzoli, E.K. Dawson. 1998. Protein, fat, and carbohydrate requirements during starvation: anaplerosis and cataplerosis. Am. J. Clin. Nutr. 68:12–34.

Pacy, P.J., G.M. Price, D. Halliday, M.R. Quevedo, and D.J. Millward. 1994. Nitrogen homeostasis in man: The diurnal responses of protein synthesis and degradation and amino acid oxidation to diets with increasing protein intakes. Clin. Sci. 86(1):103–116.

Pannemans, D.L.E., A.J.M. Wagenmakers, A.E. Jeukendrup, A.P. Gijsen, J.M.G. Senden, and W.H.M. Saris. 1997. The effect of prolonged moderate intensity exercise on protein metabolism of trained men. Med. Sci. Sport Exerc. 29: S224.

Pannier, J.L., J.J. Bouckaert, and R.A. Lefebvre. 1995. The antiserotonin agent pizotifen does not increase endurance performance in humans. Eur. J. Appl. Physiol. 12:175–178.

Pardridge, W.M. 1977. Regulation of amino acid availability to the brain. Pp. 141–190 in Nutrition and the Brain, Vol. 1, R.J. Wurtman and J.J. Wurtman, eds. New York: Raven Press.

Parry-Billings M., J. Evans, P.C. Calder, and E.A. Newsholme. 1990. Does glutamine contribute to immunosuppression after major burns? Lancet 336:523–525.

Parry-Billings M., R. Budgett, Y. Koutedakis, E. Blomstrand, S. Brooks, C. Williams, P.C. Calder, S. Pilling, R. Baigrie, and E.A. Newsholme. 1992. Plasma amino acid concentrations in the overtraining syndrome: Possible effects on the immune system. Med. Sci. Sports. Exerc. 24:1353–1358.

Patterson, B.W., T. Nguyen, E. Pierre, D.N. Herndon, R.R. Wolfe. 1997. Urea and protein metabolism in burned children: effect of dietary protein intake. Metabolism 46(5):573–578.

Paul, G.L. 1989. Dietary protein requirements of physically active individuals. Sports Med. 8:154–176.

Pearlstone, D.B., R.F. Wolf, R.S. Berman, M. Burt, M.F. Brennan. 1994. Effect of systemic insulin on protein kinetics in postoperative cancer patients. Ann. Surg. Oncol. 1(4):321–332.

Pelletier, V., L. Marks, D.A. Wagner, R.A. Hoerr, and V.R. Young. 1991a. Branched-chain amino acid interactions with reference to amino acid requirements in adult men: Valine metabolism at different leucine intakes. Am. J. Clin. Nutr. 54:395–401.

Pelletier, V., L. Marks, D.A. Wagner, R.A. Hoerr, and V.R. Young. 1991b. Branched-chain amino acid interactions with reference to amino acid requirements in adult men: Leucine metabolism at different valine and isoleucine intakes. Am. J. Clin. Nutr. 54:402–407.

Phillips, S.M., S.A. Atkinson, M.A. Tarnopolsky, and J.D. MacDougal. 1993. Gender differences in leucine kinetics and nitrogen balance in endurance athletes. J. Appl. Physiol. 75:2134–2141.

Picou, D., and T. Taylor-Roberts. 1969. The measurement of total protein synthesis and catabolism and nitrogen turnover in infants in different nutritional states and receiving different amounts of dietary protein.. Clin. Sci. 36:283–296

Pietrobelli, A., D. Gallagher, R. Baumgartner, R. Ross, S.B. Heymsfield. R. 1998. Lean value for DXA Two-component Soft-Tissue Model: Influence of Age and Tissue/Organ Type. Appl. Radiat. Isot. 49(5-6):743–744.

Pineda, O., B. Torun, F.E. Viteri, and G. Arroyave. 1981. Protein quality in relation to estimates of essential amino acids requirements. Pp. 131–139 in Protein Quality in Humans: Assessment and *In Vitro* Estimation, C.E. Bodwell, ed. Westport, Conn.: AVI Publishing Company Inc.

Pozefsky, T., P. Felig, J.D. Tobin, J.S. Soeldner, and G.F. Cahill. 1969. Amino acid balance across tissue of forearm in postabsorptive man. Effect of insulin at two dose levels. J. Clin. Invest. 48:2273–2282.

Price, G.M., D. Halliday, P.J. Pacy, M.R. Quevedo, and D.J. Millward. 1994. Nitrogen homeostasis in man: 1. Influence of protein intake on the amplitude of diurnal cycling of body nitrogen. Clin. Sci. 86:91–102.

Price, S.R., J.L. Bailey, X. Wang, C. Jurkovitz, B.K. England, X. Ding, L. S. Phillips, and W.E. Mitch. 1996. Muscle wasting in insulinopenic rats results from activation of the ATP-dependent, ubiquitin–proteosome proteolytic pathway by a mechanism including gene transcription. J. Clin. Invest. 98: 1703–1708.

Puche, R.C., and S. Feldman. 1992. Relative importance of urinary sulfate and net acid excretion as determinants of calciuria in normal subjects. Medicina (Buenos Aires) 52:220–224.

Puche, R.C., A.F. Carlomagno, A. Gonzalez, and A. Sanchez. 1987. A correlation and path coefficient analysis of components of calciuria in normal subjects and idiopathic stone formers. Bone Miner. 2:405–411.

Quevedo, M.R., G.M. Price, D. Halliday, P.J. Pacy, and D.J. Millward. 1994. Nitrogen homeostasis in man: 3. Diurnal changes in nitrogen excretion, leucine oxidation and whole body leucine kinetics during a reduction from a high to a moderate protein intake. Clin. Sci. 86:185–193.

Rand, W.M., V.R. Young, and N.S. Scrimshaw. 1976. Change of urinary nitrogen excretion in response to low protein diets in adults. Am. J. Clin. Nutr. 29:639–644.

Rand, W.M., N.S. Scrimshaw, and V.R. Young. 1979. An analysis of temporal patterns in urinary nitrogen excretion of young adults receiving constant diets at two nitrogen intakes for 8 to 11 weeks. Am. J. Clin. Nutr. 32:1408–1414.

Rand, W.M., N.S. Scrimshaw, and V.R. Young. 1981. Conventional ("long-term") nitrogen balance studies for protein quality evaluation in adults: Rationale and limitations. Pp. 61–94 in Protein Quality in Humans: Assessment and In Vitro Estimation, C.E. Bodwell, J.S. Adkin, and D.T. Hopkins, eds. Westport, Conn.: AVI Publishing.

Rand, W.M., N.S. Scrimshaw, and V.R. Young. 1985. A retrospective analysis of long-term metabolic balance studies: Implications for understanding dietary nitrogen and energy utilization. Am. J. Clin. Nutr. 42:1329–1350.

Rao, P.N., V. Prendiville, A. Buxton, D.G. Moss, and N.J. Blacklock. 1982. Dietary management of urinary risk factors in renal stone formers. Br. J. Urol. 54:578–583.

Rauch, T.M., and H.R. Lieberman. 1990. Pre-treatment with tyrosine reverses hypothermia induced behavioral depression. Brain Res. Bull. 24:147–150.

Reddy, V. 1971. Lysine supplementation of wheat and nitrogen retention in children. Am. J. Clin. Nutr. 24:1246–1249.

Reed, M.J., J.T. Brozinick, M.C. Lee, and J.L. Ivy. 1989. Muscle glycogen storage postexercise: Effect of mode of carbohydrate administration. J. Appl. Physiol. 66:720–726.

Reed, R.L., L. Pearlmutter, K. Yochum, E.E. Meredith, and A.D. Mooradian. 1991. The relationship between muscle mass and muscle strength in the elderly. J. Am. Geratr. Soc. 39:555–561.

Reeds, P.J. 1990. Amino acid needs and protein scoring patterns. Proc. Nutr. Soc. 49:489–497.

Reeds, P.J., and P.R. Becket. 1996. Protein and amino acids. Pp. 67–86 in Present Knowledge in Nutrition, 7th ed., E.E. Ziegler and L.J. Filer, eds. Washington, D.C.: ILSI Press.

Reeds, P.J., M.F. Fuller, and B.A. Nicholson. 1985. Metabolic basis of energy expenditure with particular reference to protein. Pp. 46–47 in Substrate and Energy Metabolism in Man, J. S. Garrow and D. Halliday, eds. London: John Libby.

Rennie, M.J. 1996. Influence of exercise on protein and amino acid metabolism. Pp. 995–1035 in American Physiological Society Handbook of Physiology on Exercise, Chapter 12, Section 12, Control of Energy Metabolism During Exercise, R. L. Terjung, ed. Bethesda, Md.: American Physiological Society.

Rennie, M.J., P. Babij, P.M. Taylor, H.S. Hundal, P. MacLennan, P.W. Watt, M.M. Jepson, and D.J. Millward. 1986. Characteristics of a glutamine carrier in skeletal muscle have important consequences for nitrogen loss in injury, infection and chronic disease. Lancet 2:1008–1012.

Roberts, S.B., M.B. Heyman, W.J. Evans, P. Fuss, R. Tsay, and V.R. Young. 1990. Dietary requirements of young adult men, determined using doubly labeled water method. Am. J. Clin. Nutr. 54:499–505.

Roberts, S.B., P. Fuss, M.B. Heyman, and V.R. Young. 1995. Influence of age on energy requirements. Am. J. Clin. Nutr. 62:1053S–1058S.

Robertson, W.G., M. Peacock, and A. Hodgkinson. 1979a. Dietary changes and the incidence of urinary calculi in the U.K. between 1958 and 1976. J. Chronic Dis. 32:469–476.

Robertson, W.G., P.J. Heyburn, M. Peacock, F.A. Hanes, and R. Swaminathan. 1979b. The effect of high animal protein intake on the risk of calcium-stone-formation in the urinary tract. Clin. Sci. 57:285–288.

Robertson, W.G., M. Peacock, P.J. Heyburn, F.A. Hanes, A. Rutherford, E. Clementson, R. Swaminthan, and P.B. Clark. 1979c. Should recurrent calcium oxalate stone formers become vegetarians? Brit J. Urol. 51:427–431.

Rocher, L.L., and R.D. Swartz. 1987. Kidney donors and protein intake [letter to the editor]. Ann. Intern. Med. 107:427.

Rohde, T., D.A. MacLean, A. Hartkopp, and B.K. Pedersen. 1996. The immune-system and serum glutamine during a triathlon. Eur. J. Appl. Physiol. Occupat. Physiol. 14:428–434.

Romijn, J.A., E.F. Coyle, L. Sidossis, A. Castaldelli, J.F. Horowitz, E. Endert, and R.R. Wolfe. 1993. Regulation of endogenous fat and carbohydrate metabolism in relation to exercise intensity. Am. J. Physiol. 265:E380–E391.

Rooyackers, O., and K.S. Nair. 1997. Hormonal regulation of human muscle protein metabolism. Ann. Rev. Nutr. 17:457–485.

Rooyackers, O.E., D.B. Adey, P.A. Ades, and K.S. Nair. 1996. Effect of age on *in vivo* synthesis rates of mitochondrial protein in human skeletal muscle. Proc. Natl. Acad. Sci. USA 93:15364–15369.

Rose, M.S. and D.E. Carlson. 1986. Effects of A Ration meals on body weight during sustained field operations. Technical Report T2-87. Natick, Mass.: U.S. Army Research Institute of Environmental Medicine.

Rose, M.S., P.C. Szlyk, R.P. Francesconi, L.S. Lester, L. Armstrong, W. Matthew, A.V. Cardello, R.D. Popper, I. Sils, G. Thomas, D. Schilling, and R. Whang. 1989. Effectiveness and acceptability of nutrient solutions in enhancing fluid intake in the heat. Technical Report No. T10-89. Natick, Mass.: U.S. Army Research Institute of Environmental Medicine.

Rose, R.W., C.J. Baker, W. Wisnaskas, J.S.A. Edwards, and M.S. Rose. 1989. Dietary assessment of U.S. Army basic trainees at Fort Jackson, South Carolina. Technical Report No. T6-89. Natick, Mass.: U.S. Army Research Institute of Environmental Medicine.

Rose, W.C. 1957. The amino acid requirements of adult man. Nutr. Abstr. Rev. 27:631–647.

Rose, W.C., A. Borman, M.J. Coon, and G.F. Lambert. 1955a. The amino acid requirements of man. X. The lysine requirement. J. Biol. Chem. 214:579–587.

Rose, W.C., B.E. Leach, M.J. Coon, and G.F. Lambert. 1955b. The amino acid requirements of man. IX. The phenylalanine requirement. J. Biol. Chem. 213:913–922.

Ross, R., H. Pedwell, J. Rissanen. 1995. Response of total and regional lean tissue and skeletal muscle to a program of energy restriction and resistance exercise. Int J Obes. 19:781–787.

Rothman, J.E., and F.T. Wieland. 1996. Protein sorting by transport vesicles. Science 272: 227–234.

Rowbottom, D.G., D. Keast, C. Goodman, and A.R. Morton 1995. The haematological, biochemical and immunological profile of athletes suffering from the overtraining syndrome. Eur. J. Appl. Physiol. 70:502–509.

Rowbottom, D.G., D. Keast, and A.R. Morton. 1996. The emerging role of glutamine as an indicator of exercise stress and overtaining. Sports Med. 11:80–97.

Rowland, I.R., T. Granli, O.C. Bockman, P.E. Key, and R.C. Massey. 1991. Endogenous N nitrosation in man assessed by measurement of apparent total N nitroso compounds in faeces. Carcinogenesis 12:1359–1401.

Rudman, D., D.E. Mattson, A.G. Feller, R. Cotter, and R.C. Johnson. 1989. Fasting plasma amino acids in elderly men. Am. J. Clin. Nutr. 49:559–566.

Rush, D., Z. Stein, and M.A. Susser. 1980. A randomized controlled trial of prenatal nutritional supplementation in New York City. Pediatrics 65:683–697.

Sahlin, K., A. Katz, and S. Broberg. 1990. Tricarboxylic acid cycle intermediates in human muscle during prolonged exercise. Am. J. Physiol. 159:C834–C841.

Sahlin, K., L. Jorfeldt, and K.G. Henriksson. 1995. Tricarboxylic acid cycle intermediates during incremental exercise in healthy subjects and in patients with McArdle's disease. Clin. Sci. 19:687–693.

Said, A.K., and D.M. Hegsted. 1970. Response of adult rats to low dietary levels of essential amino acids. J. Nutr. 100:1363–1376.

Sakurai, Y., and S. Klein. 1998. Metabolic alteration in patients with cancer: Nutritional implications. Surg. Today 28(3):247–257

Sakurai, Y., A. Aarsland, D.N. Herndon, D. L. Chinkes, E. Pierre, T.T. Nguyen, B.W. Patterson, and R.R. Wolfe. 1995. Stimulation of muscle protein synthesis by long-term insulin infusion in severely burned patients. Ann. Surg. 222(3):283–297.

Salter, C. 1989. Dietary tyrosine as an aid to stress resistance among troops. Milit. Med. 154:144–146.

Samuels, J.P., R.P. McDevitt, M.C. Bollman, W. Maclinn, L.M. Richardson, and L.G. Voss. 1947. Ration Development: A Report of Wartime Problems in Subsistence Research and Development, Vol. 12. Chicago, Ill.: Quartermaster Food and Container Institute for the Armed Forces, Research and Development Branch, Office of the Quartermaster General.

Sandage, B.W., Jr., L. Sabounjian, R. White, and R.J. Wurtman. 1992. Choline citrate may enhance athletic performance. Paper presented at the American Physiological Society Conference on Integrative Biology of Exercise, Colorado Springs, Colo.

Santos, A.A., M.R. Scheltinga, E. Lynch, E.F. Brown, P. Lawton, E. Chambers, J. Browning, and C.A. Dinarello. 1993. Receptor antagonist is not attenuated by glucocorticoids after endotoxemia. Arch. Surg. 128(2):138–143.

Saris, W.H.M. 1997. Limits of human endurance: Lessons from the Tour de France. Pp. 451–462 in Physiology, Stress, and Malnutrition: Functional Correlates, Nutritional Intervention, J.M. Kinney and H.N. Tucker, eds. New York: Lippincott-Raven Publishers.

Sayegh, R., I. Schiff, J. Wurtman, P. Spiers, J. McDermott, and R.J. Wurtman. 1995. The effect of a carbohydrate-rich beverage on mood, appetite, and cognitive function in women with premenstrual syndrome. Obstet. Gynecol. 86:520–528.

Scheltinga, M.R., L.S. Young, K. Benfell, R.L. Bye, T.R. Ziegler, A.A. Santos, J.H. Antin, P.R. Schloerb, and D.W. Wilmore. 1991. Glutamine-enriched intravenous feedings attenuate extracellular fluid expansion after a standard stress. Ann. Surg. 214:385–393.

Schimmel, P. 1993. GTP hydrolysis in protein synthesis: Two for tu? Science 259:1264–1265.

Schneider, W., and E. Menden. 1988. [The effect of long-term increased protein administration on mineral metabolism and kidney function in the rat. I. Renal and enteral metabolism of calcium, magnesium, phosphorus, sulfate and acid]. Z. Ernährungwiss. 27:170–185.

Schuette, S.A., and H.M. Linkswiler. 1982. Effects on Ca and P metabolism in humans by adding meat, meat plus milk, or purified proteins plus Ca and P to a low protein diet. J. Nutr. 112:338–349.

Schuette, S.A., M. Hegsted, M.B. Zemel, and H.M. Linkswiler. 1981. Renal acid, urinary cyclic AMP, and hydroxyproline excretion as affected by the level of protein, sulfur amino acid, and phosphorus intake. J. Nutr. 111:2106–2116.

Schwartz, R.S. 1995. Trophic factor supplementation: effect on the age-associated changes in body composition. J. Gerontol. A. Biol. Sci. Med. Sci. 50:151–156.

Scrimshaw, N.S., M.A. Hussein, E. Murray, W.M. Rand, and V.R. Young. 1972. Protein requirements of man: Variations in obligatory urinary and fecal nitrogen losses in young men. J. Nutr. 102:1595–1604.

Scrimshaw, N.S., Y. Taylor, and V.R. Young. 1973. Lysine supplementation of wheat gluten at adequate and restricted energy intakes in young men. Am. J. Clin. Nutr. 26:965–972.

Segura, R., and J.L. Ventura. 1988. Effect of l-tryptophan supplementation on exercise performance. Int. J. Sports Med. 9:301–305.

Seltzer, S., D. Dewart, R. L. Pollack, and E. Jackson. 1983. The effects of dietary tryptophan on chronic maxillofacial pain and experimental pain tolerance. J. Psychiat. Res. 17(2):181–186.

Shapses, S.A., S.P. Robins, E.I. Schwartz, and H. Chowdhury. 1995. Short-term changes in calcium but not protein intake alter the rate of bone resorption in healthy subjects as assessed by urinary pyridinium cross-link excretion. J. Nutr. 125:2814–2821.

Sharp, T., S.R. Bramwell, and D.G. Grahame-Smith. 1992. Effect of acute administration of L-tryptophan on the release of 5-HT in rat hippocampus in relation to serotoninergic neuronal activity: An *in vivo* microdialysis study. Life Sci. 50:1215–1223.

Shippee, R., E.W. Askew, M. Mays, B. Fairbrother, K. Friedl, J. Vogel, R. Hoyt, L. Marchitelli, B. Nindl, P. Frykman, L. Martinez-Lopez, E. Bernton, M. Kramer, R. Tulley, J. Rood, J. DeLany, D. Jezior, and J. Arsenault. 1994. Nutritional and immunological assessment of Ranger students with increased caloric intake. Technical Report No. T95-5. Natick, Mass.: U.S. Army Research Institute of Environmental Medicine.

Shippee, R.L., S. Wood, P. Anderson, T.R. Kramer, M. Neita, and K. Wolcott. 1995. Effects of glutamine supplementation on immunological responses of soldiers during the Special Forces Assessment and Selection Course [abstract]. FASEB J. 9:731.

Shukitt-Hale, B., M.J. Stillman, and H.R. Lieberman. 1996. Tyrosine administration prevents hypoxia-induced decrements in learning and memory. Physiol. Behav. 59:867–871.

Shurtleff, D., J.R. Thomas, S.T. Ahlers, and J. Schrot. 1993. Tyrosine ameliorates a cold-induced delayed matching-to-sample performance decrement in rats. Psychopharmacol. 112:228–232.

Shurtleff, D., J.R. Thomas, J. Schrot, K. Kowalski, and R. Harford. 1994. Tyrosine reverses a cold-induced working memory deficit in humans. Pharmacol. Biochem. Behav. 47(4):935–941.

Sierakowski, R., B. Finlayson, R.R. Landes, C.D. Finlayson, and N. Sierakowski. 1978. The frequency of urolithiasis in hospital discharge diagnoses in the United States. Invest. Urol. 15:438–441.

Simat, T., B. van Wickern, H. Eulitz, and H. Steinhart. 1996. Contaminants in biotechnologically manufactured l-tryptophan. J. Chromatogr. B. Biomed. 685:41–51.

Singh, P.P., F. Hussain, R.C. Gupta, A.K. Pendse, R. Kiran, and R. Ghosh. 1993. Effect of dietary methionine and inorganic sulfate with and without calcium supplementation on urinary calcium excretion of guinea pigs (*Cavia porcellus*). Indian J. Exp. Biol. 31:96–97.

Sipila S., H. Suominen. 1991. Ultrasound imaging of the quadriceps muscle in elderly athletes and untrained men. Muscle and Nerve 14:527–533.

Sjöström, L.A. 1991. A computer-tomography based multicompartment body composition technique and anthropometric predictions of lean body mass, total and subcutaneous adipose tissue. Int J Obesity. 15:19–30.

Smith, S.E., R.O. Pihl, S.N. Young, and F.R. Ervin. 1987. A test of possible cognitive and environmental influences on the mood lowering effect of tryptophan depletion in normal males. Psychopharmacology. 91:451–457.

Snyder W.S., M.J. Cook, E.S. Nasset, L.R. Karhausen, G.P. Howells, I.H. Tipton. 1975. Report of the Task Group on Reference Man. Oxford: Pergamon Press.

Soares, M.J., L.S. Piers, P.S. Shetty, S. Robinson, A.A. Jackson, and J.C. Waterlow. 1991. Basal metabolic rate, body composition and whole body protein turnover in Indian men with differing nutritional status. Clin. Sci. Colch. 81(3):419–425.

Solomon, F., R.C. Cuneo, R. Hesp, and P.H. Sonksen. 1989. The effects of treatment with recombinant human growth hormone on body composition and metabolism in adults with growth hormone deficiency. New Engl. J. Med. 321:1797.

Souba, W.W. 1991. Glutamine: A key substrate for the splanchnic bed. Ann. Rev. Nut. 11:285–308.

Souba, W.W., and D.W. Wilmore. 1994. Diet and nutrition in the care of the patient with surgery, trauma, and sepsis. Pp. 1207–1240 in Modern Nutrition in Health and Disease, 8th e., M.E. Shils, J.A. Olson, and M. Shike, eds. Philadelphia: Lea and Febiger.

Spencer, H., and L. Kramer. 1986. The calcium requirements and factors causing calcium loss. Fed. Proc. 45(12):2758–2762.

Spencer, H., L. Kramer, and D. Osis. 1988. Do protein and phosphorus cause calcium loss? J. Nutr. 118:657–660.

Spring, B. 1984. Recent research on the behavioral effects of tryptophan and carbohydrate. Nutr. Health 3:55–67.

Starnes, H.F., R.S. Warren, M. Jeevanandam, J.L. Gabrilove, W. Larchian, H.F. Oettgen, and M.F. Brennan. 1988. Tumor necrosis factor and the acute metabolic response to tissue injury in man. J. Clin. Invest. 82:1321–1325.

Stein, T.P., R.W. Hoyt, M.O. Toole, M.J. Leskiw, and M.D. Schluter. 1989. Protein and energy metabolism during prolonged exercise in trained athletes. Int. J. Sports Med. 10:311–316.

Stensrud, T., F. Ingjer, H. Holm, and S.B. Strømme. 1992. l-Tryptophan supplementation does not improve running performance. Int. J. Sports Med. 13:481–485.

Stephen, A.M., and J.H. Cummings. 1979. The influence of dietary fibre on fecal nitrogen excretion in man. Proc. Nutr. Soc. 38:141A.

Stone, E. A. 1975. Stress and catecholamines. Pp. 31–71 in Catecholamines and Behavior, A.J. Freidhoff, ed. New York: Plenum Press.

Stroud, M.A., A.A. Jackson, and J.C. Waterlow. 1996. Protein turnover rates of two human subjects during an unassisted crossing of Antarctica. Br. J. Nutr. 16:165–174.

Struder, H.K., W. Hollmann, P. Platen, J. Duperly, H.G. Fischer, and K. Weber. 1996. Alterations in plasma-free tryptophan and large neutral amino acids. Int. J. Sports Med. 17:73–79.

Stucky, W.P., and A.E. Harper. 1962. Effects of altering indispensable to dispensable amino acids in diets for rats. J. Nutr. 78:278–286.

Svanberg, E., A.C. Moller-Loswick, D.E. Mathews, U. Korner, M. Andersson, and K. Lundholm. 1996. Effects of amino acids on synthesis and degradation of skeletal muscle proteins in humans. Am. J. Physiol. 271:E718–E724.

Szeto, E.G., D.E. Carlson, T.B. Dugan, and J.C. Buchbinder. 1987. A comparison of nutrient intakes between a Ft. Riley contractor-operated and a Ft. Lewis military-operated garrison dining facility. Technical Report No. T2-88. Natick, Mass.: U.S. Army Research Institute of Environmental Medicine.

Szeto, E.G., T.B. Dugan, and J.A. Gallo. 1988. Assessment of habitual diners' nutrient intakes in a military-operated garrison dining facility, Ft. Devens I. Technical Report No. T3-89. Natick, Mass.: U.S. Army Research Institute of Environmental Medicine.

Szeto, E.G., J.A. Gallo, and K.W. Samonds. 1989. Passive nutrition intervention in a military-operated garrison dining facility, Ft. Devens II. Technical Report No. T7-89. Natick, Mass: U.S. Army Research Institute of Environmental Medicine.

Tanaka, N., K. Kubo, K. Shiraki, H. Koishi, and H. Yoshimura. 1980. A pilot study on protein metabolism in the Papuan New Guinea highlanders. J. Nutr. Sci. Vitaminol. 26:247–259.

Tapp, D.C., S. Kobayashi, G. Fernandes, and M.A. Venkatachalam. 1989a. Protein restriction or calorie restriction? A critical assessment of the influence of selective calorie restriction on the progression of experimental renal disease [review]. Seminars in Nephrol. 9(4):343–353.

Tapp, D.C., W.G. Wortham, J.F. Addison, D.N. Hammonds, J.L. Barnes, and M.A. Venkatachalam. 1989b. Food restriction retards body growth and prevents end-stage renal pathology in remnant kidney of rats regardless of protein intake. Lab. Invest. 60:184–195.

Tarnopolsky, M.A., J.D. Mac Dougal, and S.A. Atkinson. 1988. Influence of protein intake and training status on nitrogen balance and lean body mass. J. Appl. Physiol. 64:187–193.

Tarnopolsky, M.A., P.W.R. Lemon, J.D. MacDougall, and J.A. Atkinson. 1990a. Effect of body building exercise on protein requirements. Can. J. Sport Sci. 15:225–226.

Tarnopolsky, L.J., J.D. MacDougall, S.A. Atkinson, M.A. Tarnopolsky, and J.R. Sutton. 1990b. Gender differences in substrate for endurance exercise. J. Appl. Physiol. 68:302–308.

Tarnopolsky, M.A., S.A. Atkinson, S.M. Phillips, and J.D. Mac Dougal. 1995. Carbohydrate loading and metabolism during exercise in men and women. J. Appl. Physiol. 78:1360–1368.

Taylor, H.L., E.R. Buskirk, J. Brozek, J.T. Anderson, and F. Grande. 1957. Performance capacity and effects of caloric restriction with hard physical work on young men. J. Appl. Physiol. 10:421–429.

Tessari, P., S.L. Nissen, J. Miles, and M.W. Haymond. 1986. Inverse relationship of leucine flux and oxidation to free fatty acid availability *in vivo*. J. Clin. Invest. 77:575–581.

Tessari, P., M. Zanetti, R. Barazonni, M. Vettore, and F. Michielan. 1996. Mechanisms of postprandial protein accretion in human skeletal muscle. Insight from leucine and phenylalanine forearm kinetics. J. Clin. Invest. 98:1361–1372.

Tharion, W.J., A.D. Cline, N. Hotson, W. Johnson, P. Niro, C.J. Baker-Fulco, S. McGraw, R.L. Shippee, T.M. Skibinski, R.W. Hoyt, J.P. Delany, R.E. Tulley, J. Rood, W.R. Santee, S.H.M. Boquist, M. Bordic, M. Kramer, S.H. Slade, and H.R. Lieberman. 1997. Nutritional challenges for field feeding in a desert environment: Use of the Unitized Group Ration (UGR) and a supplemental carbohydrate beverage. Technical Report No. T97-9. Natick, Mass.: U.S. Army Research Institute of Environmental Medicine.

Thomas, C.D., K.E. Friedl, M.Z. Mays, S.H. Mutter, R.J. Moore, D.A. Jezior, C.J. Baker-Fulco, L.J. Marchitelli, R.T. Tulley, and E.W. Askew. 1995. Nutrient intakes and nutritional status of soldiers consuming the Meal, Ready-To-Eat (MRE XIII) during a 30-day field training exercise. Technical Report No. T95-6. Natick, Mass.: U.S. Army Research Institute of Environmental Medicine.

Thompson, G.N., P.J.H. Pacy, H. Merritt, G.C. Ford, M.A. Read, K.N. Cheng, and D. Halliday. 1989. Rapid measurement of whole body and forearm protein turnover using a [^{2}H$_5$]phenylalanine model. Am. J. Physiol. 256:E631–E639.

Tiao, G., S. Hobler, J.J. Wang, T.A. Meyer, F.A. Luchette, J.E. Fischer, and P-0. Hasselgren. 1997. Sepsis is associated with increased mRNAs of ubiquitin–proteasome proteolytic pathway in human skeletal muscle. J. Clin. Invest. 99:163–168.

Tipton, K.D., and R.R. Wolfe. 1998. Exercise-induced changes in protein metabolism. Acta Physiol. Scand. 162(3): 377–387.

Tobin, J., and D. Spector. 1986. Dietary protein has no effect on future creatinine clearance. Gerontologist 25:59A.

Todd, K.S., G.E. Butterfield, and D.H. Calloway. 1984. Nitrogen balance in men with adequate and deficient energy intake at three levels of work. J. Nutr. 114:2107–2118.

Tolbert, B., and J.H. Watts. 1963. Phenylalanine requirements of women consuming a minimal tyrosine diet and the sparing effect of tyrosine on the phenylalanine requirement. J. Nut. 80, 111–117.

Tollardona, D., C.I. Harris, E. Milne, and M.F. Fuller. 1994. The contribution of intestinal microflora to amino acid requirements in pigs. Pp. 245–248 in Proceedings of the 6th International Symposium on Digestive Physiology in Pigs, W.B. Souffran, and H. Hagemeister, eds. Dummerstorf, Germany: Forsch. Inst. Biol. Landwirtsch. Nutztiere.

Tom, K., V.R. Young, T. Chapman, T. Masud, L. Akpele, and B.J. Maroni. 1995. Long-term adaptive responses to dietary restriction in chronic renal failure. Am. J. Physiol. 268:E668–E677.

Tomlinson B.E., J.N. Walton, J.J. Rebeiz. 1969. The effects of aging and of cachexia upon skeletal muscle. A histopathological study. J Neurol Sci. 9:321–346.

Trilok, G., and H.H. Draper. 1989. Sources of protein-induced endogenous acid production and excretion by human adults. Calcif. Tissue Int. 44:335–338.

Trinchieri, A., A. Mandressi, P. Luongo, and E. Pisani. 1991. The influence of diet on urinary risk factors for stones in healthy subjects and idiopathic renal calcium stone formers. Brit. J. Urol. 67:230–236.

Tschope, W., and E. Ritz. 1985. Sulfur-containing amino acids are the major determinant of urinary calcium. Mineral Electrolyte Metab. 11:137–139.

Ungerstedt, U., C. Forster, M. Herrera-Marschitz, I. Hoffman, U. Jungnelius, U. Tossman, and T. Zetterstrom. 1982. Brain dialysis: A new *in vivo* technique for studying neurotransmitter release and metabolism. Neuroscience Lett. 10(Suppl.):S493.

Urban, R.J., Y.H. Bodenburg, C. Gilkison, J. Foxworth, R.R. Wolfe, and A. Ferrado. 1995. Testosterone administration to healthy elderly men increases muscle strength and protein synthesis. Am. J. Physiol. 296:E820–E826.

Urivetzky, M., J. Motola, S. Braverman, and A.D. Smith. 1987. Dietary protein levels affect the excretion of oxalate and calcium in patients with absorptive hypercalciuria type II. J. Urol. 137:690–692.

USACDEC/USARIEM (U.S. Army Combat Developments and Experimentation Center and U.S. Army Research Institute of Environmental Medicine). 1986. Combat Field Feeding System-Force Development Test and Experimentation (CFFS-FDTE) Technical Report CDEC-TR-85-006A. Vol. 1, Basic Report; vol. 2, Appendix A; vol. 3, Appendixes B through L. Fort Ord, Calif.: U.S. Army Combat Developments and Experimentation Center.

U.S. Army Natick Research, Development and Engineering Center. 1992. Operational Rations of the Department of Defense. Natick PAM 30-2. Natick, Mass: U.S. Army Natick Research, Development and Engineering Center.

U.S. Department of the Army. 1947. Army Regulation 40-250. Nutrition. Washington, D.C.

U.S. Departments of the Army, the Navy, and the Air Force. 1985. Army Regulation 40-25/Naval Command Medical Instruction 10110.1/Air Force Regulation 160-95. Nutritional Allowances, Standards, and Education. May 15. Washington, D.C.

U.S. War Department. 1944. Nutrition. War Department Circular No. 98. Washington, D.C.

Van Acker, B.A.C, K.W.E. Hulsewé, A.J.M. Wagenmakers, N.E.P. Deutz, B.K. Van Kreel, P.B. Soeters, and M.F. Von Meijenfeldt. 1996. Measurement of glutamine metabolism in gastrointestinal cancer patients. Clin. Nutr. 15(Suppl. 1):1.

Van der Hulst, R.R., B.K. van Kreel, M.F. von Meyenfeldt, R.J. Brummer, J.W. Arends, N.E. Deutz, and P.B. Soeters. 1993. Glutamine and the preservation of gut integrity. Lancet 341(8857):1363–1365.

Van Hall, G. 1996. Amino acids, ammonia and exercise in man. Thesis, Maastricht University, The Netherlands.

Van Hall, G., J.S.H. Raaymakers, W.H.M. Saris, and A.J.M. Wagenmakers. 1995a. Ingestion of branched chain amino acids and tryptophan during sustained exercise in man: Failure to affect performance. J. Physiol. Lond. 186:789–794.

Van Hall, G., B. Saltin, G.J. Van der Vusse, K. Söderlund, and A.J.M. Wagenmakers. 1995b. Deamination of amino acids as a source for ammonia production in human skeletal muscle during prolonged exercise. J. Physiol. 489:251–261.

Van Hall, G., D.A. MacLean, B. Saltin, and A.J.M. Wagenmakers. 1996. Mechanisms of activation of muscle branched-chain α-keto acid dehydrogenase during exercise in man. J. Physiol. 494:899–905.

Varnier, M., G.P. Leese, J. Thompson, and M.J. Rennie. 1995. Stimulatory effect of glutamine on glycogen accumulation in human skeletal muscle. Am. J. Physiol. 269:E309–E315.

Varnier, M., P. Sarto, D. Martines, L. Lora, F. Carmignoto, G.P. Leese, and R. Naccarato. 1994. Effect of infusing branched-chain amino acids during incremental exercise with reduced muscle glycogen content. Eur. J. Appl. Physiol. 69:26–31.

Vinnars, E., J. Bergström, and P. Fürst. 1975. Influence of the postoperative state on the intracellular free amino acids in human muscle tissue. Annals Surg. 182:665–671.

Wagenmakers, A.J.M. 1996. Muscle glycogen depletion and fatigue. The Physiologist 39:A68.

Wagenmakers, A.J.M., and P.B. Soeters. 1995. Metabolism of branched-chain amino acids. Pp. 67–83 in Amino Acid Metabolism and Therapy in Health and Nutritional Disease, L.A. Cynober, ed. New York, CRC Press.

Wagenmakers, A.J.M., H.J.M. Salden, and J.H. Veerkamp. 1985. The metabolic fate of branched-chain amino acids and 2-oxo acids in rat muscle homogenates and diaphragms. Int. J. Biochem. 17:957–965.

Wagenmakers, A.J.M., J.H. Coakley, and R.H.T. Edwards. 1990. Metabolism of branched-chain amino acids and ammonia during exercise: Clues from McArdle's disease. Int. J. Sports Med. 11:S101–S113.

Wagenmakers, A.J.M., E.J. Beckers, F. Brouns, H. Kuipers, P.B. Soeters, G.J. van der Vusse, and W.H.M. Saris. 1991. Carbohydrate supplementation, glycogen depletion, and amino acid metabolism during exercise. Am. J. Physiol. 260:E883–E890.

Wagenmakers, A.J.M., G. Van Hall, and B. Saltin. 1996. Excessive muscle proteolysis during one leg exercise is exclusively attended by increased *de novo* synthesis of glutamine, not of alanine. Clin. Nutr. 15(Suppl. 1):1.

Wagenmakers, A.J.M., G. Van Hall, P.A.I. Van de Schoor, and W.H.M. Saris. 1997. Glutamine does not stimulate glycogen resynthesis in human skeletal muscle. Med. Sci. Sport Ex. 29:S280

Wallace, W.M. 1959. Nitrogen content of the body and its relation to retention and loss of nitrogen. Fed. Proc. 18:1125–1130.

Walser, M. 1961a. Calcium clearance as a function of sodium clearance in the dog. Am. J. Physiol. 200:1099–1104.

Walser, M. 1961b. Ion association. VII. Dependence of calciuresis on natriuresis during sulfate infusion. Am. J. Physiol. 201:769–773.

Walser, M. 1986. The roles of urea production, ammonium excretion, and amino acid oxidation in acid–base balance. Am. J. Physiol. 250:F181–F188.

Walser, M. 1992. Dietary proteins and their relationship to kidney disease. Pp. 168–178 in Dietary Proteins in Health and Disease, G.U. Liepa, ed. Champaign, Ill.: American Oil Chemists' Society.

Walser, M., and A.A. Browder. 1959. Ion association. III. The effect of sulfate infusion on calcium excretion. J. Clin. Invest. 38:1404–1411.

Walser, M., and J.R. Trounce. 1961. The effect of diuresis and diuretics upon the renal tubular transport of alkaline earth cations. Biochem. Pharmacol. 8:157.

Walters, J.K., M. Davis, M.H. Sheard. 1979. Tryptophan-free diet: effects on the acoustic startle reflex in rats. Psychopharmacology (Berl) 62(2):103–109.

Wang, H., and P.A. Murphy. 1994. Isoflavone composition of American and Japanese soybeans in Iowa: Effects of variety, crop year and location. J. Agric. Food Chem. 42:1674–1677.

Wang, T.C., and M.F. Fuller. 1989. The optimum dietary amino acid pattern for growing pigs. 1. Experiments by amino acid deletion. Br. J. Nutr. 62(1):77–89.

Wang Z., R.N. Pierson Jr, and S.B. Heymsfield. 1992. The five-level model: A new approach to organizing body composition research. Am. J. Clin. Nutr. 56:19–28.

Wang Z., S. Heshka, R. Pierson, S.B. Heymsfield. 1995. Systematic organization of body-composition methodology: An overview with emphasis on component-based methods. Am J Clin Nutr. 61: 457–65.

Wang, Z., D. Gallagher, M. Nelson, D. Matthews, S.B. Heymsfield. 1996a. Total body skeletal muscle mass: Evaluation of 24 hour urinary creatinine excretion method by computerized axial tomography. Am. J. Clin. Nutr. 63:863–869.

Wang, Z., M. Visser, R. Ma, R. Baumgartner, D. Kotler, D. Gallagher, and S.B. Heymsfield. 1996b. Skeletal muscle mass: Evaluation of neutron activation and dual-energy x-ray absorptiometry methods. J. Appl. Physiol. 80(3):824–831.

Wang, Z., P. Deurenberg, D.E. Matthews, and S.B. Heymsfield. 1998. Urinary 3-methylhistidine excretion: Association with total body skeletal muscle mass by computerized axial tomography. J. Parenter. Enteral Nutr. 22(2): 82–86.

Wannamacher, R.W., Jr. 1975. Protein metabolism (applied biochemistry). P. 133 in Total Parenteral Nutrition: Premises and Promises, H. Ghadimi, ed. New York: John Wiley.

Warber, J.P., F.M. Kramer, S.M. McGraw, L.L. Lesher, W. Johnson, and A.D. Cline. 1996. The Army Food and Nutrition Survey, 1995–97. Technical Report. Natick, Mass.: U.S. Army Research Institute of Environmental Medicine.

Waterlow, J.C. 1996. The requirements of adult man for indispensable amino acids. Eur. J. Clin. Nutr. 50:S151–176.

Waterlow, J.C., and D.J. Millward. 1989. Energy cost of turnover of protein and other cellular constituents. Pp. 277–282 in Energy Transformations in Cells and Organisms, W. Wieser and E. Gnaiger, eds. Stuttgart: Georg Thieme Verlag.

Waterlow, J.C., P.J. Garlick, and D.J. Millward. 1978. Protein Turnover in Mammalian Tissues and the Whole Body. Amsterdam: North Holland Elsevier.

Watters, J.M., P.Q. Bessey, C.A. Dinarello, S.M. Wolf, and D.W. Wilmore. 1986. Both inflammatory and endocrine mediators stimulate host responses to sepsis. Arch. Surg. 121:179–190.

Welle, S., C. Thornton, R. Jozefowicz, and M. Statt. 1993. Myofibrillar protein synthesis in young and old men. Am. J. Physiol. 264:E693–E698.

Welle, S., C. Thornton, M. Statt, and B. McHenry. 1996. Growth hormone increases muscle mass and strength but does not rejuvenate myofibrillar protein synthesis in healthy subjects over 60 years old. J. Clin. Endocrinol. Metab. 81:3239–3243.

Weller, L.A., D.H. Calloway, S. Margen. 1971. Nitrogen balance of men fed amino acid mixtures based on Rose's requirements, egg white protein, and serum free amino acid patterns. J. Nutr. 101(11):1499–1507.

Wibom, R., and E. Hultman. 1990. ATP production rate in mitochondria isolated from microsamples of human muscle. Am. J. Physiol. 159:E204–E209.

Williams, C. 1995. Macronutrients and performance. J. Sports Sci. 13(Spec. No.):S1–10.

Williams, H.H., A.E. Harper, D.M. Hegsted, G. Arroyave, L.E. Hold, Jr. 1974. Nitrogen and amino acid requirements. Pp. 23–63 in Improvement of Protein Nutritive Value. Institute of Medicine. Washington, D.C.: National Academy Press.

Wilmore, D.W. 1977. The Metabolic Management of the Critically Ill. New York: Plenum Medical.

Wilmore, D.W. 1991. Catabolic illness: Strategies for enhancing recovery. N. Engl. J. Med. 325(10):695–702.
Wilmore, D.W. 1997a. Glutamine saves lives! What does it mean? Nutrition 13(4):375–376.
Wilmore, D.W. 1997b. Homeostasis: Bodily changes in trauma and surgery. Pp. 55–67 in Textbook of Surgery, 15th ed., D.C. Sabiston, ed. Philadelphia: W.B. Saunders.
Wilmore, D.W., J.A. Moylan, B.F. Bristow, H.D. Mason, Jr., and B.A. Pruitt, Jr. 1974. Anabolic effects of growth hormone and high caloric feedings following thermal injury. Surg. Gynecol. Obstet. 138:875–884.
Wilmore, D.W., C.W. Goodwin, L.H. Aulick, M.C. Powanda, A.D. Mason, Jr., and B.A. Pruitt, Jr. 1980. Effect of injury and infection on visceral metabolism and circulation. Ann. Surg. 192:491–504.
Windmueller, H.G., and A.E. Spaeth. 1974. Uptake and metabolism of plasma glutamine by the small intestine. J. Biol. Chem. 249:5070–5079.
Wolf, R.F., D.B. Pearlstone, E. Newman, M.J. Heslin, A. Gonenne, M.E. Burt, and M.F. Brennan. 1992. Growth hormone and insulin reverse net whole body and skeletal muscle protein catabolism in cancer patients. Ann. Surg. 216:280–288.
Wood, S., J. Mortola, Y.F. Chan, F. Moossazadeh, and S. Yen. 1992. Treatment of premenstrual syndrome with fluoxetine: A double-blind, placebo-controlled, crossover study. Obstet. Gynecol. 80:339–344.
Working Group (G. Clugston, K.G. Dewey, C. Fjeld, J. Millward, P. Reeds, N.S. Scrimshaw, K. Tontisirin, J.C. Waterlow, and V.R. Young). 1996. Report of the working group on protein and amino acid requirements. Eur. J. Clin. Nutr. 50(Suppl. 1):S193–S195.
Wurtman, J.J., and J.D. Fernstrom. 1979. Free amino acid, protein and fat contents of breast milk from Guatemalan mothers consuming a corn-based diet. Early Human Development 3:67–77.
Wurtman, R.J., F. Hefti, and E. Melamed. 1981. Precursor control of neurotransmitter synthesis. Pharmacol. Rev. 32:315–335.
Yanovski, S.Z., V.S. Hubbard, S.B. Heymsfield, H.C. Lukasksi. 1996. Bioelectrical Impedance Analysis in Body Composition Measurement. Proceedings of a National Institutes of Health Technology Assessment Conference. Bethesda, Maryland, December 12–14, 1994. Am. J. Clin Nutr. 64(3S):387–532S.
Yarasheski, K.E., J.J. Zachwieja, T.J. Angelopoulos, and D.M. Bier DM. 1993a. Short term growth hormone treatment does not increase muscle protein synthesis in experienced weight lifters. J. Appl. Physiol. 74:3073–3076.
Yarasheski, K.E., J.J. Zachwieja, and D.M. Bier. 1993b. Acute effects of resistance exercise on muscle protein synthesis rate in young and elderly men and women. Am. J. Physiol. 265:E210–E214.
Yarasheski, K.E., J.J. Zachwieja, J.A. Campell, and D.M. Bier. 1995. Effect of growth hormone and resistance training on muscle growth and strength in older men. Am. J. Physiol. 268:E268–E276.
Yates, F.E., 1983. Contribution of statistics to ethics of science. Am. J. Physiol 13:R3–R5.
Yeboah, N., E. Al-Sing, A. Bodalloo, T. Forrester, A. Jackson, and D.J. Millward. 1996. Transfer of ^{15}N from urea to the circulating lysine pool in the human infant. Proc. Nutr. Soc. 55:37A.
Yeghiayan, S.K., C. Amendola, T.H. Wu, T.J. Maher, and H.R. Lieberman. In press. Hyperthermia and tyrosine administration affect behavioral performance but not hippocampal catecholamines. Society for Neuroscience Abstracts.
Yoshida, A. 1983. Specificity of amino acids for the nutritional evaluation of proteins. Pp. 163–182 in Proceedings of the International Association of Cereal Chemists Symposium on Amino Acid Composition and Biological Value of Cereal Proteins, R. Lasztity and M. Hidvegi, eds. Budapest: Akademiai Kiado.

Yoshida, A. 1986. Nutritional aspects of amino acid oxidation. Pp. 378–382 in Proceedings of the XIII International Congress of Nutrition, T.G. Taylor and N.K. Jenkins, eds. London: John Libbey and Co., Ltd.
Young, S. 1993. Use of diet and dietary components. J. Psychiatry Neurosci. 18(5):235–244.
Young, S.N. 1996. Behavioral effects of dietary neurotransmitter precursors: Basic and clinical aspects. Neurosci. Behav. Rev. 20(2):313–323.
Young, V.R. 1986. Nutritional balance studies: Indicators of human requirements or adaptive mechanisms. J. Nutr. 116:700–703.
Young, V.R. 1987. McCollum Award Lecture: Kinetics of human amino acid metabolism: Nutritional implications and some lessons. Am. J. Clin. Nutr. 46:709–725.
Young, V.R. 1991a. Nutrient interactions with reference to amino acid and protein metabolism in nonruminants: Particular emphasis on protein–energy relations in man. Z. Ernahrungswiss 30:239–267.
Young, V.R. 1991b. Soy protein in relation to human and amino acid nutrition. J. Am. Diet. Assoc. 91:828–835.
Young, V.R. 1992. Protein and amino acid requirements in humans: Metabolic basis and current recommendations. Scand. J. Nutr. Naringsforskning 36:47–56.
Young, V.R. 1994. Adult amino acid requirement: The case for a major revision in current recommendations. J. Nutr. 124:1517S–1523S.
Young, V.R. 1997. Paper 5. Dietary protein standards can be halved (Chittenden, 1904). J. Nutr. 127:1025S–1027S.
Young, V.R., and A. E. El-Khoury. 1995a. Can amino acid requirements for nutritional maintenance in adult humans be approximated from the amino acid composition of body mixed proteins? Proc. Natl. Acad. Sci. 921:300–304.
Young, V.R., and A.E. El-Khoury. 1995b. The notion of the nutritional essentiality of amino acids, revisited, with a note on the indispensable amino acid requirements in adults. Pp. 191–232 in Amino Acid Metabolism and Therapy in Health and Nutritional Disease, L.A. Cynober, ed. Boca Raton, Fla.: CRC Press.
Young, V.R., and A.E. El-Khoury. 1996. Human amino acid requirements: A re-evaluation. Food Nutr. Bull. 17:191–203.
Young, V.R., and J.S. Marchini. 1990. Mechanisms and nutritional significance of metabolic responses to altered intakes of protein and amino acids, with reference to nutritional adaptation in humans. Am. J. Clin. Nutr. 51:270–289.
Young, V.R., and N.S. Scrimshaw. 1968. Endogenous nitrogen metabolism and plasma amino acids in young adults given a "protein-free" diet. Br. J. Nutr. 22:9–20.
Young, V.R., and P.L. Pellett. 1985. Wheat proteins in relation to protein requirements and availability of amino acids. Am. J. Clin. Nutr. 41:1077–1090.
Young, V.R., and P.L. Pellett. 1990. Current concepts concerning indispensable amino acid needs in adults and their implications for international nutrition planning. Food. Nutr. Bull. 12:289–300.
Young, V.R., and Y.M. Yu. 1996. Protein and amino acid metabolism. Pp. 159–200 in Nutrition and Metabolism in the Surgical Patient, 2nd edition, Josef E. Fischer, ed. Boston: Little, Brown and Company.
Young, V.R., Y.S.M. Taylor, W.R. Rand, and N.S. Scrimshaw. 1973. Protein requirements of man: Efficiency of egg protein utilization at maintenance and sub-maintenance levels in young men. J. Nutr. 103:1164–1174.
Young, V.R., L. Fajardo, E. Murray, W.M. Rand, and N.S. Scrimshaw. 1975. Protein requirements of man: Comparative nitrogen balance response within the submaintenance-to-maintenance range of intakes of wheat and beef proteins. J. Nutr. 105:534–544.

Young, V.R., M. Puig, E. Queiroz, N.S. Scrimshaw, and W.M. Rand. 1984. Evaluation of the protein quality of an isolated soy protein in young men: Relative nitrogen requirements and effect of methionine supplementation. Am. J. Clin. Nutr. 39:16–24.

Young, V.R., D.M. Bier, and P.L. Pellet. 1989. A theoretical basis for increasing current estimates of the amino acid requirements in adult man with experimental support. Am. J. Clin. Nutr. 50:80–92.

Young, V.R., N.S. Scrimshaw, and P.L. Pellett. 1998. Significance of dietary protein source in human nutrition: Animal or plant proteins. Pp. 205–212 in Feeding a World Population of More than Eight Billion People: A Challenge to Science, J.C. Waterlow, D.G. Armstrong, L. Fowden and R. Riley, eds. Oxford: Oxford University Press.

Zawadzki, K.M., B.B. Yaspelkis, and J.L. Ivy. 1992. Carbohydrate–protein complex increases the rate of muscle glycogen storage after exercise. J. Appl. Physiol. 72:1854–1859.

Zello, G.A., P.B. Pencharz, and R.O. Ball. 1990. Phenylalanine flux, oxidation and conversion to tyrosine in humans studied with 1-[1-^{13}C]phenylalanine. Am. J. Physiol. 259:E835–E843.

Zello, G.A., P.B. Pencharz, and R.O. Ball. 1992. Lysine requirement in young adult males. Am. J. Physiol. 264:E677–E685.

Zello, G.A., P.B. Pencharz, and R.O. Ball. 1993. Dietary lysine requirement of young adult males determined by oxidation of 1-[1-^{13}C]phenylalanine. Am. J. Physiol. 264:E677–E685.

Zello, G.A., L.J. Wykes, R.O. Ball, and B. Pencharz. 1995. Recent advances in methods of assessing dietary amino acid requirements for adult humans. J. Nutr. 125:2907–2915.

Zemel, M.B. 1988. Calcium utilization: Effect of varying level and source of dietary protein. Am. J. Clin Nutr. 48:880–883.

Zemel, M.B., S.A. Schuette, M. Hegsted, and H.M. Linkswiler. 1981. Role of sulfur-containing amino acids in protein-induced hypercalciuria in man. J. Nutr. 111:545–552.

Zhao, X.H., Z.M. Wen, C.N. Meredith, D.E. Matthews, D.M. Bier, and V.R. Young. 1986. Threonine kinetics at graded threonine intakes in young men. Am. J. Clin. Nutr. 43:795–802.

Ziegler, T.R., L.S. Young, K. Benfell, M. Scheltinga, K. Hortos, R. Bye, F.D. Morrow, D.O. Jacobs, R.J. Smith, J.H. Antin, and D.W. Wilmore. 1992. Clinical and metabolic efficacy of glutamine-supplemented parenteral nutrition after bone marrow transplantation. A randomized, double-blind, controlled study. Ann. Intern. Med. 116(10):821–828.

Ziegler, T.R., R.L. Bye, R.L. Persinger, L.S. Young, J.H. Antin, and D.W. Wilmore. 1994. Glutamine-enriched parenteral nutrition increases circulating lymphocytes after bone marrow transplantation. J. Parenter. Enteral Nutr. 18:17S.

Ziegler, T.R., M.P. Mantell, J.C. Chow, J.H. Rombeau, and R.J. Smith. 1996. Gut adaptation and the insulin-like growth factor system: Regulation by glutamine and IGF-1 administration. Am. J. Physiol. 271:G866–G875.

Ziegler, T.R., R.L. Bye, R.L. Persinger, L.S. Young, J.H. Antin, and D.W. Wilmore. 1998. Effects of glutamine supplementation on circulating lymphocytes after bone marrow transplantation: A pilot study. Am. J. Med. Sci. 315:4–10.

Zurlo, F., K. Larson, C. Bogardus, and E. Ravussin. 1990. Skeletal muscle is a major determinant of resting energy expenditure. J. Clin. Invest. 86:1423–1427.

E

Protein and Energy Content of Selected Operational Rations

Table E-1 shows the protein and energy content of the operational rations currently in use by the military. The information in this table is based on the pamphlet "Operational Rations of the Department of Defense."

Table E-1 Protein and Energy Content of Operational Rations

Ration	Total Energy (kcal/d)	Protein (g/%)	Protein/CHO
B	4300	140/13	0.24
T	4260[a]	174/16	0.30
Unitized Group Ration (UGR)	4350	152/14	0.26
Meal, Ready-to-Eat (MRE)	3900	127/13	0.25
Supplement	200	6/12	
Go to War (GTW)	3900[b]	156/16	0.27
Ration, Cold Weather (RCW)	4500	90/8	0.13
Long-Range Patrol (LRP)	1560[c]	59/15	0.30

[a]Based on consumption of one breakfast and two lunch or dinner menus; the cold-weather supplement adds 1020 kcal.

[b]Designed to sustain an individual during early stages of mobilization—three meals may not be available each day.

[c]Designed to be an extended-life ration to sustain personnel during initial assault, special operations, and long-range missions.

SOURCE: U.S. Army Natick Research, Development and Engineering Center, 1992

Index

Acute renal failure
  protein intake enhancement of, 138, 144
  See also Diseases and disorders; Renal function
Adaptation
  amino acid metabolism, adaptive component of, 180–184, 221–224
Adenosine 5´-triphosphate (ATP), 111, 112, 113, 115, 121–122, 123, 124, 125, 244, 245
Age and aging
  dual-energy x-ray absorptiometry models, influence of age, 272
  glomerulosclerosis development role of, 144, 145, 146, 147
  muscle mass changes over time, monitoring considerations, 273–275
  protein Recommended Dietary Allowances by age group, 6, 37
  sarcopenia of aging, muscle mass/function effects of, 48, 123–126
Alaska, 87, 98, 99–100
Amino acid metabolism
  adaptive component of, 180–184, 221–224
  diurnal cycling role in, 184–186, 224–225
  energy-dependent processes of, 110–112
  exercise and catabolism, 29–30, 244–247
  exercise, endurance, and, 310–311
  infection/injury effects on muscle transmembrane transport, 280–281
  influences on, 199–202
  physical activity and, 243, 244, 250–251
  physical activity, increased, role of energy supply and, 244, 248–249
  See also Amino acids (AA); Indispensable amino acids (IAA); Obligatory oxidative amino acid loss (OOL)
Amino acid requirements
  animal data for composition/requirement patterns, 178
  brain function and, 289–293
  controversy over, 3–4, 25–28, 170–173
  definition, inherent difficulties, 169–202

413

estimation feasibility and necessity, 217–237
FAO/WHO/UNU requirements, 218, 219, 220, 229, 230, 231, 236, 237
FAO/WHO/UNU values, 3–4, 24–28, 171
FAO/WHO/UNU values adjustment, 192–193
isotope, stable, use for estimating, considerations, 193–197, 232–234
military operations and, CMNR conclusions/recommendations, 77–79
MIT Requirement Pattern proposal and physical activity, considerations, 235–237
MIT Requirement Pattern proposal, overview, 218–221
nitrogen balance use for estimating, limitations, 22–24, 186–202, 226–231
obligatory oxidative loss patterns, use for predicting, 175
Amino acids (AA), 19, 133, 160, 162
Army's interest in role of, 2–3, 20–21
blood-brain barrier, overview of, 290–293
central nervous system transport mechanisms for, 291
cognitive performance effects of military lightweight rations, field studies results, 301–303
functions of, 110, 111
intake adequacy, functional indicators of, 198–202
intake adequacy, protein turnover as indicator of, 198–199
intake of, metabolic studies of long-distance cyclists, 52
military limited-use operational rations, for optimal cognitive/physical function, research needs, 86–87
neurotransmitter system products of, 292, 293
performance role of, 20
Amino acid supplements
Army personnel use of, 39, 42
benefits/health risks of, 7–11, 39, 42–63
branched-chain amino acids, endurance performance and metabolism effects of, 53–54, 310, 312–317
branched-chain amino acids, possible benefits, 249
cognitive performance effects of, 8, 42–45
glutamine, endurance performance and metabolism effects of, 310, 317–320
immune function effects of, 9, 54–56
metabolism and performance effects of, rationale for, 309–322
military performance optimization using, CMNR conclusions/recommendations, 79–81
military personnel reported use, by gender/military specialty, 103–104
performance and, study considerations, 334
physical performance effects of, 10, 52–54
pregnancy effects of, 11, 62–63
safety of, 11, 13, 14, 61–63, 79–81, 332–334
tryptophan, and cognitive performance, 43–44
tyrosine, and cognitive performance, 44–45
See also Amino acids (AA); Food supplements; Indispensable amino acids (IAA); Protein supplements
Aminotransferase
branched-chain amino acid reaction with, interaction with tricarboxylic acid in muscle, 315–317
Ammonia
branched-chain amino acids ingestion, effects on plasma concentrations/muscle production of, during exercise, 317
Anabolic hormones
injuries/infection treatment with, effects of, CMNR conclusions/recommendations, 12, 14, 78, 79
muscle protein effects of, 49–51, 125, 127, 128, 129
muscle protein metabolism, effects of following infection/injury, 281–282
See also Growth hormone (GH); Hormones; Insulin; Insulin-like growth factor-I (IGF-I)

INDEX

Animal studies
amino acid composition and requirement patterns for selected animal, data, 178
dogs, calcium clearance determinants for, 139, 140, 141
ileal indispensable amino acid losses, animal data, 180
obligatory metabolic demand pattern, animal data for, 175–180
protein-free diet, responses to, data on selected animals, 177

Anorexia
infection/injury effects on whole body protein metabolism, 160

Antarctic
walk across, protein turnover rates during, 36, 249

Anthropometry
muscle mass/composition measurement, methods overview, 45, 46, 258, 259–261, 273–274
See also Measurement techniques

Arginine, 13, 14
See also Amino acids (AA)

Army Food and Nutrition Survey I, 97, 99, 103

Army Regulation 40-25 (AR 40-25), 95

Army Regulation 40-250 (AR 40-250), 102

Athletics
Antarctic, walk across, protein turnover rates during, 36, 249
Boston Marathon and other athletes, endurance performance studies, 336
endurance athletes, amino acid/protein metabolism in, 310–312
overtrained athletes, glutamine plasma concentrations in, 318–319
overtraining syndrome, glutamine and immune function, 249–250
Tour de France cyclists, amino acid/protein metabolism in, 52, 311
Tour de France cyclists, glutamine plasma concentrations in, 318–319
Yale university athletes, reduced-protein diet study, 32, 201, 251
See also Physical activity

ATP. *See* Adenosine 5′-triphosphate (ATP)

Auditory Consonants Trigrams Recognition test, 336

Behavioral function. *See* Cognitive performance

Bioelectric impedance analysis (BIA)
muscle mass/composition measurement, methods overview, 45, 46, 258, 259, 262–263, 264, 274
See also Measurement techniques

Biological value (BV)
plant proteins biological value in nitrogen balance trials, 190–192

Blood-brain barrier (BBB)
nutritional status of brain, determination by, overview of, 290–293
See also Central nervous system (CNS)

Body composition, 256, 258
infection/injury effects on, 160, 161
See also Body weight (BW); Skeletal muscle (SM); Visceral organs

Body measurements. *See* Anthropometry

Body weight (BW), 37, 161, 256

Bolivia
U.S. military protein/energy intake studies, 40, 99, 101

Boston Marathon
endurance performance studies, 336

Brain. *See* Blood-brain barrier (BBB); Central nervous system (CNS); Cognitive performance

Branched-chain amino acids (BCAA), 53, 54, 245, 246
aminotransferase reaction with, interaction with tricarboxylic acid cycle in muscle, 315–317
ammonia, plasma concentrations and muscle production of during exercise, 317
central fatigue theory and, 53–54, 312–313
endurance performance/metabolism effects of ingestion, studies results, 310, 313–315
supplementation of, possible benefits, 249
See also Amino acids (AA)

Breastfeeding, 37, 39
  protein military Recommended Dietary Allowance for, CMNR conclusions/recommendations, 7, 39, 79
  protein Recommended Dietary Allowance for, 6–7, 37, 39
  See also Women
Bursailleus, Jacques, 19, 109–110

Calcium
  clearance determinants, 139–142
  protein supplementation and status of, 10, 60–61
  renal stone disease and intake/excretion of, 138–143
Camp Carson, 96, 97
Camp Parks, California, 41, 99, 102
Carbohydrates (CHO), 58, 244, 248, 250
  glycogen resynthesis following exercise, supplementation effects, 310, 320–322
  military operational rations energy content, by ration type, 411
  premenstrual syndrome management using supplements of, 335–336
Catabolic hormones
  muscle protein effects of, 52, 131
Catecholamines. *See* Dopamine; Epinephrine; Norepinephrine
Central fatigue theory
  branched-chain amino acids and, 53–54, 312–313
Central nervous system (CNS)
  amino acid/protein requirements and function of, 289–303
  blood-brain barrier determination of nutritional status of, 290–293
  stress and amino acids changes during field studies, effects on, 301–303
  tryptophan/serotonin effects on brain function, 293–294, 296, 297
  tyrosine/norepinephrine effects on brain function, 289, 290, 292, 297–300
  See also Blood-brain barrier (BBB)
Children
  preschool, indispensable amino acid requirement patterns, FAO/WHO/UNU, 171, 219
  protein/indispensable amino acid requirements, FAO/WHO/UNU values, 3–4, 171
Chocolate Mountain Desert Gunnery Range, California, 98, 100
Choline
  dietary supplementation and endurance performance, 336
Cognitive performance
  amino acid changes in military lightweight rations, effects of, field studies results, 301–303
  military limited-use operational rations, for optimal function, research needs, 86–87
  military performance optimization, use of protein/amino acid supplements, CMNR conclusions/recommendations, 13, 14, 79–81
  protein/amino acid requirements, role in, 20, 289–303
  protein/amino acid supplements effects on, 8, 42–45
  protein intake timing, effects on, 57–58
  stress and amino acid changes, effects on, 87, 298–300, 301–303
  tryptophan/serotonin effects on, 43–44, 87, 289, 290, 292, 293–297
  tyrosine/norepinephrine effects on, 44–45, 87, 289, 290, 292, 297–300
  See also Physical performance
Committee on Military Nutrition Research (CMNR), 2, 3, 6, 8, 21, 35, 36, 54, 55, 61, 86, 87, 88, 110
  endurance exercise, protein requirements in, conclusions/recommendations, 77, 78
  injuries/infection effects of anabolic hormones treatment, conclusions/recommendations, 12, 14, 78, 79
  operational rations and Military Recommended Dietary Allowance, conclusions/recommendations, 12, 14, 79
  operational rations protein/energy content, conclusions/recommendations, 12–13, 14, 79
  pregnancy/lactation and protein Military Recommended Dietary Allow-

ance, conclusions/recommendations, 12, 14, 79
protein/amino acid requirements effects of military operational stressors, conclusions/recommendations, 11–12, 14, 77–79
protein/amino acid supplementation, safety of, conclusions/recommendations, 12–13, 14, 79–81
protein/amino acid supplementation to optimize military performance, conclusions/recommendations, 13, 14, 79–81
Computerized tomography (CT), 266, 267
muscle mass/composition measurement, methods overview, 46, 47–48, 258, 259, 268–269, 274
See also Imaging methods; Measurement techniques
Cortisol, 131, 133
See also Catabolic hormones; Hormones
Creatinine (Cr), 145, 146, 147
urinary metabolite muscle mass/composition measurement, methods overview, 45, 46, 258, 259, 263, 265–267, 274
Cytochrome *c* oxidase (COX), 126

Defense Women's Health Research Program, 88
Demographics. *See* Age and aging; Children; Gender
Depression. *See* Mood
Dietary Supplement Health and Education Act (DSHEA), 333
Diseases and disorders
muscle protein degradation, altering conditions, 115
protein requirements effects of illness, 5, 20, 33–34
renal disease development, protein intake effects on, 137–147
See also Acute renal failure; Eosinophilia-myalgia syndrome (EMS); Glomerulosclerosis; Infectious diseases and disorders; Injuries; Renal stone disease

Diurnal cycling
amino acid, indispensable, requirements and, 184–186, 224–225
Dopamine, 290, 292, 293
cognitive performance effects of, 297, 298
See also Neurotransmitters
Drugs
renal stone disease treatment, 143
Dual-energy x-ray absorptiometry (DXA), 262, 263
muscle mass/composition measurement, methods overview, 46, 47–48, 259, 269–272, 274–275
See also Measurement techniques

Endurance
branched-chain amino acids/tryptophan ingestion, effects on performance, studies results, 310, 313–315
choline dietary supplementation and performance, 336
exercise, endurance, and protein/amino acid metabolism, 310–311
exercise, endurance, and protein requirements, CMNR conclusions/recommendations, 77, 78
glutamine ingestion, effects on performance, studies results, 310, 317–320
See also Physical performance
Energy
deficit of, protein/amino acid supplementation effects on, CMNR conclusions/recommendations, 79–81
intake of, metabolic studies of long-distance cyclists, 52
military operations protein requirements, energy intake effects, CMNR conclusions/recommendations, 11, 14, 77, 78
military operational rations content of, by ration type, 94–95, 411
military operational rations content of, CMNR conclusions/recommendations, 12–13, 14, 79
military operational rations, historic intake of, World War II studies, 96–97
military operational rations, requirements of, determination, 102–103

physical activity increases, and interaction with protein metabolism, 244, 248–249
physical activity restriction of, effects on protein requirements, 4–5, 29–33
protein breakdown, costs of, 114–116
protein metabolism use of, energy-dependent processes associated with, 110–112
protein requirements and balance of, 31–33
proteins, regulatory, turnover, energy-dependent processes in, 114
protein synthesis and regulation, costs of, 112–114
Eosinophilia-myalgia syndrome (EMS) L-tryptophan-associated, 294, 331–332, 335
Epinephrine, 131
cognitive performance effects of, 297
See also Catabolic hormones; Hormones; Neurotransmitters
Exercise. *See* Athletics; Physical activity

Federation of the American Societies for Experimental Biology (FASEB)
amino acids as dietary supplements, study on safety, 332–334
Field operations
protein intake studies, 38, 40–41
protein supplemental intake by personnel, overview, 93–104
rations amino acid changes, lightweight v. standard, cognitive performance effects, 301–303
rations, operational, protein/energy current intake, 97–102
rations, operational, protein/energy historic intake, World War II studies, 96–97
See also Military operations; Military personnel
Fluid Doctrine, 13, 14, 81
Food
Minnesota bread study, 191, 192, 231
North Carolina wheat study, 191–192, 231
See also Food and drug safety; Food supplements; Military rations; Plants and vegetation; Soybeans; Wheat

Food and Agriculture Organization (FAO), 24, 25, 26, 27, 57, 170, 172, 173, 174, 175, 186, 192, 194, 195, 199
nitrogen balance use for estimating amino acid requirements, 186–202
protein/amino acid obligatory oxidative losses, FAO/WHO/UNU values, 171
protein/amino acid requirements, 3–4, 24–28, 171, 218, 219, 220, 229, 230, 231, 236, 237
protein/amino acid requirements adjustment, 192–193
Food and Drug Administration (FDA), 332, 334
"Current Good Manufacturing Practice in Manufacturing, Packing, or Holding Dietary Supplements," 334
Food and drug safety
amino acid supplements, toxicity of, 11, 61–63, 331, 332–334
protein/amino acid supplements safety, CMNR conclusions/recommendations, 13, 14, 79–81
protein supplements, health risks of, 7–11, 39, 58–63
supplements, dietary, for enhancing performance, efficacy/safety considerations, 331–337
Food supplements
choline, and endurance performance, 336
military training nutritional intervention for immune function, research needs, 88
performance enhancement using, efficacy/safety considerations, 331–337
premenstrual syndrome management using, 335–336
See also Amino acid supplements; Protein powders (PP); Protein supplements
Fort Chaffee, Arkansas, 41, 98, 100
Fort Devens, Massachusetts, 40, 98, 100–101
Fort Jackson, South Carolina, 40, 99, 102
Fort Hood, Texas, 40, 99, 101–102
Fort Lewis, Washington, 40, 98, 100–101
Fort Polk, Louisiana, 98, 101, 103
Fort Riley, Kansas, 98, 100–101

INDEX 419

Fort Sam Houston, Texas, 41, 99, 102
Fractional synthesis rate (FSR), 125, 126, 130
Free fatty acids (FFA), 53, 312–313

Gender
  protein/amino acids supplemental use by military personnel, by sex, 103–104
  protein Recommended Dietary Allowance by sex, 6, 37
Glomerular filtration rate (GFR), 59, 139, 140, 144, 145, 147
Glomerulosclerosis
  aging and development of, 144, 145, 146, 147
  protein intake and development of, 138, 144–147
  See also Diseases and disorders; Renal function
Glucagon, 131, 133
  See also Catabolic hormones
Glutamine
  endurance performance/metabolism effects of, 310, 317–320
  glycogen resynthesis following exercise, supplementation effects, 310, 321–322
  immune response effects of supplementation, CMNR conclusions/recommendations, 13, 14
  immune system, link to, 319–320
  muscle concentrations during exercise, 244–246
  overtraining syndrome and immune function, role of, 249–250
  plasma concentrations of, 318–319
  See also Amino acids (AA)
Glycogen
  resynthesis following exercise, effects of amino acid/protein supplementation, 310, 320–322
Grains. *See* Plants and vegetation
Growth hormone (GH), 50–51, 52
  muscle protein effects of, 50–51, 125, 128–129, 133
  muscle protein metabolism following infection/injury, effects of, 281
  See also Anabolic hormones

Hormones, 19, 20
  infection/injury effects on whole body protein metabolism, regulators of, 161–162
  muscle protein effects of, 127–131
  See also Amino acids (AA); Anabolic hormones; Catabolic hormones; Growth hormone (GH); Insulin; Insulin-like growth factor-I (IGF-I); Sex steroids
Hunter Army Airfield, Georgia, 98, 101, 103

Imaging methods
  muscle mass/composition measurement, methods overview, 46, 47–48, 258, 259, 268–269, 274
  See also Computerized tomography (CT); Magnetic resonance imaging (MRI); Measurement techniques
Immune function
  amino acid supplementation effects on, 9, 13, 14, 54–56
  glutamine, effects on, 319–320
  glutamine, overtraining syndrome, and, 249–250
  military training/stressors effects on, nutritional intervention research needs, 88
India
  Anura Kurpad, amino acid requirements studies, 234
  Bangalore amino acid requirements studies, 223, 234
  nitrogen loss studies, 223
  St. John's Medical College, amino acid requirements studies, 223
Indispensable amino acids (IAA)
  composition/requirement patterns in, animal data, 178
  ileal losses, animal data, 180
  metabolic demand for, adaptive component, 180–184, 221–224
  metabolic demand for, diurnal cycling role in, 184–186, 224–225
  metabolic demand for, summary of, 186, 187
  nitrogen balance use for estimating FAO requirement patterns in, 22–24, 186–202

obligatory metabolic demand pattern for, animal data for, 175–180
obligatory oxidative losses, FAO/WHO/UNU values, 171
plant proteins biological value in nitrogen balance trials, 190–192
protein-free diet, responses to, animal data, 177
protein quality issues, 56–57
requirements for, debate over, 3–4, 25–28, 170–173
requirements for, FAO/WHO/UNU values, 3–4, 24–28, 171
requirements for, FAO/WHO/UNU values adjustment, 192–193
requirements for, feasibility and necessity of estimates, 217–237
supplementation of, cognitive performance effects of, 8, 42–45
See also Amino acids (AA); Leucine; Lysine; Tryptophan; Tyrosine
Infectious diseases and disorders
anabolic hormones treatment, effects on, CMNR conclusions/recommendations, 12, 14, 78, 79
muscle protein metabolism effects of, 20, 279–282
nitrogen loss/translocation following, 156, 158, 159–160
protein requirements effects of, 5, 20, 33–34
protein translocation during, regulators of, 160–162
whole body protein metabolism, effects of, 155–159
See also Diseases and disorders
Injuries
anabolic hormones treatment, effects on, CMNR conclusions/recommendations, 12, 14, 78, 79
muscle protein metabolism effects of, 20, 279–282
nitrogen loss/translocation following, 156, 158, 159–160
protein requirements effects of, 5, 20, 33–34
protein translocation during, regulators of, 160–162
whole body protein metabolism, effects of, 155–159

Insulin
infection/injury, and effects on muscle protein metabolism, 282
muscle protein effects of, 49–50, 125, 127–128, 133
See also Anabolic hormones
Insulin-like growth factor-I (IGF-I), 20, 50–51, 52
muscle protein effects of, 50–51, 125, 128–129, 133
See also Anabolic hormones
In vivo neutron activation analysis (IVNA)/whole body counting muscle mass/composition measurement, methods overview, 47, 258, 259, 272–273, 275
See also Measurement techniques

Japan
nitrogen loss studies, 223
renal stone disease incidence, 138
women breast cancer incidence, relationship to diet, 334–335

Kidney. *See* Acute renal failure; Renal function; Renal stone disease

Lactation. *See* Breastfeeding
Large neutral amino acids (LNAA), 42, 43, 44, 335
See also Amino acids (AA); Tryptophan; Tyrosine
Leucine, 219, 221–224, 225, 233, 234, 246
See also Amino acids (AA); Indispensable amino acids (IAA)
Life Sciences Research Office (LSRO), 332–334
Luzon, Phillipines
U.S. military personnel, protein intake studies, 96, 97
Lysine, 219, 221–224, 225, 233–234
nitrogen balance use for estimating requirements of, limitations, 226–231
Toronto Break Point stable isotope studies on, limitations of, 197–198
wheat flour content of, 229, 230
See also Amino acids (AA); Indispensable amino acids (IAA)

INDEX

Magnetic resonance imaging (MRI), 31, 260, 262, 267
  muscle mass/composition measurement, methods overview, 46, 47, 48, 258, 259, 268–269, 274
  See also Imaging methods; Measurement techniques
Massachusetts Institute of Technology (MIT), 25, 88, 175, 176, 189, 194, 197, 199, 223
Massachusetts Institute of Technology Amino Acid Requirement Pattern (MIT-AARP), 26, 27, 170, 172–173, 229, 230, 231
  FAO/WHO/UNU patterns comparison, 219
  overview of, 218–221
  physical activity and, considerations, 235–237
Maximal oxygen uptake (VO$_{2max}$), 124, 126
Meal, Ready-to-Eat (MRE), 20–21, 36, 38, 86, 87
  lightweight rations comparison, cognitive performance effects of amino acid changes, field studies results, 301–303
  nutrient intake of, field study, 301
  See also Military rations
Measurement techniques
  descriptive methods, for muscle mass, 45–46
  in vivo methods, for muscle mass, 46, 47, 258–275
  model-based methods, for muscle mass, 46–48
  See also Anthropometry; Bioelectric impedance analysis (BIA); Computerized tomography (CT); Dual-energy x-ray absorptiometry (DXA); Imaging methods; In vivo neutron activation analysis (IVNA)/whole body counting; Magnetic resonance imaging (MRI); Stable isotopes; Ultrasound; Urinary metabolites
Medicine. *See* Drugs
Metabolic demand (MD)
  amino acids, indispensable, adaptive component of, 180–184, 221–224
  amino acids, indispensable, diurnal cycling role in, 184–186, 224–225
  amino acids, indispensable, summary of, 186, 187
  See also Amino acid metabolism; Obligatory metabolic demand (OMD); Protein metabolism
Metabolism. *See* Amino acid metabolism; Metabolic demand (MD); Protein metabolism
Methionine
  urinary excretion effects of, 141, 142
Methodology, 3
  isotope, stable, amino acid requirements prediction using, limitations, 193–197
  muscle mass/composition measurement, methods organization, 257–259
  nitrogen balance studies, amino acid requirements prediction using, limitations, 186–192
  protein requirements assessment, 22–24
  Toronto Break Point stable isotope studies on phenylalanine/lysine, limitations of, 197–198
3-Methylhistidine
  urinary metabolite measurement of muscle mass/composition, 45, 46, 258, 259, 263, 265, 267, 274
Military operations
  protein/amino acids requirements during, CMNR conclusions/recommendations, 77–79
  protein/amino acids role in physiological optimization during, research needs, 85–89
  protein/amino acid supplementation to optimize performance, CMNR conclusions/recommendations, 79–81
  rations, limited-use, for optimal cognitive/physical function, research needs, 86–87
  rations protein/energy content, by ration type, 94–95, 411
  rations protein/energy content, CMNR conclusions/recommendations, 12–13, 14, 79
  rations protein/energy current intake, 97–102
  rations protein/energy historic intake, World War II studies, 96–97

rations protein/energy requirements, determination of, 102–103
rations protein Military Recommended Dietary Allowance, CMNR conclusions/recommendations, 79
stressors during, effects on protein requirements, 35–36
urea formation and recycling during, research needs, 89
Military personnel
Army protein/amino acid supplements use, 39, 42
protein/amino acids supplement use, by gender/military specialty, 103–104
protein/energy historic ration intake, World War II studies, 96–97
protein/energy operational intake, recent studies, 98, 99–101
protein intake studies of, 38, 40–41
protein requirements effects of various stressors, 3–6
training and stress-induced changes in immune function, nutritional intervention research needs, 88
See also Field operations; Military operations; Military rations; Military Recommended Dietary Allowance (MRDA); Women in military; U.S. Air Force (USAF); U.S. Army; U.S. Army Rangers; U.S. Army Reserves; U.S. Marine Corps; U.S. Special Forces
Military rations, 38
field/operational protein intake, overview, 93–104
garrison protein intake, overview, 93–104
limited-use, for optimal cognitive/physical function, research needs, 86–87
operational, protein/energy content, by ration type, 94–95, 411
operational, protein/energy content, CMNR conclusions/recommendations, 12–13, 14, 79
operational, protein/energy current intake in, 97–102
operational, protein/energy historic intake in, World War II studies, 96–97

operational, protein/energy requirements for, determination of, 102–103
operational, protein Military Recommended Dietary Allowance for, CMNR conclusions/recommendations, 79
protein requirements for, crucial questions, 250
protein supplemental intake, overview, 93–104
World War II studies of, overview of, 93–97, 102
See also Meal, Ready-to-Eat (MRE); Military personnel; Ration, Light Weight-30 (RLW-30)
Military Recommended Dietary Allowance (MRDA), 35, 36, 37–39, 78, 79, 87, 100, 101
pregnancy/lactation, protein allowance, CMNR conclusions/recommendations, 7, 12, 14, 39, 79
protein allowance for operational rations, CMNR conclusions/recommendations, 6–7, 12, 14, 37–39, 79
See also Military personnel; Recommended Dietary Allowance (RDA)
Minnesota bread study, 191, 192, 231
Models and modeling, 46–48, 257, 258, 260, 265, 272
Mood
tryptophan/serotonin effects on, 293–294, 296
Mulder, Gerardus Johannes, 19–20, 109–110

Natick Research, Development and Engineering Center (NRDEC), 93–94
National Center for Health Statistics (NCHS), 191
National Health and Nutrition Examination Survey III (NHANES III), 37, 103
Naval Medical Research Institute, 87
Nephrolithiasis. *See* Renal stone disease
Neurotransmitters, 19, 289, 290, 292, 293
See also Amino acids (AA); Dopamine; Epinephrine; Norepinephrine; Serotonin

NHANES III. *See* National Health and Nutrition Examination Survey III (NHANES III)
Nitric oxide (NO), 55–56
Nitrogen, 225
  balance of, use for estimating indispensable amino acid requirements, limitations, 22–24, 172, 186–202, 226–231
  plant proteins biological value in nitrogen balance trials, 190–192
  translocation/loss associated with infection/injury, 156, 157, 158, 159–160, 161
  See also Obligatory nitrogen loss (ONL)
Norepinephrine, 290, 292, 293
  cognitive performance effects of, 297–299
  stress exposure and cognitive performance, effects of, 297–299
  See also Neurotransmitters
North Carolina wheat study, 191–192, 231
Nurses' Health Study, 61
Nutrients
  infection/injury effects on whole body protein metabolism, regulators of, 162
  military standard field and lightweight ration groups, intakes of, 301
Nutrition
  brain status determination by blood-brain barrier, overview of, 290–293
  intervention for immune function changes during military training/stresses, research needs, 88
  muscle mass/function, implications of, 132
  undernutrition and military lightweight rations, cognitive performance effects of amino acid changes, field studies results, 301–303
  See also Amino acid metabolism; Amino acids (AA); Amino acid supplements; Energy; Food supplements; Metabolic demand (MD); Nutrients; Protein; Protein metabolism; Protein supplements

Obligatory metabolic demand (OMD)
  animal data for pattern of, 175–180
  obligatory oxidative losses and, 174–175
  overview of, 173–174
  protein/amino acids, overview of, 173–174
  See also Amino acid metabolism; Metabolic demand (MD); Protein metabolism
Obligatory nitrogen loss (ONL), 4, 170, 171, 172, 174, 175, 180
  See also Amino acid metabolism; Nitrogen; Protein metabolism
Obligatory oxidative amino acid loss (OOAL/OOL), 4, 170, 175, 176, 218, 219, 221, 223
  amino acids requirements patterns, adjustment of FAO/WHO/UNU values for, 192–193
  amino acids requirements patterns, prediction using, 175
  FAO/WHO/UNU values for, 171
  metabolic demands, obligatory, and, 174–175
  See also Amino acid metabolism; Protein metabolism

Pacific Islands
  U.S. military energy/protein intake studies, 96, 97
Phenylalanine, 128, 219
  isotope, stable, estimates of requirement values, 232–233
  nitrogen balance use for estimating requirements of, 227
  Toronto Break Point stable isotope studies on, limitations of, 197–198
  See also Amino acids (AA)
Physical activity
  branched-chain amino acids ingestion effects during exercise, 53–54, 317
  contractile activity, muscle protein turnover effects of, 30–31, 244, 247–248
  energy restriction and amino acid/protein requirements, 29–33
  exercise and amino acid catabolism and, 29–30, 244–247
  exercise effects of protein intake timing, 57–58

exercise, endurance, and amino acid/
  protein metabolism, 310–311
exercise, endurance, and protein requirements, CMNR conclusions/
  recommendations, 77, 78
exercise, prolonged, plasma glutamine
  concentrations following, 318
glycogen resynthesis following exercise, effects of amino acid/protein
  supplementation, 310, 320–322
increased, energy supply and protein/
  amino acid metabolism interactions
  in, 248–249
MIT amino acid requirement pattern
  and, considerations, 235–237
protein/amino acid requirements effects of high workload, CMNR conclusions/recommendations, 77–79
protein/amino acid supplementation
  effects on performance during high
  workload, CMNR conclusions/recommendations, 79–81
protein metabolism/requirements and,
  20, 243–244, 250–251
See also Athletics
Physical performance
  branched-chain amino acids/tryptophan ingestion, endurance effects,
    studies results, 53–54, 310, 313–315
  dietary supplements use to enhance,
    safety/efficacy considerations, 331–337
  endurance exercise, protein/amino acid
    supplementation effects on, rationale
    for, 309–322
  endurance performance and choline
    dietary supplementation, 336
  military limited-use operational rations, for optimal function, research
    needs, 86–87
  protein/amino acid supplementation
    effects on, 10, 52–54, 309–332
  protein/amino acid supplementation to
    optimize military performance,
    CMNR conclusions/recommendations, 79–81
  protein, dietary, and amino acids supplementation and, study considerations, 334
  See also Cognitive performance; Endurance

Plants and vegetation
  proteins in, biological value in nitrogen balance trials, 190–192
  proteins in, potential benefits of, 57
  See also Food; Soybeans; Wheat
Plasma
  ammonia concentrations, branched-chain amino acids ingestion effects
    during exercise, 317
  glutamine concentrations following
    prolonged exercise, 318
  glutamine concentrations in overtrained athletes and Tour de France
    cyclists, 318–319
Population characteristics. *See* Age and
  aging; Gender
Porsolt swim test, 299, 300
Pregnancy, 37, 39
  amino acid/protein supplementation
    effects on, 62–63
  protein Military Recommended Dietary Allowance for, CMNR conclusions/recommendations, 7, 39, 79
  protein Recommended Dietary Allowance for, 6–7, 37, 39
Premenstrual syndrome (PMS)
  dietary supplementation for management of, studies on, 335–336
Profile of Mood States (POMS), 295, 336
Progesterone. *See* Sex steroids
Prophylactics. *See* Drugs
Protein
  acute renal failure, intake role in, 138,
    144
  Army's interest in role of, 2–3, 20–21
  calcium status and intake of, 10, 60–61
  cognitive performance effects of
    changes during stress, 301–303
  functions of, 110, 111
  glomerulosclerosis development, intake role in, 138, 144–147
  intake adequacy, functional indicators
    of, 198–202
  intake of, metabolic studies of long-distance cyclists, 52
  intake timing, effects on exercise and
    cognitive function, 57–58
  military operational rations content, by
    ration type, 94–95, 411

INDEX 425

military operational rations content, CMNR conclusions/recommendations, 12, 14, 79
military operational rations current intake of, 97–102
military operational rations historic intake of, World War II studies, 96–97
military personnel intake of, recent studies, 38, 40–41
Military Recommended Dietary Allowance for, 6–7, 12, 14, 37–39
nonsecretory, targeting costs, 113
obligatory oxidative losses, FAO/WHO/UNU values, 171
origins of term, 19–20, 109–110
plant/legume sources of, potential benefits of, 57
plant versus animal sources of, efficacy/safety considerations, 334–335
quality of, issues, 56–57
Recommended Dietary Allowances for, 6–7, 37, 39
renal disease development, intake role in, 137–147
renal function, intake role in, 10, 59–60, 137–138
renal stone disease development, intake role in, 59–60, 138–143
secretory, targeting costs, 113
total body, distribution of, 256
urea formation and recycling of, research needs for military operations, 89
wheat flour content, 230
See also Amino acids (AA); Protein metabolism; Protein powders (PP); Protein requirements; Protein supplements
Protein-digestibility corrected amino acid score (PDCAAS), 170, 172
Protein metabolism
amino acid/protein supplements, effects on, rationale for, 309–322
anorexia response to infection/injury, 160
energy costs of breakdown/degradation, 114–116
energy costs of synthesis and regulation, 112–114

energy-dependent processes associated with, summary, 110–112
energy supply and, interactions during increased physical activity, 4–5, 29–33, 244, 248–249
environmental stresses influence on, 6, 20, 35–36
exercise, endurance, and, 310–311
hormonal response pattern following infection/injury, 161
hormones influence on, 20, 49–52
infection/injury effects on, overview of, 20, 155–159
muscle protein degradation, altering conditions, 115
muscle protein metabolism, hormonal interactions with, 49–52
muscle protein turnover effects of contractile activity, 30, 244, 247–248
nitrogen translocation following infection/injury, 157, 158, 159–160, 161
overview of, 21–22, 173–186
physical activity influence and, 20, 243–244, 250–251
translocation of following infection/injury, regulators of, 160–162
turnover, protein, as indicator of amino acid intake adequacy, limitations, 198–199
turnover, regulatory proteins, energy-dependent processes in, 114
urea formation/recycling during military operations, research needs, 89
Protein powders (PP)
supplements use reported by military personnel, by gender/military specialty, 103–104
See also Food supplements; Protein supplements
Protein requirements
assessment methods for, 22–24
brain function and, 289–303
cognitive performance, effects of stress and, 298–300
conclusions/recommendations of CMNR on, 77–79
debate over, 3–4, 25–28, 170–173
endurance exercise and, CMNR conclusions/recommendations, 77, 78
energy balance and, 11, 14, 31–33

environmental stresses influence on, 6, 20, 35–36
FAO/WHO/UNU values, 3–4, 24–28, 171
FAO/WHO/UNU values adjustment, 192–193
infection/injury/illness effects on, 5, 20, 33–34
military limited-use operational rations, for optimal cognitive/physical function, research needs, 86–87
military women and, research needs, 88–89
obligatory oxidative amino acid loss pattern use to predict, 175
operational rations, determination for, 102–103
physical activity and, 4–5, 20, 29–33, 243–244, 250–251
stressors effects on, 3–6, 11–12, 38–36
Protein supplements
amino acid/protein, effects on metabolism, rationale for, 309–322
Army personnel use of, 39, 42
cognitive performance effects of, 8, 42–45
benefits/health risks of, 7–11, 39, 49–63
hydrolysates, effects on glycogen resynthesis following exercise, 310, 322
metabolism and performance effects of, rationale for, 309–322
military performance optimization, use for, CMNR conclusions/recommendations, 12–13, 14, 79–81
military personnel reported use, by gender/military specialty, 103–104
muscle building, use for, CMNR conclusions/recommendations, 12, 14
performance and, study considerations, 334
physical performance effects of, 10, 52–54
plant versus animal sources of, efficacy of, 334–335
pregnancy, effects of high-protein diets and amino acid supplements, 11, 62–63
safety/efficacy considerations, 13, 14, 79–81, 334

timing of intake, effects, 57–58
See also Amino acid supplements; Food supplements; Protein; Protein powders (PP); Protein requirements
Ration, Light Weight-30 (RLW-30)
amino acid changes from standard rations, cognitive performance effects, field studies results, 301–303
nutrient intake of, field study, 301
Recommended Dietary Allowance (RDA), 78, 94, 102, 173
pregnant/lactating women, protein allowance, 6–7, 37, 39, 71
protein allowance by age group and sex, 6, 37
protein requirements of FAO/WHO/UNU and, 24–28
See also Military Recommended Dietary Allowance (MRDA)
Renal function
disease development, protein intake effects on, 137–147
protein intake effects on, 10, 59–60, 137–138
See also Acute renal failure; Glomerulosclerosis
Renal stone disease
calcium clearance determinants, 139–142
calcium excretion and calcium-containing stones, 138–143
calcium intake and, 138
protein intake role in, 59–60, 138–143
treatment, prophylactic, 143
uric acid stones, 138, 143
urinary and excretion patterns in stone formers, 142–143
See also Diseases and disorders
Reproductive diseases and disorders. *See* Premenstrual syndrome (PMS)
Research
immune function changes during military training/stressors, nutritional intervention research needs, 88
injuries/infection effects of anabolic hormones treatment, CMNR conclusions/recommendations, 79
military limited-use operational rations for optimal cognitive/physical function, research needs, 86–87

military operations and protein/amino acids role in physiological optimization, research needs, 85–89
military women's protein requirements, research needs, 88–89
urea formation and recycling, research needs for military operations, 89

Sarcopenia of aging. *See* Age and aging
Serotonin
  cognitive performance role of, 43–44, 87, 290, 292, 293–297
  See also Neurotransmitters
Sex steroids
  muscle protein effects of, 125, 129–131
  testosterone effects on muscle protein metabolism, 51, 282
Skeletal muscle (SM)
  aging effects on mass/function regulation, 123–126
  ammonia production of, branched-chain amino acids ingestion effects during exercise, 317
  anabolic hormones effects on, 49–51
  anabolic hormones effects on following injury/infection, 281–282
  anthropometric prediction methods for, overview, 45–46, 258, 259–261, 273–274
  bioelectric impedance analysis prediction of, methods overview, 45, 46, 258, 259, 262–263, 264, 274
  building, use of protein supplements for, CMNR conclusions/recommendations, 12, 14
  catabolic hormones effects on, 52, 131
  descriptive measurement methods for, 45–46
  dual-energy x-ray absorptiometry prediction of, methods overview, 46, 47–48, 259, 269–272, 274–275
  endurance of, 49
  glycogen resynthesis in following exercise, effects of amino acid/protein supplementation, 310, 320–322
  growth hormone effects on, 128–129, 133, 281
  hormonal effects on muscle protein, 127–131
  imaging methods for predicting, overview, 46–47, 258, 259, 268–269, 274
  infection/injury effects on body composition, 160, 161
  infection/injury effects on protein metabolism in, 279–281
  insulin effects on, 127, 128, 133, 282
  insulin-like growth factor-1 (IGF-I), effects on, 128–129, 133
  in vivo neutron activation/whole body counting as predictors of, methods overview, 46, 258, 259, 272–273, 275
  mass and function regulation, overview of, 48–49, 121–123
  measurement of, methods overview, 45–48, 259–273
  metabolism of, 49
  model-based measurement methods for, 46–48
  nitrogen translocation to visceral organs following infection/injury, 156, 159–160
  protein degradation, altering conditions, 115
  protein turnover effects of contractile activity, 30, 244, 247–248
  strength of, 48–49
  substrates and nutrition, implications for, 132
  testosterone effects on, 282
  time-related changes in mass and composition, monitoring considerations, 273–275
  tricarboxylic acid cycle in muscle, branched-chain amino acid-aminotransferase reaction, interaction with, 315–317
  ultrasound prediction of, methods overview, 45, 258, 259, 261–262, 273–274
  urinary metabolites as indices of, methods overview, 45, 46, 258, 259, 263, 265, 274
Sodium sulfate
  calcium clearance role of, 140, 141, 143
Soybeans
  protein alternative sources, efficacy/safety considerations, 80, 334–335

Special Forces Assessment and Selection (SFAS), 88
  See also U.S. Special Forces
Stable isotopes
  amino acid, indispensable, requirement values, use for, 193–197, 232–234
  Toronto Break Point studies on phenylalanine/lysine, limitations of, 197–198
  See also Measurement techniques
Stressors
  altitude, high, protein requirements effects of, 35–36, 101
  immune function stress-induced alterations, nutritional intervention research needs, 88
  protein/amino acid requirements and, cognitive performance effects, 86, 87, 289, 290
  protein/amino acid requirements and military operational stressors, effects of, CMNR conclusions/recommendations, 3–6, 35–36, 77–79
  protein/amino acid supplementation to optimize military performance during, CMNR conclusions/recommendations, 79–81
  protein requirements, effects of, 3–6, 11–12, 20, 28–36
  tyrosine role in effects on cognitive performance, 44–45, 87, 298–300
Stroop Task, 300
Substrates
  muscle mass/function, implications for, 132

Temperature
  cold stress and cognitive performance, tyrosine role in, 87, 298–299
  cold stress, protein requirements effects of, 35–36
  heat stress and cognitive performance, tyrosine role in, 300
  heat stress, protein requirements effects of, 35
Thyroid hormone, 131
  See also Catabolic hormones; Hormones
Toronto Break Point studies
  phenylalanine/lysine stable isotope data, limitations of, 197–198

Total body nitrogen (TBN), 272, 273, 275
Total body potassium (TBK), 272, 273, 275
Tour de France, 52
  cyclists, amino acid/protein metabolism during, 52, 311
  cyclists, glutamine plasma concentrations in, 318–319
Tricarboxylic acid (TCA), 24, 53–54
  branched-chain amino acid-aminotransferase reaction, interaction with in muscle, 315–317
Tryptophan
  cognitive performance/mood role of, 43–44, 87, 289, 290, 292, 293–297
  endurance performance effects of ingestion, studies results, 310, 313–315
  L-tryptophan, 331–332, 335
  military lightweight rations changes in, cognitive performance effect of, 301–303
Tyrosine, 232–233
  cognitive performance/brain function role of, 44–45, 87, 289, 290, 292, 297–300
  military lightweight rations changes in, cognitive performance effect of, 301–303
  stress response, role of, 87, 298–300

Ultrasound
  muscle mass/composition measurement, methods overview, 45, 258, 259, 261–262, 273–274
  See also Measurement techniques
United Nations University (UNU), 24, 25, 174, 175, 192
  amino acid/protein obligatory oxidative losses, FAO/WHO/UNU values, 171
  amino acid/protein requirement patterns of FAO/WHO/UNU, 3–4, 24–28, 171, 218, 219, 220, 229, 230, 231, 236, 237
Urinary excretion
  calcium clearance determinants, 139–142
  protein/nitrogen loss following infection/injury, 157, 159

renal stone formers, patterns in, 138, 142–143
urea formation and recycling, research needs for military operations, 89
Urinary metabolites
muscle mass/composition measurement, methods overview, 45, 46, 258, 259, 263, 265–267, 274
See also Measurement techniques
Urolithiasis. See Renal stone disease
U.S. Air Force (USAF), 96
U.S. Army, 7, 39, 96, 97, 99, 100, 102, 104
Air Defense Artillery company, 103
Military Nutrition Division, 2
protein/amino acid requirement questions of, CMNR responses, 77–81
protein/amino acid supplements use by personnel, 39, 42
protein, dietary, and protein balance, interest in, 2–3, 20–21
Special Operations, 101
See also Natick Research, Development and Engineering Center (NRDEC)
U.S. Army Medical Research and Materiel Command (USAMRMC), 2
U.S. Army Rangers, 20–21, 55, 86, 87, 88, 101, 103, 110
U.S. Army Research Institute of Environmental Medicine (USARIEM), 20, 301
U.S. Army Reserves, 101
U.S. Army Surgeon General, 93
U.S. Marine Corps, 45, 87, 100
U.S. Military Academy (USMA), 61
U.S. Special Forces, 2, 39, 103
See also Special Forces Assessment and Selection (SFAS)

Visceral organs
infection/injury effects on body composition, 160, 161
nitrogen translocation from muscle mass following infection/injury, 156, 159–160
See also Body composition

Visual Analog Mood Scale (VAMS), 295
Voit, Carl, 156, 186–187

Wheat
amino acid diet restriction, animal studies, 179
flour amino acid/protein content, 229, 230
North Carolina wheat study, 191–192, 231
See also Plants and vegetation
Whole body counting. See In vivo neutron activation analysis (IVNA) analysis/whole body counting
Women
college women nitrogen balance estimation of lysine requirements, 226
Japan women breast cancer incidence, relationship to diet, 334–335
See also Breastfeeding; Pregnancy; Premenstrual syndrome (PMS)
Women in military
protein/amino acids supplemental use by, 103–104
protein/energy operational intakes, recent studies, 99, 101–102
protein intake studies, 38, 40–41
protein requirements, research needs, 88–89
World Health Organization (WHO), 24, 25, 26, 57, 174, 175, 192
amino acid/protein obligatory oxidative losses, FAO/WHO/UNU values, 171
amino acid/protein requirement patterns of FAO/WHO/UNU, 3–4, 24–28, 171, 218, 219, 220, 229, 230, 231, 236, 237
World War II, 37, 93, 138, 261
operational rations protein/energy recommendations during, 94, 102
operational rations protein/energy content, studies during, 94–95
operational rations protein/energy intake, studies during, 96–97